AF734953

Advances in Complex Geometry

CONTEMPORARY MATHEMATICS

735

Advances in Complex Geometry

JHU-UMD Complex Geometry Seminar
2015–2018
Johns Hopkins University, Baltimore, Maryland
and
University of Maryland, College Park, Maryland

Yanir A. Rubinstein
Bernard Shiffman
Editors

AMS AMERICAN MATHEMATICAL SOCIETY

EDITORIAL COMMITTEE

Dennis DeTurck, Managing Editor

Michael Loss Kailash Misra Catherine Yan

2010 *Mathematics Subject Classification.* Primary 53-06, 32-06.

Library of Congress Cataloging-in-Publication Data

Names: Rubinstein, Yanir A., editor. | Shiffman, Bernard, editor.
Title: Advances in complex geometry / Yanir A. Rubinstein, Bernard Shiffman, editors.
Description: Providence, Rhode Island: American Mathematical Society, [2019] | Series: Contemporary mathematics; volume 735 | "2015-2018 JHU-UMD complex geometry seminars, Johns Hopkins University, Baltimore Maryland, University of Maryland, College Park, Maryland." | A selection of papers from the seminars. | Includes bibliographical references.
Identifiers: LCCN 2019011687 | ISBN 9781470443337 (alk. paper)
Subjects: LCSH: Geometry, Differential–Congresses. | Geometry, Algebraic–Congresses. | Geometry–Congresses. | AMS: Differential geometry – Proceedings, conferences, collections, etc. msc | Several complex variables and analytic spaces – Proceedings, conferences, collections, etc. msc
Classification: LCC QA641 .A5795 2019 | DDC 516.3/6–dc23
LC record available at https://lccn.loc.gov/2019011687
DOI: https://doi.org/10.10190/conm/735

Copying and reprinting. Individual readers of this publication, and nonprofit libraries acting for them, are permitted to make fair use of the material, such as to copy select pages for use in teaching or research. Permission is granted to quote brief passages from this publication in reviews, provided the customary acknowledgment of the source is given.

Republication, systematic copying, or multiple reproduction of any material in this publication is permitted only under license from the American Mathematical Society. Requests for permission to reuse portions of AMS publication content are handled by the Copyright Clearance Center. For more information, please visit `www.ams.org/publications/pubpermissions`.

Send requests for translation rights and licensed reprints to `reprint-permission@ams.org`.

© 2019 by the American Mathematical Society. All rights reserved.
The American Mathematical Society retains all rights
except those granted to the United States Government.
Printed in the United States of America.

∞ The paper used in this book is acid-free and falls within the guidelines
established to ensure permanence and durability.
Visit the AMS home page at `https://www.ams.org/`

10 9 8 7 6 5 4 3 2 1 24 23 22 21 20 19

Contents

Preface

There is a rich history of research in complex geometry and complex analysis of one and several variables at Johns Hopkins University (JHU; Baltimore, Maryland) and at the University of Maryland (UMD; College Park, Maryland). W.-L. Chow came to Hopkins in 1948, remaining there throughout his career and bringing K. Kodaira to Hopkins from 1962 to 1965. At College Park, an international conference in several complex variables convened in 1970, the proceedings of which are published in a two-part volume[1] edited by J. Horváth. Then in 1973, the University of Maryland held a Special Year in Complex Function Theory; some of the talks during that year are published in a volume[2] edited by W. E. Kirwan and L. Zalcman. As the editors of that volume wrote, the common denominator for the contributions was "a certain emphasis, in point of view or in method, on problems having concrete geometric content". Subsequently, C. Berenstein and the second editor of the present proceedings organized a Hopkins–Maryland joint seminar in complex analysis, which ran through 1989. Seminars on complex geometry then continued at Hopkins through 2012, organized by the second editor of this proceedings together with V. V. Shokurov and S. Zucker during 1991–2004 and with R. Wentworth and S. Zelditch during 2004–10. The complex geometry seminars at Hopkins were also co-organized by J. Noguchi in 1998 and by the first editor of the present proceedings in 2008–09. Additional activity was provided by conferences involving complex geometry sponsored by the Japan–U.S. Mathematics Institute (JAMI)[3] at Johns Hopkins in 1991, 1998, 2004, and 2019.

In 2012, the editors of this proceedings revived the Hopkins–Maryland joint seminar by establishing the JHU-UMD Complex Geometry Seminar, with the assistance of R. Wentworth, S. Wolpert and Y. Yuan. Topics presented at the seminar include geometric flows, canonical metrics, geometric stability, pluripotential theory, the Monge–Ampère equation, zeros of random holomorphic sections, L^2 extension, deformation theory, Bergman kernel, special Lagrangians, and Gromov–Witten theory, as well as topics on the interface of complex geometry with convex geometry, symplectic geometry, algebraic geometry, and topology.

This volume contains contributions from speakers at the seminar during three academic years from 2015 to 2018. The volume begins with a survey by T. Darvas of recent developments in pluripotential theory and its application to Kähler–Einstein metrics. The next article is a survey by S. Dinew of recent advances in the theory of local regularity of plurisubharmonic functions and the complex Monge–Ampère

---

[1] *Several Complex Variables I, II*, Lecture Notes in Mathematics, Vols. 155, 185, Springer, 1970, 1971.

[2] *Advances in Complex Function Theory*, Lecture Notes in Mathematics, Vol. 505, Springer, 1976.

[3] https://mathematics.jhu.edu/events/jami/.

equation. The article by F. R. Harvey and H. B. Lawson concerns properties of pluriharmonic functions arising in the generalized potential theories associated to subequations. Next, C. Li and G. Tian discuss orbifold regularity for weak Kähler–Einstein metrics of certain singular Fano varieties via resolution of singularities. The article by Z. Lu and H. Xu concerns the spectrum and self-adjointness properties of the Laplacian on the moduli space of polarized Calabi-Yau manifolds. L. Ni and F. Zheng review recent progress in the study of compact Kähler manifolds with positive orthogonal Ricci curvature. D. H. Phong, S. Picard, and X. Zhang describe the long-time behavior of the Anomaly flow on unimodular Lie groups. The volume concludes with an article by Z. Slodkowski on the pseudoconcave decomposition of the core of a complex manifold.

We would like to thank T. Darvas, H. J. Hein, J. Martinez-Garcia, V. P. Pingali, R. Wentworth, S. Wolpert, and H. Xu for their assistance in the organization of the seminar during 2015–2018. We are also grateful for the financial and logistical support of the mathematics departments at Johns Hopkins University and the University of Maryland. Finally, we thank the speakers for their inspiring contributions to the mathematical activities of our two institutions.

Yanir A. Rubinstein
Bernard Shiffman

List of seminars

2015–2016

Gábor Székelyhidi (University of Notre Dame)
The J-flow on toric manifolds

Ruadhaí Dervan (University of Cambridge)
K-stability of finite covers

Gang Tian (Princeton University)
K-stability implies CM-stability

Blaine Lawson (Stony Brook University)
Differential inequalities and generalized pluripotential theories

Zhiqin Lu (University of California at Irvine)
On the L^2 estimates on moduli space of Calabi-Yau manifolds

Tamás Darvas (University of Maryland)
Infinite-dimensional geometry on the space of Kähler metrics and applications to canonical Kähler metrics

Joaquim Ortega-Cerdà (Universitat de Barcelona)
Sampling polynomials in algebraic varieties

Dror Varolin (Stony Brook University)
Berndtsson's Convexity Theorem and the L^2 Extension Theorem

2016–2017

Ben Weinkove (Northwestern University)
Monge-Ampère equations on complex and almost complex manifolds

Hao Xu (University of Pittsburgh)
Asymptotic expansion of Bergman and heat kernels

Mu-Tao Wang (Columbia University)
Lagrangian curvature flows in cotangent bundles of spheres

Jake Solomon (Hebrew University of Jerusalem)
Point-like bounding chains in open Gromov-Witten theory

Duong Phong (Columbia University)
Supersymmetric vacua of superstrings and geometric flows

Xiaofeng Sun (Lehigh University)
Deformation of Fano manifolds

Xiaojun Huang (Rutgers University)
Bergman-Einstein metrics on strongly pseudoconvex domains of $\mathbb{C}^n$.

Mattias Jonsson (University of Michigan)
A variational approach to the Yau-Tian-Donaldson conjecture

2017–2018

Zbigniew Slodkowski (University of Illinois at Chicago)
Pseudoconcave decompositions in complex manifolds

Sebastien Picard (Columbia University)
The anomaly flow and the Hull-Strominger system

Sławomir Dinew (Jagiellonian University, Kraków)
Singular sets of plurisubharmonic functions

Thomas Bloom (University of Toronto)
Universality for zeros of random polynomials

Lei Ni (University of California at San Diego)
Metric characterizations of the projectivity

Jeff Streets (University of California at Irvine)
Generalized Kähler-Ricci flow in the commuting case

Contemporary Mathematics
Volume **735**, 2019
https://doi.org/10.1090/conm/735/14822

Geometric pluripotential theory on Kähler manifolds

Tamás Darvas

ABSTRACT. Finite energy pluripotential theory accommodates the variational theory of equations of complex Monge–Ampère type arising in Kähler geometry. Recently it has been discovered that many of the potential spaces involved have a rich metric geometry, effectively turning the variational problems in question into problems of infinite dimensional convex optimization, yielding existence results for solutions of the underlying complex Monge–Ampère equations. The purpose of this survey is to describe these developments from basic principles.

2010 *Mathematics Subject Classification.* Primary 32Q15; 32Q20, 53C25.
This research was partially supported by NSF grant DMS-1610202 and BSF grant 2012236.

©2019 American Mathematical Society

Contents

Preface

A circle of problems, going back to Calabi [**29**], asks to find Kähler metrics with special curvature properties on a compact Kähler manifold (X, ω). Of special interest are the *Kähler–Einstein* (KE) metrics $\tilde{\omega}$, that are cohomologous to ω, and whose Ricci curvature is proportional to the metric tensor, i.e.,

$$\operatorname{Ric}\tilde{\omega} = \lambda\tilde{\omega}.$$

Existence of such metrics on (X, ω) is only possible under cohomological restrictions, in particular the first Chern class needs to be a scalar multiple of the Kähler class $[\omega]$:

$$(\star) \qquad c_1(X) = \lambda[\omega].$$

When $\lambda \leq 0$, by the work of Aubin and Yau it is always possible to find a unique KE metric on (X, ω) [**1, 113**]. The case of Fano manifolds, structures that satisfy $(\star)$ with $\lambda > 0$, is much more intricate. In particular, the problem of finding KE metrics in this case is equivalent with solving the following global scalar equation of complex Monge–Ampère type on X:

$$(\star\star) \qquad (\omega + i\partial\bar{\partial}u)^n = e^{-\lambda u + f_0}\omega^n,$$

where f_0 is a fixed smooth function on X. The solution u belongs to $\mathcal{H}_\omega$, the set of smooth functions (potentials) that satisfy $\omega + i\partial\bar{\partial}u > 0$, which is an open subset of $C^\infty(X)$. As is well known, this equation does not always admit a solution, and our desire is to characterize Fano manifolds that admit KE metrics.

Switching point of view, the KE problem has a very rich *variational* theory as well. Indeed, Mabuchi and Ding [**64, 88**] introduced functionals

$$\mathcal{K} : \mathcal{H}_\omega \to \mathbb{R} \text{ and } \mathcal{F} : \mathcal{H}_\omega \to \mathbb{R}$$

whose minimizers are exactly the KE potentials, the solutions of $(\star\star)$ (see (4.6) and (4.46) for precise definition of these functionals). As a result, KE metrics exist if and only if the minimizer set of $\mathcal{K}$ (or $\mathcal{F}$) is non–empty. Along these lines we ask ourselves: what conditions guarantee existence/uniqueness of minimizers? Our source of inspiration will be the following elementary finite–dimensional result, which will allow to turn the variational approach into a problem of infinite–dimensional *convex optimization*:

THEOREM. *Suppose $F : \mathbb{R}^n \to \mathbb{R}$ is a strictly convex functional. If F has a minimizer it has to be unique. Regarding the existence, the following are equivalent:*
(i) F has a (unique) minimizer $x_0 \in \mathbb{R}$.
(ii) F is proper, in the sense that there exists $C, D > 0$ such that $F(x) \geq C|x| - D$, $x \in \mathbb{R}$.

We give a sketch of the elementary proof. Uniqueness of minimizers is a consequence of strict convexity. That properness of F implies existence of a minimizer follows from the fact that a bounded F–minimizing sequence subconverges to some $x_0 \in \mathbb{R}$, and that convex functions are continuous. That existence of a (unique) minimizer $x_0 \in \mathbb{R}$ implies properness of F follows from the fact that the unit sphere $\mathbb{S}^{n-1}(x_0, 1)$ is compact, hence $C := \inf_{\mathbb{S}^{n-1}(x_0,1)}(F(x) - F(x_0)) = F(y) - F(x_0)$ for some $y \in \mathbb{S}^{n-1}(x_0, 1)$. Uniqueness of minimizers implies that $C > 0$. Using convexity of F, one concludes that $F(x) \geq C|x| - D$ for some $D > 0$.

The above simple argument already sheds light on what needs to be accomplished in our infinite dimensional setting to obtain an analogous result for existence/uniqueness of KE metrics. First, we need to understand the convexity of $\mathcal{K}$ and $\mathcal{F}$. Second, even in the above short argument we have used twice that $\mathbb{R}^n$ is complete. Consequently, an adequate metric structure needs to be chosen on $\mathcal{H}_\omega$, and its completion needs to be understood. Third, (pre)compactness of spheres/balls in this new metric geometry needs to be explored.

Regarding convexity, unfortunately $\mathcal{K}$ and $\mathcal{F}$ are not convex along the straight line segments of $\mathcal{H}_\omega$. In order to address this, Mabuchi, Semmes and Donaldson independently introduced a non-positively curved Riemannian L^2 type metric on $\mathcal{H}_\omega$ that produces geodesics along which $\mathcal{K}$ and $\mathcal{F}$ are indeed convex [**67**, **88**, **99**]. Inspired from this, a careful analysis of infinite dimensional spaces led Bando–Mabuchi to prove uniqueness of KE metrics [**2**], and later Berman–Berndtsson to discover even more general uniqueness results [**8**].

On the other hand, there is strong evidence to suggest that the L^2 geometry of Mabuchi–Semmes–Donaldson alluded to above does not have the right compactness properties to allow for a characterization of existence of KE metrics. In order to address this, one needs to introduce more general L^p type Finsler metrics on $\mathcal{H}_\omega$ and compute the metric completion of the related path length metric spaces $(\mathcal{H}_\omega, d_p)$ [**48**, **49**].

After sufficient metric theory is developed, it is apparent that the L^1 geometry of $\mathcal{H}_\omega$ will be the one we should focus on. Indeed, sublevel sets of $\mathcal{K}$ restricted to spheres/balls are d_1–precompact, allowing to establish an equivalence between existence of KE metrics and d_1–properness of $\mathcal{K}$ and $\mathcal{F}$. Lastly, d_1–properness can be expressed using simple analytic means. This allowed the author and Y.A. Rubinstein to verify numerous related conjectures of Tian [**58**] going back to the nineties [**107**, **108**].

Structure of the survey. The aim of this work is to give a self contained introduction to special Kähler metrics using pluripotential theory/infinite–dimensional geometry.

In Part 1, we give a very brief introduction to Orlicz spaces that are generalizations of the classical L^p spaces. Our treatment will be rather minimalistic and we refer to [**98**] for a complete treatment.

In Part 2, we develop some background in finite energy pluripotential theory, necessary for later developments, closely following the original treatises of Guedj–Zeriahi and collaborators [**11**, **12**, **73**, **74**], that were inspired by work of Cegrell [**33**] in the local case. We refer to these works for a comprehensive treatment, as well as the recent excellent textbook [**75**].

Part 3 contains the main technical machinery presented in this work. Here we introduce the L^p Finsler geometry of the space of Kähler potentials $\mathcal{H}_\omega$, and compute the metric completion of this space with respect to the corresponding path length metrics d_p. The d_p–completions of $\mathcal{H}_\omega$ will be identified with $\mathcal{E}_p(X, \omega)$, the finite energy spaces of Guedj–Zeriahi described in the previous part (Theorem 3.36). In particular, we can endow these spaces with a rich *metric geometry*, inspiring the title of this work.

In Part 4 we discuss applications to existence/uniqueness of KE metrics on Fano manifolds. First we describe an abstract properness/existence principle (Theorem 4.7) that adapts the above finite–dimensional Theorem to our infinite–dimensional setting. As we verify the assumptions of this principle, we will present self contained proofs of the Bando–Mabuchi uniqueness theorem (Theorem 4.23) [**2**] and the Matsushima theorem about reductivity of the automorphism group of a KE manifold (Proposition 4.25) [**89**]. After this, in Theorem 4.11 and Theorem 4.36 we resolve different versions of Tian's conjectures [**107, 108**] characterizing existence of KE metrics in terms of energy properness, following [**58**].

Prerequisites. An effort has been made to keep prerequisites at a minimum. However due to size constraints, such requirements on part of the reader are inevitable. We assume that our reader is familiar with the basics of Bedford–Taylor theory of the complex Monge–Ampère operator. Mastery of [**20**, Chapters I-III] or [**61**, Chapter I and Chapter III.1-3] is more then sufficient, and for a thorough treatment we highly recommend the recent textbook [**75**]. We also assume that our reader is familiar with the basics of Kähler geometry, though we devote a section in the appendix to introduce our terminology, and recall some of the essentials. For a comprehensive introduction into Kähler geometry we refer the reader to the recent textbook [**103**], as well as [**61, 114**].

Though our main focus is the pluripotential theoretic point of view, some results in this survey rest on important regularity theorems regarding equations of complex Monge–Ampère type. Due to space constraints we cannot present a detailed proof of these theorems, but we will isolate their statements and keep them at a minimum, while providing precise references at all times.

Relation to other works. As stressed above, our focus in this work is on self–contained treatment of the chosen topics. On the down side, we could not devote enough space to the vast historical developments of the subject, and for such a treatment we refer to the survey [**96**], that discusses similar topics using a more chronological approach.

Without a doubt the choice of topics represent our bias and limitations, and many important recent developments could not be surveyed. In particular, recent breakthroughs on K–stability (the work of Chen–Donaldson–Sun [**38–40**], Tian [**109**], Chen–Wang–Sun [**42**], Berman–Boucksom–Jonsson [**10**]) could not be presented, and we refer to [**69, 110**] for recent surveys on this topic. For results about the quantization of the geometry of the space of Kähler metrics, we refer to the original papers [**18, 43, 56, 68, 91, 92**], as well as the survey [**93**] along with references therein. Geometric flows could not be discussed either, and we refer to [**12, 13, 26, 53, 101, 102**] for work on the Ricci and Calabi flows that uses the theoretical machinery described in this survey.

The relation with constant scalar curvature Kähler (csck) metrics is also not elaborated. A preliminary version of this memoir appeared on the website of the author in the early months of 2017. Since then a number of important works have appeared building on the topics presented in this work: Chen–Cheng cracked the PDE theory of the csck equation [**36, 37**], allowing to fully prove a converse of a theorem by Berman–Darvas–Lu [**14**]. Very recently He–Li pointed out that the contents of this survey generalize to Sasakian manifolds [**79**], paving the way to existence theorems for canonical metrics in that context as well.

Acknowledgments. I thank W. He, L. Lempert, J. Li, C.H. Lu, Y.A. Rubinstein, K. Smith, V. Tosatti and the anonymous referees for their suggested corrections, careful remarks, and precisions. Also, I thank the students of MATH868D at the University of Maryland for their intriguing questions and relentless interest throughout the Fall of 2016. Part 3 is partly based on work done as a graduate student at Purdue University, and I am indebted to L. Lempert for encouragement and guidance. Part 4 surveys to some extent joint work with Y. Rubinstein, and I am grateful for his mentorship over the years.

Part 1. A primer on Orlicz spaces

Plainly speaking, Orlicz spaces are generalizations of L^p Banach spaces. As we will see in our later study, we will prefer working with Orlicz norms over L^p norms, since a careful choice of weight makes Orlicz norms smooth away from the origin. The same cannot be said about L^p norms. We give here a brief and self-contained introduction, only touching on aspects that will be needed later. For a more thorough treatment we refer to [**98**].

Suppose (Ω, Σ, μ) is a measure space with $\mu(\Omega) = 1$ and (χ, χ^*) is a complementary pair of Young weights. This means that $\chi : \mathbb{R} \to \mathbb{R}^+ \cup \{\infty\}$ is convex, even, lower semi-continuous (lsc) and satisfies the normalizing conditions

$$\chi(0) = 0, \quad 1 \in \partial\chi(1).$$

Recall that $\partial\chi(l) \subset \mathbb{R}$ is the set of subgradients to χ at l, i.e., $v \in \partial\chi(l)$ if and only if $\chi(l) + vh \leq \chi(l + h)$, $h \in \mathbb{R}$. As described, χ is simply a *normalized Young weight.* The complement χ^* is the Legendre transform of χ:

$$\chi^*(h) = \sup_{l\in\mathbb{R}}(lh - \chi(l)). \tag{1.1}$$

Using convexity of χ and the above identity, one can verify that χ^* is also a normalized Young weight. Additionally (χ, χ^*) satisfies the *Young identity* and *inequality*:

$$\chi(a) + \chi^*(\chi'(a)) = a\chi'(a), \ \chi(a) + \chi^*(b) \geq ab, \ a, b \in \mathbb{R}, \ \chi'(a) \in \partial\chi(a), \tag{1.2}$$

in particular, due to our normalization: $\chi(1) + \chi^*(1) = 1$.

The most typical example to keep in mind is the pair $\chi_p(l) = |l|^p/p$ and $\chi_p^*(l) = |l|^q/q$, where $p, q > 1$ and $1/p + 1/q = 1$. Let $L^\chi(\mu)$ be the following space of measurable functions:

$$L^\chi(\mu) = \Big\{f : \Omega \to \mathbb{R} \cup \{\infty, -\infty\} : \ \exists r > 0 \text{ s.t. } \int_\Omega \chi(rf)d\mu < \infty\Big\}.$$

One can introduce the following norm on $L^\chi(\mu)$:

$$\|f\|_{\chi,\mu} = \inf\Big\{r > 0 : \int_\Omega \chi\Big(\frac{f}{r}\Big)d\mu \leq \chi(1)\Big\}. \tag{1.3}$$

The set $\{f \in L^\chi(\mu) : \int_\Omega \chi(f)d\mu \leq \chi(1)\}$ is convex and symmetric in $L^\chi(\mu)$, hence $\|\cdot\|_{\chi,\mu}$ is nothing but the Minkowski seminorm of this set. This is the content of the following lemma:

LEMMA 1.1. *Suppose $f, g \in L^\chi(\mu)$. Then $\|f + g\|_{\chi,\mu} \leq \|f\|_{\chi,\mu} + \|g\|_{\chi,\mu}$.*

PROOF. Suppose $\int_\Omega \chi(f/r_1)d\mu \leq \chi(1)$, $\int_\Omega \chi(g/r_2)d\mu \leq \chi(1)$ for some $r_1, r_2 > 0$. Convexity of χ implies that

$$\int_\Omega \chi\Big(\frac{f+g}{r_1+r_2}\Big)d\mu \leq \frac{r_1}{r_1+r_2}\int_\Omega \chi\Big(\frac{f}{r_1}\Big)d\mu + \frac{r_2}{r_1+r_2}\int_\Omega \chi\Big(\frac{g}{r_2}\Big)d\mu \leq \chi(1). \tag{1.4}$$

Hence $\|f + g\|_{\chi,\mu} \leq r_1 + r_2$, finishing the argument. □

Together with the previous one, the next lemma implies that $(L^\chi(\mu), \|\cdot\|_{\chi,\mu})$ is a normed space:

LEMMA 1.2. *Suppose $f \in L^\chi(\mu)$. Then $\|f\|_{\chi,\mu} = 0$ implies that $f = 0$ a.e. with respect to μ.*

PROOF. As $1 \in \partial\chi(1)$ and $\chi \geq 0$, it follows that

$$l \leq \chi(1) + l \leq \chi(1+l) \text{ for all } l \geq 0. \tag{1.5}$$

As a consequence of this inequality, $L^\chi(\mu) \subset L^1(\mu)$. Also, since $\|f\|_{\chi,\mu} = 0$, it follows that $\int_\Omega \chi(nf)d\mu \leq \chi(1)$ for all $n \in \mathbb{N}$. By (1.5) we can write:

$$\int_{\{n|f|>1\}} \Big(|f| - \frac{1}{n}\Big)d\mu \leq \int_{\{n|f|>1\}} \frac{\chi(n|f|)}{n}d\mu \leq \int_\Omega \frac{\chi(nf)}{n}d\mu \leq \frac{\chi(1)}{n}.$$

Applying the dominated convergence theorem to this inequality gives $\int_{|f|>0} |f|d\mu = 0$, finishing the proof. □

Though we will not make use of it, one can also show that $(L^\chi(\mu), \|\cdot\|_{\chi,\mu})$ is complete, hence it is a Banach space (see [**98**, Theorem 3.3.10]). The reason we work with a complementary pair of Young weights is because in this setting the Hölder inequality holds:

PROPOSITION 1.3. *For $f \in L^\chi(\mu)$ and $g \in L^{\chi^*}(\mu)$ we have*

$$\int_\Omega fgd\mu \leq \|f\|_{\chi,\mu}\|g\|_{\chi^*,\mu}, \quad f \in L^\chi(\mu), \ g \in L^{\chi^*}(\mu). \tag{1.6}$$

PROOF. Let $r_1 > \|f\|_{\chi,\mu}$ and $r_2 > \|g\|_{\chi^*,\mu}$. Using both the Young inequality and the identity (1.2), (1.6) follows in the following manner:

$$\int_\Omega \frac{fg}{r_1 r_2}d\mu \leq \int_\Omega \chi\Big(\frac{f}{r_1}\Big)d\mu + \int_\Omega \chi^*\Big(\frac{g}{r_2}\Big)d\mu \leq \chi(1) + \chi^*(1) = 1.$$ □

Orlicz spaces can be quite general and in our study we will be interested in spaces whose normalized Young weight is finite and satisfies the growth estimate

$$l\chi'(l) \leq p\chi(l), \ l > 0, \tag{1.7}$$

for some $p \geq 1$. To clarify, $\chi'(l)$ is just an arbitrary subgradient of χ at l. For such weights we write $\chi \in \mathcal{W}_p^+$, following the notation of [**74**]. As it turns out, weights χ that satisfy (1.7) can be thought of as distant cousins of the homogeneous L^p weight $|l|^p/p \in \mathcal{W}_p^+$:

PROPOSITION 1.4. *For $\chi \in \mathcal{W}_p^+$, $p \geq 1$ and $0 < \varepsilon < 1$ we have*

$$\varepsilon^p\chi(l) \leq \chi(\varepsilon l) \leq \varepsilon\chi(l), \ l > 0. \tag{1.8}$$

PROOF. The second estimate follows from convexity of χ. For the first estimate we notice that for any $\delta > 0$ the weight $\chi_\delta(l) := \chi(l) + \delta|l|$ also satisfies (1.7). As $\chi_\delta(h) > 0$ for $h > 0$, we can integrate $\chi'_\delta(h)/\chi_\delta(h) \leq p/h$ from εl to l to obtain:

$$\varepsilon^p\chi_\delta(l) \leq \chi_\delta(\varepsilon l).$$

Letting $\delta \to 0$ the desired estimate follows. □

Estimate (1.8) immediately implies that for $f \in L^\chi(\mu)$ the function $l \to \int_\Omega \chi(lf)d\mu$ is continuous, hence we have

$$\|f\|_{\chi,\mu} = \alpha > 0 \text{ if and only if } \int_\Omega \chi\Big(\frac{f}{\alpha}\Big)d\mu = \chi(1). \tag{1.9}$$

To simplify future notation, we introduce the increasing functions

$$M_p(l) = \max\{l, l^p\}, \quad m_p(l) = \min\{l, l^p\},$$

for $p > 0, l \geq 0$. Observe that $M_p \circ m_{1/p}(l) = m_p \circ M_{1/p}(l) = l$. As a consequence of (1.8) and (1.9), we immediately obtain the following estimates, characterizing Orlicz norm convergence:

PROPOSITION 1.5. *If $\chi \in \mathcal{W}_p^+$ and $f \in L^\chi(\mu)$ then*

$$m_p(\|f\|_{\chi,\mu}) \leq \frac{\int_\Omega \chi(f)d\mu}{\chi(1)} \leq M_p(\|f\|_{\chi,\mu}). \tag{1.10}$$

$$m_{1/p}\Big(\frac{\int_\Omega \chi(f)d\mu}{\chi(1)}\Big) \leq \|f\|_{\chi,\mu} \leq M_{1/p}\Big(\frac{\int_\Omega \chi(f)d\mu}{\chi(1)}\Big). \tag{1.11}$$

As a result, for a sequence $\{f_j\}_{j\in\mathbb{N}}$ we have $\|f_j\|_{\chi,\mu} \to 0$ if and only if $\int_\Omega \chi(f_j)d\mu \to 0$. Also, $\|f_j\|_{\chi,\mu} \to N$ for $N > 0$ if and only if $\int_\Omega \chi(f_j/N)d\mu \to \chi(1)$.

Later in this survey we will need to approximate certain Orlicz norms with Orlicz norms having *smooth* $\mathcal{W}_p^+$-weights. The following two approximation results, which are by no means optimal, will be useful in our treatment:

PROPOSITION 1.6. *Suppose $\chi \in \mathcal{W}_p^+$ and $\{\chi_k\}_{k\in\mathbb{N}}$ is a sequence of normalized Young weights that converges uniformly on compacts to χ. Let f be a bounded μ-measurable function on X. Then $f \in L^\chi(\mu), L^{\chi_k}(\mu),\ k \in \mathbb{N}$ and we have that*

$$\lim_{k\to+\infty} \|f\|_{\chi_k,\mu} = \|f\|_{\chi,\mu}.$$

PROOF. Suppose $N = \|f\|_{\chi,\mu}$. If $N = 0$, then $f = 0$ a.e. with respect to μ implying that $\|f\|_{\chi_k,\mu} = \|f\|_{\chi,\mu} = 0$. So we assume that $N > 0$. As $\chi \in \mathcal{W}_p^+$, by (1.9), for any $\varepsilon > 0$ there exists $\delta > 0$ such that

$$\int_X \chi\Big(\frac{f}{(1+\varepsilon)N}\Big)d\mu < \chi(1) - 2\delta < \chi(1) < \chi(1) + 2\delta < \int_X \chi\Big(\frac{f}{(1-\varepsilon)N}\Big)d\mu.$$

As χ_k tends uniformly on compacts to χ, f is bounded and $\mu(X) = 1$, it follows from the dominated convergence theorem that for k big enough we have

$$\int_X \chi_k\Big(\frac{f}{(1+\varepsilon)N}\Big)d\mu < \chi_k(1) - \delta < \chi_k(1) < \chi_k(1) + \delta < \int_X \chi_k\Big(\frac{f}{(1-\varepsilon)N}\Big)d\mu.$$

This implies that $(1-\varepsilon)N \leq \|f\|_{\chi_k,\mu} \leq (1+\varepsilon)N$, from which the conclusion follows. □

PROPOSITION 1.7. *Given $\chi \in \mathcal{W}_p^+$, there exists $\chi_k \in \mathcal{W}_{p_k}^+ \cap C^\infty(\mathbb{R}),\ k \in \mathbb{N}$, with $\{p_k\}_k$ possibly unbounded, such that $\chi_k \to \chi$ uniformly on compacts.*

PROOF. There is great freedom in constructing the sequence χ_k. First we smoothen and normalize χ, introducing the sequence $\tilde{\chi}_k$ in the process:

$$\tilde{\chi}_k(l) = (\delta_k \star \chi)(h_k l) - (\delta_k \star \chi)(0),\ l \in \mathbb{R}, k \in \mathbb{N}^*.$$

Here δ is a typical choice of bump function that is smooth, even, has support in $(-1,1)$, and $\int_\mathbb{R} \delta(t)dt = 1$. Accordingly, $\delta_k(\cdot) := k\delta(k(\cdot))$, and $h_k > 0$ is chosen in such a way that $\tilde{\chi}_k$ becomes normalized ($\tilde{\chi}_k(0) = 0$ and $\tilde{\chi}_k'(1) = 1$). As χ is normalized it follows that $\tilde{\chi}_k'(1-\frac{1}{k}) \leq 1 \leq \tilde{\chi}_k'(1+\frac{1}{k})$. In particular, $1-1/k \leq h_k \leq 1+1/k$, hence the $\tilde{\chi}_k$ are normalized Young weights that converge to χ uniformly on compacts.

As χ is strictly increasing on $\mathbb{R}^+$, so is each $\tilde{\chi}_k$, but our construction does not seem to guarantee that $\tilde{\chi}_k \in \mathcal{W}_{p_k}^+$ for some $p_k \geq 1$. This can be fixed by changing smoothly the values of $\tilde{\chi}_k$ on the sets $|x| < 1/k$ and $|x| > k$ in such a manner that the altered weights χ_k satisfy the estimate

$$l\chi_k'(l) \leq p_k\chi_k(l), \ l > 0,$$

for some $p_k \geq 1$. Since $\tilde{\chi}_k'(t) > 0, \ t > 0$ and $\tilde{\chi}_k'(0) = 0$, this can be done easily. The sequence $\{\chi_k\}_k$ thus obtained satisfies the required properties. □

Part 2. Finite energy pluripotential theory on Kähler manifolds

In this section we introduce the basics of finite energy pluripotential theory on Kähler manifolds. Our treatment is by no means comprehensive and we refer the readers to [**75**, Chapter 8-10] for a more elaborate study. Our short introduction closely follows [**74**] and [**48**]. More precisely, Sections 2.1–2.3 are based on [**74**], and Section 2.4 is based on [**48**].

The *plurifine* topology of an open set $U \subset \mathbb{C}^n$ is the coarsest topology making all plurisubharmonic (psh) functions on U continuous. Clearly, the resulting topology on U is finer then the Euclidean one. The motivating result behind this notion is that of Bedford–Taylor [**3**], according to which the complex Monge–Ampère operator is local with respect to the plurifine topology, i.e., if $\phi, \psi \in \mathrm{PSH} \cap L^\infty_{\mathrm{loc}}(U)$ and $V \subset U$ is plurifine open such that $\phi|_V = \psi|_V$, then

$$\mathbb{1}_V (i\partial\bar\partial\phi)^n = \mathbb{1}_V (i\partial\bar\partial\psi)^n. \tag{2.1}$$

When dealing with a n dimensional Kähler manifold (X, ω), the largest class of potentials one can consider is that of *ω-psh* functions:

$$\mathrm{PSH}(X,\omega) = \{u \in L^1(X),\ u \text{ is usc and } \omega_u := \omega + i\partial\bar\partial u \geq 0 \text{ as currents}\}.$$

In practical terms, $u \in \mathrm{PSH}(X,\omega)$ means that $g + u|_U \in \mathrm{PSH}(U)$, for all open sets $U \subset X$ on which the metric ω can be written as $\omega = i\partial\bar\partial g,\ g \in C^\infty(U)$.

Given the local nature of (2.1), it generalizes to Kähler manifolds in a straightforward manner, in particular we have the following identity for $u, v \in \mathrm{PSH}(X,\omega) \cap L^\infty$ that we will use numerous times below:

$$\mathbb{1}_{\{u>v\}} \omega^n_{\max(u,v)} = \mathbb{1}_{\{u>v\}} \omega^n_u. \tag{2.2}$$

2.1. Full mass ω-psh functions

One of the cornerstones of Bedford–Taylor theory is associating a complex Monge–Ampère measure to bounded psh functions. Their construction generalizes to elements of $\mathrm{PSH}(X,\omega) \cap L^\infty$ in a straightforward manner.

As it turns out, it is possible to generalize the Bedford–Taylor construction to all elements of $\mathrm{PSH}(X,\omega)$, as we describe now. Let $v \in \mathrm{PSH}(X,\omega)$ and by $v_h = \max(v, -h) \in \mathrm{PSH}(X,\omega) \cap L^\infty,\ h \in \mathbb{R}$ we denote the *canonical cutoffs* of v. If $h_1 < h_2$ then (2.2) implies that

$$\mathbb{1}_{\{v>-h_1\}} \omega^n_{v_{h_1}} = \mathbb{1}_{\{v>-h_1\}} \omega^n_{v_{h_2}} \leq \mathbb{1}_{\{v>-h_2\}} \omega^n_{v_{h_2}},$$

hence $\{\mathbb{1}_{\{v>-h\}} \omega^n_{v_h}\}_h$ is an increasing sequence of Borel measures on X. This leads to the following natural defintion of ω^n_v, that generalizes the Bedford–Taylor construction:

$$\omega^n_v := \lim_{h\to\infty} \mathbb{1}_{\{v>-h\}} \omega^n_{v_h}. \tag{2.3}$$

This definition means that $\int_B \omega^n_v = \lim_{h\to\infty} \int_B \mathbb{1}_{\{v>-h\}} \omega^n_{v_h}$ for all Borel sets $B \subset X$, hence it is stronger then simply saying that ω^n_v is the weak limit of $\mathbb{1}_{\{v>-h\}} \omega^n_{v_h}$.

Before we examine this construction more closely, let us recall a few facts about smooth elements of $\mathrm{PSH}(X,\omega)$. Of great importance in this work is the set of smooth *Kähler potentials*:

$$\mathcal{H}_\omega = \{u \in C^\infty(X),\ \omega + i\partial\bar\partial u > 0\}. \tag{2.4}$$

Clearly, $\mathcal{H}_\omega \subset \mathrm{PSH}(X,\omega)$ and in fact any element of $\mathrm{PSH}(X,\omega)$ can be approximated by a decreasing sequence in $\mathcal{H}_\omega$:

THEOREM 2.1 ([**23**]). *Given $u \in \mathrm{PSH}(X,\omega)$, there exists a decreasing sequence $\{u_k\}_k \subset \mathcal{H}_\omega$ such that $u_k \searrow u$.*

We give a proof of this result in Appendix A.2. Among other things, this theorem implies that the total volume of X is the same for all currents ω_u with bounded potential:

LEMMA 2.2. *if $v \in \mathrm{PSH}(X,\omega) \cap L^\infty$, then $\int_X \omega_v^n = \int_X \omega^n =: \mathrm{Vol}(X)$.*

PROOF. By an application of Stokes theorem, the statement holds for $v \in \mathcal{H}_\omega$. The general result follows after we approximate $v \in \mathrm{PSH}(X,\omega) \cap L^\infty$ with a decreasing sequence $v_k \in \mathcal{H}_\omega$ (Theorem 2.1), and we use Bedford–Taylor theory ([**20**, Theorem 2.2.5]) to conclude that $\omega_{v_k}^n \to \omega_v^n$ weakly. □

In contrast with the above, given our definition (2.3), it is clear that we only have $\int_X \omega_v^n \leq \mathrm{Vol}(X)$ for $v \in \mathrm{PSH}(X,\omega)$. This leads to the natural definition of ω-psh functions of *full mass*:

$$\mathcal{E}(X,\omega) := \{v \in \mathrm{PSH}(X,\omega) \text{ s.t. } \int_X \omega_v^n = \mathrm{Vol}(X)\}.$$

Though $\mathrm{PSH}(X,\omega) \cap L^\infty \subsetneq \mathcal{E}(X,\omega)$ (see the examples of Section 2.3), many of the properties that hold for bounded ω-psh functions, still hold for (unbounded) elements of $\mathcal{E}(X,\omega)$ as well, like the *comparison principle*:

PROPOSITION 2.3. [**74**, Theorem 1.5] *Suppose $u, v \in \mathcal{E}(X,\omega)$. Then*

$$\int_{\{v<u\}} \omega_u^n \leq \int_{\{v<u\}} \omega_v^n. \tag{2.5}$$

PROOF. First we show (2.5) for u, v bounded. Using (2.2) we can write:

$$\int_{\{v<u\}} \omega_u^n = \int_{\{v<u\}} \omega_{\max(u,v)}^n = \int_X \omega_{\max(u,v)}^n - \int_{\{u\leq v\}} \omega_{\max(u,v)}^n$$
$$\leq \int_X \omega_{\max(u,v)}^n - \int_{\{u<v\}} \omega_{\max(u,v)}^n.$$

Using Lemma 2.2 and (2.2) we can continue to write:

$$\int_{\{v<u\}} \omega_u^n \leq \int_X \omega_v^n - \int_{\{u<v\}} \omega_v^n = \int_{\{v\leq u\}} \omega_v^n.$$

Noticing that $\{v<u\} = \cup_{\varepsilon>0}\{v+\varepsilon<u\} = \cup_{\varepsilon>0}\{v+\varepsilon\leq u\}$, we can replace v with $v+\varepsilon$ in the above inequality and let $\varepsilon \to 0$ to obtain (2.5) for bounded potentials.

In general, let $u_l = \max(u,-l), v_k = \max(u,-k)$, $l,k \in \mathbb{N}$ be the canonical cutoffs of u, v. For these, by the above we have $\int_{\{v_l<u_k\}} \omega_{u_k}^n \leq \int_{\{v_l<u_k\}} \omega_{v_l}^n$. This inequality together with the inclusions $\{v_l<u\} \subset \{v_l<u_k\} \subset \{v<u_k\}$ gives

$$\int_{\{v_l<u\}} \omega_{u_k}^n \leq \int_{\{v<u_k\}} \omega_{v_l}^n.$$

We can let $l \to \infty$, and using the definition of $\omega_{v_l}^n$ from (2.3) it follows that $\int_{\{v<u\}} \omega_{u_k}^n \leq \int_{\{v<u_k\}} \omega_v^n$. Now letting $k \to \infty$ and using again (2.3) we obtain

$$\int_{\{v<u\}} \omega_u^n \leq \int_{\{v\leq u\}} \omega_v^n.$$

As in the bounded case, replacing v with $v+\varepsilon$ in the above inequality, and then taking the limit $\varepsilon \to 0$ yields (2.5). □

Of great importance in what follows will be the *monotonicity property* of $\mathcal{E}(X,\omega)$:

PROPOSITION 2.4. [**74**, Proposition 1.6] *Let $u \in \mathcal{E}(X,\omega)$ and $v \in \mathrm{PSH}(X,\omega)$. If $u \le v$ then $v \in \mathcal{E}(X,\omega)$.*

PROOF. Set $\psi := v/2$. We can assume without loss of generality that $u \le v < -2$, hence $\psi < -1$. This choice of normalization allows to write the following inclusions for the canonical cutoffs u_j, v_j, ψ_j:

$$\{\psi \le -j\} = \{\psi_j \le -j\} \subset \{u_{2j} < \psi_j - j + 1\} \subset \{u_{2j} \le -j\}.$$

Using these inclusions and the comparison principle (Proposition 2.3) we conclude:

$$\int_{\{\psi_j \le -j\}} \omega^n_{\psi_j} \le \int_{\{u_{2j}<\psi_j-j+1\}} \omega^n_{\psi_j} \le \int_{\{u_{2j}<\psi_j-j+1\}} \omega^n_{u_{2j}} \le \int_{\{u_{2j}\le -j\}} \omega^n_{u_{2j}}.$$

Using Lemma 2.2 and (2.2) we get $\int_{\{u_{2j}\le -j\}} \omega^n_{u_{2j}} = \int_{\{u_j \le -j\}} \omega^n_{u_j} = \int_{\{u\le -j\}} \omega^n_{u_j}$, hence $\int_{\{\psi_j\le -j\}} \omega^n_{\psi_j} \to 0$ as $j \to \infty$, implying that $\psi = v/2 \in \mathcal{E}(X,\omega)$.

Now we argue that in fact $v \in \mathcal{E}(X,\omega)$. For the canonical cutoffs we have the identity $v_{2j}/2 = \psi_j$, $j \in \mathbb{N}$, hence we get the estimate $\omega_{\psi_j} = \omega/2 + \omega_{v_{2j}}/2 \ge \omega_{v_{2j}}/2$. This allows to conclude that

$$\int_{\{v\le -2j\}} \omega^n_{v_{2j}} \le 2^n \int_{\{v \le -2j\}} \omega^n_{\psi_j} = 2^n \int_{\{\psi \le -j\}} \omega^n_{\psi_j},$$

hence the left hand integral converges to zero as $j \to \infty$, since $\psi \in \mathcal{E}(X,\omega)$. Using (2.3), this implies that $\int_X \omega_v^n = \mathrm{Vol}(X)$, i.e., $v \in \mathcal{E}(X,\omega)$ as desired. □

Lastly we note that formula (2.2) generalizes to full mass ω-psh functions as well:

LEMMA 2.5. *For $u, v \in \mathcal{E}(X,\omega)$ we have $\mathbb{1}_{\{u>v\}}\omega^n_{\max(u,v)} = \mathbb{1}_{\{u>v\}}\omega^n_u$.*

PROOF. Proposition 2.4 implies that $\alpha := \max(u,v) \in \mathcal{E}(X,\omega)$. For the canonical cutoffs we observe that $\max(u_j, v_{j+1}) = \max(u,v,-j) = \alpha_j$. As the cutoffs are bounded we have

$$\mathbb{1}_{\{u_j > v_{j+1}\}}\omega^n_{\alpha_j} = \mathbb{1}_{\{u_j>v_{j+1}\}}\omega^n_{u_j}. \tag{2.6}$$

From (2.3) we have that $\mathbb{1}_{\{u>v\}}\omega^n_{u_j} \to \mathbb{1}_{\{u>v\}}\omega^n_u$ and $\mathbb{1}_{\{u>v\}}\omega^n_{\alpha_j} \to \mathbb{1}_{\{u>v\}}\omega^n_\alpha$. Since $\{u>v\} \subset \{u_j > v_{j+1}\}$ and $\{u_j > v_{j+1}\} \setminus \{u>v\} \subset \{u \le -j\}$, it follows that

$$0 \le (\mathbb{1}_{\{u>v\}} - \mathbb{1}_{\{u_j>v_{j+1}\}})\omega^n_{u_j} \le \mathbb{1}_{\{u\le -j\}}\omega^n_{u_j} \to 0.$$

Since $\{u_j > v_{j+1}\} \setminus \{u>v\} \subset \{\max(u,v) \le -j\}$ we also obtain that $0 \le (\mathbb{1}_{\{u>v\}} - \mathbb{1}_{\{u_j>v_{j+1}\}})\omega^n_{\alpha_j} \le \mathbb{1}_{\{\alpha\le -j\}}\omega^n_{\alpha_j} \to 0$. Combining these last facts with (2.6), and taking the limit, we arrive at the desired result. □

2.2. Finite energy classes

By considering weights $\chi \in \mathcal{W}_p^+$, one can introduce various finite energy subclasses of $\mathcal{E}(X,\omega)$, important in our later geometric study:

$$\mathcal{E}_\chi(X,\omega) := \{u \in \mathcal{E}(X,\omega) \text{ s.t. } E_\chi(u) < \infty\}, \tag{2.7}$$

where E_χ is the χ*-energy* defined by the expression:

$$E_\chi(u) := \int_X \chi(u)\omega_u^n.$$

Recall from Part 1 that the condition $E_\chi(u) < \infty$ is equivalent to $u \in L^\chi(\omega_u^n)$. Of special importance are the weights $\chi_p(t) := |t|^p/p$ and the associated finite energy classes

$$\mathcal{E}_p(X,\omega) := \mathcal{E}_{\chi_p}(X,\omega). \tag{2.8}$$

By the next result, to test membership in $\mathcal{E}_\chi(X,\omega)$ it is enough to test the finiteness condition $E_\chi(u) < \infty$ on the canonical cutoffs:

PROPOSITION 2.6. [**74**, Proposition 1.4] *Suppose $u \in \mathcal{E}(X,\omega)$ with canonical cutoffs $\{u_k\}_{k\in\mathbb{N}}$. If $h : \mathbb{R}^+ \to \mathbb{R}^+$ is continuous and increasing then*

$$\int_X h(|u|)\omega_u^n < \infty \text{ if and only if } \limsup_k \int_X h(|u_k|)\omega_{u_k}^n < \infty.$$

If the above condition holds, then $\int_X h(|u|)\omega_u^n = \lim_{k\to\infty} \int_X h(|u_k|)\omega_{u_k}^n$.

PROOF. We can assume that $u \leq 0$. If $\limsup_k \int_X h(|u_k|)\omega_{u_k}^n < C \in \mathbb{R}^+$ then the family of Borel measures $\{h(|u_k|)\omega_{u_k}^n\}_k$ is precompact, hence one can extract a weakly converging subsequence $\{h(|u_{k_j}|)\omega_{u_{k_j}}^n\}_{k_j} \to \mu$ with $\mu(X) < C$. By the definition of ω_u^n (2.3) it follows that $\omega_{u_{k_j}}^n \to \omega_u^n$. Since $\{h(|u_{k_j}|)\}_{k_j}$ is a sequence of lsc functions increasing to $h(|u|)$, a standard measure theoretic lemma implies that $\int_X h(|u|)\omega_u^n \leq \mu(X)$ (see [**20**, Lemma A2.2]).

Now assume that $\int_X h(|u|)\omega_u^n \leq C$. By (2.2) and the definition of ω_u^n it follows that $\mathbb{1}_{\{u>-k\}}\omega_u^n = \mathbb{1}_{\{u>-k\}}\omega_{u_k}^n$, implying that $\int_{\{u\leq -k\}} \omega_{u_k}^n = \int_{\{u\leq -k\}} \omega_u^n$. As a consequence, we can write the following:

$$\begin{aligned}\Big|\int_X h(|u_k|)\omega_{u_k}^n - \int_X h(|u|)\omega_u^n\Big| &\leq \int_{\{u\leq -k\}} h(k)\omega_{u_k}^n + \int_{\{u\leq -k\}} h(|u|)\omega_u^n\\ &= h(k)\int_{\{u\leq -k\}} \omega_u^n + \int_{\{u\leq -k\}} h(|u|)\omega_u^n\\ &\leq 2\int_{\{u\leq -k\}} h(|u|)\omega_u^n.\end{aligned}$$

Since $\int_{\{u=-\infty\}} h(|u|)\omega_u^n = 0$, it follows that $\int_X h(|u_k|)\omega_{u_k}^n$ is bounded above and moreover $\int_X h(|u_k|)\omega_{u_k}^n \to \int_X h(|u|)\omega_u^n$. □

With the aid of the previous proposition, we can prove our next result, sometimes called the *"fundamental estimate"*:

PROPOSITION 2.7. [**74**, Lemma 3.5] *Suppose $\chi \in \mathcal{W}_p^+$ and $u, v \in \mathcal{E}_\chi(X,\omega)$ satisfies $u \leq v \leq 0$. Then*

$$E_\chi(v) \leq (p+1)^n E_\chi(u). \tag{2.9}$$

PROOF. First assume that $u, v \in \mathrm{PSH}(X,\omega) \cap L^\infty$. We first show the following estimate:

$$\int_X \chi(u)\omega_v^{j+1} \wedge \omega_u^{n-j-1} \leq (p+1) \int_X \chi(u)\omega_v^j \wedge \omega_u^{n-j}. \tag{2.10}$$

To show this estimate, integration by parts gives

$$\int_X \chi(u)\omega_v^{j+1} \wedge \omega_u^{n-j-1} = \int_X \chi(u)\omega \wedge \omega_v^j \wedge \omega_u^{n-j-1} + \int_X vi\partial\bar\partial\chi(u) \wedge \omega_v^j \wedge \omega_u^{n-j-1}. \tag{2.11}$$

The first integral on the right hand side is bounded above by $\int_X \chi(u)\omega_v^j \wedge \omega_u^{n-j}$. Indeed, as $u \leq 0$ we have $\int_X i\chi'(u)\partial u \wedge \bar\partial u \wedge \omega_v^j \wedge \omega_u^{n-j-1} \leq 0$.

Concerning the second integral on the right hand side of (2.11) we notice that $i\partial\bar\partial\chi(u) = i\chi''(u)\partial u \wedge \bar\partial u + \chi'(u)i\partial\bar\partial u \geq \chi'(u)\omega_u$. Since $v \leq 0$ we can write

$$\begin{aligned}\int_X vi\partial\bar\partial\chi(u) \wedge \omega_v^j \wedge \omega_u^{n-j-1} &\leq \int_X v\chi'(u)\omega_v^j \wedge \omega_u^{n-j} = \int_X |v|\chi'(|u|)\omega_v^j \wedge \omega_u^{n-j} \\ &\leq \int_X |u|\chi'(|u|)\omega_v^j \wedge \omega_u^{n-j} \leq p\int_X \chi(u)\omega_v^j \wedge \omega_u^{n-j},\end{aligned}$$

where in the last inequality we have used that $\chi \in \mathcal{W}_p^+$. Combining the above with (2.11) yields (2.10). Iterating (2.10) n times and using the fact that $u \leq v$ gives (2.9) for bounded potentials.

In case $u, v \in \mathcal{E}_\chi(X,\omega)$, we know that $E_\chi(u_k) \leq (p+1)^n E_\chi(v_k)$ for the canonical cutoffs u_k, v_k. Proposition 2.6 allows to take the limit $k \to \infty$, to obtain (2.9). □

As a corollary of this result, we obtain the *monotonicity property* for $\mathcal{E}_\chi(X,\omega)$:

COROLLARY 2.8. *Suppose $u \in \mathcal{E}_\chi(X,\omega)$ and $v \in \mathrm{PSH}(X,\omega)$. If $u \leq v$ then $v \in \mathcal{E}_\chi(X,\omega)$.*

PROOF. We can assume without loss of generality that $u \leq v \leq 0$. Proposition 2.4 implies that $v \in \mathcal{E}(X,\omega)$. Also for the canonical cutoffs v_k we have $u \leq v_k$, hence $E_\chi(v_k) \leq (p+1)^n E_\chi(u)$ for all $k \in \mathbb{N}$. Proposition 2.6 now gives that $E_\chi(v) \leq (p+1)^n E_\chi(u)$, finishing the proof. □

The next result says that if $u, v \in \mathcal{E}_\chi(X,\omega)$, then in fact $u \in L^\chi(\omega_v^n)$:

PROPOSITION 2.9. [**74**, Proposition 3.6] *Suppose $u, v \in \mathcal{E}_\chi(X,\omega)$, $\chi \in \mathcal{W}_p^+$. If $u, v \leq 0$ then*

$$\int_X \chi(u)\omega_v^n \leq p2^p\big(E_\chi(u) + E_\chi(v)\big)$$

PROOF. We first show that $\chi'(2t) \leq p2^{p-1}\chi'(t),\ t > 0$. For this, after possibly adding $\delta|t|$ to $\chi(t)$, we can momentarily assume that $\chi'(t), \chi(t) > 0$ for any $t > 0$.

By convexity we have $\chi(t)/t \leq \chi'(t)$, and (1.8) gives $\chi(2t)/\chi(t) \leq 2^p$, hence we can write:

$$\frac{\chi'(2t)}{\chi'(t)} = \frac{2t\chi'(2t)}{\chi(2t)} \cdot \frac{\chi(2t)}{2\chi(t)} \cdot \frac{\chi(t)}{t\chi'(t)} \leq p2^{p-1}. \tag{2.12}$$

Consequently we have the following sequence of inequalities:

$$\int_X \chi(u)\omega_v^n = \int_0^\infty \chi'(t)\omega_v^n\{|u| > t\}dt \leq 2^p p\int_0^\infty \chi'(t)\omega_v^n\{|u| > 2t\}dt$$

Noticing that $\{u < -2t\} \subset \{u < -t + v\} \cup \{v < -t\}$ we can continue to write:

$$\begin{aligned}
\int_X \chi(u)\omega_v^n &\leq 2^p p\Big(\int_0^\infty \chi'(t)\omega_v^n\{u < -t+v\}dt + \int_0^\infty \chi'(t)\omega_v^n\{v < -t\}dt\Big) \\
&\leq 2^p p\Big(\int_0^\infty \chi'(t)\omega_u^n\{u < -t+v\}dt + E_\chi(v)\Big) \\
&\leq 2^p p\Big(\int_0^\infty \chi'(t)\omega_u^n\{u < -t\}dt + E_\chi(v)\Big) = 2^p p\big(E_\chi(u) + E_\chi(v)\big),
\end{aligned}$$

where in the second line we have used Proposition 2.3 and in the last line we have used that $\{u < -t + v\} \subset \{u < -t\}$. □

Before we can establish the continuity property of the Monge–Ampère operator along monotonic sequences of potentials, we need to establish the following auxiliary result, which states that if $u \in \mathcal{E}_\chi(X, \omega)$, then it is possible to find a $\tilde{\chi}$ with bigger growth than χ such that $u \in \mathcal{E}_{\tilde{\chi}}(X, \omega)$ still holds:

LEMMA 2.10. *Suppose $u \in \mathcal{E}_\chi(X, \omega)$, $\chi \in \mathcal{W}_p^+$. Then there exists $\tilde{\chi} \in \mathcal{W}^+_{2p+1}$ such that $\chi(t) \leq \tilde{\chi}(t)$, $\chi(t)/\tilde{\chi}(t) \searrow 0$ as $t \to \infty$, and $u \in \mathcal{E}_{\tilde{\chi}}(X, \omega)$.*

PROOF. The weight $\tilde{\chi} : \mathbb{R}^+ \to \mathbb{R}^+$ will be constructed as an increasing limit of weights $\tilde{\chi}^j \in \mathcal{W}^+_{2p+1}, j \in \mathbb{N}$ constructed below. We set $\tilde{\chi}_0(t) = \chi$. Let $t_0 \in \mathbb{R}^+$, to be specified later. We define $\tilde{\chi}_1 : \mathbb{R}^+ \to \mathbb{R}^+$ by the formula

$$\tilde{\chi}_1(t) = \begin{cases} \tilde{\chi}_0(t), & \text{if } t \leq t_1 \\ \tilde{\chi}_0(t_1) + 2(\tilde{\chi}_0(t) - \tilde{\chi}_0(t_1)), & \text{if } t > t_1. \end{cases}$$

Notice that $\tilde{\chi}_1$ satisfies $\chi \leq \tilde{\chi}_1$ and the following also hold

$$\sup_{t>0} \frac{|t\tilde{\chi}_1'(t)|}{|\tilde{\chi}_1(t)|} \leq \sup_{t>0} \frac{2|t\tilde{\chi}_0'(t)|}{|\tilde{\chi}_0(t)|} < 2p + 1, \tag{2.13}$$

$$\lim_{t\to\infty} \frac{|t\tilde{\chi}_1'(t)|}{|\tilde{\chi}_1(t)|} = p. \tag{2.14}$$

We can choose t_1 to be big enough such that $E_{\tilde{\chi}_1}(u) < E_\chi(u) + 1$.

Now pick $t_2 > t_1$, again specified later. One defines $\tilde{\chi}_2 : \mathbb{R}^+ \to \mathbb{R}^+$ in a similar manner:

$$\tilde{\chi}_2(t) = \begin{cases} \tilde{\chi}_1(t), & \text{if } t \leq t_2 \\ \tilde{\chi}_1(t_2) + 2(\tilde{\chi}_1(t) - \tilde{\chi}_1(t_2)), & \text{if } t > t_2. \end{cases}$$

As (2.14) holds for $\tilde{\chi}_1$, it is possible to choose $t_2 > t_1$ big enough so that the $\tilde{\chi}_2$-analogs of (2.13),(2.14) are satisfied and $E_{\tilde{\chi}_2}(u) < E_\chi(u) + 1$.

We define $\tilde{\chi}_k, k \in \mathbb{N}$, following the above procedure. As $\lim_{t\to\infty} \tilde{\chi}_k(t)/\chi(t) = 2^k$, the limit weight $\tilde{\chi}(t) = \lim_{k\to\infty} \tilde{\chi}_k(t)$ is seen to satisfy the requirements of the lemma. □

The following result, allowing to take weak limits of certain measures, will be used in many different contexts throughout the survey:

PROPOSITION 2.11. *Assume that $\{\phi_k\}_{k\in\mathbb{N}}, \{\psi_k\}_{k\in\mathbb{N}}, \{v_k\}_{k\in\mathbb{N}} \subset \mathcal{E}_\chi(X, \omega)$ decrease (increase a.e.) to $\phi, \psi, v \in \mathcal{E}_\chi(X, \omega)$ respectively. Suppose the following hold:*
(i) $\psi_k \leq \phi_k$ and $\psi_k \leq v_k$.

(ii) $h : \mathbb{R} \to \mathbb{R}$ *is continuous with* $\limsup_{|l|\to\infty} |h(l)|/\chi(l) \leq C$ *for some* $C > 0$. *Then* $h(\phi_k - \psi_k)\omega^n_{v_k} \to h(\phi - \psi)\omega^n_v$ *weakly.*

We will apply this porposition mostly for $h = \chi$ and $h = 1$. In the latter case this proposition simply tells that the complex Monge–Ampère measures, as defined in (2.3), converge weakly along monotonic sequences of $\mathcal{E}_\chi(X,\omega)$. Though we will not use it here, this also holds more generally for sequences inside $\mathcal{E}(X,\omega)$, as shown in [**74**].

PROOF. We can suppose without loss of generality that all the functions involved are negative. First we suppose that there exists $L > 1$ such that $-L < \phi, \phi_k, \psi, \psi_k, v, v_k < 0$ are prove the theorem under this assumption. Given $\varepsilon > 0$ one can find an open $O \subset X$ such that $\mathrm{Cap}_X(O) < \varepsilon$ and $\phi, \phi_k, \psi, \psi_k, v, v_k$ are all continuous on $X \setminus O$ ([**20**, Theorem 2.2] or [**73**, Definition 2.4, Corollary 2.8]). We have

$$\int_X h(\phi_k - \psi_k)\omega^n_{v_k} - \int_X h(\phi - \psi)\omega^n_{v_k} = \int_O + \int_{X\setminus O} [h(\phi_k - \psi_k) - h(\phi - \psi)]\omega^n_{v_k}. \tag{2.15}$$

The integral on O is bounded by $2\varepsilon L^n|h(L)|$. The second integral tends to 0 as on the closed set $X \setminus O$ we have $\phi_k \to \phi$ and $\psi_k \to \psi$ uniformly. We also have

$$\int_X h(\phi - \psi)\omega^n_{v_k} - \int_X h(\phi - \psi)\omega^n_v \to 0, \tag{2.16}$$

as the function $h(\phi - \psi)$ is quasi–continuous and bounded. Indeed, quasi–continuity and boundedness implies again that for all $\varepsilon > 0$ one can find an open $\tilde{O} \subset X$ such that $\mathrm{Cap}_X(\tilde{O}) < \varepsilon$ and $h(\phi - \psi)$ is continuous on $X \setminus \tilde{O}$. Furthermore, by Tietze's extension theorem we can extend $h(\phi - \psi)|_{X\setminus\tilde{O}}$ to a continuous function α on X. As $\mathrm{Cap}_X(\tilde{O}) < \varepsilon$, we have that $\int |h(\phi - \psi) - \alpha|\omega^n_{v_k} \leq C\varepsilon$ and $|h(\phi - \psi) - \alpha|\omega^n_v \leq C\varepsilon$. On the other hand $\int_X \alpha\omega^n_{v_k} \to \int_X \alpha\omega^n_v$ by Bedford-Taylor theory (see [**20**, Theorem 2.2.5]). Putting these facts together we get (2.16). Finally, (2.15) and (2.16) together give the proposition for bounded potentials.

Now we argue that the result also holds when $\phi, \phi_k, \psi, \psi_k, v, v_k$ are unbounded. For this we show that

$$\int_X h(\phi_k - \psi_k)\omega^n_{v_k} - \int_X h(\phi^L_k - \psi^L_k)\omega^n_{v^L_k} \to 0 \tag{2.17}$$

$$\int_X h(\phi - \psi)\omega^n_v - \int_X h(\phi^L - \psi^L)\omega^n_{v^L} \to 0 \tag{2.18}$$

as $L \to \infty$, uniformly with respect to k, where $v^L = \max(v, -L)$ and $v^L_k, \psi^L_k, \psi^L, \phi^L_k$, ϕ^L are defined similarly.

Now we pick $\tilde{\chi} \in \mathcal{W}^+_{2p+1}$. If $\psi_k \searrow \psi$ then choose $\tilde{\chi}$ in such a way that it satisfies the assumptions of the previous lemma for ψ, i.e., $\psi \in \mathcal{E}_{\tilde{\chi}}(X,\omega)$. In case $\psi_k \nearrow \psi$, choose $\tilde{\chi}$ in such a way that it satisfies the assumptions of the previous lemma for ψ_1, i.e. $\psi_1 \in \mathcal{E}_{\tilde{\chi}}(X,\omega)$. By Corollary 2.8 we obtain in both cases that $\phi_k, \phi, \psi_k, \psi, v_k, v \in \mathcal{E}_{\tilde{\chi}}(X,\omega), k \in \mathbb{N}$.

For the rest of the proof assume that the sequences ϕ_k, ψ_k, v_k are decreasing. The case of increasing sequences is argued similarly.

Since $\{\phi_k \leq -L\} \cup \{v_k \leq -L\} \subseteq \{\psi_k \leq -L\}$, and for big enough L we have $|h(t)| \leq (C+1)\chi(t)$ for $|t| \geq L$, using the definition of the complex Monge–Ampère measure (2.3) we can start writing:

$$
\begin{aligned}
&\Big|\int_X \big[h(\phi_k - \psi_k)\omega^n_{v_k} - h(\phi^L_k - \psi^L_k)\omega^n_{v^L_k}\big]\Big| = \\
&\quad = \Big|\int_{\{\psi_k \leq -L\}} \big[h(\phi_k - \psi_k)\omega^n_{v_k} - h(\phi^L_k - \psi^L_k)\omega^n_{v^L_k}\big]\Big| \\
&\quad \leq (C+1)\int_{\{\psi_k \leq -L\}} \chi(\psi_k)\omega^n_{v_k} + (C+1)\int_{\{\psi_k \leq -L\}} \chi(\psi^L_k)\omega^n_{v^L_k} \\
&\quad \leq \frac{(C+1)\chi(L)}{\tilde\chi(L)}\Big(\int_{\{\psi_k \leq -L\}} \tilde\chi(\psi_k)\omega^n_{v_k} + \int_{\{\psi_k \leq -L\}} \tilde\chi(\psi^L_k)\omega^n_{v^L_k}\Big) \\
&\quad \leq \frac{(C+1)\chi(L)}{\tilde\chi(L)}\Big(\int_X \tilde\chi(\psi_k)\omega^n_{v_k} + \int_X \tilde\chi(\psi^L_k)\omega^n_{v^L_k}\Big) \\
&\quad \leq \frac{C(p)\chi(L)}{\psi(L)}\Big(E_{\tilde\chi}(\psi_k) + E_{\tilde\chi}(v_k) + E_{\tilde\chi}(\psi^L_k) + E_{\tilde\chi}(v^L_k)\Big) \leq \frac{C(p)\chi(L)}{\tilde\chi(L)} E_{\tilde\chi}(\psi),
\end{aligned}
$$

where in the last line we have used Proposition 2.9 and Proposition 2.7. As $\chi(L)/\tilde\chi(L) \to 0$, this justifies (2.17) and (2.18) is established the same way, finishing the proof. □

Later, when showing that $\mathcal{E}_\chi(X,\omega)$ admits a complete metric structure, we will make use of the following corollary:

COROLLARY 2.12. [**74**, Proposition 5.6] *Suppose* $\chi \in \mathcal{W}^+_p$ *and* $\{u_k\}_{k\in\mathbb{N}} \subset \mathcal{E}_\chi(X,\omega)$ *is a sequence decreasing to* $u \in \text{PSH}(X,\omega)$. *If* $\sup_k E_\chi(u_k) < \infty$ *then* $u \in \mathcal{E}_\chi(X,\omega)$ *and*

$$E_\chi(u) = \lim_{k\to\infty} E_\chi(u_k).$$

PROOF. Let $u^l_k = \max(u,-l)$ and $u^l = \max(u,-l)$ be the canonical cutoffs. As $-l \leq u^l \leq u^k_l$, by the previous proposition and Proposition (2.7) we get that $E_\chi(u^l) = \lim_k E_\chi(u^l_k) \leq (p+1)^n C := (p+1)^n \limsup_k E_\chi(u_k)$.

Finally, we apply Proposition 2.6 to obtain that $E_\chi(u) = \lim_{l\to\infty} E_\chi(u^l) \leq (p+1)^n C$, i.e., $u \in \mathcal{E}_\chi(X,\omega)$. Another application of the previous proposition gives that $E_\chi(u) = \lim_{k\to\infty} E_\chi(u_k)$. □

2.3. Examples and singularity type of finite energy potentials

It is clear that $\text{PSH}(X,\omega) \cap L^\infty \subset \mathcal{E}_p(X,\omega)$ for all $p \geq 1$. In this short section we will show that this inclusion is always strict, as one can construct unbounded elements of $\mathcal{E}_p(X,\omega)$. However the content of our first result is that the singularity type of full mass potentials is always mild, even when they are unbounded.

Given $u \in \text{PSH}(X,\omega)$ and $x_0 \in X$ it is possible to measure the local singularity of u at x_0 using the *Lelong number* $\mathcal{L}(u,x_0)$, whose definition we now recall:

(2.19) $$\mathcal{L}(u,x_0) := \sup\{r \geq 0 \mid u(x) \leq r\log|x| + C_r \ \forall x \in U_r \ \text{ for some } C_r > 0\},$$

where $U_r \subset \mathbb{C}^n$ is some coordinate neighborhood of x_0 that identifies x_0 with $0 \in \mathbb{C}^n$. Roughly speaking, $\mathcal{L}(u,x_0)$ measures the extent to which the singularity of u at x_0 is logarithmic. For an extensive treatment of Lelong numbers we refer to [**75**, Section 2.3].

To begin, we observe that elements of $\mathcal{E}(X,\omega)$ have singularity so mild that Lelong numbers can not detect them:

PROPOSITION 2.13. [**74**, Corollary 1.8] *If $u \in \mathcal{E}(X,\omega)$ then all Lelong numbers of u are zero.*

PROOF. Let $x_0 \in X$ and $U \subset X$ a coordinate neighborhood of x_0, identifying x_0 with $0 \in \mathbb{C}^n$ via a biholomorphism $\varphi : B(0,2) \to U$. Let $v \in \mathrm{PSH}(X,\omega)$ such that v is smooth on $X \setminus \{x_0\}$ and $v \circ \varphi|_{B(0,1)} = c\log|x|$ for some $c > 0$. Using a partition of unity, such v can be easily constructed.

If $\mathcal{L}(u,x_0) > 0$, then for some $\varepsilon > 0$ we will have $u \leq \varepsilon v + 1/\varepsilon$. The monotonicity property (Proposition 2.4) now implies that $\varepsilon v \in \mathcal{E}(X,\omega)$. However this cannot happen, yielding a contradiction. Indeed, by the definition of $\omega^n_{\varepsilon v}$ (see (2.3)) we have that

$$\begin{aligned}\int_X \omega^n_{\varepsilon v} &= \mathrm{Vol}(X) - \lim_{k\to\infty}\int_{\{\varepsilon v \leq -k\}} \omega^n_{\max(\varepsilon v,-k)}\\ &\leq \mathrm{Vol}(X) - \lim_{k\to\infty}\int_{B(0,1)} i\big(\partial\bar\partial \max(c\varepsilon\log|x|,-k)\big)^n\\ &= \mathrm{Vol}(X) - c^n\varepsilon^n(2\pi)^n,\end{aligned}$$

where in the last identity we have used the fact that $i\partial\bar\partial\big(\max(\log|z|,\log r)\big)^n = (2\pi)^n d\sigma_{\partial B(0,r)}$, where $d\sigma_{\partial B(0,r)}$ is the Euclidean surface measure of $\partial B(0,r)$ (see the exercise following [**20**, Corollary 2.2.7]). □

The purpose of our next proposition is to give a flexible construction for unbounded elements of $\mathcal{E}_p(X,\omega)$:

PROPOSITION 2.14. [**74**, Example 2.14], [**21**, Proposition 5] *Suppose that $u \in \mathrm{PSH}(X,\omega)$, $u < -1$ and $\alpha \in (0,1/2)$. Then $v = -(-u)^\alpha \in \mathcal{E}_p(X,\omega)$ for all $p \in [1,(1-\alpha)/\alpha)$.*

PROOF. First assume that $u \in \mathcal{H}_\omega$ and $u \leq -1$. For $v = -(-u)^\alpha$ we have

$$\begin{aligned}\omega_v &= \alpha(1-\alpha)(-u)^{\alpha-2} i\partial u \wedge \bar\partial u + \alpha(-u)^{\alpha-1}\omega_u + (1-\alpha(-u)^{\alpha-1})\omega\\ &\leq (-u)^{\alpha-2} i\partial u\wedge\bar\partial u + (-u)^{\alpha-1}\omega_u + \omega.\end{aligned}$$

Consequently $v \in \mathcal{H}_\omega$, and for some $C := C(\alpha) > 1$ we have

$$\omega_v^n \leq C\Big(\sum_{k=0}^{n-1}(-u)^{\alpha-2+k(\alpha-1)} i\partial u\wedge\bar\partial u\wedge\omega_u^k\wedge\omega^{n-1-k} + \sum_{k=0}^{n}(-u)^{k(\alpha-1)}\omega_u^k\wedge\omega^{n-k}\Big). \tag{2.20}$$

For arbitrary $a > 0$, integration by parts gives the following estimates

$$\begin{aligned}a\int_X (-u)^{-a-1} i\partial u\wedge\bar\partial u\wedge\omega_u^k\wedge\omega^{n-k-1} &= \int_X i\partial(-u)^{-a}\wedge\bar\partial u\wedge\omega_u^k\wedge\omega^{n-k-1}\\ &= -\int_X (-u)^{-a} i\partial\bar\partial u\wedge\omega_u^k\wedge\omega^{n-k-1}\\ &\leq \int_X (-u)^{-a}\omega_u^k\wedge\omega^{n-k} \leq \mathrm{Vol}(X),\end{aligned}$$

where in the last inequality we have used that $(-u)^{-a} \leq 1$. This estimate and (2.20) implies that for any $b \in (0, 1-\alpha)$ we have

$$\int_X |v|^{\frac{b}{\alpha}} \omega_v^n = \int_X |u|^b \omega_v^n \leq C(\omega, \alpha, b)\Big(1 + \int_X |u|\omega^n\Big).$$

Returning to the general case $u \in \mathrm{PSH}(X, \omega)$, let $\{u_k\}_k \subset \mathcal{H}_\omega$ be a sequence of smooth potentials decreasing to u (Theorem 2.1). We can assume that $u_k < -1$, hence the above inequality implies that $\int_X |v_k|^{b/\alpha} \omega_{v_k}^n$ is uniformly bounded for $v_k := -(-u_k)^\alpha$. Consequently, Corollary 2.12 implies that $v \in \mathcal{E}_p(X, \omega)$ for any $p \in (0, (1-\alpha)/\alpha)$. □

2.4. Envelopes of finite energy classes

Envelope constructions are ubiquitous throughout pluripotential theory. In our setting, given an usc function $f : X \to [-\infty, \infty)$, the simplest envelope one can consider is

$$P(f) := \sup\{u \in \mathrm{PSH}(X, \omega) \text{ s.t. } u \leq f\}. \tag{2.21}$$

As we know, the supremum of a family of ω-psh functions may not be ω-psh, as $P(f)$ may not be usc to begin with. However by [**20**, Theorem 1.2.3] the usc regularization $P(f)^*$ is indeed ω-psh. As f is usc, we obtain that $P(f)^* \leq f^* = f$, immediately giving that $P(f)^*$ is a candidate in the definition of $P(f)$, hence $P(f) = P(f)^*$, i.e., $P(f) \in \mathrm{PSH}(X, \omega)$.

Slightly generalizing the above concept, for usc functions $\{f_1, f_2, \ldots, f_k\}$ we introduce the *rooftop envelope*

$$P(f_1, f_2, \ldots, f_k) := P(\min(f_1, f_2, \ldots, f_k)).$$

When f is smooth (or just continuous), as $P(f)$ is usc, we obtain that the *non-contact set* $\{f > P(f)\} \subset X$ is open. A classical Perron type argument (see [**3**, Corollary 9.2] or [**20**, Proposition 1.4.10]). yields that

$$\omega_{P(f)}^n(\{f > P(f)\}) = 0 \tag{2.22}$$

This observation suggests that $P(f)$ can have at most bounded but not continuous second derivatives. This is mostly confirmed by the next result, whose proof is provided in the appendix:

THEOREM 2.15 (Theorem 5.12, [**57**, Theorem 2.5]). *Given $f_1, \ldots, f_k \in C^\infty(X)$, then $P(f_1, f_2, \ldots, f_k) \in C^{1,\alpha}(X)$, $\alpha \in (0,1)$. More precisely, the following estimate holds:*

$$\|P(f_1, f_2, \ldots, f_k)\|_{C^{1,\bar{1}}} \leq C(X, \omega, \|f_1\|_{C^{1,\bar{1}}}, \|f_2\|_{C^{1,\bar{1}}}, \ldots, \|f_k\|_{C^{1,\bar{1}}}).$$

By a bound on the $C^{1,\bar{1}}$ norm of $P(f)$ we mean a uniform bound on all mixed second order derivatives $\partial^2 P(f)/\partial z_j \bar{\partial} z_k$. Since $P(f)$ is ω-psh, this is equivalent to saying that $\Delta^\omega P(f)$ is bounded, and by the Calderon–Zygmund estimate [**76**, Chapter 9, Lemma 9.9], we automatically obtain that $P(f_1, f_2, \ldots, f_k) \in C^{1,\alpha}(X), \alpha < 1$. We introduce the following subspace of $\mathrm{PSH}(X, \omega)$:

$$\mathcal{H}_\omega^{1,\bar{1}} := \{u \in \mathrm{PSH}(X, \omega) \text{ s.t. } \|u\|_{C^{1,\bar{1}}} < \infty\}. \tag{2.23}$$

When $k = 1$, the above result was first proved in [**15**] using the Kiselman technique for attenuation of singularities. An independent "PDE proof" has been given by Berman [**6**], and we present this in the appendix (see Theorem 5.7). The proof of the general case $k \geq 1$ was given in the paper [**57**], that provided a detailed regularity analysis for the rooftop envelopes introduced above. By reduction to the case $k = 1$, we prove this more general result in the appendix as well (see Theorem 5.12). Lastly, we mention that very recently it was shown by Tosatti [**112**] and Chu-Zhou [**45**] that in fact it is possible to bound the full Hessian of $P(f_1, \dots, f_k)$ in terms of the Hessians of $f_1, \dots, f_k$.

With the above introduced notation, we will use Theorem 2.15 in the following form:

COROLLARY 2.16. *If* $u_0, u_1, \dots, u_k \in \mathcal{H}_\omega^{1,\bar{1}}$ *then* $P(u_0, u_1, \dots, u_k) \in \mathcal{H}_\omega^{1,\bar{1}}$.

PROOF. Let $f_j^i \in C^\infty(X)$ be such that $f_j^i \to u_j$ uniformly, and the mixed second derivatives of f_j^i are uniformly bounded. By the previous theorem, the mixed second derivatives of $P(f_0^i, f_1^i, \dots, f_k^i)$ are uniformly bounded as well. Since $P(f_0^i, f_1^i, \dots, f_k^i) \to P(u_0, u_1, \dots, u_k)$ uniformly, the result follows. □

As a consequence of the above corollary, we get a volume partition formula for $\omega^n_{P(u_0,u_1)}$:

PROPOSITION 2.17. [**48**, Proposition 2.2] *For* $u_0, u_1 \in \mathcal{H}_\omega^{1,\bar{1}}$, *we introduce the contact sets* $\Lambda_{u_0} = \{P(u_0, u_1) = u_0\}$ *and* $\Lambda_{u_1} = \{P(u_0, u_1) = u_1\}$. *Then the following partition formula holds:*

$$\omega^n_{P(u_0,u_1)} = \mathbb{1}_{\Lambda_{u_0}} \omega^n_{u_0} + \mathbb{1}_{\Lambda_{u_1} \setminus \Lambda_{u_0}} \omega^n_{u_1}. \tag{2.24}$$

PROOF. As pointed out in (2.22), $\omega^n_{P(u_0,u_1)}$ is concentrated on the coincidence set $\Lambda_{u_0} \cup \Lambda_{u_1}$. Having bounded Laplacian implies that all second order partials of $P(u_0, u_1)$ are in any $L^p(X)$, $p < \infty$ [**76**, Chapter 9, Lemma 9.9]. It follows from [**76**, Chapter 7, Lemma 7.7] that on Λ_{u_0} all the second order partials of $P(u_0, u_1)$ and u_0 agree a.e., and the analogous statement holds on Λ_{u_1}. Hence, using [**20**, Proposition 2.1.6] one can write:

$$\omega^n_{P(u_0,u_1)} = \mathbb{1}_{\Lambda_{u_0} \cup \Lambda_{u_1}} \omega^n_{P(u_0,u_1)} = \mathbb{1}_{\Lambda_{u_0}} \omega^n_{u_0} + \mathbb{1}_{\Lambda_{u_1} \setminus \Lambda_{u_0}} \omega^n_{u_1},$$

finishing the proof. □

The partition formula (2.24) is at the core of many theorems presented later in this survey. Interestingly, it fails to hold evein in the slightly more general case of Lipschitz potentials

$$u_0, u_1 \in \mathcal{H}_\omega^{0,1} := \mathrm{PSH}(X, \omega) \cap C^{0,1}(X).$$

For a counterexample, suppose $\dim X = 1$ and g_x is the $\omega-$Green function with pole at $x \in X$. Such function is characterized by the property $\int_X g_x \omega = 0$ and $\omega + i\partial\bar{\partial} g_x = \delta_x$. We choose $u_0 = \max\{g_x, 0\}$ and $u_1 = 0$. In this case $P(u_0, u_1) = 0$, $\Lambda_{u_0} = \{g_x \leq 0\}$ and $\Lambda_{u_1} \setminus \Lambda_{u_0} = X \setminus \Lambda_{u_0} \neq \emptyset$. As $\mathrm{Vol}(X) = \int_{\Lambda_{u_0}} \omega^n_{u_0} = \int_X \omega^n_{P(u_0,u_1)}$, it is seen that the right hand side of (2.24) has total integral greater then the left hand side, hence they can not equal.

As $\int_X \omega^n_{u_1}$ is finite, it follows that $\omega^n_{u_1}(\{u_0 = u_1 + \tau\}) > 0$ only for a countable number of values $\tau \in \mathbb{R}$. Consequently, (2.24) implies the following observation:

REMARK 2.18. *Given* $u_0, u_1 \in \mathcal{H}_\omega^{1,\bar{1}}$, *for any* $\tau \in \mathbb{R}$ *outside a countable set we have:*

$$\omega^n_{P(u_0,u_1+\tau)} = \mathbb{1}_{\Lambda_{u_0}} \omega^n_{u_0} + \mathbb{1}_{\Lambda_{u_1+\tau}} \omega^n_{u_1}.$$

In particular, $\mathrm{Vol}(X) = \int_{\{P(u_0,u_1+\tau)=u_0\}} \omega^n_{u_0} + \int_{\{P(u_0,u_1+\tau)=u_1+\tau\}} \omega^n_{u_1}$.

Corollary 2.16 simply says that the operation $(u_0, u_1) \to P(u_0, u_1)$ is closed inside the class $\mathcal{H}^{1,\bar{1}}_\omega$. The next proposition tells that the same holds inside the finite energy classes as well:

PROPOSITION 2.19. [**48**, Lemma 3.4] *Suppose $\chi \in \mathcal{W}^+_p$ and $u_0, u_1 \in \mathcal{E}_\chi(X, \omega)$. Then $P(u_0, u_1) \in \mathcal{E}_\chi(X, \omega)$, and if $u_0, u_1 \leq 0$ then following estimate holds:*

$$E_\chi(P(u_0, u_1)) \leq (p+1)^n (E_\chi(u_0) + E_\chi(u_1)). \tag{2.25}$$

PROOF. As $P(u_0 - c, u_1 - c) = P(u_0, u_1) - c$ for $c \in \mathbb{R}$, it follows that without loss of generality we can assume that $u_0, u_1 < 0$.

By Theorem 2.1, it is possible to find negative potentials $u^j_0, u^j_1 \in \mathcal{H}_\omega$ that decrease to u_0, u_1. Furthermore, by Proposition 2.17 we can also assume that $P(u^j_0, u^j_1) \in \mathcal{H}^{1,\bar{1}}_\omega$ satisfies

$$\omega^n_{P(u^j_0, u^j_1)} \leq \mathbb{1}_{\Lambda_{u^j_0}} \omega^n_{u^j_0} + \mathbb{1}_{\Lambda_{u^j_1}} \omega^n_{u^j_1}. \tag{2.26}$$

Using this formula we can write:

$$\begin{aligned} E_\chi(P(u^j_0, u^j_1)) &= \int_X \chi(P(u^j_0, u^j_1)) \omega^n_{P(u^j_0, u^j_1)} \\ &\leq \int_{\{P(u^j_0, u^j_1) = u^j_0\}} \chi(u^j_0) \omega^n_{u^j_0} + \int_{\{P(u^j_0, u^j_1) = u^j_1\}} \chi(u^j_1) \omega^n_{u^j_1} \\ &\leq \int_X \chi(u^j_0) \omega^n_{u^j_0} + \int_X \chi(u^j_1) \omega^n_{u^j_1} = E_\chi(u^j_0) + E_\chi(u^j_1) \\ &\leq (p+1)^n (E_\chi(u_0) + E_\chi(u_1)), \end{aligned}$$

where in the last line we have used Proposition 2.7. As $P(u^j_0, u^j_1)$ decreases to $P(u_0, u_1)$, by Corollary 2.12 we have $P(u_0, u_1) \in \mathcal{E}_\chi(X, \omega)$, and (2.25) holds. □

Observe that $tu_0 + (1-t)u_1 \geq P(u_0, u_1)$ for any $t \in [0, 1]$, hence as a consequence of the previous proposition and the monotonicity property of $\mathcal{E}_\chi(X, \omega)$ (Corollary 2.8) we obtain that $\mathcal{E}_\chi(X, \omega)$ is convex:

COROLLARY 2.20. *If $u_0, u_1 \in \mathcal{E}_\chi(X, \omega)$ then $tu_0 + (1-t)u_1 \in \mathcal{E}_\chi(X, \omega)$ for any $t \in [0, 1]$.*

Lastly, we prove the *domination principle* for the class $\mathcal{E}_1(X, \omega)$ (recall (2.8)). We mention that these results also hold more generally for the class $\mathcal{E}(X, \omega)$, by a theorem of S. Dinew [**24**, **66**]. The short proof below was pointed out to us by C.H. Lu, and it is based on the arguments of [**52**].

PROPOSITION 2.21. *Let $\phi, \psi \in \mathcal{E}_1(X, \omega)$. If $\psi \leq \phi$ almost everywhere with respect to ω^n_ϕ then $\psi \leq \phi$.*

Since $\mathcal{E}_p(X, \omega) \subset \mathcal{E}_1(X, \omega)$, $p \geq 1$, we obtain that the domination principle trivially holds for $\phi, \psi \in \mathcal{E}_p(X, \omega)$, $p \geq 1$ as well.

PROOF. We can assume without loss of generality that $\phi, \psi < 0$. As $\phi, \psi \in \text{PSH}(X, \omega)$, it suffices to prove that $\psi \leq \phi$ a.e. with respect to ω^n. This will then imply that $\psi \leq \phi$ globally.

Suppose that $\omega^n(\{\phi < \psi\}) > 0$. By the next lemma, this implies existence of $u \in \mathcal{E}_1(X, \omega)$ such that $u \leq \psi$ and $\omega^n_u(\{\phi < \psi\}) > 0$.

By Corollary 2.20 we have that $tu + (1-t)\psi \in \mathcal{E}(X,\omega)$ for any $t \in [0,1]$. By the comparison principle (Proposition 2.3) we can write:

$$t^n \int_{\{\phi < tu+(1-t)\psi\}} \omega_u^n \leq \int_{\{\phi < tu+(1-t)\psi\}} \omega_{tu+(1-t)\psi}^n$$
$$\leq \int_{\{\phi < tu+(1-t)\psi\}} \omega_\phi^n = \int_{\{\phi<\psi\}} \omega_\phi^n = 0.$$

We conclude that $0 = \omega_u^n(\{\phi < tu + (1-t)\psi\}) \nearrow \omega_u^n(\{\phi < \psi\})$, as $t \searrow 0$. As a result, $\omega_u^n(\{\phi < \psi\}) = 0$, a contradiction. □

LEMMA 2.22. *Suppose that $U \subset X$ is a Borel set with non–zero Lebesgue measure and $\psi \in \mathcal{E}_1(X,\omega)$. Then there exists $u \in \mathcal{E}_1(X,\omega)$ such that $u \leq \psi$ and $\omega_u^n(U) > 0$.*

PROOF. We can assume that $\psi < 0$. We fix $C > 0$ and let $\psi_C := P(\psi + C, 0)$. Let $u_j \in \mathcal{H}_\omega$ be a decreasing approximating sequence of ψ, that exists by Theorem 2.1. We can assume that $u_j < 0$. The partition formula of Proposition 2.17 gives:

$$\omega_{P(u_j+C,0)}^n \leq \mathbb{1}_{\{u_j \leq -C\}}\omega_{u_j}^n + \omega^n \leq -\frac{u_j}{C}\omega_{u_j}^n + \omega^n.$$

Notice that $P(u_j + C, 0) \searrow \psi_C$ as $j \to \infty$. Taking the limit, Proposition 2.11 gives us the estimate of measures $\omega_{\psi_C}^n \leq -\frac{\psi}{C}\omega_\psi^n + \omega^n$. Using this we can write:

$$\omega_{\psi_C}^n(X \setminus U) \leq \frac{1}{C}\int_{X\setminus U} |\psi|\omega_\psi^n + \omega^n(X \setminus U) < \frac{1}{C}\int_X |\psi|\omega_\psi^n + \omega^n(X \setminus U).$$

As $\omega_{\psi_C}^n(X) = \omega^n(X)$, for C big enough we get that $\omega_{\psi_C}^n(U) > 0$. The choice $u := \psi_C - C$ satisfies the requirements of the lemma. □

Part 3. The Finsler geometry of the space of Kähler potentials

As follows from the definition (2.4), the space of Kähler potentials $\mathcal{H}_\omega$ is a convex open subset of $C^\infty(X)$, hence one can think of it as a trivial Fréchet manifold. As such, one can introduce on $\mathcal{H}_\omega$ a collection of L^p type Finsler metrics whose geometry we will study in this section. If $u \in \mathcal{H}_\omega$ and $\xi \in T_u\mathcal{H}_\omega \simeq C^\infty(X)$, then the L^p-length of ξ is given by the following expression:

$$\|\xi\|_{p,u} = \Big(\frac{1}{\mathrm{Vol}(X)}\int_X |\psi|^p\omega_u^n\Big)^{\frac{1}{p}}, \tag{3.1}$$

where $\mathrm{Vol}(X) = \int_X \omega^n$. In case $p = 2$, we recover the Riemannian geometry of Mabuchi [**88**] (independently discovered by Semmes [**99**] and Donaldson [**67**]). Though these metrics will be our primary object of study, unfortunately the associated weights $\chi_p(l) = l^p/p$ are not twice differentiable for $1 \leq p < 2$. This will cause problems as we shall see, and an approximation with more regular weights will be necessary to carry out even the most basic geometric arguments. For this reason, one needs to first work with even more general Finsler metrics on $\mathcal{H}_\omega$ with smooth weights in $\mathcal{W}_p^+$ (see (1.7)) and then use approximation via Propositions 1.6 and 1.7 to return to the L^p metrics in the end.

Following our discussion above, we introduce the Orlicz–Finsler length of ξ for any weight $\chi \in \mathcal{W}_p^+$:

$$\|\xi\|_{\chi,u} = \inf\Big\{r > 0 : \frac{1}{\mathrm{Vol}(X)}\int_X \chi\Big(\frac{\xi}{r}\Big)\omega_u^n \leq \chi(1)\Big\}. \tag{3.2}$$

The above expression introduces a norm on each fiber of $T\mathcal{H}_\omega$, and the length of a smooth curve $[0,1] \ni t \to \alpha_t \in \mathcal{H}_\omega$ is computed by the usual formula:

$$l_\chi(\alpha_t) = \int_0^1 \|\dot\alpha_t\|_{\chi,\alpha_t} dt. \tag{3.3}$$

To clarify, smoothness of $t \to \alpha_t$ simply means that the map $\alpha(t,x) = \alpha_t(x)$ is smooth as a map from $[0,1] \times X$ to $\mathbb{R}$. Furthermore, the distance $d_\chi(u_0,u_1)$ between $u_0, u_1 \in \mathcal{H}_\omega$ is the infimum of the l_χ-length of smooth curves joining u_0 and u_1:

$$d_\chi(u_0,u_1) = \inf\{l_\chi(\gamma_t) : \ t \to \gamma_t \ \text{ is smooth and } \ \gamma_0 = u_0, \gamma_1 = u_1\}. \tag{3.4}$$

The distance d_χ is a pseudo–metric (the triangle inequality holds), but $d_\chi(u,v) = 0$ may not imply $u = v$, as our setting is infinite–dimensional. We will see in Theorem 3.6 below that d_χ is a bona fide metric, but this will require a careful analysis of our Finsler structures.

When dealing with the L^p metric structures (3.1), the associated curve length and path length metric will be denoted by l_p and d_p respectively.

As we will see, the different weights in $\mathcal{W}_p^+$ induce different geometries on $\mathcal{H}_\omega$, however the equation for the shortest length curves between points of $\mathcal{H}_\omega$, the so called geodesics, will be essentially the same. Motivated by this, in the next section we will focus on the L^2 Riemannian geometry first, in which case we can get an explicit equation for these geodesics.

3.1. Riemannian geometry of the space of Kähler potentials

When $p = 2$ the metric of (3.1) is induced by the following non-degenerate inner product:

$$\langle \phi, \psi \rangle_u = \frac{1}{\text{Vol}(X)} \int_X \phi\psi\omega_u^n, \quad u \in \mathcal{H}_\omega, \ \phi, \psi \in T_u\mathcal{H}_\omega. \tag{3.5}$$

This Riemannian structure was first studied by Mabuchi [**88**] and later independently by Semmes [**99**] and Donaldson [**67**]. For another introductory survey on the Mabuchi geometry of $\mathcal{H}_\omega$ we refer to [**22**].

Let us compute the Levi–Civita connection of this metric. For this we choose a smooth curve $[0, 1] \ni t \to u_t \in \mathcal{H}_\omega$ and $[0, 1] \ni t \to \phi_t, \psi_t \in C^\infty(X)$, two vector fields along $t \to u_t$. In the future, when working with time derivatives, we will use the notation $\dot{u}_t = du_t/dt$, $\ddot{u}_t = d^2u_t/dt^2$, etc.

We will identify the Levi–Civita connection $\nabla_{(\cdot)}(\cdot)$, using the fact that it is torsion free and satisfies the following product rule:

$$\frac{d}{dt}\langle \phi_t, \psi_t\rangle_{u_t} = \langle \nabla_{\dot{u}_t}\phi_t, \psi_t\rangle_{u_t} + \langle \phi_t, \nabla_{\dot{u}_t}\psi_t\rangle_{u_t}. \tag{3.6}$$

We know that $\frac{d}{dt}\omega^n_{u_t} = ni\partial\bar{\partial}\dot{u}_t \wedge \omega^{n-1}_{u_t} = \frac{1}{2}\Delta^{\omega_{u_t}}\dot{u}_t\omega^n_{u_t}$ and $-\int_X \langle \nabla^{\omega_{u_t}} f, \nabla^{\omega_{u_t}} g\rangle \omega^n_{u_t} = \int_X f\Delta^{\omega_{u_t}} g\omega^n_{u_t}$ (see the discussion following (5.4) in the appendix), hence can start writing:

$$\begin{aligned}
\frac{d}{dt}\langle \phi_t, \psi_t\rangle_{u_t} &= \int_X (\dot{\phi}_t\psi_t + \phi_t\dot{\psi}_t + \frac{1}{2}\phi_t\psi_t\Delta^{\omega_{u_t}}\dot{u}_t)\omega^n_{u_t} \\
&= \int_X (\dot{\phi}_t\psi_t + \phi_t\dot{\psi}_t - \frac{1}{2}\langle\nabla^{\omega_{u_t}}(\phi_t\psi_t), \nabla^{\omega_{u_t}}\dot{u}_t\rangle)\omega^n_{u_t} \\
&= \int_X (\dot{\phi}_t - \frac{1}{2}\langle\nabla^{\omega_{u_t}}\phi_t, \nabla^{\omega_{u_t}}\dot{u}_t\rangle)\psi_t\omega^n_{u_t} + \int_X \phi_t(\dot{\psi}_t - \frac{1}{2}\langle\nabla^{\omega_{u_t}}\psi_t, \nabla^{\omega_{u_t}}\dot{u}_t\rangle)\omega^n_{u_t}.
\end{aligned}$$

Comparing with (3.6), this line of calculation suggests the following formula for the Levi–Civita connection, and it is easy to see that the resulting connection is indeed torsion free:

$$\nabla_{\dot{u}_t}\phi_t = \dot{\phi}_t - \frac{1}{2}\langle\nabla^{\omega_{u_t}}\dot{u}_t, \nabla^{\omega_{u_t}}\phi_t\rangle, \ t \in [0, 1]. \tag{3.7}$$

This immidiately implies that $t \to u_t$ is a geodesic if and only if $\nabla_{\dot{u}_t}\dot{u}_t = 0$, or equivalently

$$\ddot{u}_t - \frac{1}{2}\langle\nabla^{\omega_{u_t}}\dot{u}_t, \nabla^{\omega_{u_t}}\dot{u}_t\rangle = 0, \ t \in [0, 1]. \tag{3.8}$$

As discovered independently by Semmes [**99**] and Donaldson [**67**], the above equation can be understood as a complex Monge–Ampère equation. For this one has to introduce the trivial complexification $u \in C^\infty(S \times X)$, using the formula

$$u(s, x) = u_{\text{Re } s}(x),$$

where $S = \{0 < \text{Re s} < 1\} \subset \mathbb{C}$ is the unit strip. We pick a coordinate patch $U \subset X$, where the metric ω_u has a potential $g_u \in C^\infty(X)$, i.e., $\omega_u = i\partial\bar{\partial}g_u = ig_{u_{j\bar{k}}}dz_j \wedge d\bar{z}_k$. Then on $[0, 1] \times U$ the geodesic equation (3.8) is seen to be equivalent to $\ddot{u} - g_u^{j\bar{k}}\dot{u}_j\dot{u}_{\bar{k}} = 0$, where $g_u^{j\bar{k}}$ is the inverse of $g_{u_{j\bar{k}}}$. By involving the complexified variable s, this identity is further seen to be equivalent to $u_{s\bar{s}} - g_u^{j\bar{k}}u_{j\bar{s}}u_{s\bar{k}} = 0$ on

$S \times U$. After multiplying with $\det(g_{u_{j\bar{k}}})$, this last equation can be written globally on $S \times X$ as:

$$(\pi^*\omega + i\partial\bar{\partial}u)^{n+1} = 0. \tag{3.9}$$

where $\pi : S \times X \to X$ is the projection map to the second component. Consequently, the problem of joining the potentials $u_0, u_1 \in \mathcal{H}_\omega$ with a smooth geodesic equates to finding a smooth solution $u \in \mathcal{C}^\infty(S \times X)$ to the following boundary value problem:

$$\begin{cases} (\pi^*\omega + i\partial\bar{\partial}u)^{n+1} = 0, \\ \omega + i\partial\bar{\partial}u\big|_{\{s\}\times X} > 0, \ s \in S, \\ u(t + ir, x) = u(t, x) \ \forall x \in X, t \in (0,1), \ r \in \mathbb{R}. \\ \lim_{s\to 0} u(s,\cdot) = u_0 \text{ and } \lim_{s\to 1} u(s,\cdot) = u_1. \end{cases} \tag{3.10}$$

To be precise, here $\lim_{s\to 0,1} u(s,\cdot) = u_{0,1}$ simply means that $u(s,\cdot)$ converges uniformly to $u_{0,1}$ as $s \to 0, 1$. Unfortunately, as detailed below, this boundary value problem does not usually have smooth solutions, but a unique weak solution (in the sense of Bedford–Taylor) does exist. Instead, one replaces (3.8) with the following equation for *ε-geodesics*:

$$\Big(\ddot{u}^\varepsilon_t - \frac{1}{2}\langle \nabla^{\omega_{u^\varepsilon_t}} \dot{u}^\varepsilon_t, \nabla^{\omega_{u^\varepsilon_t}} \dot{u}^\varepsilon_t\rangle\Big)\omega^n_{u^\varepsilon_t} = \varepsilon\omega^n, \ t \in [0,1]. \tag{3.11}$$

By an elementary calculation, similar to the one giving (3.9), the associated boundary value problem for this equation becomes:

$$\begin{cases} (\pi^*\omega + i\partial\bar{\partial}u^\varepsilon)^{n+1} = \frac{\varepsilon}{4}(ids \wedge d\bar{s} + \pi^*\omega)^{n+1}, \\ u^\varepsilon(t + ir, x) = u^\varepsilon(t, x) \ \forall x \in X, t \in (0,1), r \in \mathbb{R}. \\ \lim_{s\to 0} u^\varepsilon(s,\cdot) = u_0 \text{ and } \lim_{s\to 1} u^\varepsilon(s,\cdot) = u_1. \end{cases} \tag{3.12}$$

Since $\omega_{u_0}, \omega_{u_1} > 0$, we see that $\pi^*\omega + i\partial\bar{\partial}u^\varepsilon > 0$ on $S \times X$. As a result the condition $\omega + i\partial\bar{\partial}u|_{\{s\}\times X} > 0, \ s \in S$ is automatically satisfied. In contrast with (3.10), this Dirichlet problem is elliptic and its solutions are smooth, moreover we have the following regularity result due to X.X. Chen [**34**] (with complements by Blocki [**22**]):

THEOREM 3.1. *The boundary value problem* (3.12) *admits a unique smooth solution $u^\varepsilon \in C^\infty(\overline{S} \times X)$ with the following bounds that are independent of $\varepsilon > 0$:*

$$\|u^\varepsilon\|_{C^0(\overline{S}\times X)}, \|u^\varepsilon\|_{C^1(\overline{S}\times X)}, \|\Delta u^\varepsilon\|_{C^0(\overline{S}\times X)} \leq C(\|u_0\|_{C^3(X)}, \|u_1\|_{C^3(X)}, X, \omega). \tag{3.13}$$

Recall from our discussion preceding (2.23) that having a bound on Δu is equivalent to bounding all mixed second order complex derivatives of u on $S \times X$. We refer to [**22**, Theorem 12] for an elaborate treatment of Theorem 3.1 (see also the survey paper [**25**]). Although we will not need it, let us mention that recently Chu–Tosatti–Weinkove have showed that one can more generally bound the Hessian of u^ε [**44**] independently of ε. Additionally, it was shown by Berman–Demailly [**15**] and He [**77**] that one can in fact bound each $\Delta^\omega u^\varepsilon_t, \ t \in [0,1]$, using bounds on $\Delta^\omega u_0$ and $\Delta^\omega u_1$.

Using the Bedford–Taylor interpretation of $(\pi^*\omega + i\partial\bar{\partial}u^\varepsilon)^{n+1}$ as a Borel measure, the boundary value problems (3.10) and (3.12) can be stated for $u, u^\varepsilon \in \mathrm{PSH}(S \times X, \pi^*\omega)$ that are only bounded and not necessarily smooth. Additionally, after pulling back by the log function, we can equivalently state (3.10) and (3.12) as boundary value problems with circle–invariant solutions on $A \times X$, where A is the annulus $\{e^0 < |z| < e^1\} \subset \mathbb{C}$. Consequently, the next result (whose proof closely follows [**22**, Theorem 21]) will assure that uniqueness of solutions to (3.10) and (3.12) holds not only for smooth solutions, but also for solutions that are merely in $\mathrm{PSH}(S \times X, \pi^*\omega) \cap L^\infty$:

THEOREM 3.2. *Let M be a k dimensional complex manifold with smooth boundary, and Kähler form η. If $u, v \in \mathrm{PSH}(M, \eta) \cap L^\infty$ with $\liminf_{x \to \partial M}(u - v)(x) \geq 0$ and $(\eta + i\partial\bar{\partial}v)^k \geq (\eta + i\partial\bar{\partial}u)^k$, then $u \geq v$ on M.*

Since $\pi^*\omega$ is only non-negative on $S \times X$, the above result is not directly applicable to our situation. This small inconvenience can be fixed by taking $\eta := \pi^*\omega + i\partial\bar{\partial}g$, where g is a smooth function on S, such that $i\partial\bar{\partial}g > 0$, $g(t+ir) = g(t)$, and $g(ir) = g(1 + ir) = 0$.

PROOF. Let $\delta > 0$ and $v_\delta := \max(u, v - \delta) \in \mathrm{PSH}(M, \eta) \cap L^\infty$. Then $v_\delta = u$ near ∂M. To conclude the proof, it is enough to show that $v_\delta = u$ on M.

From Bedford–Taylor theory (see [**20**, Theorem 2.2.10]) it follows that

$$\eta^k_{v_\delta} \geq \mathbb{1}_{\{u \geq v - \delta\}} \eta^k_u + \mathbb{1}_{\{u < v - \delta\}} \eta^k_v \geq \eta^k_u.$$

As v_δ and u agree near the boundary and $v_\delta \geq u$, we can integrate by parts [**20**, Theorem 1.3.4] and write:

$$0 \leq \int_M (v_\delta - u)(\eta^k_{v_\delta} - \eta^k_u) = -\sum_{j=0}^{k-1} \int_M i\partial(v_\delta - u) \wedge \bar{\partial}(v_\delta - u) \wedge \eta^j_u \wedge \eta^{k-j-1}_{v_\delta}.$$

As each of the summands above is non-negative, it follows that

$$\int_M i\partial(u - v_\delta) \wedge \bar{\partial}(u - v_\delta) \wedge \eta^j_u \wedge \eta^{k-j-1}_{v_\delta} = 0, \ j \in \{0, 1, \dots, k-1\}. \tag{3.14}$$

The equality $v_\delta = u$ will follow, if we can show that $\int_M i\partial(u - v_\delta) \wedge \bar{\partial}(u - v_\delta) \wedge \eta^{k-1} = 0$. Indeed, this would imply that all Sobolev derivatives of $u - v_\delta$ are zero. This last identity will be established as the last step in an inductive argument showing that

$$\int_M i\partial(u - v_\delta) \wedge \bar{\partial}(u - v_\delta) \wedge \eta^j_u \wedge \eta^{k-j-1} = 0, \ j \in \{0, 1, \dots, k-1\}. \tag{3.15}$$

This identity holds for $j = k - 1$ by the above. As all the steps are carried out similarly, we only show that

$$\int_M i\partial(u - v_\delta) \wedge \bar{\partial}(u - v_\delta) \wedge \eta^{k-2}_u \wedge \eta = 0. \tag{3.16}$$

Denote $f = u - v_\delta$. Using (3.15) and integration by parts we can write:

$$\begin{aligned}
\int_M i\partial f \wedge \bar{\partial} f \wedge \eta^{k-2}_u \wedge \eta &= \int_M i\partial f \wedge \bar{\partial} f \wedge \eta^{k-1}_u - \int_M i\partial f \wedge \bar{\partial} f \wedge i\partial\bar{\partial}u \wedge \eta^{k-2}_u \\
&= -\int_M i\partial f \wedge i\partial\bar{\partial}u \wedge \bar{\partial} f \wedge \eta^{k-2}_u \\
&= -\int_M i\partial f \wedge \bar{\partial}u \wedge (\eta_u - \eta_{v_\delta}) \wedge \eta^{k-2}_u \\
&= -\int_M i\partial f \wedge \bar{\partial}u \wedge \eta^{k-1}_u + \int_M i\partial f \wedge \bar{\partial}u \wedge \eta_{v_\delta} \wedge \eta^{k-2}_u. \qquad (3.17)
\end{aligned}$$

Using the Cauchy–Schwarz inequality of Bedford–Taylor theory [**20**] we can write

$$\Big|\int_M i\partial f \wedge \bar\partial u \wedge \eta_u^{k-1}\Big|^2 \leq \int_M i\partial u \wedge \bar\partial u \wedge \eta_u^{k-1} \cdot \int_M i\partial f \wedge \bar\partial f \wedge \eta_u^{k-1},$$

hence by (3.14), it follows that the first term in (3.17) is zero. It can be shown similarly that the second term in (3.17) is zero as well, implying (3.16) and finishing the proof. □

Now we return to (3.10) and (3.12). As each u^ε is invariant in the $i\mathbb{R}$ direction, plurisubharmonicity of u^ε implies that $t \to u^\varepsilon(t,x)$ is convex for each $x \in X$. As each u^ε solves the boundary value problem (3.12), we additionally obtain that

$$u^\varepsilon(s,x) \leq (1 - \text{Re } s)u_0(x) + \text{Re } s u_1(x). \tag{3.18}$$

On top of proving uniqueness of general solutions to (3.10) and (3.12), the above theorem also shows that the family $\{u^\varepsilon\}_\varepsilon$ increases as $\varepsilon \searrow 0$. Using this, (3.18) and the continuity of the complex Monge–Ampère operator along increasing sequences [**20**, Theorem 2.2.5] it follows that solutions of (3.12) approximate solutions of (3.10), i.e.,

$$\lim_{\varepsilon\to 0} u^\varepsilon = u. \tag{3.19}$$

From here, an application of the Arzela–Ascoli compactness theorem yields that the estimates of (3.13) also hold for solutions of (3.10):

THEOREM 3.3. *For the unique solution* $u \in \text{PSH}(S \times X, \pi^*\omega)$ *of* (3.10) *the following estimates hold:*

$$\|u\|_{C^0(\overline{S}\times X)}, \|u\|_{C^1(\overline{S}\times X)}, \|\Delta u\|_{C^0(\overline{S}\times X)} \leq C(\|u_0\|_{C^3(X)}, \|u_1\|_{C^3(X)}, X, \omega). \tag{3.20}$$

By the examples of [**54, 85**] this regularity result is essentially optimal, as no higher order estimates are possible for weak solutions. Often we will refer to the unique solution of (3.10) as the *weak* $C^{1,\bar 1}$*–geodesic* joining $u_0, u_1 \in \mathcal{H}_\omega$, and it will be denoted as

$$[0,1] \ni t \to u_t \in \mathcal{H}^{1,\bar 1}_\omega. \tag{3.21}$$

Moving on, as follows from standard elliptic PDE theory, given a smooth family of boundary data for (3.12) one obtains a corresponding smooth family of solutions. As another simple consequence of the uniqueness theorem, we obtain a uniform bound on the rate of change of this family as the parameter changes:

COROLLARY 3.4 ([**34**]). *Suppose* $[0,1] \ni \rho \to u_0(\cdot,\rho), u_1(\cdot,\rho) \in \mathcal{H}_\omega$ *is smooth family of boundary data for* (3.12). *For the smooth family of solutions* $[0,1] \ni \rho \to u^\varepsilon(\cdot,\cdot,\rho) \in C^\infty(\overline{S} \times X)$ *we have the following estimate for all* $\varepsilon > 0$*:*

$$\Big\|\frac{du^\varepsilon}{d\rho}\Big\|_{C^0(\overline{S}\times X\times[0,1])} \leq \max\Big(\Big\|\frac{du_0}{d\rho}\Big\|_{C^0(X\times[0,1])}, \Big\|\frac{du_1}{d\rho}\Big\|_{C^0(X\times[0,1])}\Big).$$

PROOF. Let $\rho_0 \in [0,1]$ and $C > \max(\|du_0/d\rho\|_{C^0(X\times[0,1])}, \|du_1/d\rho\|_{C^0(X\times[0,1])})$. Then Theorem 3.2 implies that

$$u^\varepsilon(s,x,\rho_0) - C\delta \leq u^\varepsilon(s,x,\rho_0+\delta) \leq u^\varepsilon(s,x,\rho_0) + C\delta, \ \ (s,x) \in \overline{S} \times X$$

for $\delta \in \mathbb{R}$ such that $\rho_0 + \delta \in [0,1]$. Consequently, $|u^\varepsilon(s,x,\rho_0+\delta) - u^\varepsilon(s,x,\rho_0)|/\delta \leq C$, implying the desired estimate. □

As a last consequence of the uniqueness theorem, we note the following concrete estimate for the tangent vectors of weak $C^{1,\bar{1}}$-geodesics:

LEMMA 3.5. *Given $u_0, u_1 \in \mathcal{H}_\omega$, let $[0,1] \ni t \to u_t \in \mathcal{H}_\omega$ be the weak $C^{1,\bar{1}}$-geodesic joining u_0, u_1. Then the following estimate holds:*

$$\|\dot{u}_t\|_{C^0(X)} \leq \|u_0 - u_1\|_{C^0(X)}, \quad t \in [0,1]. \tag{3.22}$$

PROOF. Let $C := \sup_X |u_0 - u_1|$. Convexity in the t variable implies that $\dot{u}_0 \leq \dot{u}_t \leq \dot{u}_1$. Hence we only need to show that $-C \leq \dot{u}_0$ and $\dot{u}_1 \leq C$.

Examining (3.8), it is clear that the curve $[0,1] \in t \to v_t := u_0 - Ct \in \mathcal{H}_\omega$ is a smooth geodesic connecting u_0 and $u_0 - C$. Hence its complexification v is a solution to (3.10).

As $u_0 - C \leq u_1$, Theorem 3.2 gives that that $v_t \leq u_t$, $t \in [0,1]$, implying that $-C \leq \dot{u}_0$.

The same trick applied to the "reverse" $C^{1,\bar{1}}$-geodesic $t \to u_{1-t}$ gives $-C \leq -\dot{u}_1$, finishing the proof. □

3.2. The Orlicz geometry of the space of Kähler potentials

As discussed in the beginning of the present chapter, in our study of L^p Finsler metrics on $\mathcal{H}_\omega$, we need to return to the full generality of Orlicz–Finsler metrics (3.2) on $\mathcal{H}_\omega$ with weight in $\mathcal{W}_p^+$. We will do this in this section, and our main theorem connects the d_χ pseudo–distance (see (3.4)) with the weak $C^{1,\bar{1}}$ geodesic of the L^2 Mabuchi geometry (see (3.21)), in the process showing that d_χ is indeed a bona fide metric:

THEOREM 3.6. [**49**, Theorem 1] *If $\chi \in \mathcal{W}_p^+$, $p \geq 1$ then $(\mathcal{H}_\omega, d_\chi)$ is a metric space and for any $u_0, u_1 \in \mathcal{H}_\omega$ the weak $C^{1,\bar{1}}$ geodesic $t \to u_t$ connecting u_0, u_1 satisfies:*

$$d_\chi(u_0, u_1) = \|\dot{u}_t\|_{\chi, u_t}, \ t \in [0,1]. \tag{3.23}$$

Although the metric spaces $(\mathcal{H}_\omega, d_\chi)$ are not quasi isometric for different χ (as we will see, they have different metric completions), it is remarkable that the *same* weak $C^{1\bar{1}}$–geodesic is "length minimzing" for *all* d_χ metric structures. Additionally, as we will see in the next sections, this same curve is an honest metric geodesic in the completion of each space $(\mathcal{H}_\omega, d_\chi)$.

In the proof of Theorem 3.6, we first show the result for Finsler metrics with smooth weight χ, and afterwards use approximation via Proposition 1.6 and Proposition 1.7 to establish the result for all metrics with weight in $\mathcal{W}_p^+$, which includes as particular case the L^p metrics.

In the particular case of the L^2 metric, Theorem 3.6 was obtained by X.X. Chen [**34**] and our proof in case of Finsler metrics with smooth weight follows his ideas and a careful differential analysis of Orlicz norms.

To ease the technical nature of future calculations, for the rest of this section we assume the normalizing condition

$$V := \mathrm{Vol}(X) = \int_X \omega^n = 1. \tag{3.24}$$

This can always be achieved by rescaling the Kähler metric ω. Given a differentiable normalized Young weight χ, the differentiability of the associated norm $\|\cdot\|_\chi$ (see

(1.3)) is well understood (see [**98**, Chapter VII]). As a first step in proving Theorem 3.6, we adapt [**98**, Theorem VII.2.3] to our setting:

PROPOSITION 3.7. [**49**, Proposition 3.1] *Suppose $\chi \in \mathcal{W}_p^+ \cap C^\infty(\mathbb{R})$. Given a smooth curve $(0,1) \ni t \to u_t \in \mathcal{H}_\omega$, and a vector field $(0,1) \ni t \to f_t \in C^\infty(X)$ along this curve with $f_t \not\equiv 0,\ t \in (0,1)$, the following formula holds:*

$$\frac{d}{dt}\|f_t\|_{\chi,u_t} = \frac{\int_X \chi'\Big(\frac{f_t}{\|f_t\|_{\chi,u_t}}\Big)\nabla_{\dot u_t} f_t \omega^n_{u_t}}{\int_X \chi'\Big(\frac{f_t}{\|f_t\|_{\chi,u_t}}\Big)\frac{f_t}{\|f_t\|_{\chi,u_t}}\omega^n_{u_t}}, \tag{3.25}$$

where $\nabla_{(\cdot)}(\cdot)$ is the covariant derivative from (3.7).

PROOF. We introduce the smooth function $F : \mathbb{R}^+ \times (0,1) \to \mathbb{R}$ given by

$$F(r,t) = \int_X \chi\Big(\frac{f_t}{r}\Big)\omega^n_{u_t}.$$

As $\chi \in \mathcal{W}_p^+$, by (1.8) we have $\chi'(l) > 0,\ l > 0$ and $\chi'(l) < 0,\ l < 0$. As $t \to f_t$ is non-vanishing, it follows that

$$\frac{d}{dr}F(r,t) = -\frac{1}{r^2}\int_X f_t\chi'\Big(\frac{f_t}{r}\Big)\omega^n_{u_t} < 0$$

for all $r > 0,\ t \in (0,1)$. Using the fact that $F(\|f_t\|_{\chi,u_t}, t) = \chi(1)$ (see (1.9)), an application of the implicit function theorem yields that the map $t \to \|f_t\|_{\chi,u_t}$ is differentiable and the following formula holds:

$$\frac{d}{dt}\|f_t\|_{\chi,u_t} = \frac{\int_X \Big[\dot f_t\chi'\Big(\frac{f_t}{\|f_t\|_{\chi,u_t}}\Big) + \frac{1}{2}\|f_t\|_{\chi,u_t}\chi\Big(\frac{f_t}{\|f_t\|_{\chi,u_t}}\Big)\Delta^{\omega_{u_t}}\dot u_t\Big]\omega^n_{u_t}}{\int_X \frac{f_t}{\|f_t\|_{\chi,u_t}}\chi'\Big(\frac{f_t}{\|f_t\|_{\chi,u_t}}\Big)\omega^n_{u_t}}.$$

Recalling the formula for the covariant derivative (3.7), an integration by parts yields (3.25). □

The estimate for ε-geodesics (see (3.11)) from the following technical lemma will be of great use in our later study:

LEMMA 3.8. *Suppose $\chi \in \mathcal{W}_p^+ \cap C^\infty(\mathbb{R})$ and $u_0, u_1 \in \mathcal{H}_\omega$. Then the ε-geodesic $[0,1] \ni t \to u_t^\varepsilon \in \mathcal{H}_\omega$ connecting u_0, u_1 satisfies the following estimate:*

$$\int_X \chi(\dot u_t^\varepsilon)\omega^n_{u_t^\varepsilon} \geq \max\Big(\int_X \chi(\min(u_1-u_0,0))\omega^n_{u_0}, \int_X \chi(\min(u_0-u_1,0))\omega^n_{u_1}\Big) - \varepsilon R > 0, \tag{3.26}$$

for all $t \in [0,1]$, where $R := R(\chi, \|u_0\|_{C^2}, \|u_1\|_{C^2})$.

PROOF. As $t \to u_t^\varepsilon(x)$ is convex for any $x \in X$, on the set $\{u_0 \geq u_1\}$ the estimate $\dot u_0^\varepsilon \leq u_1 - u_0 \leq 0$ holds, hence

$$\int_X \chi(\dot u_0^\varepsilon)\omega^n_{u_0} \geq \int_X \chi(\min(u_1 - u_0, 0))\omega^n_{u_0}.$$

We can similarly deduce that

$$\int_X \chi(\dot u_1^\varepsilon)\omega^n_{u_1} \geq \int_X \chi(\min(u_0 - u_1, 0))\omega^n_{u_1}.$$

For $t \in [0,1]$, using the Riemannian connection $\nabla_{(\cdot)}(\cdot)$ (see (3.7)) and the fact that $t \to u^\varepsilon_t$ is an ε–geodesic (see (3.11)), we can write:

$$\Big|\frac{d}{dt}\int_X \chi(\dot u^\varepsilon_t)\omega^n_{u^\varepsilon_t}\Big| = \Big|\int_X \chi'(\dot u^\varepsilon_t)\nabla_{\dot u^\varepsilon_t}\dot u^\varepsilon_t \omega^n_{u^\varepsilon_t}\Big| = \varepsilon\Big|\int_X \chi'(\dot u^\varepsilon_t)\omega^n\Big|$$
$$\leq \varepsilon R(\chi, \|u_0\|_{C^2}, \|u_1\|_{C^2}),$$

where in the last estimate we have used that $\dot u^\varepsilon_t$ is uniformly bounded in terms of $\|u_0\|_{C^2}, \|u_1\|_{C^2}$ (Theorem 3.1). After putting together the last three estimates, (3.26) follows. □

As a consequence of the previous two results we obtain the following corollary:

COROLLARY 3.9. *Suppose $\chi \in \mathcal{W}^+_p \cap C^\infty(\mathbb{R})$ and $u_0, u_1 \in \mathcal{H}_\omega$, $u_0 \neq u_1$. Then there exists $\varepsilon_0 > 0$ dependent on upper bounds for $\|u_0\|_{C^2(X)}, \|u_1\|_{C^2(X)}$ and lower bounds for $\|\chi(u_1 - u_0)\|_{L^1(\omega^n)}, \omega^n_{u_0}/\omega^n$ and $\omega^n_{u_1}/\omega^n$, such that for all $\varepsilon \in (0, \varepsilon_0)$ the ε–geodesic $[0,1] \ni t \to u^\varepsilon_t \in \mathcal{H}_\omega$ of (3.12), connecting u_0, u_1 satisfies:*

$$\frac{d}{dt}\|\dot u^\varepsilon_t\|_{\chi, u^\varepsilon_t} = \varepsilon \frac{\int_X \chi'\Big(\frac{\dot u^\varepsilon_t}{\|\dot u^\varepsilon_t\|_{\chi,u^\varepsilon_t}}\Big)\omega^n}{\int_X \frac{\dot u^\varepsilon_t}{\|\dot u^\varepsilon_t\|_{\chi,u^\varepsilon_t}}\chi'\Big(\frac{\dot u^\varepsilon_t}{\|\dot u^\varepsilon_t\|_{\chi,u^\varepsilon_t}}\Big)\omega^n_{u^\varepsilon_t}}, \quad t \in [0,1]. \tag{3.27}$$

PROOF. This is a simple application of the formula of Proposition 3.7 for the ε–geodesic $t \to u^\varepsilon_t$ and the vector field $t \to f_t := \dot u_t$. Indeed, by the previous lemma, one can choose $\varepsilon_0 > 0$ as indicated, so that $\dot u_t$ is non-vanishing for all $t \in [0,1]$, hence the assumptions of Proposition 3.7 are satisfied. □

Continuing to focus on smooth weights χ, we establish concrete bounds for the χ–length of tangent vectors along the ε–geodesics and their derivatives. This is the analog of [**22**, Lemma 13] in our more general setting:

PROPOSITION 3.10. *Suppose $\chi \in \mathcal{W}^+_p \cap C^\infty(\mathbb{R})$ and $u_0, u_1 \in \mathcal{H}_\omega$, $u_0 \neq u_1$. Then there exists $\varepsilon_0 > 0$ such that for any $\varepsilon \in (0, \varepsilon_0)$ the ε–geodesic $[0,1] \ni t \to u^\varepsilon_t \in \mathcal{H}_\omega$ connecting u_0, u_1 satisfies:*
(i)$\|\dot u^\varepsilon_t\|_{\chi,u^\varepsilon_t} > R_0$, $t \in [0,1]$,
(ii)$\big|\frac{d}{dt}\|\dot u^\varepsilon_t\|_{\chi,u^\varepsilon_t}\big| \leq \varepsilon R_1$, $t \in [0,1]$,
where ε_0, R_0, R_1 depend on upper bounds for $\|u_0\|_{C^2(X)}$, $\|u_1\|_{C^2(X)}$ and lower bounds for $\|\chi(u_1 - u_0)\|_{L^1(\omega^n)}, \omega^n_{u_0}/\omega^n$ and $\omega^n_{u_1}/\omega^n$.

PROOF. The estimate of (i) follows from (3.26) and Proposition 1.5. To establish (ii) we shrink ε_0 enough to satisfy the requirements of Corollary 3.9. Using the Young identity (1.2) we can write:

$$\Big|\frac{d}{dt}\|\dot u^\varepsilon_t\|_{\chi,u^\varepsilon_t}\Big| = \varepsilon\frac{\Big|\int_X \chi'\Big(\frac{\dot u^\varepsilon_t}{\|\dot u^\varepsilon_t\|_{\chi,u^\varepsilon_t}}\Big)\omega^n\Big|}{\int_X \frac{\dot u^\varepsilon_t}{\|\dot u^\varepsilon_t\|_{\chi,u^\varepsilon_t}}\chi'\Big(\frac{\dot u^\varepsilon_t}{\|\dot u^\varepsilon_t\|_{\chi,u^\varepsilon_t}}\Big)\omega^n_{u^\varepsilon_t}} \tag{3.28}$$
$$= \varepsilon\frac{\Big|\int_X \chi'\Big(\frac{\dot u^\varepsilon_t}{\|\dot u^\varepsilon_t\|_{\chi,u^\varepsilon_t}}\Big)\omega^n\Big|}{\chi(1) + \int_X \chi^*\Big(\chi'\Big(\frac{\dot u^\varepsilon_t}{\|\dot u^\varepsilon_t\|_{\chi,u^\varepsilon_t}}\Big)\Big)\omega^n_{u^\varepsilon_t}}$$
$$\leq \frac{\varepsilon}{\chi(1)}\int_X \chi'\Big(\frac{\dot u^\varepsilon_t}{\|\dot u^\varepsilon_t\|_{\chi,u^\varepsilon_t}}\Big)\omega^n.$$

Using (i) and the fact that $\dot{u}^\varepsilon_t$ is uniformly bounded in terms of $\|u_0\|_{C^2}, \|u_1\|_{C^2}$ (Theorem 3.1) the estimate of (ii) follows. □

We can now establish the main geometric estimate for ε–geodesics, which generalizes the corresponding statement for the L^2–metric ([**22**, Theorem 14]):

PROPOSITION 3.11 ([**49**]). *Suppose $\chi \in \mathcal{W}^+_p \cap C^\infty(\mathbb{R})$, $[0,1] \ni s \to \psi_s \in \mathcal{H}_\omega$ is a smooth curve, $\phi \in \mathcal{H}_\omega \setminus \psi([0,1])$ and $\varepsilon > 0$. We denote by $u^\varepsilon \in C^\infty([0,1]\times[0,1]\times X)$ the smooth function for which $[0,1] \ni t \to u^\varepsilon_t(\cdot,s) := u^\varepsilon(t,\cdot,s) \in \mathcal{H}_\omega$ is the ε–geodesic connecting ϕ and ψ_s. There exists $\varepsilon_0(\phi,\psi) > 0$ such that for any $\varepsilon \in (0,\varepsilon_0)$ the following holds:*

$$l_\chi(u^\varepsilon_t(\cdot,0)) \leq l_\chi(\psi) + l_\chi(u^\varepsilon_t(\cdot,1)) + \varepsilon R,$$

for some $R(\phi,\psi,\chi,\varepsilon_0) > 0$ independent of $\varepsilon > 0$.

PROOF. Whenever it will not cause confusion, we will drop sub/superscripts when addressing the ε–geodesic $t \to u^\varepsilon_t(\cdot,s)$. To avoid cumbersome notation, derivatives in the t–direction will be denoted by dots, derivatives in the s–direction will be denoted by d/ds, and sometimes we also omit dependence on (t,s).

Fix $s \in [0,1]$. By Proposition 3.7 and Proposition 3.10(i) there exists $\varepsilon_0(\phi,\psi) > 0$ such that for $\varepsilon \in (0,\varepsilon_0)$ the following holds:

$$\frac{d}{ds} l_\chi(u_t(\cdot,s)) = \int_0^1 \frac{d}{ds}\|\dot{u}(t,\cdot,s))\|_{\chi,u(t,s)} dt = \int_0^1 \frac{\int_X \chi'\Big(\frac{\dot{u}}{\|\dot{u}\|_{\chi,u}}\Big)\nabla_{\frac{du}{ds}} \dot{u}\omega_u^n}{\int_X \chi'\Big(\frac{\dot{u}}{\|\dot{u}\|_{\chi,u}}\Big)\frac{\dot{u}}{\|\dot{u}\|_{\chi,u}}\omega_u^n} dt$$

Using the Young identity (1.2) and the fact that $\nabla_{(\cdot)}(\cdot)$ is a Riemannian connection, we can continue:

$$= \int_0^1 \frac{\int_X \chi'\Big(\frac{\dot{u}}{\|\dot{u}\|_{\chi,u}}\Big)\nabla_{\frac{du}{ds}} \dot{u}\omega_u^n}{\chi(1) + \int_X \chi^*\Big(\chi'\Big(\frac{\dot{u}}{\|\dot{u}\|_{\chi,u}}\Big)\Big)\omega_u^n} dt$$

$$= \int_0^1 \frac{\int_X \chi'\Big(\frac{\dot{u}}{\|\dot{u}\|_{\chi,u}}\Big)\nabla_{\dot{u}} \frac{du}{ds}\omega_u^n}{\chi(1) + \int_X \chi^*\Big(\chi'\Big(\frac{\dot{u}}{\|\dot{u}\|_{\chi,u}}\Big)\Big)\omega_u^n} dt$$

$$(3.29)\qquad = \int_0^1 \frac{\frac{d}{dt}\int_X \chi'\Big(\frac{\dot{u}}{\|\dot{u}\|_{\chi,u}}\Big)\frac{du}{ds}\omega_u^n - \int_X \frac{du}{ds}\nabla_{\dot{u}}\Big(\chi'\Big(\frac{\dot{u}}{\|\dot{u}\|_{\chi,u}}\Big)\Big)\omega_u^n}{\chi(1) + \int_X \chi^*\Big(\chi'\Big(\frac{\dot{u}}{\|\dot{u}\|_{\chi,u}}\Big)\Big)\omega_u^n} dt.$$

We make the following side computation:

$$(3.30)\qquad \nabla_{\dot{u}}\Big(\chi'\Big(\frac{\dot{u}}{\|\dot{u}\|_{\chi,u}}\Big)\Big)\omega_u^n = \chi''\Big(\frac{\dot{u}}{\|\dot{u}\|_{\chi,u}}\Big)\Big(\frac{\nabla_{\dot{u}}\dot{u}}{\|\dot{u}\|_{\chi,u}} - \frac{\dot{u}}{\|\dot{u}\|^2_{\chi,u}}\frac{d}{dt}\|\dot{u}\|_{\chi,u}\Big)\omega_u^n.$$

After possibly further shrinking $\varepsilon_0(\phi,\psi) > 0$, from Proposition 3.10(i)(ii) and (3.11) it follows that $\|\dot{u}\|_{\chi,u}$ is uniformly bounded away from zero and both $\nabla_{\dot{u}}\dot{u}\omega_u^n$ and $\frac{d}{dt}\|\dot{u}\|_{\chi,u}$ are of the form εR, where R is an uniformly bounded quantity for $\varepsilon < \varepsilon_0(\phi,\psi)$.

Furthermore, it follows from Theorem 3.1 that $\dot{u}$ is uniformly bounded independently of ε. All of this implies that the quantity of (3.30) is also of the form εR. Lastly, Corollary 3.4 implies that du/ds is uniformly bounded as well, hence putting the above together it follows that the second term in the numerator of (3.29) is also of the form εR, and we can continue to write:

$$= \int_0^1 \frac{\frac{d}{dt}\int_X \chi'\Big(\frac{\dot u}{\|\dot u\|_{\chi,u}}\Big)\frac{du}{ds}\omega_u^n}{\chi(1)+\int_X \chi^*\Big(\chi'\Big(\frac{\dot u}{\|\dot u\|_{\chi,u}}\Big)\Big)\omega_u^n}dt + \varepsilon R$$

As χ^* is the Legendre transform of χ, it follows that $\chi^{*\prime}(\chi'(l)) = l,\ l \in \mathbb{R}$. Using this, our prior observations and the chain rule, we obtain that the expression

$$\frac{d}{dt}\Big(\chi(1)+\int_X \chi^*\Big(\chi'\Big(\frac{\dot u}{\|\dot u\|_{\chi,u}}\Big)\Big)\omega_u^n\Big) = \int_X \frac{\dot u}{\|\dot u\|_{\chi,u}}\chi''\Big(\frac{\dot u}{\|\dot u\|_{\chi,u}}\Big)\nabla_{\dot u}\Big(\frac{\dot u}{\|\dot u\|_{\chi,u}}\Big)\omega_u^n$$

is again of magnitude εR, hence in our sequence of calculations we can write

$$\begin{aligned}
&= \int_0^1 \frac{d}{dt}\frac{\int_X \chi'\Big(\frac{\dot u}{\|\dot u\|_{\chi,u}}\Big)\frac{du}{ds}\omega_u^n}{\chi(1)+\int_X \chi^*\Big(\chi'\Big(\frac{\dot u}{\|\dot u\|_{\chi,u}}\Big)\Big)\omega_u^n}dt + \varepsilon R\\
&= \frac{\int_X \chi'\Big(\frac{\dot u(1,s)}{\|\dot u(1,s)\|_{\chi,\psi}}\Big)\frac{d\psi(s)}{ds}\omega_\psi^n}{\chi(1)+\int_X \chi^*\Big(\chi'\Big(\frac{\dot u(1,s)}{\|\dot u(1,s)\|_{\chi,\psi}}\Big)\Big)\omega_\psi^n} + \varepsilon R
\end{aligned}$$

$$\geq -\Big\|\frac{d\psi(s)}{ds}\Big\|_{\chi,\psi} + \varepsilon R, \tag{3.31}$$

where in the last line we have used the Young inequality (1.2) in the following manner:

$$\begin{aligned}
\frac{\int_X \chi'\Big(\frac{\dot u(1,s)}{\|\dot u(1,s)\|_{\chi,\psi}}\Big)\frac{d\psi(s)}{ds}\omega_\psi^n}{\|d\psi/ds\|_{\chi,\psi}} &\geq -\int_X \Big[\chi\Big(\frac{d\psi/ds}{\|d\psi/ds\|_{\chi,\psi}}\Big) + \chi^*\Big(\chi'\Big(\frac{\dot u(1,s)}{\|\dot u(1,s)\|_{\chi,\psi}}\Big)\Big)\Big]\omega_\psi^n\\
&= -\Big[\chi(1)+\int_X \chi^*\Big(\chi'\Big(\frac{\dot u(1,s)}{\|\dot u(1,s)\|_{\chi,\psi}}\Big)\Big)\omega_\psi^n\Big].
\end{aligned}$$

Integrating estimate (3.31) with respect to s yields the desired inequality. □

With the previous result established, there is no more need to differentiate expressions involving the l_χ length of curves, hence we can return to general Finsler metrics on $\mathcal{H}_\omega$, with possibly non–smooth weight $\chi \in \mathcal{W}_p^+$, and prove Theorem 3.6 in the process:

Proof of Theorem 3.6. First we show the following identity for the weak $C^{1,\bar 1}$–geodesic joining u_0, u_1:

$$d_\chi(u_0,u_1) = l_\chi(u_t). \tag{3.32}$$

We can assume that $u_0 \neq u_1$. By Theorem 3.1 and (3.19),(3.20), the smooth ε–geodesics u^ε connecting u_0, u_1 $C^{1,\alpha}$–converge to the weak $C^{1,\bar 1}$ geodesic u, connecting u_0, u_1. As a result, $\dot u_t^\varepsilon$ converges uniformly to $\dot u_t$. Next we argue that the lengths of these tangent vectors converge as well:

Claim 3.12. *$\|\dot u_t^\varepsilon\|_{\chi,u_t^\varepsilon} \to \|\dot u_t\|_{\chi,u_t}$ as $\varepsilon \to 0$.*

From Proposition 3.10(i) and Theorem 3.1 it follows that there exists $C_2 > C_1 > 0$ such that for small enough $\varepsilon > 0$ we have

$$0 < C_1 \leq \|\dot u_t^\varepsilon\|_{\chi,u_t^\varepsilon} \leq C_2.$$

In particular, we only have to argue that all cluster points of the set $\{\|\dot{u}^\varepsilon_t\|_{\chi,u^\varepsilon_t}\}_\varepsilon$ are equal to $\|\dot{u}_t\|_{\chi,u_t}$. Let N be such a cluster point, and after taking a subsequence, we can assume that $\|\dot{u}^\varepsilon_t\|_{\chi,u^\varepsilon_t} \to N$ as $\varepsilon \to 0$. By uniform convergence of tangent vectors, we obtain that $\dot{u}^\varepsilon_t/\|\dot{u}^\varepsilon_t\|_{\chi,u^\varepsilon_t}$ converges to $\dot{u}_t/N$ uniformly as well. Since $\omega^n_{u^\varepsilon_t} \to \omega^n_{u_t}$ weakly, this allows to conclude that

$$\chi(1) = \int_X \chi\Big(\frac{\dot{u}^\varepsilon_t}{\|\dot{u}^\varepsilon_t\|_{\chi,u^\varepsilon_t}}\Big)\omega^n_{u^\varepsilon_t} \to \int_X \chi\Big(\frac{\dot{u}_t}{N}\Big)\omega^n_{u_t}, \text{ as } \varepsilon \to 0.$$

Using (1.9) we get that $N = \|\dot{u}_t\|_{\chi,u_t}$, finishing the proof of the claim.

Using the claim, we can apply the dominated convergence theorem to conclude that

$$\lim_{\varepsilon\to 0} l_\chi(u^\varepsilon_t) = l_\chi(u_t), \tag{3.33}$$

hence $d_\chi(u_0,u_1) \leq l_\chi(u_t)$. To prove the reverse inequality, and with that establishing (3.32), we assume first that $\chi \in \mathcal{W}^+_p \cap C^\infty$. We have to prove that

$$l_\chi(\phi_t) \geq l_\chi(u_t) \tag{3.34}$$

for all smooth curves $[0,1] \ni t \to \phi_t \in \mathcal{H}$ connecting u_0, u_1. We can assume that $u_1 \notin \phi[0,1)$ and let $h \in [0,1)$. Letting $\varepsilon \to 0$ in the estimate of the previous result, by (3.33) we obtain that

$$l_\chi(u_{1-t}) \leq l_\chi(\phi_t|_{[0,h]}) + l_\chi(w^h_t),$$

where $[0,1] \ni t \to u_{1-t}, w^h_t \in \mathcal{H}^{1,\bar{1}}_\omega$ are the weak $C^{1,\bar{1}}$-geodesic segments joining u_1, u_0 and u_1, ϕ_h respectively. As $h \to 1$, by Lemma 3.5 we have $l_\chi(w^h_t) \to 0$ and we obtain (3.34) for smooth weights χ.

For general $\chi \in \mathcal{W}^+_p$ by Proposition 1.7 there exists a sequence $\chi_k \in \mathcal{W}^+_{p_k} \cap C^\infty(\mathbb{R})$ such that χ_k converges to χ uniformly on compacts. From what we just proved it follows that

$$\int_0^1 \|\dot{\phi}_t\|_{\chi_k,\phi_t}dt = l_{\chi_k}(\phi_t) \geq l_{\chi_k}(u_t) = \int_0^1 \|\dot{u}_t\|_{\chi_k,u_t}dt.$$

Using Proposition 1.6 and the dominated convergence theorem ($\dot{\phi}_t, \dot{u}_t$ are uniformly bounded), we can take the limit in this last estimate to conclude (3.34), which gives(3.32). Formula (3.23) follows now from the fact that $l_\chi(u_t) = \int_0^1 \|\dot{u}_l\|_{\chi,u_l}dl$ and Lemma 3.13 below.

Finally, if $u_0 \neq u_1$ then after taking the limit $\varepsilon \to 0$ in the estimate of Lemma 3.8 we obtain that $\dot{u}_0 \not\equiv 0$, hence $d_\chi(u_0,u_1) = \|\dot{u}_0\|_{\chi,u_0} > 0$. This implies that $(\mathcal{H}, d_\chi)$ is a metric space, as claimed. □

According to the last lemma of this section, the χ-length of tangent vectors along a $C^{1,\bar{1}}$-geodesic is always constant. This parallels a similar result of Berndtsson [**18**, Proposition 2.2].

LEMMA 3.13. *Given $u_0, u_1 \in \mathcal{H}_\omega$, let $[0,1] \ni t \to u_t \in \mathcal{H}^{1,\bar{1}}_\omega$ be the weak $C^{1,\bar{1}}$-geodesic connecting u_0, u_1. Then for any $\chi \in \mathcal{W}^+_p$ and $t_0, t_1 \in [0,1]$ the following hold:*

$$\|\dot{u}_{t_0}\|_{\chi,u_{t_0}} = \|\dot{u}_{t_1}\|_{\chi,u_{t_1}}. \tag{3.35}$$

PROOF. As we already argued in the proof of the previous result, for the ε–geodesics $t \to u_t^\varepsilon$ joining u_0, u_1 we have $\|\dot u^\varepsilon_{t_0}\|_{\chi,u^\varepsilon_{t_0}} \to \|\dot u_{t_0}\|_{\chi,u_{t_0}}$ and $\|\dot u^\varepsilon_{t_1}\|_{\chi,u^\varepsilon_{t_1}} \to \|\dot u_{t_1}\|_{\chi,u_{t_1}}$ as $\varepsilon \to 0$. Furthermore Proposition 3.10(ii) implies that $|\|\dot u^\varepsilon_{t_1}\|_{\chi,u^\varepsilon_{t_1}} - \|\dot u^\varepsilon_{t_0}\|_{\chi,u^\varepsilon_{t_0}}| \leq |t_1 - t_0|\varepsilon R_1$. Putting all of this together, and letting $\varepsilon \to 0$, (3.35) follows. □

3.3. The weak geodesic segments of PSH(X, ω)

As noted after Theorem 3.3, it is not possible to join smooth potentials u_0, u_1 with a geodesic staying inside $\mathcal{H}_\omega$. The main point of the present section is to show that a similar phenomenon does not occur if u_0, u_1 is allowed to be more singular. Indeed, as we will see, if u_0, u_1 are bounded, or they are from a finite energy space, then it is possible to define a weak geodesic connecting them that is also bounded or stays inside the finite energy space respectively. Our study will connect properties of weak geodesics with that of envelopes, and we will make good use of the results of Section 2.4.

When considering the boundary value problem (3.10), we constructed the (weak) solution u as the limit of solutions to the family of elliptic problems (3.12). Moving away from this idea, as noted by Berndtsson [**19**, Section 2.1], it possible to describe the (weak) solution u in another way, using a slight generalization of the classical Perron–Bremmerman envelope from the local theory.

The advantage of this approach is that one can consider very general boundary data in (3.10). Indeed, to begin, let $u_0, u_1 \in \text{PSH}(X, \omega)$. In the future, we will refer to $i\mathbb{R}$–invariant elements of PSH$(S \times X, \pi^*\omega)$ as *weak subgeodesics* (recall that $S = \{0 < \text{Re } z < 1\} \subset \mathbb{C}$). This name is justified by the following formula:

$$u = \sup_{v \in \mathcal{S}} v, \tag{3.36}$$

where $\mathcal{S}$ is the following family of subgeodesics:

$$\mathcal{S} = \{(0,1) \ni t \to v_t \in \text{PSH}(X, \omega) \text{ is a subgeodesic with } \lim_{t \to 0,1} v_t \leq u_{0,1}\}.$$

As we know, the supremum of a family of $\pi^*\omega$-psh functions may not be $\pi^*\omega$-psh, and the first step is to show that u, as defined in (3.36), is $\pi^*\omega$-psh nonetheless. Indeed, by convexity in the t variable, each member of $w_t \in \mathcal{S}$ satisfies $w_t \leq (1-t)u_0 + tu_1$, hence this also holds for the supremum u:

$$u_t \leq (1-t)u_0 + tu_1. \tag{3.37}$$

By taking the usc regularization of the above inequality, we conclude that the same inequality holds with u^* in place of u:

$$u_t^* \leq (1-t)u_0 + tu_1.$$

We obtain that $u^* \in \mathcal{S}$, hence $u^* \leq u$ by (3.36). Trivially $u \leq u^*$, and this implies that $u = u^* \in \text{PSH}(S \times X, \pi^*\omega)$.

We will call the curve $[0,1] \ni t \to u_t \in \text{PSH}(X, \omega)$ resulting from the construction of (3.36) the *weak geodesic* connecting u_0, u_1. This terminology is justified by the following result, which says that (3.36) gives the unique solution to (3.10) for boundary data that is merely bounded:

LEMMA 3.14. *When $u_0, u_1 \in \text{PSH}(X, \omega) \cap L^\infty$ then the unique bounded $\pi^*\omega$-psh solution of* (3.10) *is given by* (3.36).

As a result of this lemma, for $u_0, u_1 \in \mathrm{PSH}(X,\omega) \cap L^\infty$, we will call the weak geodesic $t \to u_t$ connecting u_0, u_1 a *bounded geodesic.*

PROOF. Let $C := \|u_1 - u_0\|_{L^\infty}$. It is easy to see that $t \to (u_0 - Ct)$ and $t \to (u_1 - C(1-t))$ are (sub)geodesics that are both members of $\mathcal{S}$, hence so is their maximum $v_t := \max(u_0 - Ct, u_1 - C(1-t)) \in \mathcal{S}$. This and (3.37) gives

$$\max(u_0 - Ct, u_1 - C(1-t)) \leq u_t \leq (1-t)u_0 + tu_1. \tag{3.38}$$

Consequently, $u \in \mathrm{PSH}(S \times X, \pi^*\omega) \cap L^\infty$ and $\lim_{t\to 0,1} u_t = u_{0,1}$. The classical Perron–Bremmerman argument can now be adapted to this setting to give $(\pi^*\omega + i\partial\bar\partial u)^{n+1} = 0$. Lastly, uniqueness of u is a consequence of Theorem 3.2. $\square$

Turning back to non-bounded endpoints u_0, u_1, it turns out that even the very general weak geodesic segment $t \to u_t$ connecting u_0, u_1 exhibits some structure, as we will see in the next two results:

PROPOSITION 3.15 ([**48**]). *Suppose $u_0^k, u_0, u_1^k, u_1 \in \mathrm{PSH}(X,\omega)$ are such that $u_0^k \searrow u_0$ and $u_1^k \searrow u_1$. Let $[0,1] \ni t \to u_t^k, u_t \in \mathrm{PSH}(X,\omega)$ be the weak geodesics connecting u_0^k, u_1^k and u_0, u_1 respectively. Then the following hold:*
(i) $u_t^k \searrow u_t$, $t \in [0,1]$.
(ii) For any $t_1, t_2 \in [0,1]$ we have that $[0,1] \ni l \to u_{(1-l)t_1 + lt_2} \in \mathrm{PSH}(X,\omega)$ is the weak geodesic joining u_{t_1} and u_{t_2}.

PROOF. By the definition of $u^k \in \mathrm{PSH}(S \times X, \pi^*\omega)$ (3.36) it is clear that u^k is decreasing in k and $v := \lim_k u_k \in \mathrm{PSH}(S \times X, \pi^*\omega)$. As u is a candidate in the definition of each u_k, it follows that $u \leq u^k$, hence also $u \leq v$. For the other direction, by (3.37) we have that $u_t^k \leq (1-t)u_0^k + tu_1^k$ hence we can take the limit to obtain that

$$v \leq (1-t)u_0 + tu_1.$$

Consequently v is a candidate for u, giving that $v \leq u$, finishing the proof of (i).

Now we turn to proving (ii). Let $u_0^k = \max(u_0, -k)$ and $u_1^k = \max(u_1, -k)$ be the canonical cutoffs and let $t \to u_t^k$ be the bounded geodesics joining u_0^k and u_1^k. Part (i) implies that $u_{t_1}^k \searrow u_{t_1}$ and $u_{t_2}^k \searrow u_{t_2}$. Hence, applying (i) again, it is enough to prove that $[0,1] \ni l \to u^k_{(1-l)t_1 + lt_2} \in \mathrm{PSH}(X,\omega)$ is the bounded/weak geodesic joining $u_{t_1}^k, u_{t_2}^k$.

Now (3.38) implies that each $t \to u_t^k$ is Lipschitz continuous in the t variable, hence $u_t \to u_{t_{1,2}}$ uniformly as $t \to t_{1,2}$. By Lemma 3.14, $[0,1] \ni l \to u^k_{(1-l)t_1 + lt_2} \in \mathrm{PSH}(X,\omega)$ is indeed the unique bounded/weak geodesic joining $u_{t_1}^k$ and $u_{t_2}^k$. $\square$

The next result connects weak geodesics to the rooftop envelopes of Section 3.3:

LEMMA 3.16 ([**48**]). *Suppose $u_0, u_1 \in \mathrm{PSH}(X,\omega)$ and $t \to u_t$ is the weak geodesic connecting u_0, u_1. Then the following holds:*

$$\inf_{t\in(0,1)} (u_t - t\tau) = P(u_0, u_1 - \tau),\ \tau \in \mathbb{R}.$$

PROOF. Notice that $t \to v_t := u_t - \tau t$ is the weak geodesic connecting $u_0, u_1 - \tau$, hence the proof of the general case reduces to the particlar case $\tau = 0$.

By definition, $P(u_0, u_1) \leq u_0, u_1$. As a result, for the constant (sub)geodesic $t \to h_t := P(u_0, u_1)$ we have $h \in \mathcal{S}$. This trivially gives $h_l \leq u_t$, $t \in [0,1]$, hence $P(u_0, u_1) \leq \inf_{t\in(0,1)} u_t$.

For the reverse inequality, we use the Kiselman minimum principle [**61**, Chapter I, Theorem 7.5], which guarantees that $w := \inf_{t\in(0,1)} u_t \in \mathrm{PSH}(X,\omega)$. Using this and (3.37) we obtain that $w \le u_0, u_1$ hence w is a candidate for $P(u_0,u_1)$, i.e., $w \le P(u_0,u_1)$, finishing the proof. □

We can now relate the super–level sets of tangent vectors along weak geodesics to contact sets of rooftop envelopes:

LEMMA 3.17 ([**48**]). *Suppose $u_0, u_1 \in \mathrm{PSH}(X,\omega)$. Let $t \to u_t$ be the weak geodesic joining u_0, u_1. Then for any $\tau \in \mathbb{R}$ we have*

$$\{\dot u_0 \ge \tau\} = \{P(u_0, u_1 - \tau) = u_0\}.$$

PROOF. By the previous result we have $\inf_{t\in[0,1]}(u_t - t\tau) = P(u_0, u_1-\tau)$. Given $x \in X$, it follows that $P(u_0, u_1 - \tau t)(x) = u_0(x)$ if and only if $\inf_{t\in[0,1]}(u_t(x) - \tau t) = u_0(x)$. Convexity in the t variable implies that this last identity is equivalent to $\dot u_0(x) \ge \tau$. □

With the aid of Lemma 3.16, we can show that weak geodesics with endpoints in finite energy classes stay inside finite energy classes:

PROPOSITION 3.18 ([**48**]). *Suppose $u_0, u_1 \in \mathcal{E}_\chi(X,\omega)$, $\chi \in \mathcal{W}_p^+$, $p \ge 1$. Then for the weak geodesic $t \to u_t$ connecting u_0, u_1 we have that $u_t \in \mathcal{E}_\chi(X,\omega)$ for all $t \in [0,1]$. In case $u_0, u_1 \le 0$ the following estimate holds:*

$$E_\chi(u_t) \le (p+1)^{2n}(E_\chi(u_0) + E_\chi(u_1)),\ t \in [0,1].$$

As a consequence of this proposition, for $u_0, u_1 \in \mathcal{E}_\chi(X,\omega)$ we will call the the curve $[0,1] \ni t \to u_t \in \mathcal{E}_\chi(X,\omega)$ the *finite energy geodesic* connecting u_0, u_1.

PROOF. To start, Proposition 2.19 implies that $P(u_0,u_1) \in \mathcal{E}_\chi(X,\omega)$. By Lemma 3.16, we have that $P(u_0,u_1) \le u_t$, hence by the monotonicity property (Corollary 2.8) it follows that $u_t \in \mathcal{E}_\chi(X,\omega)$, $t \in [0,1]$.

When $u_0, u_1 \le 0$, (3.37) implies that $u_t \le 0$, $t \in [0,1]$. To finish the proof, by Proposition 2.19 and Proposition 2.7 we have the following estimates:

$$E_\chi(u_t) \le (p+1)^n E_\chi(P(u_0,u_1)) \le (p+1)^{2n}(E_\chi(u_0) + E_\chi(u_1)).$$

□

3.4. Extension of the L^p metric structure to finite energy spaces

For the rest of this chapter we will focus only various L^p Finsler geometries of $\mathcal{H}_\omega$. Most of the results we present also have analogs for the more general Orlicz Finsler structures discussed in the previous sections (see [**49**]). Having later applications in mind, we do not seek the greatest generality, and we leave it to the interested reader to adapt our argument to more general metrics. As done it previously, we will assume the volume normalization condition (3.24) throughout this section as well.

Given $u_0, u_1 \in \mathcal{E}_p(X,\omega)$, by Theorem 2.1 there exists decreasing sequences $u_0^k, u_1^k \in \mathcal{H}_\omega$ such that $u_0^k \searrow u_0$ and $u_1^k \searrow u_1$. We propose to define the distance $d_p(u_0,u_1)$ by the formula:

$$d_p(u_0,u_1) = \lim_{k\to\infty} d_p(u_0^k, u_1^k). \tag{3.39}$$

We will show that the above limit exists and it is also independent of the approximating sequences. Developing this further, our main result of this section is the following:

THEOREM 3.19 ([**49**]). *$(\mathcal{E}_p(X,\omega), d_p)$ is a geodesic pseudo–metric space extending $(\mathcal{H}_\omega, d_p)$. Additionally, for $u_0, u_1 \in \mathcal{E}_p(X,\omega)$ the finite energy geodesic $t \to u_t$ joining u_0, u_1 (given by Proposition 3.18) is a d_p-geodesic.*

Recall that a pseudo–metric is just a metric that may not satisfy the non-degeneracy condition. Also, given a pseudo–metric space (M, d), we say that a curve $[0,1] \ni t \to \gamma_t \in M$ is a *d–geodesic* if

$$d(\gamma_{t_1}, \gamma_{t_2}) = |t_1 - t_2| d(\gamma_0, \gamma_1), \quad t_1, t_2 \in [0,1]. \tag{3.40}$$

A *geodesic pseudo–metric space* (M, d) is pseudo–metric space in which any two points can be joining by a d–geodesic.

In the next section we will show that in fact d_p is in fact a bona fide metric, but this will require additional machinery. Finally, as the last major theorem of this chapter, we will prove that the resulting metric space $(\mathcal{E}_p(X,\omega), d_p)$ is the completion of $(\mathcal{H}_\omega, d_p)$.

The proof of Theorem 3.19 will be split into a sequence of lemmas and propositions. Our first one is an estimate for "comparable" potentials:

PROPOSITION 3.20. *Suppose $u, v \in \mathcal{H}_\omega$ with $u \leq v$. Then we have:*

$$\frac{1}{2^{n+p}} \max\Big(\int_X |v-u|^p \omega_u^n, \int_X |v-u|^p \omega_v^n\Big) \leq d_p(u,v)^p \leq \int_X |v-u|^p \omega_u^n. \tag{3.41}$$

PROOF. Suppose $[0,1] \ni t \to w_t \in \mathcal{H}_\omega^{1,\bar{1}}$ is the $C^{1,\bar{1}}$-geodesic segment joining $u_0 = u$ and $u_1 = v$. By (3.23) we have

$$d_p(u,v)^p = \int_X |\dot{w}_0|^p \omega_u^n = \int_X |\dot{w}_1|^p \omega_v^n.$$

Since $u \leq v$, we have that $u \leq w_t$, as follows from (3.36) (or Theorem 3.2). Since $(t,x) \to w_t(x)$ is convex in the t variable, we get $0 \leq \dot{w}_0 \leq v - u \leq \dot{w}_1$, and together with the above identity we obtain part of (3.41):

$$\int_X |v-u|^p \omega_v^n \leq d_p(u,v)^p \leq \int_X |v-u|^p \omega_u^n. \tag{3.42}$$

Now we prove the rest of (3.41). Using $\omega_u^n \leq 2^n \omega_{(u+v)/2}^n$ we obtain that

$$\frac{1}{2^{n+p}} \int_X |v-u|^p \omega_u^n \leq \int_X \Big|u - \frac{u+v}{2}\Big|^p \omega_{(u+v)/2}^n.$$

Since $u \leq (u+v)/2$, the first estimate of (3.42) allows us to continue and write:

$$\frac{1}{2^{n+p}} \int_X |v-u|^p \omega_u^n \leq d_p\Big(\frac{u+v}{2}, u\Big)^p.$$

The lemma below implies that $d_p((u+v)/2, u) \leq d_p(v,u)$, giving the remaining estimate in (3.41). □

LEMMA 3.21. *Suppose $u, v, w \in \mathcal{H}_\omega$ and $u \geq v \geq w$. Then $d_p(v,w) \leq d_p(u,w)$ and $d_p(u,v) \leq d_p(u,w)$.*

PROOF. We introduce the $C^{1,\bar{1}}$ geodesics $[0,1] \ni t \to \alpha_t, \beta_t \in \mathcal{H}^{1,\bar{1}}_\omega$ connecting $\alpha_0 := w, \alpha_1 := v$ and $\beta_0 := w, \beta_1 := u$ respectively. From (3.36) (or Theorem 3.2) it follows that both of these curves are increasing in t. Additionally, $\alpha \leq \beta$ by Theorem 3.2. As $\alpha_0 = \beta_0$, it follows that $0 \leq \dot{\alpha}_0 \leq \dot{\beta}_0$. Using this and Theorem 3.6 we obtain that $d_p(w,v) \leq d_p(w,u)$. The estimate $d_p(u,v) \leq d_p(u,w)$ is proved similarly. □

Next we turn our attention to smooth approximants of finite energy potentials:

LEMMA 3.22. *Suppose $u \in \mathcal{E}_p(X,\omega)$ and $\{u_k\}_k \subset \mathcal{H}_\omega$ is a sequence decreasing to u. Then $d_p(u_l, u_k) \to 0$ as $l,k \to \infty$.*

PROOF. We can suppose that $l \leq k$. Then $u_k \leq u_l$, hence by Proposition 3.20 we have:

$$d_p(u_l,u_k)^p \leq \int_X |u_k - u_l|^p \omega^n_{u_k}.$$

We clearly have $u - u_l, u_k - u_l \in \mathcal{E}_p(X, \omega_{u_l})$ and $u - u_l \leq u_k - u_l \leq 0$. Hence, applying Proposition 2.7 for the class $\mathcal{E}_p(X,\omega_{u_l})$ we obtain that

$$d_p(u_l,u_k)^p \leq (p+1)^n \int_X |u - u_l|^p \omega^n_u. \tag{3.43}$$

As u_l decreases to $u \in \mathcal{E}_p(X,\omega)$, it follows from the dominated convergence theorem that $d_p(u_l,u_k) \to 0$ as $l,k \to \infty$. □

Our next lemma confirms that the way we proposed to extend the d_p to $\mathcal{E}_p(X,\omega)$ (see (3.39)) does not have inconsistencies:

LEMMA 3.23. *Given $u_0, u_1 \in \mathcal{E}_p(X,\omega)$, the limit in (3.39) is finite and independent of the approximating sequences $u_0^k, u_1^k \in \mathcal{H}_\omega$.*

In particular, this result implies that for $u_0, u_1 \in \mathcal{H}_\omega$ the distance $d_p(u_0,u_1)$ will be the same according to both (3.39) and our original definition in (3.4). Lastly, the triangle inequality will also hold, hence d_p is a pseudo–metric on $\mathcal{E}_p(X,\omega)$, as claimed in Theorem 3.19.

PROOF. By the triangle inequality and Lemma 3.22 we can write:

$$|d_p(u_0^l,u_1^l) - d_p(u_0^k,u_1^k)| \leq d_p(u_0^l,u_0^k) + d_p(u_1^l,u_1^k) \to 0, \ l,k \to \infty,$$

proving that $d_p(u_0^k,u_1^k)$ is indeed convergent.

Now we prove that the limit in (3.39) is independent of the choice of approximating sequences. Let $v_0^l, v_1^l \in \mathcal{H}_\omega$ be different approximating sequences. By adding small constants if necessary, we can arrange that the sequences u_0^l, u_1^l, respectively v_0^l, v_1^l, are strictly decreasing to u_0, u_1.

Fixing k for the moment, the sequence $\{\max\{u_0^{k+1}, v_0^j\}\}_{j\in\mathbb{N}}$ decreases pointwise to u_0^{k+1}. By Dini's lemma the convergence is uniform, hence there exists $j_k \in \mathbb{N}$ such that for any $j \geq j_k$ we have $v_0^j < u_0^k$. By repeating the same argument we can also assume that $v_1^j < u_1^k$ for any $j \geq j_k$. By the triangle inequality again

$$|d_p(u_0^j,u_1^j) - d_p(v_0^k,v_1^k)| \leq d_p(u_0^j,v_0^k) + d_p(u_1^j,v_1^k), \ j \geq j_k.$$

From (3.43) it follows that for k big enough the quantities $d(u_0^j,v_0^k), d(u_1^j,v_1^k), \ j \geq j_k$ are arbitrarily small, hence $d_p(u_0,u_1)$ is independent of the choice of approximating sequences.

When $u_0, u_1 \in \mathcal{H}_\omega$, then it is possible to approximate with a constant sequence, hence our argument implies that the restriction of d_p (as extended in (3.39)) to $\mathcal{H}_\omega$ coincides with the original definition in (3.4). □

By the previous result, the triangle inequality is inherited by the extension of d_p to $\mathcal{E}_p(X,\omega)$, making $(\mathcal{E}_p(X,\omega), d_p)$ a pseudo–metric space. In the last part of this section we turn our attention to showing that finite energy geodesic segments of Proposition 3.18 are in fact d_p–geodesics:

LEMMA 3.24. *Suppose $u_0, u_1 \in \mathcal{E}_p(X,\omega)$ and $[0,1] \ni t \to u_t \in \mathcal{E}_p(X,\omega)$ is the finite energy geodesic segment connecting u_0, u_1. Then $t \to u_t$ is a d_p-geodesic.*

PROOF. First we prove that

$$d_p(u_0, u_l) = l d_p(u_0, u_1), \quad l \in [0,1]. \tag{3.44}$$

By Theorem 2.1, suppose $u_0^k, u_1^k \in \mathcal{H}_\omega$ are strictly decreasing approximating sequences of u_0, u_1 and let $[0,1] \ni t \to u_t^k \in \mathcal{H}_\omega^{1,\bar{1}}$ be the decreasing sequence of $C^{1,\bar{1}}$ geodesics connecting u_0^k, u_1^k. By the definition of (3.39) and Theorem 3.6 we can write:

$$d_p(u_0, u_1)^p = \lim_{k\to\infty} d_p(u_0^k, u_1^k)^p = \lim_{k\to\infty} \int_X |\dot{u}_0^k|^p \omega^n_{u_0^k}.$$

By Proposition 3.15(i), the geodesic segments $[0,1] \ni t \to u_t^k \in \mathcal{H}_\omega^{1,\bar{1}}$ are decreasing pointwise to $[0,1] \ni t \to u_t \in \mathcal{E}_p(X,\omega)$. In particular, this implies that $u_l^k \searrow u_l$. We want to find a decreasing sequence $\{w_l^k\}_k \subset \mathcal{H}_\omega$ such that $u_l^k \leq w_l^k$, $w_l^k \searrow u_l$ and

$$l^p d_p(u_0^k, u_1^k)^p - d_p(u_0^k, w_l^k)^p = l^p \int_X |\dot{u}_0^k|^p \omega^n_{u_0^k} - \int_X |\dot{w}_0^k|^p \omega^n_{u_0^k} \to 0 \text{ as } k \to \infty, \tag{3.45}$$

where $t \to w_t^k$ is the $C^{1,\bar{1}}$–geodesic segment connecting u_0^k and w_l^k. By the definition of d_p, letting $k \to \infty$ in (3.45) would give us (3.44).

Finding such sequence w_l^k is always possible by an application of Lemma 3.25 applied to $v_0 := u_0^k$ and $v_1 := u_l^k$, as we observe that the bounded geodesic segment connecting u_0^k and u_l^k is exactly $t \to u_{lt}^k$ (see Proposition 3.15(ii)).

Finally, to finish the proof, we argue that for $t_1, t_2 \in [0,1]$, $t_1 \leq t_2$ we have

$$d_p(u_{t_1}, u_{t_2}) = (t_2 - t_1) d_p(u_0, u_1). \tag{3.46}$$

Let $h_0 = u_{t_2}$ and $h_1 = u_0$. From Proposition 3.15(ii) it follows that $[0,1] \ni t \to h_t := u_{t_2(1-t)} \in \mathcal{E}_p(X,\omega)$ is the finite energy geodesic connecting h_0, h_1. Applying 3.44 to $t \to h_t$ and $l = 1 - t_1/t_2$ we obtain

$$(1 - t_1/t_2) d_p(u_{t_2}, u_0) = d_p(u_{t_2}, u_{t_1}).$$

Now applying (3.44) for $t \to u_t$ and $l = t_2$ we have

$$d_p(u_0, u_{t_2}) = t_2 d_p(u_0, u_1).$$

Putting these last two formulas together we obtain (3.46), finishing the proof. □

LEMMA 3.25. *Suppose* $v_0, v_1 \in \mathrm{PSH}(X,\omega) \cap L^\infty$ *and* $\{v_1^j\}_{j \in \mathbb{N}} \subset \mathrm{PSH}(X,\omega) \cap L^\infty$ *is sequence decreasing to* v_1. *By* $[0,1] \ni t \to v_t, v_t^j \in \mathrm{PSH}(X,\omega) \cap L^\infty$ *we denote the bounded geodesics connecting* v_0, v_1 *and* v_0, v_1^j *respectively. By convexity in the* t *variable, we can introduce* $\dot{v}_0 = \lim_{t \to 0}(v_t - v_0)/t$ *and* $\dot{v}_0^j = \lim_{t\to 0}(v_t^j - v_0)/t$, *and the following holds:*

$$\lim_{j\to\infty} \int_X |\dot{v}_0^j|^p \omega_{v_0}^n = \int_X |\dot{v}_0|^p \omega_{v_0}^n.$$

PROOF. By (3.38) there exists $C > 0$ such that $\|\dot{v}_0\|_{L^\infty(X)}, \|\dot{v}_0^j\|_{L^\infty(X)} \leq C$. We also have $v \leq v^j$, $j \in \mathbb{N}$ by (3.36) (or Theorem 3.2). As we have convexity in the t variable and all our bounded geodesics share the same starting point, it also follows that $\dot{v}_0^j \searrow \dot{v}_0$ pointwise. Consequently, the lemma follows now from Lebesgue's dominated convergence theorem. □

3.5. The Pythagorean formula and applications

In this section we explore the geometry of the operator $(u,v) \to P(u,v)$ restricted to the spaces $\mathcal{E}_p(X,\omega)$. The main focus will be on the following result, establishing a metric relationship between the vertices of the "triangle" $(u, v, P(u,v))$. This will help in proving that d_p is non–degenerate, that $v \to P(u,v)$ is a d_p–contraction, and a number of other properties. Again, throughout this section we will assume the volume normalization condition (3.24) holds. The main result of this section is the Pythagorean formula of [**49**]:

THEOREM 3.26 (Pythagorean formula, [**49**, Corollary 4.14]). *Given* $u_0, u_1 \in \mathcal{E}_p(X,\omega)$, *we have* $P(u_0,u_1) \in \mathcal{E}_p(X,\omega)$ *and*

$$d_p(u_0,u_1)^p = d_p(u_0, P(u_0,u_1))^p + d_p(P(u_0,u_1),u_1)^p.$$

The name of the above formula comes from the particular case $p = 2$, in which case it suggests that u_0, u_1 and $P(u_0,u_1)$ form a right triangle with hypotenuse $t \to u_t$.

Before we give the proof of this result, we argue how it implies the non–degeneracy of d_p, giving that $(\mathcal{E}_p(X,\omega), d_p)$ is a metric space:

PROPOSITION 3.27. *Given* $u_0, u_1 \in \mathcal{E}_p(X,\omega)$ *if* $d_p(u_0,u_1) = 0$ *then* $u_0 = u_1$.

PROOF. From Theorem 3.26 it follows that $d_p(u_0, P(u_0,u_1)) = 0$ and also $d_p(u_1, P(u_0,u_1)) = 0$. By the first estimate of the next lemma, which generalizes Proposition 3.20, it follows that $u_0 = P(u_0,u_1)$ a.e. with respect to $\omega^n_{P(u_0,u_1)}$, and similarly, $u_1 = P(u_0,u_1)$ a.e. with respect to $\omega^n_{P(u_0,u_1)}$. By the domination principle (Proposition 2.21) it follows that $u_0 \leq P(u_0,u_1)$ and $u_1 \leq P(u_0,u_1)$. As the reverse inequalities are trivial, we obtain that $u_0 = P(u_0,u_1) = u_1$. □

LEMMA 3.28. *Suppose* $u, v \in \mathcal{E}_p(X,\omega)$ *with* $u \leq v$. *Then we have:*

$$\max\Big(\frac{1}{2^{n+p}}\int_X |v-u|^p\omega_u^n, \int_X |v-u|^p \omega_v^n\Big) \leq d_p(u,v)^p \leq \int_X |v-u|^p\omega_u^n. \tag{3.47}$$

PROOF. By Proposition 2.9, all the integrals in (3.47) are finite. By Theorem 2.1, we can choose $u_k, v_k \in \mathcal{H}_\omega$ such that $v_k \searrow v$ and $u_k \searrow u$ and $u_k \leq v_k$. By Proposition 3.20, estimate (3.47) holds for u_k, v_k. Using Proposition 2.11 and definition (3.39) we can take the limit $k \to \infty$ in the estimates for u_k, v_k, and obtain (3.47). □

By the next corollary we will only need to prove Theorem 3.26 for smooth potentials u_0, u_1:

COROLLARY 3.29. *If $\{w_k\}_{k \in \mathbb{N}} \subset \mathcal{E}_p(X, \omega)$ decreases (increases a.e.) to $w \in \mathcal{E}_p(X, \omega)$ then $d_p(w_k, w) \to 0$.*

PROOF. By the previous lemma, we have $d_p(w, w_k)^p \leq \int_X |w - w_k|^p (\omega^n_{w_k} + \omega^n_w)$. We can use Proposition 2.11 again to conclude that $d_p(w, w_k) \to 0$. □

As $P(u_0, u_1) \in \mathcal{H}^{1,\bar{1}}_\omega$ for $u_0, u_1 \in \mathcal{H}_\omega$ (Corollary 2.16), we need to generalize Theorem 3.6 for endpoints in $\mathcal{H}^{1,\bar{1}}_\omega$, before we can prove Theorem 3.26. The first step is the next lemma:

LEMMA 3.30. *Suppose $u_0, u_1 \in \mathcal{H}^{1,\bar{1}}_\omega$ and $t \to u_t$ is the bounded geodesic connecting them. Then the following holds:*

$$\int_X |\dot{u}_0|^p \omega^n_{u_0} = \int_X |\dot{u}_1|^p \omega^n_{u_1} \tag{3.48}$$

PROOF. To obtain (3.48) we prove:

$$\int_{\{\dot{u}_0 > 0\}} |\dot{u}_0|^p \omega^n_{u_0} = \int_{\{\dot{u}_1 > 0\}} |\dot{u}_1|^p \omega^n_{u_1}, \tag{3.49}$$

$$\int_{\{\dot{u}_0 < 0\}} |\dot{u}_0|^p \omega^n_{u_0} = \int_{\{\dot{u}_1 < 0\}} |\dot{u}_1|^p \omega^n_{u_1}. \tag{3.50}$$

Using Remark 2.18 and Lemma 3.17 multiple times we can write:

$$\begin{aligned}
\int_{\{\dot{u}_0 > 0\}} |\dot{u}_0|^p \omega^n_{u_0} &= p \int_0^\infty \tau^{p-1} \omega^n_{u_0}(\{\dot{u}_0 \geq \tau\}) d\tau \\
&= p \int_0^\infty \tau^{p-1} \omega^n_{u_0}(\{P(u_0, u_1 - \tau) = u_0\}) d\tau \\
&= p \int_0^\infty \tau^{p-1} (\operatorname{Vol}(X) - \omega^n_{u_1}(\{P(u_0, u_1 - \tau) = u_1 - \tau\})) d\tau \\
&= p \int_0^\infty \tau^{p-1} \omega^n_{u_1}(\{P(u_0, u_1 - \tau) < u_1 - \tau\}) d\tau \\
&= p \int_0^\infty \tau^{p-1} \omega^n_{u_1}(\{P(u_0 + \tau, u_1) < u_1\}) d\tau \\
&= p \int_0^\infty \tau^{p-1} \omega^n_{u_1}(\{\dot{u}_1 > \tau\}) d\tau \\
&= \int_{\{\dot{u}_1 > 0\}} |\dot{u}_1|^p \omega^n_{u_1},
\end{aligned}$$

where in the second line we used Lemma 3.17, in the third line we used Remark 2.18 and in the sixth line we used Lemma 3.17 again, but this time for the "reversed" geodesic $t \to u_{1-t}$. Formula (3.50) follows if we apply (3.49) to the bounded geodesic $t \to u_{1-t}$. □

PROPOSITION 3.31. *Suppose $u_0, u_1 \in \mathcal{H}^{1,\bar{1}}_\omega$ and $t \to u_t$ is the bounded geodesic connecting them. Then we have:*

$$d_p(u_0,u_1)^p = \int_X |\dot{u}_0|^p \omega^n_{u_0} = \int_X |\dot{u}_1|^p \omega^n_{u_1}. \tag{3.51}$$

PROOF. As usual, let $u_0^k, u_1^k \in \mathcal{H}_\omega$ be a sequence of potentials decreasing to u_0, u_1. Let $[0,1] \ni t \to u_t^{kl} \in \mathcal{H}^{1,\bar{1}}_\omega$ be the $C^{1,\bar{1}}$–geodesic joining u_0^k, u_1^l. By Theorem 3.6 we have

$$d_p(u_0^k, u_1^l)^p = \int_X |\dot{u}_0^{kl}|^p \omega^n_{u_0^k}.$$

If we let $l \to \infty$, by Lemma 3.25 and (3.39) we obtain that that

$$d_p(u_0^k, u_1)^p = \int_X |\dot{u}_0^k|^p \omega^n_{u_0^k},$$

where $t \to u_t^k$ is bounded geodesic connecting u_0^k with u_1. Using the previous lemma we can write:

$$d_p(u_0^k, u_1)^p = \int_X |\dot{u}_1^k|^p \omega^n_{u_1}.$$

Letting $k \to \infty$, another application of Lemma 3.25 yields (3.51) for $t = 1$, and the case $t = 0$ follows by symmetry. □

PROOF OF THEOREM 3.26. By Corllary 3.29, it is enough to prove the Pythgaorean formula for $u_0, u_1 \in \mathcal{H}_\omega$. According to Corollary 2.16 we have $P(u_0,u_1) \in \mathcal{H}^{1,\bar{1}}_\omega$. Suppose $[0,1] \ni t \to u_t \in \mathcal{H}^{1,\bar{1}}_\omega$ is the $C^{1,\bar{1}}$–geodesic connecting u_0, u_1. By Theorem 3.6 we have:

$$d_p(u_0,u_1)^p = \int_X |\dot{u}_0|^p \omega^n_{u_0}.$$

To complete the argument we will prove the following:

$$d_p(u_1, P(u_0,u_1))^p = \int_{\{\dot{u}_0 > 0\}} |\dot{u}_0|^p \omega^n_{u_0}, \tag{3.52}$$

$$d_p(u_0, P(u_0,u_1))^p = \int_{\{\dot{u}_0 < 0\}} |\dot{u}_0|^p \omega^n_{u_0}. \tag{3.53}$$

We prove now (3.52). Using Lemma 3.17 we can write:

$$\begin{aligned}\int_{\{\dot{u}_0 > 0\}} |\dot{u}_0|^p \omega^n_{u_0} &= p\int_0^\infty \tau^{p-1}\omega^n_{u_0}(\{\dot{u}_0 \geq \tau\})d\tau \\ &= p\int_0^\infty \tau^{p-1}\omega^n_{u_0}(\{P(u_0, u_1 - \tau) = u_0\})d\tau.\end{aligned}$$

Suppose $t \to \tilde{u}_t$ is the bounded geodesic connecting $P(u_0,u_1), u_1$. Since $P(u_0,u_1) \leq \tilde{u}_t$, $t \in [0,1]$, the correspondence $(t,x) \to \tilde{u}_t(x)$ is increasing in the t variable, hence

$\ddot{u}_0 \geq 0$. By (3.51), Lemma 3.17 and Proposition 2.17 we can write:

$$\begin{aligned}
d_p(P(u_0,u_1),u_1)^p &= \int_X |\dot{\tilde{u}}_0|^p \omega^n_{P(u_0,u_1)} = \int_{\{\dot{\tilde{u}}_0>0\}} |\dot{\tilde{u}}_0|^p \omega^n_{P(u_0,u_1)} \\
&= p\int_0^\infty \tau^{p-1}\omega^n_{P(u_0,u_1)}(\{\dot{\tilde{u}}_0 \geq \tau\})d\tau \\
&= p\int_0^\infty \tau^{p-1}\omega^n_{P(u_0,u_1)}(\{P(P(u_0,u_1),u_1-\tau) = P(u_0,u_1)\})d\tau \\
&= p\int_0^\infty \tau^{p-1}\omega^n_{P(u_0,u_1)}(\{P(u_0,u_1,u_1-\tau) = P(u_0,u_1)\})d\tau \\
&= p\int_0^\infty \tau^{p-1}\omega^n_{P(u_0,u_1)}(\{P(u_0,u_1-\tau) = P(u_0,u_1)\})d\tau \\
&= p\int_0^\infty \tau^{p-1}\omega^n_{u_0}(\{P(u_0,u_1-\tau) = P(u_0,u_1) = u_0\})d\tau \\
&= p\int_0^\infty \tau^{p-1}\omega^n_{u_0}(\{P(u_0,u_1-\tau) = u_0\})d\tau,
\end{aligned}$$

where in the third line we have used Lemma 3.17, in the sixth line we have used Proposition 2.17 together with the fact that $\{P(u_0,u_1)=u_1\}\cap\{P(u_0,u_1-\tau) = P(u_0,u_1)\}$ is empty for $\tau>0$.

Comparing our above calculations (3.52) follows. One can conclude (3.53) from (3.52) after reversing the roles of u_0,u_1 and then using (3.49). □

As another application of Theorem 3.26 we will show that the d_p metric is comparable to a concrete analytic expression:

THEOREM 3.32 ([**49**]). *For any $u_0,u_1 \in \mathcal{E}_p(X,\omega)$ we have*

(3.54)
$$\frac{1}{2^{p-1}}d_p(u_0,u_1)^p \leq \int_X |u_0-u_1|^p\omega^n_{u_0} + \int_X |u_0-u_1|^p\omega^n_{u_1} \leq 2^{2n+3p+3}d_p(u_0,u_1)^p.$$

PROOF. To obtain the first estimate we use the triangle inequality and Lemma 3.28:

$$\begin{aligned}
d_p(u_0,u_1)^p &\leq (d_p(u_0,\max(u_0,u_1)) + d_p(\max(u_0,u_1),u_1))^p \\
&\leq 2^{p-1}(d_p(u_0,\max(u_0,u_1))^p + d_p(\max(u_0,u_1),u_1)^p) \\
&\leq 2^{p-1}\Big(\int_X |u_0-\max(u_0,u_1)|^p\omega^n_{u_0} + \int_X |\max(u_0,u_1)-u_1|^p\omega^n_{u_1}\Big) \\
&= 2^{p-1}\Big(\int_{\{u_1>u_0\}} |u_0-u_1|^p\omega^n_{u_0} + \int_{\{u_0>u_1\}} |u_0-u_1|^p\omega^n_{u_1}\Big) \\
&\leq 2^{p-1}\Big(\int_X |u_0-u_1|^p\omega^n_{u_0} + \int_X |u_0-u_1|^p\omega^n_{u_1}\Big).
\end{aligned}$$

Now we deal with the second estimate in (3.54). By the next result result and Theorem 3.26 we can write

$$\begin{aligned}
2^{n+p+1}d_p(u_0,u_1)^p &\geq d_p\Big(u_0,\frac{u_0+u_1}{2}\Big)^p \geq d_p\Big(u_0,P\Big(u_0,\frac{u_0+u_1}{2}\Big)\Big)^p \\
&\geq \int_X \Big|u_0 - P\Big(u_0,\frac{u_0+u_1}{2}\Big)\Big|^p\omega^n_{u_0}.
\end{aligned}$$

By a similar reasoning as above, and the fact that $2^n\omega^n_{(u_0+u_1)/2} \geq \omega^n_{u_0}$ we can write:

$$\begin{aligned}
2^{n+p+1}d_p(u_0,u_1)^p &\geq d_p\Big(u_0,\frac{u_0+u_1}{2}\Big)^p \geq d_p\Big(\frac{u_0+u_1}{2},P\Big(u_0,\frac{u_0+u_1}{2}\Big)\Big)^p\\
&\geq \int_X \Big|\frac{u_0+u_1}{2}-P\Big(u_0,\frac{u_0+u_1}{2}\Big)\Big|^p\omega^n_{(u_0+u_1)/2}\\
&\geq \frac{1}{2^n}\int_X \Big|\frac{u_0+u_1}{2}-P\Big(u_0,\frac{u_0+u_1}{2}\Big)\Big|^p\omega^n_{u_0}.
\end{aligned}$$

Adding the last two estimates, and using the convexity of $t\to t^p$ we obtain:

$$\begin{aligned}
2^{2n+p+2}d_p(u_0,u_1)^p &\geq \int_X\Big|u_0-P\Big(u_0,\frac{u_0+u_1}{2}\Big)\Big|^p+\Big|\frac{u_0+u_1}{2}-P\Big(u_0,\frac{u_0+u_1}{2}\Big)\Big|^p\omega^n_{u_0}\\
&\geq \frac{1}{2^{2p}}\int_X|u_0-u_1|^p\omega^n_{u_0}.
\end{aligned}$$

By symmetry we also have $2^{2n+3p+2}d_p(u_0,u_1)^p \geq \int_X |u_0-u_1|^p\omega^n_{u_1}$, and adding these last two estimates together the second inequality in (3.54) follows. □

LEMMA 3.33. *Suppose $u_0,u_1\in\mathcal{E}_p(X,\omega)$. Then we have*

$$d_p\Big(u_0,\frac{u_0+u_1}{2}\Big)^p \leq 2^{n+p+1}d_p(u_0,u_1)^p.$$

PROOF. Using Theorem 3.26 and Lemma 3.21 we can start writing:

$$\begin{aligned}
d_p\Big(u_0,\frac{u_0+u_1}{2}\Big)^p &= d_p\Big(u_0,P\Big(u_0,\frac{u_0+u_1}{2}\Big)\Big)^p+d_p\Big(\frac{u_0+u_1}{2},P\Big(u_0,\frac{u_0+u_1}{2}\Big)\Big)^p\\
&\leq d_p(u_0,P(u_0,u_1))^p+d_X\Big(\frac{u_0+u_1}{2},P(u_0,u_1)\Big)^p\\
&\leq \int_X |u_0-P(u_0,u_1)|^p\omega^n_{P(u_0,u_1)}+\int_X\Big|\frac{u_0+u_1}{2}-P(u_0,u_1)\Big|^p\omega^n_{P(u_0,u_1)}\\
&\leq \frac{3}{2}\int_X |u_0-P(u_0,u_1)|^p\omega^n_{P(u_0,u_1)}+\frac{1}{2}\int_X|u_1-P(u_0,u_1)|^p\omega^n_{P(u_0,u_1)}\\
&\leq 3\cdot 2^{n+p-1}d_p(u_0,P(u_0,u_1))^p+2^{n+p-1}d_p(u_1,P(u_0,u_1))^p\\
&\leq 2^{n+p+1}d_p(u_0,u_1)^p,
\end{aligned}$$

where in the second line we have used Lemma 3.21 and the fact that $P(u_0,u_1)\leq P(u_0,(u_0+u_1)/2)$, in the third and fifth line Lemma 3.28, and in the sixth line we have used Theorem 3.26 again. □

3.6. The complete metric spaces $(\mathcal{E}_p(X,\omega),d_p)$

To show completeness of $(\mathcal{E}_p(X,\omega),d_p)$ we have to argue that limits of Cauchy sequences land in $\mathcal{E}_p(X,\omega)$. According to our first result, which extends Corollary 3.29, monotone d_p-bounded sequences are Cauchy, and their limit is in $\mathcal{E}_p(X,\omega)$:

LEMMA 3.34. *Suppose $\{u_k\}_{k\in\mathbb{N}}\subset\mathcal{E}_p(X,\omega)$ is a decreasing/increasing d_p-bounded sequence. Then $u=\lim_{k\to\infty}u_k\in\mathcal{E}_p(X,\omega)$ and additionally $d_p(u,u_k)\to 0$.*

PROOF. First we assume that u_k is decreasing. Without loss of generality, we can also suppose that $u_k<0$. By Lemma 3.28 it follows that

$$\max\Big(\int_X|u_j|^p\omega^n_{u_j},\int_X|u_j|^p\omega^n\Big)\leq 2^{n+p}d_p(u_j,0)^p\leq D. \tag{3.55}$$

As $\int_X |u_j|^p \omega^n$ is uniformly bounded, it follows that the limit function $u = \lim_k u_k$ exists and $u \in \text{PSH}(X, \omega)$. Since $E_p(u_j) = \int_X |u_j|^p \omega_{u_j}^n$ is uniformly bounded we can use Corollary 2.12, to obtain that $u \in \mathcal{E}_p(X, \omega)$ and Corollary 3.29 gives that $d_p(u_k, u) \to 0$.

Now we turn to the case when u_k is increasing. We know that $d_p(u_k, 0)$ is uniformly bounded, hence using Theorem 3.32 we obtain that $\int_X |u_k|^p \omega^n$ is uniformly bounded as well. By the L^1–compactness of subharmonic functions [**61**, Chapter I, Proposition 4.21] we obtain that $u_k \nearrow u \in \text{PSH}(X, \omega)$ a.e. with respect to ω^n. By the monotonicity property (Corollary 2.8) we obtain that $u \in \mathcal{E}_p(X, \omega)$ and another application of Corollary 3.29 gives that $d_p(u_k, u) \to 0$. □

As another consequence of the Pythagorean identity, we note the contractivity of the operator $v \to P(u, v)$, when restricted to finite energy spaces. This result will be essential in proving that $(\mathcal{E}_p(X, \omega), d_p)$ is a complete metric space.

PROPOSITION 3.35. *Given $u, v, w \in \mathcal{E}_p(X, \omega)$ we have*

$$d_p(P(u, v), P(u, w)) \leq d_p(v, w).$$

PROOF. Using Corollary 3.29 we can assume that $u, v, w \in \mathcal{H}_\omega^{1,\bar{1}}$ and also $P(u, v), P(u, w) \in \mathcal{H}_\omega^{1,\bar{1}}$ (see Corollary 2.16).

First we assume that $v \leq w$. Let $t \to \phi_t$ be the $C^{1,\bar{1}}$ geodesic connecting v, w, and $t \to \psi_t$ be the bounded geodesic connecting $P(u, v), P(u, w)$. By Proposition 3.31 we have to argue that

$$\int_X |\dot{\psi}_0|^p \omega_{P(u,v)}^n \leq \int_X |\dot{\phi}_0|^p \omega_v^n. \tag{3.56}$$

Proposition 2.17 implies that

$$\int_X |\dot{\psi}_0|^p \omega_{P(u,v)}^n \leq \int_{\{P(u,v)=u\}} |\dot{\psi}_0|^p \omega_u^n + \int_{\{P(u,v)=v\}} |\dot{\psi}_0|^p \omega_v^n.$$

We argue that the first term in this sum is zero. As $P(u, v) \leq P(u, w) \leq u$, by (3.36) (or Theorem 3.2), it is clear that $P(u, v) \leq \psi_t \leq u$, $t \in [0, 1]$. Hence, if $x \in \{P(u, v) = u\}$ then $\psi_t(x) = u(x)$, $t \in [0, 1]$, implying $\dot{\psi}_0\big|_{\{P(u,v)=u\}} \equiv 0$.

At the same time, using Theorem 3.2 again, it follows that $\psi_t \leq \phi_t$, $t \in [0, 1]$. This implies that $0 \leq \dot{\psi}_0\big|_{\{P(u,v)=v\}} \leq \dot{\phi}_0\big|_{\{P(u,v)=v\}}$, which in turn implies (3.56).

The general case follows now from an application of the Pythagorean formula (Theorem 3.26) and what we just proved above:

$$\begin{aligned} d_p(P(u, v), P(u, w))^p &= d_p(P(u, v), P(u, v, w))^p + d_p(P(u, w), P(u, v, w))^p \\ &= d_p(P(u, v), P(u, P(v, w)))^p + d_p(P(u, w), P(u, P(v, w)))^p \\ &\leq d_p(v, P(v, w))^p + d_p(w, P(v, w))^p \\ &= d_p(v, w)^p. \end{aligned}$$

□

We arrive to the main result of this section. Using the previous proposition we will show that any Cauchy sequence of $(\mathcal{E}_p, d_p)$ is equivalent to a monotonic Cauchy sequence. By the Lemma 3.34, such sequences have limit in $\mathcal{E}_p(X, \omega)$, showing that this space is complete with respect to d_p:

THEOREM 3.36 ([**49**]). *$(\mathcal{E}_p(X,\omega),d_p)$ is a geodesic metric space, which is the metric completion of $(\mathcal{H}_\omega,d_p)$. Additionally, the finite energy geodesic $t \to u_t$ connecting $u_0,u_1 \in \mathcal{E}_p(X,\omega)$ is a d_p-geodesic.*

PROOF. By Theorem 2.1 and Corollary 3.29, the set $\mathcal{H}_\omega$ is a d_p–dense subset of $\mathcal{E}_p(X,\omega)$. The statement about geodesics was addressed in Theorem 3.19, hence we only need to argue completeness.

Suppose $\{u_k\}_{k\in\mathbb{N}} \subset \mathcal{E}_p(X,\omega)$ is a d_p–Cauchy sequence. We will prove that there exists $v \in \mathcal{E}_p(X,\omega)$ such that $d_p(u_k,v)\to 0$. After passing to a subsequence we can assume that

$$d_p(u_l,u_{l+1}) \leq 1/2^l,\ l\in\mathbb{N}.$$

By Proposition 2.19 we can introduce $v^k_l = P(u_k,u_{k+1},\dots,u_{k+l}) \in \mathcal{E}_p(X,\omega),\ l,k\in\mathbb{N}$. We argue first that each decreasing sequence $\{v^k_l\}_{l\in\mathbb{N}}$ is d_p–Cauchy. We observe that $v^k_{l+1} = P(v^k_l,u_{k+l+1})$ and $v^k_l = P(v^k_l,u_{k+l})$. Using this and Proposition 3.35 we can write:

$$d_p(v^k_{l+1},v^k_l) = d_p(P(v^k_l,u_{k+l+1}),P(v^k_l,u_{k+l})) \leq d_p(u_{k+l+1},u_{k+l}) \leq \frac{1}{2^{k+l}}.$$

By Lemma 3.34, it follows now that each sequence $\{v^k_l\}_{l\in\mathbb{N}}$ is d_p–convergening to some $v^k\in\mathcal{E}_p(X,\omega)$. Using the same trick as above, we can write:

$$\begin{aligned}
d_p(v^k,v^{k+1}) &= \lim_{l\to\infty} d_p(v^k_{l+1},v^{k+1}_l) = \lim_{l\to\infty} d_p(P(u_k,v^{k+1}_l),P(u_{k+1},v^{k+1}_l))\\
&\leq d_p(u_k,u_{k+1}) \leq \frac{1}{2^k},\\
d_p(v^k,u_k) &= \lim_{l\to\infty} d_p(v^k_l,u_k) = \lim_{l\to\infty} d_p(P(u_k,v^{k+1}_{l-1}),P(u_k,u_k))\\
&\leq \lim_{l\to\infty} d_p(v^{k+1}_{l-1},u_k) = \lim_{l\to\infty} d_p(P(u_{k+1},v^{k+2}_{l-2}),u_k)\\
&\leq \lim_{l\to\infty} d_p(P(u_{k+1},v^{k+2}_{l-2}),u_{k+1}) + d_p(u_{k+1},u_k)\\
&\leq \lim_{l\to\infty} \sum_{j=k}^{l+k} d_p(u_j,u_{j+1}) \leq \frac{1}{2^{k-1}}.
\end{aligned}$$

Consequently, $\{v^k\}_{k\in\mathbb{N}}$ is an increasing d_p–bounded d_p–Cauchy sequence that is equivalent to $\{u_k\}_{k\in\mathbb{N}}$. By Lemma (3.34) there exists $v\in\mathcal{E}_p(X,\omega)$ such that $d_p(v_k,v)\to 0$, which in turn implies that $d_p(u_k,v)\to 0$, finishing the proof. □

3.7. Special features of the L^1 Finsler geometry

When it comes to applications of measure theory, the most important L^p spaces are the L^1 space, its dual L^∞, and the Hilbert space L^2. A similar pattern can be observed with the L^p Finsler geometries on $\mathcal{H}_\omega$. One can show that the d_∞ metric is a multiple of the usual L^∞ metric. As follows from the work of Calabi–Chen [**30**], the completion of $(\mathcal{H}_\omega,d_2)$ is non–postively curved and as such it provides fertile ground to the study of geometric gradient flows explored in [**13, 101, 102**].

In this section we will focus exclusively on the L^1 Finsler geometry of $\mathcal{H}_\omega$, whose path length metric structure has a number of interesting properties making it suitable in our study of canonical Kähler metrics, detailed in the next chapter.

The starting point is the Monge–Ampère energy $I : \text{PSH}(X,\omega) \cap L^\infty \to \mathbb{R}$ (sometimes called Aubin–Yau, or Aubin–Mabuchi energy):

$$I(u) := \frac{1}{(n+1)V} \sum_{j=0}^{n} \int_X u\omega_u^j \wedge \omega^{n-j}. \tag{3.57}$$

The following lemma explains the choice of name for the I energy, as it turns out that the first order variation of this functional is exactly the complex Monge–Ampère operator:

LEMMA 3.37. *Suppose $[0,1] \ni t \to v_t \in \mathcal{H}_\omega$ is a smooth curve. Then $t \to I(v_t)$ is differentiable and*

$$\frac{d}{dt} I(v_t) = \frac{1}{V}\int_X \dot{v}_t \omega_{v_t}^n.$$

PROOF. Using the definition of I we can calculate the differential of $t \to I(v_t)$ directly:

$$\begin{aligned}
\frac{d}{dt} I(v_t) &= \frac{1}{(n+1)V}\Big(\sum_{j=0}^{n}\int_X \dot{v}_t \omega_{v_t}^j \wedge \omega^{n-j} + \sum_{j=0}^{n} j\int_X v_t i\partial\bar{\partial}\dot{v}_t \wedge \omega_{v_t}^{j-1}\wedge \omega^{n-j}\Big)\\
&= \frac{1}{(n+1)V}\Big(\sum_{j=0}^{n}\int_X \dot{v}_t \omega_{v_t}^j \wedge \omega^{n-j} + \sum_{j=0}^{n} j\int_X \dot{v}_t i\partial\bar{\partial}v_t \wedge \omega_{v_t}^{j-1}\wedge \omega^{n-j}\Big)\\
&= \frac{1}{(n+1)V}\Big(\sum_{j=0}^{n}\int_X \dot{v}_t \omega_{v_t}^j \wedge \omega^{n-j} + \sum_{j=0}^{n} j\int_X \dot{v}_t (\omega_{v_t}-\omega) \wedge \omega_{v_t}^{j-1}\wedge \omega^{n-j}\Big)\\
&= \frac{1}{V}\int_X \dot{v}_t \omega_{v_t}^n.
\end{aligned}$$

□

Based on the last lemma, we deduce a number of properties of the Monge–Ampère energy:

PROPOSITION 3.38. *Given $u, v \in \text{PSH}(X,\omega) \cap L^\infty$, the following hold:*

$$I(u) - I(v) = \frac{1}{(n+1)V}\sum_{j=0}^{n}\int_X (u-v)\omega_u^j \wedge \omega_v^{n-j}, \tag{3.58}$$

$$\frac{1}{V}\int_X (u-v)\omega_u^n \leq I(u) - I(v) \leq \frac{1}{V}\int_X (u-v)\omega_v^n. \tag{3.59}$$

PROOF. First we show (3.58) for $u, v \in \mathcal{H}_\omega$. Let $[0,1] \ni t \to h_t := (1-t)v + tu \in \mathcal{H}_\omega$ be the smooth affine curve joining v and u. By the previous lemma and the binomial theorem we can write

$$\begin{aligned}
I(u) - I(v) &= \int_0^1 \frac{d}{dt} I(h_t)dt = \frac{1}{V}\int_X (u-v)\int_0^1 ((1-t)\omega_v + t\omega_u)^n dt\\
&= \frac{1}{V}\sum_{j=0}^{n}\binom{n}{j}\cdot \int_0^1 t^j(1-t)^{n-j}dt \cdot \int_X (u-v)\omega_u^j \wedge \omega_v^{n-j} dt
\end{aligned}$$

After an elementary calculation involving integration by parts we obtain that

$$\int_0^1 t^j(1-t)^{n-j}dt = \frac{j!(n-j)!}{(n+1)!},$$

giving (3.58) for $u, v \in \mathcal{H}_\omega$. For the general case $u, v \in \mathrm{PSH}(X,\omega) \cap L^\infty$ by Theorem 2.1 we can find $u_k, v_k \in \mathcal{H}_\omega$ decreasing to u and v respectively. We can use the Bedford–Taylor theorem on the continuity of the complex Monge–Ampère measures ([**20**, Theorem 2.2.5]) to conclude that $I(u_k) \to I(u)$, $I(v_k) \to I(v)$ and $\int_X (u_k - v_k)\omega_{u_k}^j \wedge \omega_{v_k}^{n-j} \to \int_X (u-v)\omega_u^j \wedge \omega_v^{n-j}$, giving (3.57) for bounded potentials.

Now we turn to the estimates in (3.59). First we prove the following estimate:

$$\begin{aligned}\int_X (u-v)\omega_u^j \wedge \omega_v^{n-j} &= \\ &= \int_X (u-v)\omega_u^{j-1} \wedge \omega_v^{n-j+1} + \int_X (u-v) i\partial\bar\partial(u-v) \wedge \omega_u^{j-1} \wedge \omega_v^{n-j} \\ &= \int_X (u-v)\omega_u^{j-1} \wedge \omega_v^{n-j+1} - \int_X i\partial(u-v) \wedge i\bar\partial(u-v) \wedge \omega_u^{j-1} \wedge \omega_v^{n-j} \\ &\leq \int_X (u-v)\omega_u^{j-1} \wedge \omega_v^{n-j+1}. \end{aligned} \tag{3.60}$$

Using this estimate inductively for the terms of (3.58) yields (3.59). □

As a consequence of (3.58) we note the monotonicity property of I:

COROLLARY 3.39. *Suppose $u, v \in \mathrm{PSH}(X,\omega) \cap L^\infty$ such that $u \geq v$. Then $I(u) \geq I(v)$.*

This results allows to extend the definition of I to $\mathrm{PSH}(X,\omega)$. Indeed, if $u \in \mathrm{PSH}(X,\omega)$ then using the canonical cutoffs $u_k = \max(u, -k)$ we can write:

$$I(u) := \lim_{k\to\infty} I(u_k). \tag{3.61}$$

Though the limit in the above definition is well defined, it may equal $-\infty$ for certain potentials u. By our next result I is finite exactly on $\mathcal{E}_1(X,\omega)$. What is more, I is d_1–continuous when restricted to this space:

PROPOSITION 3.40. *Let $u \in \mathrm{PSH}(X,\omega)$. Then $I(u) > -\infty$ if and only if $u \in \mathcal{E}_1(X,\omega)$. Also, the following holds:*

$$|I(u_0) - I(u_1)| \leq d_1(u_0, u_1), \quad u_0, u_1 \in \mathcal{E}_1(X,\omega). \tag{3.62}$$

PROOF. By definition, $I(u+c) = I(u) + c$ for all $c \in \mathbb{R}$, hence we can assume that $u \leq 0$. Consequently, by (3.59) for the canonical cutoffs $u_k = \max(u, -k)$ we have:

$$-\frac{E_1(u_k)}{V} = \frac{1}{V}\int_X u_k \omega_{u_k}^n \leq I(u_k) \leq \frac{1}{(n+1)V}\int_X u_k \omega_{u_k}^n = -\frac{E_1(u_k)}{(n+1)V}.$$

If $u \in \mathcal{E}_1(X,\omega)$ then by the fundamental estimate (Proposition 2.7) we have that $E_1(u_k)$ is uniformly bounded, hence $I(u) = \lim_k I(u_k)$ is finite. On the other hand, if $I(u)$ is finite then $I(u_k)$ is uniformly bounded, hence so is $E_1(u_k)$. By Corollary 2.12 we obtain that $u \in \mathcal{E}_1(X,\omega)$.

Now we turn to (3.62). By Corollary 3.29 and (3.61) it follows that it is enough to prove (3.62) for bounded potentials. Furthermore, using Theorem 2.1 it is actually enough to prove (3.62) for potentials in $\mathcal{H}_\omega$. For $u_0, u_1 \in \mathcal{H}_\omega$ let $t \to u_t$

be the $C^{1,\bar{1}}$ geodesic connecting u_0, u_1. By (3.63), proved in the next lemma, we can finish our argument:

$$|I(u_1) - I(u_0)| = \Big| \int_0^1 \frac{d}{dt} I(u_t) dt \Big| \leq \frac{1}{V} \int_0^1 \int_X |\dot{u}_t| \omega_{u_t}^n dt = d_1(u_0, u_1),$$

where in the last equality we have used Theorem 3.6. □

LEMMA 3.41. *Suppose $t \to u_t$ is the $C^{1,\bar{1}}$ geodesic connecting $u_0, u_1 \in \mathcal{H}_\omega$. Then $t \to I(u_t)$ is differentiable, moreover*

$$\frac{d}{dt} I(u_t) = \frac{1}{V} \int_X \dot{u}_t \omega_{u_t}^n. \tag{3.63}$$

PROOF. As $t \to u_t$ is $C^{1,\bar{1}}$ it follows that $(u_{t+\delta} - u_t)/\delta \to \dot{u}_t$ uniformly as $\delta \to 0$. Using this and (3.58) we can write:

$$\frac{d}{dt} I(u_t) = \lim_{\delta \to 0} \frac{I(u_{t+\delta}) - I(u_t)}{\delta} = \frac{1}{(n+1)V} \sum_{j=0}^{n} \lim_{\delta \to 0} \int_X \frac{u_{t+\delta} - u_t}{\delta} \omega_{u_{t+\delta}}^j \wedge \omega_{u_t}^{n-j}$$

$$= \frac{1}{V} \int_X \dot{u}_t \omega_{u_t}^n, \tag{3.64}$$

where we have used that $\omega_{u_{t+\delta}}^j \wedge \omega_{u_t}^{n-j} \to \omega_{u_t}^n$ weakly. □

Further linking the Monge–Ampère energy to the L^1 Finsler geometry is the fact that I is affine along finite energy geodesics:

PROPOSITION 3.42. *Suppose $u_0, u_1 \in \mathcal{E}_1(X, \omega)$ and $t \to u_t$ is the finite energy geodesic connecting u_0, u_1. Then $t \to I(u_t)$ is affine: $I(u_t) = (1-t)I(u_0) + tI(u_1)$, $t \in [0,1]$.*

PROOF. First we show that $t \to I(u_t)$ is affine for the $C^{1,\bar{1}}$ geodesic connecting $u_0, u_1 \in \mathcal{H}_\omega$. Let $[0,1] \ni t \to u_t^\varepsilon \in \mathcal{H}_\omega$ be the smooth ε–geodesic connecting u_0, u_1 (see (3.12)). By Lemma 3.37 we have that

$$\frac{d}{dt} I(u_t^\varepsilon) = \frac{1}{V} \int_X \dot{u}_t^\varepsilon \omega_{u_t^\varepsilon}^n = \frac{1}{V} \int_X \dot{u}_t^\varepsilon \omega_{u_t^\varepsilon}^n = \langle 1, \dot{u}_t^\varepsilon \rangle_{u_t^\varepsilon}.$$

As we take one more derivative of the above formula, we can use the L^2 Mabuchi Levi–Civita connection (see (3.7)) and (3.11) to obtain:

$$\frac{d^2}{dt^2} I(u_t^\varepsilon) = \langle 1, \nabla_{\dot{u}_t^\varepsilon} \dot{u}_t^\varepsilon \rangle_{u_t^\varepsilon} = \frac{1}{V} \int_X \nabla_{\dot{u}_t^\varepsilon} \dot{u}_t^\varepsilon \omega_{u_t^\varepsilon}^n = \frac{1}{V} \int_X \varepsilon \omega^n = \varepsilon. \tag{3.65}$$

For fixed $t \in [0,1]$ we have that $u_t^\varepsilon \nearrow u_t$ uniformly, hence (3.58) gives that $I(u_t^\varepsilon) \to I(u_t)$. By an elementary argument, (3.65) implies that $t \to I(u_t) = \lim_{\varepsilon \to 0} I(u_t^\varepsilon)$ is affine.

Next we show the result for $u_0, u_1 \in \mathcal{E}_1(X, \omega)$. In this case let $u_0^k, u_1^k \in \mathcal{H}_\omega$ be decreasing approximating sequences that exist by Theorem 2.1. Let $t \to u_t^k$ be the $C^{1,\bar{1}}$ geodesics joining u_0^k, u_1^k, and $t \to u_t$ be the finite energy geodesic joining u_0, u_1. By Proposition 3.15 it follows that for fixed t we have $u_t^k \searrow u_t$. Corollary 3.29 now gives that $d_1(u_t^k, u_t) \to 0$. We can now use Proposition 3.40 to conclude that $I(u_t^k) \to I(u_t)$. Since $t \to I(u_t^k)$ is affine, so is $t \to I(u_t)$. □

By revisiting the Pythagorean formula of the L^1 geometry (Theorem 3.26) we obtain the following concrete formula for the d_1 metric in terms of the Monge–Ampère energy:

PROPOSITION 3.43. *If $u, v \in \mathcal{E}_1(X,\omega)$ then*

$$d_1(u,v) = I(u) + I(v) - 2I(P(u,v)). \tag{3.66}$$

In particular, if $u \geq v$ then $d_1(u,v) = I(u) - I(v)$.

PROOF. We show that $d_1(u,v) = I(u) - I(v)$ if $u \geq v$. Then the Pythagorean formula (Theorem 3.26) will imply (3.66). By Theorem 2.1 , Corollary 3.29 and Proposition 3.40 it is enough to show $d_1(u,v) = I(u) - I(v)$ for $u, v \in \mathcal{H}_\omega$. Let $[0,1] \ni t \to u_t \in \mathcal{H}^{1,\bar{1}}_\omega$ be the $C^{1,\bar{1}}$ geodesic joining $u_0 := v$ and $u_1 := u$. As $u_1 \geq u_0$, by Theorem 3.2 we obtain that $u_0 \leq u_t$. Using convexity in the t variable, we obtain that $0 \leq \dot{u}_0 \leq \dot{u}_t$. Since $\dot{u}_t \geq 0$, by Theorem 3.6 we can conclude that

$$d_1(u,v) = \frac{1}{V}\int_0^1 \int_X \dot{u}_t \omega^n_{u_t} dt = \int_0^1 \frac{d}{dt} I(u_t)dt = I(u_1) - I(u_0) = I(u) - I(v),$$

where in the second equality we have used (3.63). □

Recall that the correspondence $u \to \omega_u$ is one-to-one up to a constant. To make the correspondence between metrics and potentials one-to-one, we need to restrict the map $u \to \omega_u$ to a hyper–surface of $\mathcal{H}_\omega$. There are many normalizations that come to mind, but the most convenient choice is to use the following space and its completion:

$$\mathcal{H}_\omega \cap I^{-1}(0) = \{u \in \mathcal{H}_\omega \text{ s.t. } I(u) = 0\} \quad \text{and} \quad \mathcal{E}_1(X,\omega) \cap I^{-1}(0). \tag{3.67}$$

As $I : \mathcal{E}_1(X,\omega) \to \mathbb{R}$ is d_1-continuous, we see that $\mathcal{E}_1(X,\omega)\cap I^{-1}(0)$ is indeed the d_1–completion of $\mathcal{H}_\omega \cap I^{-1}(0)$. We focus on the preimage of I due to the conclusion of Proposition 3.42. Indeed, according to this result, if $u_0, u_1 \in \mathcal{E}_1(X,\omega)\cap I^{-1}(0)$ then $t \to u_t$, the finite energy geodesic connecting u_0, u_1, satisfies $u_t \in \mathcal{E}_1(X,\omega)\cap I^{-1}(0)$, hence the "hypersurface" $\mathcal{E}_1(X,\omega) \cap I^{-1}(0)$ is totally geodesic.

The J energy is closely related to the Monge–Ampère energy and is given by the following formula:

$$J(u) = \frac{1}{V}\int_X u\omega^n - I(u). \tag{3.68}$$

By (3.59) it follows that $J(u) \geq 0$ for all $u \in \mathcal{E}_1(X,\omega)$.

For $u \in \mathcal{E}_1(X,\omega)\cap I^{-1}(0)$, the J energy is given by the especially simple formula $J(u) = \frac{1}{V}\int_X u\omega^n$. Additionally, on $\mathcal{E}_1(X,\omega) \cap I^{-1}(0)$ the growth of the d_1 metric and the J energy is closely related:

PROPOSITION 3.44 ([**49**, **58**]). *] There exists $C = C(X,\omega) > 1$ such that*

$$\frac{1}{C}J(u) - C \leq d_1(0,u) \leq CJ(u) + C, \quad u \in \mathcal{E}_1(X,\omega) \cap I^{-1}(0). \tag{3.69}$$

PROOF. Let $u \in \mathcal{E}_1(X,\omega) \cap I^{-1}(0)$. By Theorem 3.32 we have

$$J(u) = \frac{1}{V}\int_X u\omega^n \leq \frac{1}{V}\int_X |u|\omega^n \leq Cd_1(0,u),$$

implying the first estimate in (3.69). For the second estimate, since $I(u) = 0$, we have that $\sup_X u \geq 0$ and

$$d_1(0,u) = -2I(P(0,u)). \tag{3.70}$$

Clearly, $u - \sup_X u \leq \min(0,u)$, so $u - \sup_X u \leq P(0,u)$. Thus, $-\sup_X u = I(u - \sup_X u) \leq I(P(0,u))$. Combined with (3.70), we obtain that $d_1(0,u) =$

$-2I(P(0,u)) \leq 2\sup_X u$. Finally, according to the next basic lemma, $\sup_X u \leq C' \int_X u\omega^n + C'$ for some $C'(X,\omega) > 1$, finishing the proof. $\square$

LEMMA 3.45. *There exists $C := C(X,\omega) > 1$ such that for any $u \in \mathrm{PSH}(X,\omega)$ we have*

$$\frac{1}{V}\int_X u\omega^n \leq \sup_X u \leq \frac{1}{V}\int_X u\omega^n + C.$$

This is a well known result in pluripotential theory. A proof using compactness can be found in [**73**, Proposition 1.7]. Below we give a constructive argument using only the sub-mean value property of psh functions.

PROOF. The first estimate is trivial. To argue the second estimate, it is enough to show that $\int_X u\omega^n$ is uniformly bounded from below, for all $u \in \mathrm{PSH}(X,\omega)$ with $\sup_X u = 0$. We fix such potential u for the rest of the proof.

We pick nested coordinate charts $U_k \subset W_k \subset X$, such that $\{U_k\}_{1\leq k\leq N}$ covers X, and there exist holomorphic diffeomorphisms $\varphi_k : B(0,4) \to W_k$ such that $\varphi_k(B(0,1)) = U_k$ and there exists $\psi_k \in C^\infty(W_k)$ such that $\omega|_{W_k} = i\partial\bar\partial\psi_k$. Notice that it is enough to show the existence of $C := C(X,\omega) < 0$ such that

$$\int_{B(0,1)} u \circ \varphi_j d\mu \geq C, \quad j \in \{1,\dots,N\}, \tag{3.71}$$

where $d\mu$ is the Euclidean measure on $\mathbb{C}^n$.

Using our setup, we obtain that $\psi_k + u \in \mathrm{PSH}(W_k)$ for all $k \in \{1,\dots,N\}$. As u is usc, its supremum is realized at some $x_1 \in X$, i.e. $u \leq u(x_1) = 0$. As $\{U_k\}_k$ covers X, $x_1 \in U_l$ for some $l \in \{1,\dots,N\}$. For simplicity, we can assume that $l = 1$.

Let $z_1 := \varphi_1^{-1}(x_1) \in B(0,1)$. As $B(z_1,2) \subset B(0,4)$, by the sub-meanvalue property of psh functions we can write:

$$\begin{aligned}\psi_1(x_1) &= u \circ \varphi_1(z_1) + \psi_1 \circ \varphi_1(z_1)\\ &\leq \frac{1}{\mu(B(z_1,2))}\int_{B(z_1,2)} (u\circ\varphi_1 + \psi_1\circ\varphi_1)d\mu.\end{aligned} \tag{3.72}$$

Since $u \leq 0$ and $B(0,1) \subset B(z_1,2)$, there exists $C_1 < 0$, independent of u, such that

$$\int_{B(0,1)} u \circ \varphi_1 d\mu \geq C_1. \tag{3.73}$$

As $\{U_k\}_k$ is a covering of X, U_1 needs to intersect at least another member of the covering. We can assume that $U_1 \cap U_2$ is non-empty.

Since $u \leq 0$ and (3.73) holds, there exists $x_2 \in U_2 \cap U_1, r_2 \in (0,1)$ and $\tilde C_1 < 0$, independent of u, such that for $z_2 = \varphi_2^{-1}(x_2)$ we have $\varphi_2(B(z_2,r_2)) \subset U_1 \cap U_2$ and

$$\frac{1}{\mu(B(z_2,r_2))}\int_{B(z_2,r_2)} (u\circ\varphi_2 + \psi_2\circ\varphi_2)d\mu \geq \tilde C_1.$$

Since $u\circ\varphi_2 + \psi_2\circ\varphi_2$ is psh in $B(0,4)$, we can further write that:

$$\begin{aligned}\frac{1}{\mu(B(z_2,2))}\int_{B(z_2,2)} (u\circ\varphi_2 + \psi_2\circ\varphi_2)d\mu &\geq \frac{1}{\mu(B(z_2,r_2))}\int_{B(z_2,r_2)} (u\circ\varphi_2 + \psi_2\circ\varphi_2)d\mu\\ &\geq \tilde C_1.\end{aligned}$$

Since $B(0,1) \subset B(z_2,2)$ and $u \leq 0$, from here we conclude the existence of $C_2 < 0$, independent of u, such that

$$\int_{B(0,1)} u \circ \varphi_2 d\mu \geq C_2. \tag{3.74}$$

Continuing the above process, after $N-2$ more steps we eventually arrive at (3.71). □

3.8. Relation to classical notions of convergence

Theorem 3.32 gave a characterization of d_p–convergence using concrete analytic expressions. When dealing with d_1, it turns out that an even more convenient characterization can be given with the help of the Monge–Ampère energy. This is the content of the next theorem, which also shows that d_1–convergence implies classical L^1 convergence of potentials and also the weak convergence of the associated complex Monge–Ampère measures:

THEOREM 3.46 ([**49**]). *Suppose $u_k, u \in \mathcal{E}_1(X,\omega)$. Then the following hold:*
(i) $d_1(u_k,u) \to 0$ if and only if $\int_X |u_k - u|\omega^n \to 0$ and $I(u_k) \to I(u)$.
(ii) If $d_1(u_k,u) \to 0$ then $\omega^n_{u_k} \to \omega^n_u$ weakly, and $\int_X |u_k - u|\omega^n_v \to 0$ for any $v \in \mathcal{E}_1(X,\omega)$.

The proof of this result is given in a number of propositions and lemmas below. The new analytic input will be given by the bi–functional $\mathcal{I}(\cdot,\cdot) : \mathcal{E}_1(X,\omega) \times \mathcal{E}_1(X,\omega) \to \mathbb{R}$ and its properties:

$$\mathcal{I}(u_0,u_1) = \int_X (u_0 - u_1)(\omega^n_{u_1} - \omega^n_{u_0}). \tag{3.75}$$

Observe that by Lemma 2.9 the above expression is indeed finite. Moreover, if $d_1(u_j,u) \to 0$ then by Theorem 3.32 we get $\mathcal{I}(u_j,u) \to 0$. As it turns out, the relationship between d_1 and $\mathcal{I}$ is much deeper then this simple observation might suggest.

When u_0, u_1 are bounded, then Bedford–Taylor theory allows to integrate by parts and obtain:

$$\mathcal{I}(u_0,u_1) = \sum_{j=0}^{n-1} \int_X i\partial(u_0 - u_1) \wedge \bar{\partial}(u_0 - u_1) \wedge \omega^j_{u_0} \wedge \omega^{n-j-1}_{u_1}.$$

In particular, $\mathcal{I}(u_0,u_1) \geq 0$ for bounded potentials. Applying the next lemma to canonical cutoffs, we deduce that this also holds in general:

LEMMA 3.47. *Suppose $u_0, u^j_0, u_1, u^j_1 \in \mathcal{E}_1(X,\omega)$ and $u^j_0 \searrow u_0$, $u^j_1 \searrow u_1$. Then the following hold:*
(i) $\mathcal{I}(u_0,u_1) = \mathcal{I}(u_0, \max(u_0,u_1)) + \mathcal{I}(\max(u_0,u_1),u_1)$.
(ii) $\lim_{j\to\infty} \mathcal{I}(u^j_0,u^j_1) = \mathcal{I}(u_0,u_1)$.

PROOF. (i) follows from the locality of the complex Monge–Ampère measure on plurifine open sets (see (2.1)). Indeed, $\mathbb{1}_{\{u_1>u_0\}}(\omega^n_{\max(u_0,u_1)} - \omega^n_{u_0}) = \mathbb{1}_{\{u_1>u_0\}}(\omega^n_{u_1} - \omega^n_{u_0})$, consequently

$$\mathcal{I}(u_0,\max(u_0,u_1)) = \int_{\{u_1>u_0\}} (u_0 - u_1)(\omega^n_{u_1} - \omega^n_{u_0}),$$

and for similar reasons, $\mathcal{I}(\max(u_0,u_1),u_1)=\int_{\{u_0>u_1\}}(u_0-u_1)(\omega^n_{u_1}-\omega^n_{u_0})$. Adding these last two identities yields (i).

To prove (ii), we start with $\mathcal{I}(u_0^j,u_1^j)=\mathcal{I}(u_0^j,\max(u_0^j,u_1^j))+\mathcal{I}(\max(u_0^j,u_1^j),u_1^j)$. Trivially, $u_0^j,u_1^j\leq\max(u_0^j,u_1^j)$, hence we can apply Proposition 2.11 to each term on the right hand side of (3.75) to conclude: $\mathcal{I}(u_0^j,\max(u_0^j,u_1^j))\to\mathcal{I}(u_0,\max(u_0,u_1))$ and $\mathcal{I}(\max(u_0^j,u_1^j),u_1^j)\to\mathcal{I}(\max(u_0,u_1),u_1)$. Another application of (i) yields (ii). ∎

The next proposition contains the main analytic properties of the $\mathcal{I}$ functional:

PROPOSITION 3.48. [**11**, Lemma 3.13, Lemma 5.8] *Let $C>0$ and $\phi,\psi,u,v\in\mathcal{E}_1(X,\omega)$ satisfying*

$$-C\leq I(\phi),I(\psi),I(u),I(v),\sup_X\phi,\sup_X\psi,\sup_X u,\sup_X v\leq C. \tag{3.76}$$

Then there exists $f_C:\mathbb{R}^+\to\mathbb{R}^+$ (only dependent on C) continuous with $f_C(0)=0$ such that

$$\Big|\int_X\phi(\omega_u^n-\omega_v^n)\Big|\leq f_C(\mathcal{I}(u,v)), \tag{3.77}$$

$$\Big|\int_X(u-v)(\omega_\phi^n-\omega_\psi^n)\Big|\leq f_C(\mathcal{I}(u,v)). \tag{3.78}$$

Before we get into the argument, we note that by the lemma following this proposition, the condition (3.76) is seen to be equivalent with

$$d_1(0,\phi),d_1(0,\psi),d_1(0,u),d_1(0,v)\leq C'. \tag{3.79}$$

PROOF. By repeated application of the dominated convergence theorem, (2.3) and Lemma 3.47(ii), one can see that it is enough to show (3.77) and (3.78) for bounded potentials.

We focus now on (3.77) and introduce the quantities $a_k:=\int_X\phi\omega_u^k\wedge\omega_v^{n-k}$. We note that (3.77) follows if we are able to prove

$$|a_{k+1}-a_k|\leq f_C(\mathcal{I}(u,v)),\quad k\in\{0,\dots,n-1\}. \tag{3.80}$$

As our potentials are bounded, we can use integration by parts and start writing:

$$\begin{aligned}a_{k+1}-a_k&=\int_X\phi i\partial\bar\partial(u-v)\wedge\omega_u^k\wedge\omega_v^{n-k-1}\\&=\int_X i\partial\phi\wedge\bar\partial(v-u)\wedge\omega_u^k\wedge\omega_v^{n-k-1},\end{aligned}$$

Consequently, by the Cauchy–Schwarz inequality we can write:

$$\begin{aligned}|a_{k+1}-a_k|^2&\leq\int_X i\partial\phi\wedge\bar\partial\phi\wedge\omega_u^k\wedge\omega_v^{n-k-1}\cdot\int_X i\partial(u-v)\wedge\bar\partial(u-v)\wedge\omega_u^k\wedge\omega_v^{n-k-1}\\&\leq\int_X i\partial\phi\wedge\bar\partial\phi\wedge\omega_u^k\wedge\omega_v^{n-k-1}\cdot\mathcal{I}(u,v).\end{aligned}$$

By the above, (3.80) would follow if we can show that $\int_X i\partial\phi \wedge \bar{\partial}\phi \wedge \omega_u^k \wedge \omega_v^{n-k-1}$ is uniformly bounded, and this exactly what we argue:

$$\begin{aligned}\int_X i\partial\phi \wedge \bar{\partial}\phi \wedge \omega_u^k \wedge \omega_v^{n-k-1} &= \int_X \phi\omega \wedge \omega_u^k \wedge \omega_v^{n-k-1} - \int_X \phi\omega_\phi \wedge \omega_u^k \wedge \omega_v^{n-k-1}\\ &\leq \int_X |\phi|\omega \wedge \omega_u^k \wedge \omega_v^{n-k-1} + \int_X |\phi|\omega_\phi \wedge \omega_u^k \wedge \omega_v^{n-k-1}\\ &\leq D\int_X |\phi|\omega^n_{\phi/4+u/8+v/8}\\ &\leq D' d_1(u/8+v/8, \phi/4+u/8+v/8).\end{aligned}$$

where in the last estimate we have used Theorem 3.32. To finish, by the triangle inequality we have to argue that $d_1(0, u/8+v/8)$ and $d_1(0, \phi/4+u/8+v/8)$ are bounded. This can be deduced from (3.79) by repeated application of Lemma 3.33 and the triangle inequality for d_1.

Now we turn to (3.78). The proof has similar philosophy, but it is slightly more intricate. We introduce $\alpha := u - v$, and the quantities $b_k := \int_X \alpha\omega_u^k \wedge \omega_\phi^{n-k}$. We will show that

$$|b_{k+1} - b_k| \leq f_C(\mathcal{I}(u,v)), \quad k \in \{0, \dots, k-1\}. \tag{3.81}$$

This will imply that $\left|\int_X (u-v)(\omega_u^n - \omega_\phi^n)\right| \leq f_C(\mathcal{I}(u,v))$, and using the symmetry in u, v, and basic properties of the absolute value, we obtain (3.78). Integration by parts yields the following:

$$b_{k+1} - b_k = \int_X \alpha i\partial\bar{\partial}(u-\phi) \wedge \omega_u^k \wedge \omega_\phi^{n-k-1} = -\int_X i\partial\alpha \wedge \bar{\partial}(u-\phi) \wedge \omega_u^k \wedge \omega_\phi^{n-k-1}$$

Introducing $\beta := (u+v)/2$ and $c_k := \int_X i\partial\alpha \wedge \bar{\partial}\alpha \wedge \omega_\beta^k \wedge \omega_\phi^{n-k-1}$, by the Cauchy–Schwarz inequality we deduce that

$$|b_{k+1} - b_k|^2 \leq \mathcal{I}(u,\phi) \int_X i\partial\alpha \wedge \bar{\partial}\alpha \wedge \omega_u^k \wedge \omega_\phi^{n-k-1} \leq 2^k \mathcal{I}(u,\phi) c_k \leq D c_k, \tag{3.82}$$

where in the last inequality we have used that $\mathcal{I}(u,\phi)$ is bounded. Indeed, this follows from (3.79), the triangle inequality for d_1, and Theorem 3.32. Consequently, to prove (3.81) it suffices to show that

$$c_k \leq f_C(\mathcal{I}(u,v)). \tag{3.83}$$

To obtain this, we integrate by parts again:

$$\begin{aligned}c_{k+1} - c_k &= \int_X i\partial\alpha \wedge \bar{\partial}\alpha \wedge i\partial\bar{\partial}(\beta-\phi) \wedge \omega_\beta^k \wedge \omega_\phi^{n-k-2}\\ &= \int_X i\partial\alpha \wedge \bar{\partial}(\beta-\phi) \wedge i\partial\bar{\partial}\alpha \wedge \omega_\beta^k \wedge \omega_\phi^{n-k-2}\\ &= \int_X i\partial\alpha \wedge \bar{\partial}(\beta-\phi) \wedge \omega_u \wedge \omega_\beta^k \wedge \omega_\phi^{n-k-2}\\ &\quad - \int_X i\partial\alpha \wedge \bar{\partial}(\beta-\phi) \wedge \omega_v \wedge \omega_\beta^k \wedge \omega_\phi^{n-k-2}.\end{aligned} \tag{3.84}$$

Using that $\omega_u \leq 2\omega_\beta$ and the Cauchy–Schwarz inequality we can estimate the first term from the right hand side in the following manner:

$$\Big|\int_X i\partial\alpha \wedge \bar\partial(\beta-\phi)\wedge\omega_u\wedge\omega_\beta^{k-1}\wedge\omega_\phi^{n-k-1}\Big|^2 \leq$$
$$\leq D'\mathcal{I}(\beta,\phi)\int_X i\partial\alpha\wedge\bar\partial\alpha\wedge\omega_\beta^{k+1}\wedge\omega_\phi^{n-k-2}$$
$$= D'\mathcal{I}\Big(\frac{u+v}{2},\phi\Big)c_{k+1} \leq D''c_{k+1},$$

where in the last inequality we used that $\mathcal{I}((u+v)/2,\phi)$ is bounded. This follows from (3.79), as repeated application of Lemma 3.33 and the triangle inequality for d_1 yields that $d_1(\phi,(u+v)/2)$ is bounded.

We can similarly estimate the other term on the right hand side of (3.84) and putting everything together we obtain that $c_k \leq c_{k+1} + 2D''c_{k+1}^{1/2}$. After a successive application of this inequality, we find that $c_k \leq f_C(c_{n-1})$, which is equivalent to (3.83). □

LEMMA 3.49. *Suppose $u \in \mathcal{E}_1(X,\omega)$ and $-C \leq I(u) \leq \sup_X u \leq C$ for some $C > 0$. Then $d_1(0,u) \leq 3C$.*

PROOF. By the triangle inequality and Proposition 3.43 we can write:

$$\begin{aligned} d_1(0,u) &\leq d_1(0,\sup_X u) + d_1(\sup_X u, u) = d_1(0,\sup_X u) + d_1(0, u - \sup_X u)\\ &= |\sup_X u| + I(0) - I(u - \sup_X u)\\ &= |\sup_X u| + \sup_X u - I(u)\\ &\leq 3C. \end{aligned}$$

□

We note the following important corollary of Proposition 3.48:

COROLLARY 3.50. *For any $C > 0$ there exists $\tilde f_C : \mathbb{R}^+ \to \mathbb{R}^+$ continuous with $\tilde f_C(0) = 0$ such that*

$$\int_X |u-v|\omega_\psi^n \leq \tilde f_C(d_1(u,v)), \tag{3.85}$$

for $u, v, \psi \in \mathcal{E}_1(X,\omega)$ satisfying $d_1(0,u), d_1(0,v), d_1(0,\psi) \leq C$.

PROOF. From Lemma 3.40 and Theorem 3.32 it follows that $d_1(0,u), d_1(0,v)$, $d_1(0,\psi) \leq C$ implies that $I(u), I(v), I(\psi)$ and $\sup_X u, \sup_X v, \sup_X \psi$ are uniformly bounded. Consequently (3.78) gives that

$$\Big|\int_X (u-v)\omega_\psi\Big| \leq f_C(\mathcal{I}(u,v)) + \int_X |u-v|\omega_u^n \leq \tilde f_C(d_1(u,v)).$$

Next, by Theorem 3.32, $d_1(u,\max(u,v))$ is uniformly bounded. By the triangle inequality, so is $d_1(0,\max(u,v))$. Consequently, since $\mathcal{I}(\max(u,v),v) \leq \mathcal{I}(u,v)$ (see Lemma (3.47)) we also have:

$$\Big|\int_X (\max(u,v)-v)\omega_\psi\Big| \leq \tilde f_C(d_1(u,v)).$$

Using $|u-v| = 2(\max(u,v)-v) - (u-v)$, we can add these last two inequalities to obtain (3.85). □

As another corollary of Proposition 3.48 we obtain the second part of Theorem 3.46:

COROLLARY 3.51. *Suppose* $u_k, u, v \in \mathcal{E}_1(X,\omega)$ *with* $d_1(u_k,u) \to 0$. *The following hold:*
(i) $\omega^n_{u_k} \to \omega^n_u$ *weakly.*
(ii) $\int_X |u_k - u|\omega^n_v \to 0$.

PROOF. Given $\phi \in C^\infty(X)$, after possibly multiplying with a small positive constant, we can assume that $\phi \in \mathcal{H}_\omega$. As $d_1(u_k,u) \to 0$ implies that $\mathcal{I}(u_k,u) \to 0$ (Theorem 3.32), we see that (3.77) gives (i). The convergence statement of (ii) follows immediately from the previous corollary. □

Now we prove the rest of Theorem 3.46:

PROPOSITION 3.52. *Suppose* $u_k, u \in \mathcal{E}_1(X,\omega)$. *Then* $d_1(u,u_k) \to 0$ *if and only if* $\int_X |u - u_k|\omega^n \to 0$ *and* $I(u_k) \to I(u)$.

PROOF. The fact that $d_1(u_k,u) \to 0$ implies $\int_X |u-u_k|\omega^n \to 0$, follows from the previous corollary. $I(u_k) \to I(u)$ follows from the d_1–continuity of I.

We focus on the reverse direction. The first step is to show that

$$\int_X u_k \omega^n_u \to \int_X u\omega^n_u. \tag{3.86}$$

By Theorem 2.1, pick $v_k \in \mathcal{H}_\omega$ decreasing to u. Consequently $\sup_X v_k$ and $I(v_k)$ is uniformly bounded, hence by (3.77) we can write:

$$\Big|\int_X u_k(\omega^n_u - \omega^n_{v_j})\Big| \leq f_C(\mathcal{I}(v_j,u)), \quad j,k \in \mathbb{N}.$$

As v_j is smooth and $\int_X |u_k - u|\omega^n \to 0$, we can write

$$\Big|\limsup_k \int_X u_k\omega^n_u - \int_X u\omega^n_{v_j}\Big| \leq f_C(\mathcal{I}(v_j,u)), \quad j \in \mathbb{N}.$$

As $v_j \searrow u$, we can use Proposition 2.11 to get $\limsup_k \int_X u_k\omega^n_u = \int_X u\omega^n_u$. The analogous statement also holds for $\liminf$, and we obtain (3.86).

Next we use the following estimate, which is a consequence of (3.58) and (3.60):

$$\frac{\mathcal{I}(u,u_k)}{(n+1)V} \leq I(u_k) - I(u) - \frac{1}{V}\int_X (u-u_k)\omega^n_u.$$

By (3.86) we have $\mathcal{I}(u,u_k) \to 0$. Using (3.86) again and (3.78) we can write

$$\int_X (u_k - u)\omega^n_{u_k} \to 0. \tag{3.87}$$

Since, $\mathcal{I}(u,u_k) \to 0$, Lemma 3.47(i) gives that $\mathcal{I}(\max(u_k,u),u) \to 0$. Also, by our assumptions, $\int_X(\max(u_k,u) - u)\omega^n \to 0$, hence (3.78) implies that

$$\int_X (\max(u_k,u) - u)\omega^n_u \to 0 \quad \text{and} \quad \int_X (\max(u_k,u) - u)\omega^n_{u_k} \to 0,$$

where in the last limit we also used the locality of the complex Monge–Ampère operator with respect to the plurifine topology (see (2.1)). Using the fact that

$|u_k - u| = 2(\max(u_k, u) - u) - (u_k - u)$, together with (3.86) and (3.87) we get that

$$\int_X |u - u_k|\omega_u^n + \int_X |u - u_k|\omega_{u_k}^n \to 0.$$

Theorem 3.32 now gives $d_1(u_k, u) \to 0$. □

For the last result of this section we return to the general d_p metric topologies. First observe that for any $u \in \mathcal{H}_\omega$ and $\xi \in T_u\mathcal{H}_\omega$ the Hölder inequality gives $\|\xi\|_{u,p'} \leq \|\xi\|_{u,p}$ for any $p' \leq p$. This in turn implies that the $L^{p'}$ length of smooth curves in $\mathcal{H}_\omega$ is shorter then their L^p length, ultimately giving that the d_p metric dominates d'_p. Consequently, all d_p metrics dominate the d_1 metric, hence Theorem (3.46)(ii) holds for d_p–convergence as well. We record this (in a slightly stronger form) in our last result:

PROPOSITION 3.53. *Suppose $v, u, u_k \in \mathcal{E}_p(X,\omega)$ and $d_p(u_k, u) \to 0$. Then $\omega_{u_k}^n \to \omega_u^n$ weakly and $\int_X |u - u_k|^p\omega_v^n \to 0$.*

PROOF. We only need to argue that $\int_X |u - u_k|^p\omega_v^n \to 0$. Given an arbitrary subsequence of u_k, there exists a sub-subsequence, again denoted by u_k, satisfying the sparsity condition:

$$d_\chi(u_k, u_{k+1}) \leq \frac{1}{2^k}, \quad k \in \mathbb{N}. \tag{3.88}$$

Using this sparsity condition and Proposition 3.35 we can write

$$\begin{aligned} &d_\chi(P(u, u_0, \dots, u_k), P(u,u_0, \dots, u_{k+1})) = \\ &\qquad = d_\chi(P(P(u, u_0, \dots, u_k), u_k), P(P(u, u_0, \dots, u_k), u_{k+1})) \\ &\qquad \leq d_\chi(u_k, u_{k+1}) \leq \frac{1}{2^k}. \end{aligned}$$

Hence, the decreasing sequence $h_k = P(u, u_0, u_1, \dots, u_k)$, $k \geq 1$ is bounded and Lemma 3.34 (or completeness) implies that the limit satisfies $h := \lim_k h_k \in \mathcal{E}_p(X,\omega)$. As $d_p(u_k, 0), d_p(u, 0)$ are bounded, by Theorem 3.32 and Lemma 3.45 there exists $M > 0$ such that $h \leq u_k, u \leq M$. Putting everything together we get

$$h - M \leq u_k - u \leq M - h. \tag{3.89}$$

By Theorem 3.46(ii) we have $\int_X |u - u_k|\omega_v^n \to 0$, hence $u_k \to u$ a.e. with respect to ω_v^n. Finally, using (3.89), the dominated convergence theorem gives $\int_X |u_k - u|^p\omega_v^n \to 0$. □

Brief historical remarks. In this chapter we put extensive focus on the particular case of the L^1 geometry. Historically however, the study of the L^2 structure was the one developed first. In particular, Calabi-Chen showed that $C^{1,1}$ geodesics of $\mathcal{H}_\omega$ satisfy the CAT(0) inequality [**30**]. As shown in [**48**], this implies that the completion $(\mathcal{E}_2(X,\omega), d_2)$ is a CAT(0) geodesic metric space. The possibility of identifying $\mathcal{E}_2(X,\omega)$ with the metric completion of $\mathcal{H}_\omega$ was conjectured by Guedj [**72**], who worked out the case of toric Kähler manifolds. As detailed in this chapter, this conjecture was confirmed in [**48**], and later generalized to the case of L^p Finsler structures [**49**].

As noticed by J. Streets [**101, 102**], the CAT(0) property allows to study of the Calabi flow [**41**] in the context of the metric completion, leading to precise convergence results for this flow in [**13**].

Endowing the space of Kähler metrics with natural geometries goes back to the work of Calabi in the 50's [**29**]. Calabi's Riemannian metric is defined in terms of the Laplacian of the potentials, and the resulting geometry differs from that of Mabuchi. The study of this structure was taken up by Calamai [**31**] and Clarke-Rubinstein [**46**]. In the latter work the completion of the Calabi path lengh metric was identified, and was compared to the Mabuchi geometry in [**51**].

It is also possible to introduce a Dirichlet type Riemannian metric on $\mathcal{H}_\omega$ in terms of the gradient of the potentials [**31, 32**]. Not much is known about the metric theory of this structure. However properties of this space seem to be closely immeresed with the study of the Kähler-Ricci flow [**16**].

Part 4. Applications to Kähler–Einstein metrics

Given a Kähler manifold (X, ω), we will be interested in picking a metric with special curvature properties from $\mathcal{H}$, the space of all Kähler metrics ω' whose de Rham cohomology class equals that of ω:

$$\mathcal{H} = \{\omega' \text{ is a Kähler metric on } X \text{ such that } [\omega'] = [\omega] \in H^2(X, \mathbb{R})\}. \tag{4.1}$$

By the $\partial\bar{\partial}$–lemma of Hodge theory, there exists $u \in \mathcal{H}_\omega$ (2.4), unique up to a constant, s.t. $\omega' = \omega_u$ (see [**22**, Theorem 3]). As follows from Stokes' theorem (see more generally Lemma 2.2), the total volume of each metric in $\mathcal{H}$ is the same, and we introduce the constant

$$V := \int_X \omega_u^n = \int_X \omega^n, \quad u \in \mathcal{H}_\omega.$$

As a result of the above observations, we will focus on the space of potentials $\mathcal{H}_\omega$ instead of $\mathcal{H}$, and our goal will be to find $u \in \mathcal{H}_\omega$ whose Ricci curvature is a multiple of the metric ω_u:

$$\text{Ric } \omega_u = \lambda \omega_u. \tag{4.2}$$

Such metrics are called *Kähler–Einstein* (KE) metrics. Recall from the Appendix that Ric ω_u is always closed. Moreover, by the well known formula (5.9) for the change of Ricci curvature

$$\text{Ric } \omega_u - \text{Ric } \omega_v = i\partial\bar{\partial} \log\Big(\frac{\omega_v^n}{\omega_u^n}\Big), \quad u, v \in \mathcal{H}_\omega \tag{4.3}$$

we deduce that the de Rham class [Ric ω_u] does not depend on the choice of $u \in \mathcal{H}_\omega$. What is more, [Ric ω_u] agrees with $c_1(X)$, the first Chern class of X (see [**114**, Section III.3] for more details). Consequently, if (4.2) holds then X needs to have special cohomological properties depending on the sign of $\lambda \in \mathbb{R}$:

(i) if $\lambda = 0$ then $c_1(X) = 0$, i.e., X is *Calabi–Yau*.
(ii) if $\lambda < 0$ then K_X is an ample line bundle, i.e., X is of *general type*.
(iii) if $\lambda > 0$ then $-K_X$ is an ample line bundle, i.e., X is *Fano*.

After close inspection it turns out that (4.2) is a fourth order PDE in terms of the derivatives of u. As we show now, with the help of (4.3) one can write down a scalar equation equivalent to (4.2), that is merely a second order PDE, greatly simplifying our subsequent treatment. Indeed, from $\lambda[\omega] = c_1(X)$ it follows that for each $u \in \mathcal{H}_\omega$ there exists a unique $f_u \in C^\infty(X)$ such that $\int_X e^{f_u}\omega_u^n = V$ and

$$\text{Ric } \omega_u = \lambda\omega + i\partial\bar{\partial} f_u. \tag{4.4}$$

Fittingly, the potential f_u is called the *Ricci potential* of ω_u, and ω_u is KE if and only if $f_u = 0$. Consequently, by (4.3), (4.2) is equivalent to

$$i\partial\bar{\partial} \log\Big(\frac{\omega_u^n}{\omega^n}\Big) = \text{Ric } \omega - \text{Ric } \omega_u = \lambda\omega + i\partial\bar{\partial} f_0 - \lambda\omega_u = i\partial\bar{\partial}(f_0 - \lambda u).$$

And now the (magical!) drop of order takes place. As there are only constants in the kernel of $i\partial\bar{\partial}$, there exists $c \in \mathbb{R}$ such that

$$\log\Big(\frac{\omega_u^n}{\omega^n}\Big) = f_0 - \lambda u + c.$$

When $\lambda \neq 0$, the constant c can be "contracted" into u. When $\lambda = 0$, then the normalization condition on f_0 implies that $c = 0$. Summarizing, we arrive at the scalar KE equation:

$$\omega_u^n = e^{-\lambda u + f_0}\omega^n, \quad u \in \mathcal{H}_\omega. \tag{4.5}$$

When $\lambda < 0$, the existence of unique solutions was proved by Aubin and Yau [**1**, **113**]. In the case $\lambda = 0$, existence and uniqueness was obtained by Yau, as a particular case of the solution of the Calabi conjecture [**113**].

When X is Fano ($\lambda > 0$) the situation is more involved. Uniqueness up to holomorphic automorpshims was shown by Bando–Mabuchi ([**2**], see also Section 4.5 below). As it turns out, on a general Fano manifold (X, ω), there are numerous obstructions to existence of KE metrics (see [**71**, **89**]). Recently the algebro–geometric notion of K–stability has been found to be equivalent with existence of KE metrics ([**38**–**40**, **109**]), with this verifying an important particular case of the Yau–Tian–Donaldson conjecture. Our goal in this chapter is to give equivalent characterizations, more in line with the variational study of partial differential equations, eventually verifying related conjectures of Tian.

As mentioned above, in the rest of this chapter we will assume that (X, ω) is Fano. Also, after possibly rescaling ω, we can also assume that $\lambda = 1$, i.e., $[\omega] = c_1(X)$. Next we introduce Ding's $\mathcal{F}$ functional [**64**]:

$$\mathcal{F}(u) = -I(u) - \log \frac{1}{V} \int_X e^{-u + f_0}\omega^n, \quad u \in \mathcal{H}_\omega, \tag{4.6}$$

where I is the Monge–Ampère energy (see (3.57)). By definition, $\mathcal{F}$ is invariant under adding constants, i.e., $\mathcal{F}(u+c) = \mathcal{F}(u)$, and as a consequence of Lemma 3.37 it follows that the critical points of $\mathcal{F}$ are exactly the KE metrics:

LEMMA 4.1. *Suppose* $[0,1] \ni t \to v_t \in \mathcal{H}_\omega$ *is a smooth curve. Then*

$$V\frac{d}{dt}\mathcal{F}(v_t) = \int_X \dot{v}_t\Big(-\omega_{v_t}^n + \frac{V}{\int_X e^{-v_t + f_0}\omega^n} e^{-v_t + f_0}\omega^n \Big).$$

Comparing with (4.5), we deduce that existence of KE potentials in $\mathcal{H}_\omega$ is equivalent with existence of critical points of $\mathcal{F}$. As we will see, the natural domain of definiton of $\mathcal{F}$ is not $\mathcal{H}_\omega$, but rather its d_1 metric completion $\mathcal{E}_1(X, \omega)$. In addition to this, $\mathcal{F}$ will be shown to be d_1–continuous and it will be convex along the finite energy geodesics of $\mathcal{E}_1(X, \omega)$. Using convexity we will deduce that critical points of $\mathcal{F}$ are exactly the minimizers of $\mathcal{F}$, and we can use properness properties of $\mathcal{F}$ to characterize existence of these minimizers. All this will be done in the forthcoming sections.

4.1. The action of the automorphism group

We continue to assume that (X, ω) is Fano with $[\omega] = c_1(X)$. Let $\mathrm{Aut}_0(X, J)$ denote the connected component of the complex Lie group of biholomorphisms of (X, J), and denote by $\mathrm{aut}(M, J)$ its complex Lie algebra of holomorphic vector fields. $\mathrm{Aut}_0(X, J)$ acts on $\mathcal{H}$ by pullback of metrics. Indeed, $f^\star \eta \in \mathcal{H}$ for any $f \in \mathrm{Aut}_0(X, J)$ and $\eta \in \mathcal{H}$.

Given the one-to-one correspondence between $\mathcal{H}$ and $\mathcal{H}_\omega \cap I^{-1}(0)$ (recall (3.67)), the group $\mathrm{Aut}_0(X, J)$ also acts on $\mathcal{H}_\omega \cap I^{-1}(0)$ and we describe this action more precisely in the next lemma:

LEMMA 4.2. [**58**, Lemma 5.8] *For* $\varphi \in \mathcal{H}_\omega \cap I^{-1}(0)$ *and* $f \in \mathrm{Aut}_0(X, J)$ *let* $f.\varphi \in \mathcal{H}_\omega \cap I^{-1}(0)$ *be the unique element such that* $f^*\omega_\varphi = \omega_{f.\varphi}$. *Then,*

$$f.\varphi = f.0 + \varphi \circ f. \tag{4.7}$$

PROOF. First we note that the right hand side of (4.7) is a Kähler potential for $f^*\omega_\varphi$. Indeed, since $f \in \mathrm{Aut}_0(X, J)$ we have $f^\star i\partial\bar\partial\varphi = i\partial\bar\partial\varphi \circ f$. The identity $I(f.0 + \varphi \circ f) = 0$ follows from (3.58) as we have:

$$I(f.0 + \varphi \circ f) = I(f.0 + \varphi \circ f) - I(f.0) = \frac{1}{(n+1)V}\int_X \varphi \circ f \sum_{j=0}^{n} f^\star\omega^{n-j} \wedge f^\star\omega_\varphi^j$$

$$= \frac{1}{(n+1)V}\int_X \varphi \sum_{j=0}^{n} \omega^{n-j} \wedge \omega_\varphi^j = I(\varphi) = 0.$$

□

With the formula of the above lemma, we show that $\mathrm{Aut}_0(X, J)$ acts on $\mathcal{H}_\omega \cap I^{-1}(0)$ by d_p–isometries:

LEMMA 4.3. *The action of* $\mathrm{Aut}_0(X, J)$ *on* $\mathcal{H}_\omega \cap I^{-1}(0)$ *is by* d_p*–isometries, for any* $p \geq 1$.

PROOF. From (4.7) it follows that $d(f.\varphi_t)/dt = \dot\varphi_t \circ f$, for any smooth curve $[0, 1] \ni t \to \varphi_t \in \mathcal{H}_\omega \cap I^{-1}(0)$. Thus, the d_p–length of $t \to f.\varphi_t$ satisfies:

$$l_p(f.\varphi_t) = \int_0^1 \Big(\frac{1}{V}\int_X |\dot\varphi_t \circ f|^p f^\star\omega^n_{\varphi_t}\Big)^{\frac{1}{p}} dt = \int_0^1 \Big(\frac{1}{V}\int_X |\dot\varphi_t|^p \omega^n_{\varphi_t}\Big)^{\frac{1}{p}} dt.$$

Since this last quantity is exactly the the d_p-length of $t \to \varphi_t$, it follows that $\mathrm{Aut}_0(X, J)$ acts by d_p–isometries. □

As a consequence of this last result (see also Lemma 4.6 below), the action of $\mathrm{Aut}_0(X, J)$ on $\mathcal{H}_\omega \cap I^{-1}(0)$ has a unique d_p–isometric extension to the d_p–metric completion $\mathcal{E}_p(X, \omega) \cap I^{-1}(0)$. In what follows we will focus on the case $p = 1$.

Recall the definition of the J functional (3.68) and let $H \leq \mathrm{Aut}_0(X, J)$ be a subgroup. The "H–dampened" functional $J_H : \big(\mathcal{E}_1 \cap I^{-1}(0)\big)/H \to \mathbb{R}$ is introduced by the formula

$$J_H(Hu) := \inf_{f \in H} J(f.u). \tag{4.8}$$

As a direct consequence of Proposition 3.44 we have the following estimates for this functional:

LEMMA 4.4. *There exists* $C := C(X, \omega) > 1$ *such that for any* $u \in \mathcal{E}_1 \cap I^{-1}(0)$ *we have*

$$\frac{1}{C}J_H(Hu) - C \leq d_{1,H}(H0, Hu) \leq CJ_H(Hu) + C, \tag{4.9}$$

where $d_{1,H}$ *is the pseudo–metric of the quotient* $\big(\mathcal{E}_1 \cap I^{-1}(0)\big)/H$ *given by the formula* $d_{1,H}(Hu, Hv) := \inf_{f \in H} d_1(u, f.v)$.

Lastly, we note that $\mathcal{F}$ is affine along one parameter subgroups of $\mathrm{Aut}_0(X, J)$, and that the map $(u, v) \to \mathcal{F}(u) - \mathcal{F}(v)$ is $\mathrm{Aut}_0(X, J)$–invariant:

LEMMA 4.5. *Suppose* $\mathbb{R} \ni t \to \rho_t \in \mathrm{Aut}_0(X,J)$ *is a one parameter subgroup. Then:*
(i) for any $u \in \mathcal{H}_\omega \cap I^{-1}(0)$ *the map* $t \to \mathcal{F}(\rho_t.u)$ *is affine.*
(ii) if $u, v \in \mathcal{H}_\omega \cap I^{-1}(0)$ *and* $g \in \mathrm{Aut}_0(X,J)$ *then* $\mathcal{F}(u) - \mathcal{F}(v) = \mathcal{F}(g.u) - \mathcal{F}(g.v)$.

PROOF. First we show that for any $u \in \mathcal{H}_\omega$ we have that

$$\frac{e^{-u+f_0}}{\int_X e^{-u+f_0}\omega^n}\omega^n = \frac{e^{f_u}}{V}\omega_u^n, \tag{4.10}$$

where f_u is the Ricci potential of u (see (4.4)). By definition, we have $\mathrm{Ric}\,\omega_u = \omega_u + i\partial\bar\partial f_u$ and $\mathrm{Ric}\,\omega = \omega + i\partial\bar\partial f_0$. Substituting this into (4.3) we see that (4.10) holds.

Now we argue (i). Let $h \in C^\infty(X)$ be the unique function such that $\int_X h\omega_u^n = 0$ and $\frac{d}{dt}\big|_{t=0}\rho_t^*\omega_u = i\partial\bar\partial h$. We claim that

$$\frac{d}{dt}\rho_t.u = h \circ \rho_t, \quad t \in \mathbb{R}. \tag{4.11}$$

Indeed, this follows from a simple calculation: $i\partial\bar\partial \frac{d}{dt}\rho_t.u = \frac{d}{dt}\omega_{\rho_t.u} = \frac{d}{dt}\rho_t^*\omega_u = \rho_t^* i\partial\bar\partial h = i\partial\bar\partial h \circ \rho_t$.

Finally, using Lemma 4.1, (4.10), (4.11), and the fact that $f_u \circ \rho_t = f_{\rho_t.u}$ we can conclude that $t \to \mathcal{F}(\rho_t.u)$ is indeed affine:

$$\begin{aligned}
V\frac{d}{dt}\mathcal{F}(\rho_t.u) &= \int_X \frac{d}{dt}\rho_t.u\Big(-\omega_{\rho_t.u}^n + \frac{V}{\int_X e^{-\rho_t.u+f_0}\omega^n}e^{-\rho_t.u+f_0}\omega^n\Big)\\
&= \int_X \frac{d}{dt}\rho_t.u\Big(-\omega_{\rho_t.u}^n + e^{f_{\rho_t.u}}\omega_{\rho_t.u}^n\Big)\\
&= \int_X h\circ\rho_t\Big(-\rho_t^*\omega_u^n + e^{f_u\circ\rho_t}\rho_t^*\omega_u^n\Big) = \int_X h\big(-1+e^{f_u}\big)\omega_u^n.
\end{aligned}$$

Now we focus on (ii). Let $[0,1] \ni t \to \gamma_t \in \mathcal{H}_\omega \cap I^{-1}(0)$ be any smooth segment connecting u, v. For example, one can take $\gamma_t := (1-t)u + tv - I((1-t)u + tv)$. Consequently, $t \to g.\gamma_t$ connects $g.u, g.v$. Using Lemma 4.1, (4.7), and (4.10) we can finish the proof:

$$\begin{aligned}
V(\mathcal{F}(g.v) - \mathcal{F}(g.u)) = V\int_0^1 \frac{d}{dt}\mathcal{F}(\gamma_t)dt &= \int_0^1\int_X \dot\gamma_t \circ g(-1+e^{f_{g.\gamma_t}})\omega_{g.\gamma_t}^n dt.\\
&= \int_0^1\int_X \dot\gamma_t\circ g(-1+e^{f_{\gamma_t}\circ g})g^*\omega_{\gamma_t}^n dt\\
&= \int_0^1\int_X \dot\gamma_t(-1+e^{f_{\gamma_t}})\omega_{\gamma_t}^n dt = V(\mathcal{F}(v) - \mathcal{F}(u)).
\end{aligned}$$

□

4.2. The existence/properness principle and relation to Tian's conjectures

Staying with a Fano manifold (X,ω), our main goal is to show that existence of KE metrics in $\mathcal{H}$ is equivalent with properness of the Ding energy (and later the K-energy). As it turns out, our proof will rest on a very general existence/properness principle for abstract metric spaces, and we describe this now as it provides the skeleton for our later arguments concerning Kähler geometry.

To begin, we introduce $(\mathcal{R}, d, F, G)$, a metric space structure with additional data satisfying the following *axioms*:

(A1) $(\mathcal{R}, d)$ is a metric space with a distinguished element $0 \in \mathcal{R}$, whose metric completion is denoted by $(\overline{\mathcal{R}}, d)$.

(A2) $F : \mathcal{R} \to \mathbb{R}$ is d–lsc. Let $F : \overline{\mathcal{R}} \to \mathbb{R} \cup \{+\infty\}$ be the largest d–lsc extension of $F : \mathcal{R} \to \mathbb{R}$:

$$F(u) = \sup_{\varepsilon > 0} \Big(\inf_{\substack{v \in \mathcal{R} \\ d(u,v) \leq \varepsilon}} F(v) \Big), \quad u \in \overline{\mathcal{R}}.$$

(A3) By $\mathcal{M}$ we denote the set of minimizers of F on $\overline{\mathcal{R}}$:

$$\mathcal{M} := \Big\{ u \in \overline{\mathcal{R}} \ : \ F(u) = \inf_{v \in \overline{\mathcal{R}}} F(v) \Big\}.$$

(A4) G is a group acting on $\mathcal{R}$ by $G \times \mathcal{R} \ni (g, u) \to g.u \in \mathcal{R}$. Denote by $\mathcal{R}/G$ the orbit space, by $Gu \in \mathcal{R}/G$ the orbit of $u \in \mathcal{R}$, and define d_G : $\mathcal{R}/G \times \mathcal{R}/G \to \mathbb{R}_+$ by

$$d_G(Gu, Gv) := \inf_{f,g \in G} d(f.u, g.v).$$

In addition to the above, our data $(\mathcal{R}, d, F, G)$ also enjoys the following *properties*:

(P1) For any $u_0, u_1 \in \mathcal{R}$ there exists a d–geodesic $[0,1] \ni t \mapsto u_t \in \overline{\mathcal{R}}$ connecting u_0, u_1 (see (3.40)) for which $t \mapsto F(u_t)$ is continuous and convex on $[0,1]$.

(P2) If $\{u_j\}_j \subset \overline{\mathcal{R}}$ satisfies $\lim_{j \to \infty} F(u_j) = \inf_{\overline{\mathcal{R}}} F$, and for some $C > 0$, $d(0, u_j) \leq C$ for all j, then there exists $u \in \mathcal{M}$ and a subsequence $\{u_{j_k}\}_k$ s.t. $d(u_{j_k}, u) \to 0$.

(P2)* If $\{u_j\}_j \subset \overline{\mathcal{R}}$ satisfies $F(u_j) \leq C$, and $d(0, u_j) \leq C$, $j \geq 0$ for some $C > 0$, then there exists $u \in \mathcal{R}$ and a subsequence $\{u_{j_k}\}_k$ s.t. $d(u_{j_k}, u) \to 0$.

(P3) $\mathcal{M} \subset \mathcal{R}$.

(P4) G acts on $\mathcal{R}$ by d-isometries.

(P5) G acts on $\mathcal{M}$ transitively.

(P6) For all $u, v \in \mathcal{R}$ and $g \in G$, $F(u) - F(v) = F(g.u) - F(g.v)$.

We make three remarks. First, by (A2),

$$\inf_{v \in \overline{\mathcal{R}}} F(v) = \inf_{v \in \mathcal{R}} F(v). \tag{4.12}$$

Second, condition (P2)* is stronger than (P2) and we will require that only one of these conditions holds.

Third, thanks to (P4) and the next lemma, the action of G, originally defined on $\mathcal{R}$ (A4), extends to an action by d–isometries on the completion $\overline{\mathcal{R}}$.

LEMMA 4.6. *Let (X, ρ) and (Y, δ) be two complete metric spaces, W a dense subset of X and $f : W \to Y$ a C-Lipschitz function, i.e.,*

$$\delta(f(a), f(b)) \leq C\rho(a, b), \quad \forall\, a, b \in W. \tag{4.13}$$

Then f has a unique C–Lipschitz continuous extension to a map $\bar{f} : X \to Y$.

PROOF. Let $w_k \in W$ be a sequence converging to some $w \in X$. Lipschitz continuity gives

$$\delta(f(w_k), f(w_l)) \leq C\rho(w_k, w_l),$$

hence $\bar{f}(w) := \lim_k f(w_k) \in Y$ exists and independent of the choice of approximating sequence w_k. Choose now another sequence $z_k \in W$ with limit $z \in X$, plugging in w_k, z_k in (4.13) and taking the limit gives that $\bar{f} : X \to Y$ is C-Lipschitz continuous. □

A d-geodesic ray $[0, \infty) \ni t \to u_t \in \overline{\mathcal{R}}$ is *G-calibrated* if the curve $t \to Gu_t$ is a d_G-geodesic with the same speed as $t \to u_t$, i.e.,

$$d_G(Gu_0, Gu_t) = d(u_0, u_t), \quad t \geq 0.$$

The next result will provide the framework that relates existence of canonical Kähler metrics to energy properness and uniform geodesic stability.

THEOREM 4.7. [**58**, Theorem 3.4] *Suppose* $(\mathcal{R}, d, F, G)$ *satisfies* (A1)–(A4) *and* (P1)–(P6)*. The following are equivalent:*

(i)(existence of minimizers) $\mathcal{M}$ *is nonempty.*

(ii)(energy properness) $F : \mathcal{R} \to \mathbb{R}$ *is G-invariant, and for some* $C, D > 0$,

$$F(u) \geq C d_G(G0, Gu) - D, \quad \text{for all } u \in \mathcal{R}. \tag{4.14}$$

If (P2)* *holds instead of* (P2)*, then the above are additionally equivalent to:*

(iii)(uniform geodesic stability) Fix $u_0 \in \mathcal{R}$ *with* $F(u_0) < +\infty$*. Then* $F : \mathcal{R} \to \mathbb{R}$ *is G-invariant, and there exists* $C > 0$ *such that for all geodesic rays* $[0, \infty) \ni t \to u_t \in \overline{\mathcal{R}}$ *we have that*

$$\limsup_{t \to \infty} \frac{F(u_t) - F(u_0)}{t} \geq C \limsup_{t \to \infty} \frac{d_G(Gu_0, Gu_t)}{t}. \tag{4.15}$$

(iv)(uniform geodesic stability) Fix $u_0 \in \mathcal{R}$ *with* $F(u_0) < +\infty$*. Then* $F : \mathcal{R} \to \mathbb{R}$ *is G-invariant, and there exists* $C > 0$ *such that for all G-calibrated geodesic rays* $[0, \infty) \ni t \to u_t \in \overline{\mathcal{R}}$ *we have that*

$$\limsup_{t \to \infty} \frac{F(u_t) - F(u_0)}{t} \geq C d(u_0, u_1). \tag{4.16}$$

Before arguing the above theorem we recall standard facts from metric geometry in the form of the following two lemmas:

LEMMA 4.8. *If* (P4) *holds, then* $(\overline{\mathcal{R}}/G, d_G)$ *and* $(\mathcal{R}/G, d_G)$ *are pseudo–metric spaces.*

PROOF. It is enough to show that $(\overline{\mathcal{R}}/G, d_G)$ is a pseudo–metric space. Since d is symmetric,

$$d_G(Gu, Gv) := \inf_{f,g \in G} d(f.u, g.v) = \inf_{f,g \in G} d(g.v, f.u) = d_G(Gv, Gu).$$

Hence d_G is also symmetric. Given $u, v, w \in \overline{\mathcal{R}}$ and $\varepsilon > 0$, there exist $f, g \in G$ such that $d_G(Gu, Gw) > d(f.u, w) - \varepsilon$ and $d_G(Gv, Gw) > d(g.v, w) - \varepsilon$. The triangle inequality for d and (P4) give

$$\begin{aligned} d_G(Gu, Gv) &\leq d(u, f^{-1}g.v) = d(f.u, g.v) \\ &\leq d(f.u, w) + d(g.v, w) < 2\varepsilon + d_G(Gu, Gv) + d_G(Gv, Gw). \end{aligned}$$

Letting $\varepsilon \to 0$ shows d_G satisfies the triangle inequality. Thus d_G is a pseudo–metric. □

LEMMA 4.9. *Suppose* (P4) *holds,* $u_0, u_1 \in \mathcal{R}$ *and* $[0,1] \ni t \to u_t \in \overline{\mathcal{R}}$ *is a d–geodesic connecting* u_0, u_1. *If* $d(u_0,u_1) - \varepsilon \leq d_G(Gu_0, Gu_1) \leq d(u_0,u_1)$ *for some* $\varepsilon > 0$ *then*

$$d(u_a,u_b) - \varepsilon \leq d_G(Gu_a, Gu_b) \leq d(u_a,u_b), \quad \forall a,b \in [0,1].$$

PROOF. The proof is by contradiction. Suppose $d_G(Gu_a,Gu_b) < d(u_a,u_b) - \varepsilon$. Since $d(u_0,u_a) + d(u_a,u_b) + d(u_b,u_1) = d(u_0,u_1)$ we can write

$$\begin{aligned} d_G(Gu_0,Gu_1) &\leq d_G(Gu_0,Gu_a) + d_G(Gu_a,Gu_b) + d_G(Gu_b,Gu_1) \\ &< d(u_0,u_a) + d(u_a,u_b) + d(u_b,u_1) - \varepsilon \\ &= d(u_0,u_1) - \varepsilon \leq d_G(Gu_0,Gu_1). \end{aligned} \tag{4.17}$$

This is a contradiction, finishing the proof. □

PROOF OF THEOREM 4.7. First we show (ii)⇒(i). If condition (ii) holds, then F is bounded from below. By (4.12), (4.14), the G–invariance of F, and the definition of d_G there exists $u_j \in \mathcal{R}$ such that $\lim_j F(u_j) = \inf_{\overline{\mathcal{R}}} F$ and $d(0,u_j) \leq d_G(G0, Gu_j) + 1 < C$ for C independent of j. By (P2), $\mathcal{M}$ is non-empty.

We now show that (i)⇒(ii). First we argue that $F : \mathcal{R} \to \mathbb{R}$ is G–invariant. Let $v \in \mathcal{M}$. By (P3), $v \in \mathcal{R}$. By (P4), $f.v \in \mathcal{M}$ for any $f \in G$. Thus, $F(v) = F(f.v)$. Consequently, $F(u) - F(v) = F(u) - F(f.v)$. By (P6), we get $F(u) - F(v) = F(f^{-1}.u) - F(v)$, so $F(u) = F(f^{-1}.u)$ for every $f \in G$, i.e., F is G–invariant.

For $v \in \mathcal{M} \subset \mathcal{R}$ we define

$$C := \inf\left\{ \frac{F(u) - F(v)}{d_G(Gv, Gu)} \ : u \in \mathcal{R}, \ d_G(Gv,Gu) \geq 2 \right\} \geq 0.$$

If $C > 0$, then we are done. Suppose $C = 0$. Then there exists $\{u^k\}_k \subset \mathcal{R}$ such that

$$(F(u^k) - F(v))/d_G(Gv, Gu^k) \to 0$$

and $d_G(Gv, Gu^k) \geq 2$. By G–invariance of F we can also assume that $d(v,u^k) - 1/k \leq d_G(Gv,Gu^k) \leq d(v,u^k)$. Thus,

$$\lim_{k\to\infty} \frac{F(u^k) - F(v)}{d(v,u^k)} = 0.$$

Using (P1), let $[0, d(v,u^k)] \ni t \mapsto u^k_t \in \overline{\mathcal{R}}$ be a unit speed d–geodesic connecting $u^k_0 := v$ and $u^k_{d(v,u^k)} := u^k$ such that $t \mapsto F(u^k_t)$ is convex. As v is a minimizer of F, by convexity we obtain

$$0 \leq F(u^k_1) - F(v) \leq \frac{F(u^k) - F(v)}{d(v,u^k)} \to 0. \tag{4.18}$$

Trivially, $d(v,u^k_1) = 1$, hence (P2) and (4.18) imply that $d(u^k_1, \tilde{v}) \to 0$ for some $\tilde{v} \in \mathcal{M}$ (after perhaps passing to a subsequence of u^k_1). By (P5), $\tilde{v} = f.v$ for some $f \in G$, hence $d_G(Gv, G\tilde{v}) = 0$.

From Lemma 4.9 we obtain $1 - 1/k \leq d_G(Gv, Gu^k_1) \leq 1$. Since $d(u^k_1,\tilde{v}) \to 0$, we also have $d_G(Gu^k_1, G\tilde{v}) \to 0$, which gives $d_G(Gv, G\tilde{v}) = 1$, a contradiction with $d_G(Gv, G\tilde{v}) = 0$. This implies that $C > 0$, finishing the proof of the implication (i)⇒(ii).

The implications (ii)⇒(iii)⇒(iv) is trivial and we finish the proof by showing that (iv)⇒(i). Suppose (i) does not hold but (iv) does. We will derive a contradiction. Since F is G-invariant there exists $\{u^k\} \subset \mathcal{R}$ such that $d(u_0,u^k) -$

$1/k \leq d_G(Gu_0, Gu^k)$ and $F(u^k)$ decreases to $\inf_{u \in \mathcal{R}} F(u)$. By (P2)* we must have $d(u_0, u^k) \to \infty$, otherwise there would exists $u \in \overline{\mathcal{R}}$ such that $F(u) = -\infty$, a contradiction.

Let $[0, d(u_0, u^k)] \ni t \to u_t^k \in \overline{\mathcal{R}}$ be the d-geodesic joining u_0, u^k from (P1). We note that by Lemma 4.9 it follows that

$$d(u_0, u_t^k) - 1/k \leq d_G(Gu_0, Gu_t^k), \quad t \in [0, d(u_0, u^k)]. \tag{4.19}$$

Fix $l \in \mathbb{Q}_+$. For big enough k using convexity of F we can write:

$$\frac{F(u_l^k) - F(u_0)}{l} \leq \frac{F(u^k) - F(u_0)}{d(u^k, u_0)} \leq 0. \tag{4.20}$$

As $d(u_l^k, u_0) = l$, we can use (P2)* and a Cantor diagonal process, to conclude the existence of a sequence $k_j \to \infty$ and $u_l \in \overline{\mathcal{R}}$ for all $l \in \mathbb{Q}_+$ such that $d(u_l^{k_j}, u_l) \to 0$. As each curve $t \to u_t^k$ is d-Lipschitz, it follows that in fact we can extend the curve $\mathbb{Q}_+ \ni l \to u_l \in \overline{\mathcal{R}}$ to a curve $[0, \infty) \ni t \to u_t \in \mathcal{R}$ such that $d(u_l^{k_t}, u_t) \to 0$.

For elementary reasons $t \to u_t$ is a d-geodesic. By (A2) and (4.20) we get that

$$\frac{F(u_l) - F(u_0)}{l} \leq d(u_l, u_0) \leq 0, \quad l \geq 0. \tag{4.21}$$

Finally, we argue that $t \to u_t$ is a G-calibrated geodesic ray, yielding a contradiction with (4.16). Let $g \in G$ be arbitrary, from (4.19) it follows that $d(u_0, u_t^{k_j}) - 1/k_j \leq d(g.u_0, u_t^{k_j})$. Letting $k_j \to \infty$, we obtain $d(u_0, u_t) \leq d(g.u_0, u_t)$ for $t \in \mathbb{Q}_+$ and by density for all $t \geq 0$. Consequently, $t \to u_t$ is G-calibrated. □

Later, when dealing with the K-energy functional, we will make use of the following observation:

REMARK 4.10. *The direction (ii)⇒(i) in the above argument only uses the compactness condition* (P2).

The next result, together with Theorem 4.36, represents the main application of Theorem 4.7:

THEOREM 4.11. [**58**, Theorem 7.1] *Suppose* (X, ω) *is Fano and set* $G := \mathrm{Aut}_0(X, J)$. *The following are equivalent:*

(i) There exists a KE metric in $\mathcal{H}$.

(ii) For some $C, D > 0$ *the following holds:*

$$\mathcal{F}(u) \geq C d_{1,G}(G0, Gu) - D, \quad u \in \mathcal{H}_\omega \cap I^{-1}(0). \tag{4.22}$$

(iii) For some $C, D > 0$ *the following holds:*

$$\mathcal{F}(u) \geq C J_G(Gu) - D, \quad u \in \mathcal{H}_\omega \cap I^{-1}(0). \tag{4.23}$$

PROOF. The equivalence between (ii) and (iii) is the content of Lemma 4.4.

For the equivalence between (i) and (ii) we wish to apply Theorem 4.7 to the data

$$\mathcal{R} = \mathcal{H}_\omega \cap I^{-1}(0), \quad d = d_1, \quad F = \mathcal{F}, \quad G := \mathrm{Aut}_0(X, J). \tag{4.24}$$

We first have to show that (4.22) implies a bit of extra information: that $\mathcal{F}$ is invariant under the action of $\mathrm{Aut}_0(X, J)$. Indeed, (4.22) implies that $\mathcal{F}$ is bounded from below. This and Lemma 4.5(i) implies that $\mathbb{R} \ni t \to \mathcal{F}(\rho_t.u) \in \mathbb{R}$ is affine and bounded for all one parameter subgroups $\mathbb{R} \ni t \to \rho_t \in \mathrm{Aut}_0(X, J)$. Consequently

$t \to \mathcal{F}(\rho_t.u)$ has to be constant equal to $\mathcal{F}(u)$. As $\text{Aut}_0(X,J)$ is connected, we obtain that $\mathcal{F}$ is invariant under the action of this group, as claimed.

To finish the proof, we go over the axioms and properties of $(\mathcal{R}, d, F, G)$, as imposed in the statement of Theorem 4.7:

(A1) By Theorem 3.36 and the fact that I is d_1–continuous (Proposition 3.40) we obtain that $\overline{(\mathcal{H}_\omega \cap I^{-1}(0), d_1)} = (\mathcal{E}_1(X,\omega) \cap I^{-1}(0), d_1)$.

(A2) That $\mathcal{F}$ is d_1–continuous on $\mathcal{H}_\omega$ will be proved in Theorem 4.17. As we will see, the d_1–continuous extension $\mathcal{F} : \mathcal{E}_1(X,\omega) \to \mathbb{R}$ is given by the original formula for smooth potentials (see (4.6)).

(A3) We choose $\mathcal{M}$ as the minimizer set of the extended functional $\mathcal{F} : \mathcal{E}_1(X,\omega) \cap I^{-1}(0) \to \mathbb{R}$.

(P1) This fact is due to Berndtsson [**19**, Theorem 1.1], and we present this result in Theorem 4.22 below.

(P2) This property will be verfied in Theorem 4.18.

(P3) That elements of $\mathcal{M}$ are in fact smooth KE potentials follows after combination of results due to Berman, Tosatti–Székelyhidi and Berman–Boucksom–Guedj–Zeriahi. We present this in Theorem 4.30 below.

(P4) This is Lemma 4.3 for $p = 1$.

(P5) This follows from (P3) and the Bando–Mabuchi uniqueness theorem that we will prove in Theorem 4.23 below. □

(P6) This is exactly the content of Lemma 4.5(ii) above.

REMARK 4.12. *The equivalence between (i) and (iii) in the above theorem verifies the analogue of a conjecture of Tian for the Ding functional $\mathcal{F}$ (see* [**108**, p. 127],[**107**, Conjecture 7.12]*).*

In Theorem 4.36 *below we will verify Tian's original conjecture for the K–energy as well, giving another characterization for existence of KE metrics on Fano manifolds. For other results of the same spirit, as well as relation to the literature, we refer to* [**14, 36, 37, 55, 58**].

4.3. Continuity and compactness properties of the Ding functional

Given a Fano manifold (X,ω), in this section we will show that the $\mathcal{F}$ functional is d_1–continuous on $\mathcal{H}_\omega$, hence naturally extends to the d_1–completion $\mathcal{E}_1(X,\omega)$ (Theorem 4.17). In addition to this, we will show that d_1–bounded sequences that are $\mathcal{F}$ minimizing are d_1–subconvergent, with this establishing an important compactness property of $\mathcal{F}$ (Theorem 4.18).

To start, we need to prove the following preliminary compactness lemma:

LEMMA 4.13. *For $B, D \in \mathbb{R}$ we consider the following subset of $\mathcal{E}_1(X,\omega)$:*

$$\mathcal{C} = \{u \in \mathcal{E}_1(X,\omega) \ : \ B \leq I(u) \leq \sup_X u \leq D\}.$$

Then $\mathcal{C}$ is compact with respect to the weak $L^1(\omega^n)$ topology of $\text{PSH}(X,\omega)$.

PROOF. Since $\sup_X u$ is bounded for any $u \in \mathcal{C}$, by Lemma 3.45 it follows that $\int_X |u|\omega^n$ is bounded as well. By the Montel property of psh functions ([**61**, Proposition I.4.21]) it follows that $\mathcal{C}$ is precompact with respect to the weak $L^1(\omega^n)$ topology.

Now we argue that $\mathcal{C}$ is L^1–closed. Let $u_k \in \mathcal{C}$ and $u \in \text{PSH}(X,\omega)$ such that $\int_X |u - u_k|\omega^n \to 0$. As $u_k \leq D$ and $u_k \to u$ a.e., it follows that $\sup_X u \leq D$.

Since $u_k \to u$ a.e., it follows that $v_k \searrow u$, where $v_k := (\sup_{j \geq k} u_j)^* \in \text{PSH}(X, \omega)$. By the monotonicity property of I we have that $B \leq I(u_k) \leq I(v_k) \leq \sup_X v_k \leq D$, $k \geq 1$. By Lemma 3.49 and Lemma 3.34 it follows that $d_1(v_k, u) \to 0$. Consequently $I(v_k) \to I(u)$ (Proposition 3.40), hence $u \in \mathcal{C}$. □

In contrast with Proposition 3.40, as a consequence of the above argument, we also obtain that I is $L^1(\omega^n)$–usc:

COROLLARY 4.14. *For* $u_k, u \in \mathcal{E}_1(X, \omega)$ *we have* $\limsup_{k\to\infty} I(u_k) \leq I(u)$.

Before we proceed, we recall Zeriahi's uniform version of the famous Skoda integrability theorem [**115**]:

THEOREM 4.15. *Suppose* $\mathcal{S} \subset \text{PSH}(X, \omega)$ *is an* $L^1(\omega^n)$*–compact family whose elements have zero Lelong numbers. Then for* $p \geq 1$ *there exists* $C := C(p, \mathcal{S}, \omega) > 1$ *such that*

$$\int_X e^{-pu} \omega^n \leq C, \quad u \in \mathcal{S}.$$

For a full account of this result we refer to [**75**, Theorem 2.50]. Since full mass potentials have zero Lelong numbers (Propostion 2.13), we obtain the following corollary:

COROLLARY 4.16. *For* $D, p \geq 1$ *there exists* $C := C(p, D, \omega) > 0$ *such that for any* $u \in \mathcal{E}_1(X, \omega)$ *with* $d_1(0, u) \leq D$ *we have*

$$\int_X e^{-pu} \omega^n \leq C.$$

PROOF. From Theorem 3.32 it follows that $\int_X u\omega^n$ is uniformly bounded. By Lemma 3.45 so is $\sup_X u$. By Proposition 3.40 it follows that $I(u)$ is bounded as well. This allows to apply Lemma 4.13, and together with Proposition 2.13 the conditions of of Theorem 4.15 are satisfied to conclude the result. □

In particular, this last corollary implies that the original definition of the $\mathcal{F}$ functional for smooth potentials (see (4.6)) extends to $\mathcal{E}_1(X, \omega)$ as well. Additionally, our next theorem shows that this extension is in fact d_1–continuous:

THEOREM 4.17. *The map*

$$\mathcal{E}_1(X, \omega) \ni u \to \mathcal{F}(u) := -I(u) - \log \frac{1}{V} \int_X e^{-u+f_0} \omega^n \in \mathbb{R}$$

is d_1*–continuous.*

PROOF. We know that $u \to I(u)$ is d_1–continuous and finite on $\mathcal{E}_1(X, \omega)$, hence we only have to argue that so is $u \to \int_X e^{-u+f_0}\omega^n$. For $u, v \in \mathcal{E}_1(X, \omega)$, using the inequality $|e^x - e^y| \leq |x - y|(e^x + e^y)$, $x, y \in \mathbb{R}$, we have the following estimates:

$$\left| \int_X e^{-u+f_0}\omega^n - \int_X e^{-v+f_0}\omega^n \right| \leq \int_X e^{f_0} \left| e^{-u} - e^{-v} \right| \omega^n \leq \int_X e^{f_0} |u-v|(e^{-u}+e^{-v})\omega^n.$$

Using the fact that f_0 is bounded, the Hölder inequality gives

$$\Big|\int_X e^{-u+f_0}\omega^n - \int_X e^{-v+f_0}\omega^n\Big|^2 \leq C\Big(\int_X (e^{-2u}+e^{-2v})\omega^n\Big)\cdot\Big(\int_X |u-v|^2\omega^n\Big). \tag{4.25}$$

Suppose $u_k \in \mathcal{E}_1(X,\omega)$ such that $d_1(u_k,u) \to 0$. Then Proposition 3.53 implies that $\int_X |u_k - u|\omega^n \to 0$. As all L^p topologies on $\mathrm{PSH}(X,\omega)$ are equivalent (see [**81**, Theorem 4.1.8]) we obtain that $\int_X |u_k - u|^2\omega^n \to 0$. Finally, (4.25) and Corollary 4.16 implies that $\int_X e^{-u_k+f_0}\omega^n \to \int_X e^{-u+f_0}\omega^n$. □

Lastly, we argue the compactness property of $\mathcal{F}$ that is a vital ingredient in the existence/properness principle of the previous section:

THEOREM 4.18. *Suppose $u_k \in \mathcal{E}_1(X,\omega)$ such that $\mathcal{F}(u_k) \to \inf_{\mathcal{E}_1(X,\omega)} \mathcal{F}$ and $d_1(u_k,0) \leq C$ for some $C > 0$. After possibly taking a subsequence, there exists $u \in \mathcal{E}_1(X,\omega)$ such that $d_1(u_k,u) \to 0$. In particular, $\mathcal{F}(u) = \inf_{\mathcal{E}_1(X,\omega)} \mathcal{F}$.*

PROOF. First we construct a candidate for the minimizer $u \in \mathcal{E}_1(X,\omega)$. From Theorem 3.32 it follows that $\int_X u_k\omega^n$ is uniformly bounded. By Lemma 3.45 so is $\sup_X u_k$. By Proposition 3.40 it follows that $I(u_k)$ is bounded as well. Now Lemma 4.13 implies that after possibly taking a subsequence we can find $u \in \mathcal{E}_1(X,\omega)$ such that $\int_X |u_k - u|\omega^n \to 0$.

We now show that u is actually a minimizer of $\mathcal{F}$. Using (4.25) in the same way as in the proof of the previous result, we obtain that $\int_X e^{-u_k+f_0}\omega^n \to \int_X e^{-u+f_0}\omega^n$. By Corollary 4.14 I is usc with respect to the weak $L^1(\omega^n)$ topology, hence we can write

$$\lim_k \mathcal{F}(u_k) \geq -\limsup_k I(u_k) - \log \frac{1}{V}\int_M e^{-u+f_0}\omega^n \geq \mathcal{F}(u). \tag{4.26}$$

As $\{u_k\}_k$ minimizes $\mathcal{F}$, it follows that all the inequalities above are equalities. Thus, u minimizes $\mathcal{F}$.

Lastly, we show that there is a subsequence of u_k that d_1-converges to u. As $\limsup_k I(u_k) = I(u)$, after possibly passing to a subsequence, $\lim_k I(u_k) = I(u)$. This together with $|u_k - u|_{L^1(\omega^n)} \to 0$ and Theorem 3.46 gives that $d_1(u_k,u) \to 0$. □

4.4. Convexity of the Ding functional

We stay with a Fano manifold (X,ω) for this section as well. Previously we extended Ding's functional to $\mathcal{E}_1(X,\omega)$. In this short section we show that this extension is convex along the finite energy geodesics of $\mathcal{E}_1(X,\omega)$ (Theorem 4.22). We start by computing the Hessian of the Ding functional with respect to the L^2 Mabuchi metric of $\mathcal{H}_\omega$:

PROPOSITION 4.19. *Suppose $u \in \mathcal{H}_\omega$ and $\phi,\psi \in C^\infty(X) \simeq T_u\mathcal{H}_\omega$. We have the following formula for the Hessian of $\mathcal{F}$ with respect to the L^2 Mabuchi metric:*

(4.27)

$$\nabla^2\mathcal{F}(u)(\phi,\psi)$$
$$= \frac{1}{V}\int_X \Big[\frac{1}{2}\langle\nabla^{\omega_u}\phi,\nabla^{\omega_u}\psi\rangle_{\omega_u} - \Big(\phi - \frac{1}{V}\int_X \phi e^{f_u}\omega_u^n\Big)\cdot\Big(\psi - \frac{1}{V}\int_X \psi e^{f_u}\omega_u^n\Big)\Big]e^{f_u}\omega_u^n.$$

PROOF. First we compute the Hessian of the Aubin–Yau energy I. From Lemma 3.37 it follows that for small $t > 0$ we have:

$$dI(u + t\psi)(\phi) = \frac{1}{V}\int_X \phi\omega^n_{u+t\psi}.$$

Using the formulas $\nabla^2 I(u+t\psi)(\phi,\psi) = \frac{d}{dt}dI(u+t\psi)(\phi) - dI(u+t\psi)(\nabla_{\frac{d}{dt}}\phi)$ and $\nabla_{\frac{d}{dt}}\phi = -\frac{1}{2}\langle\nabla^{\omega_u}\phi, \nabla^{\omega_u}\psi\rangle_{\omega_u}$, we can write:

$$\begin{aligned}\nabla^2 I(u)(\phi,\psi) &= \frac{n}{V}\int_X \phi i\partial\bar\partial\psi\wedge\omega_u^{n-1} + \frac{1}{2V}\int_X \langle\nabla^{\omega_u}\phi,\nabla^{\omega_u}\psi\rangle_{\omega_u}\omega_u^n \\ &= \frac{1}{2V}\int_X \phi(\Delta^{\omega_u}\psi)\omega_u^n + \frac{1}{2V}\int_X \langle\nabla^{\omega_u}\phi,\nabla^{\omega_u}\psi\rangle_{\omega_u}\omega_u^n = 0,\end{aligned} \tag{4.28}$$

where in the last line we have used (5.4) in the appendix and the formula below it. Next we introduce $\mathcal{B} : \mathcal{H}_\omega \to \mathbb{R}$ by the formula

$$\mathcal{B}(u) = -\log\frac{1}{V}\int_X e^{-u+f_0}\omega^n, \ u \in \mathcal{H}_\omega. \tag{4.29}$$

Using (4.10) and $d\mathcal{B}(u+t\psi)(\phi) = \int_X \phi e^{-u-t\psi+f_0}\omega^n / \int_X e^{-u-t\psi+f_0}\omega^n$, we obtain that $d\mathcal{B}(u+t\psi)(\phi) = \frac{1}{V}\int_X \phi e^{f_u}\omega_u^n$. Another differentiation gives

$$\begin{aligned}\frac{d}{dt}\Big|_{t=0} d\mathcal{B}(u+t\psi)(\phi) &= -\frac{\int_X \phi\psi e^{-u+f_0}\omega^n}{\int_X e^{-u+f_0}\omega^n} + \frac{\int_X \phi e^{-u+f_0}\omega^n}{\int_X e^{-u+f_0}\omega^n}\cdot\frac{\int_X \psi e^{-u+f_0}\omega^n}{\int_X e^{-u+f_0}\omega^n} \\ &= -\frac{1}{V}\int_X \phi\psi e^{f_u}\omega_u^n + \frac{1}{V^2}\int_X \phi e^{f_u}\omega_u^n\cdot\int_X \psi e^{f_u}\omega_u^n.\end{aligned}$$

where in the last line we have used (4.10) again. Reorganizing terms in this identity and using $\nabla^2\mathcal{B}(u+t\psi)(\phi,\psi) = \frac{d}{dt}d\mathcal{B}(u+t\psi)(\phi) - d\mathcal{B}(u+t\psi)(\nabla_{\frac{d}{dt}}\phi)$ we conclude that

$$\begin{aligned}&\nabla^2\mathcal{B}(u)(\phi,\psi) = \\ &= \frac{1}{V}\int_X\Big[\frac{1}{2}\langle\nabla^{\omega_u}\phi,\nabla^{\omega_u}\psi\rangle_{\omega_u} - \Big(\phi - \frac{1}{V}\int_X \phi e^{f_u}\omega_u^n\Big)\cdot\Big(\psi - \frac{1}{V}\int_X \psi e^{f_u}\omega_u^n\Big)\Big]e^{f_u}\omega_u^n.\end{aligned}$$

The proof is finished after we subtract (4.28) from this last formula. □

Next we will show that $\nabla^2\mathcal{F}(u)(\cdot,\cdot)$ is positive semi–definite for all $u \in \mathcal{H}_\omega$. Before this we introduce and study the following complex valued weighted complex Laplacian

$$L^{f_u}h = \partial^*\partial h - \langle\partial h, \partial f_u\rangle_{\omega_u}, \quad h \in C^\infty(X,\mathbb{C}). \tag{4.30}$$

To clarify, by $\langle \partial h, \partial f_u \rangle_{\omega_u}$ we mean the quantity $g_u^{j\bar{k}} h_j f_{u\bar{k}}$ (expressed in local coordinates). Also ∂^* is the Hermitian L^2 adjoint of ∂ with respect to ω_u. For $g, h \in C^\infty(X, \mathbb{C})$ integration by parts gives the following:

$$\int_X (L^{f_u} g)\bar{h} e^{f_u} \omega_u^n = \int_X \langle \partial g, \partial h \rangle_{\omega_u} e^{f_u} \omega_u^n = \int_X g \overline{(L^{f_u} h)} e^{f_u} \omega_u^n. \tag{4.31}$$

Consequently L^{f_u} is a self–adjoint elliptic operator with respect to the Hermitian inner product

$$\langle \alpha, \beta \rangle = \int_X \alpha \bar{\beta} e^{f_u} \omega_u, \quad \alpha, \beta \in C^\infty(X, \mathbb{C}).$$

We conclude that $L^2(e^{f_u}\omega_u^n)$ has an orthonormal base composed of eigenfunctions corresponding to the eigenvalues $\lambda_0 < \lambda_1 < \ldots$ of L^{f_u}. As another application of (4.31) we see that $\lambda_0 = 0$ and the eigenspace of this eigenvalue is composed by the constant functions. Moreover, we have the following general result about the eigenfunctions of L^{f_u} due to Futaki:

PROPOSITION 4.20 ([**71**]). *Suppose $v \in C^\infty(X, \mathbb{R})$ and $u \in \mathcal{H}_\omega$ such that $\int_X v e^{f_u} \omega_u^n = 0$. In addition, let h be an eigenfunction L^{f_u}, i.e., $L^{f_u} h = \lambda h$. Then the following hold:*

$$\lambda \int_X \langle \partial h, \partial h \rangle_{\omega_u} e^{f_u} \omega_u^n = \int_X \langle \partial h, \partial h \rangle_{\omega_u} e^{f_u} \omega_u^n + \int_X \langle \mathcal{L} h, \mathcal{L} h \rangle_{\omega_u} e^{f_u} \omega_u^n, \tag{4.32}$$

$$\begin{aligned} \int_X |v|^2 e^{f_u} \omega_u^n \leq \frac{1}{2} \int_X \langle \nabla^{\omega_u} v, \nabla^{\omega_u} v \rangle_{\omega_u} e^{f_u} \omega_u^n &= \int_X \langle \partial v, \partial v \rangle_{\omega_u} e^{f_u} \omega_u^n \\ &= \int_X (L^{f_u} v) \bar{v} e^{f_u} \omega_u^n, \end{aligned} \tag{4.33}$$

where $\mathcal{L}$ is the Lichnerowitz operator (see (5.12) in the appendix). Also, one has equality in (4.33) if and only if $\mathcal{L} v = 0$, or equivalently, $\nabla_{0,1}^{\omega_u} v \in T_{\mathbb{C}}^{1,0} X$ is a holomorphic vector field.

Recall from the discussion following (5.12) in the appendix that the condition $\mathcal{L} v = 0$ is indeed equivalent with $\nabla_{0,1}^{\omega_u} v \in T_{\mathbb{C}}^{1,0} X$ being holomorphic.

PROOF. By the the discussion preceding the result, the condition $\int_X v e^{f_u} \omega_u^n = 0$ simply means that v is orthogonal to the eigenspace of λ_0, hence to prove (4.33) we only need to argue (4.32). Indeed, the rightmost term in (4.32) is nonnegative hence either $\lambda = \lambda_0 = 0$ (in which case h is a constant) or $\lambda \geq \lambda_1 \geq 1$. Consequently the inequality between the first and last term of (4.33) follows after expressing v using the orthonormal base of $L^2(e^{f_u}\omega_u^n)$ composed of eigenfunctions of L^{f_u}.

Lastly, we note that the middle identities of (4.33) are simply a consequence of (4.31) and the formula following (5.4), since v is real valued.

We now argue (4.32). To ease notation, we will drop the subscript of f_u and ω_u in the rest of the proof. Also, recall that by a choice of normal coordinates identifying a neighborhood of $x \in X$ with that of $0 \in \mathbb{C}^n$, we can assure that locally $\omega = i\partial\bar{\partial} g$, with $g_{j\bar{k}}(0) = \delta_{jk}$ and $g_{j\bar{k}l}(0) = g_{j\bar{k}\bar{l}}(0) = 0$ (see Proposition 5.2 below). With such a choice of coordinates we also have $\text{Ric}\, \omega_{j\bar{k}}|_x = -g_{j\bar{k}a\bar{a}}(0)$

(see (5.11)). Making use of this and integrating by parts multiple times we get

$$\begin{aligned}
\lambda \int_X \langle \partial h, \partial h \rangle e^f \omega^n &= \int_X \langle \partial(L^f h), \partial h \rangle e^f \omega^n = \int_X g^{j\bar{k}} (L^f h)_j \bar{h}_{\bar{k}} e^f \omega^n \\
&= \int_X g^{j\bar{k}} (-g^{a\bar{b}} h_{a\bar{b}} - g^{a\bar{b}} h_a f_{\bar{b}})_j \bar{h}_{\bar{k}} e^f \omega^n \\
&= \int_X (g_{\bar{a}j b} h_{a\bar{b}} \bar{h}_{\bar{j}} - h_{a\bar{a}j} \bar{h}_{\bar{j}} - h_{aj} f_{\bar{a}} \bar{h}_{\bar{j}} - h_a f_{\bar{a}j} \bar{h}_{\bar{j}}) e^f \omega^n \\
&= \int_X (g_{\bar{a}jb} h_{a\bar{b}} \bar{h}_{\bar{j}} + h_{aj} \bar{h}_{\bar{a}\bar{j}} + h_{aj} f_{\bar{a}} \bar{h}_{\bar{j}} - h_{aj} f_{\bar{a}} \bar{h}_{\bar{j}} - h_a f_{\bar{a}j} \bar{h}_{\bar{j}}) e^f \omega^n \\
&= \int_X (g_{\bar{a}jb} h_{a\bar{b}} \bar{h}_{\bar{j}} + h_{aj} \bar{h}_{\bar{a}\bar{j}} - h_a f_{\bar{a}j} \bar{h}_{\bar{j}}) e^f \omega^n \\
&= \int_X (-g_{\bar{a}jb\bar{b}} h_a \bar{h}_{\bar{j}} + h_{aj} \bar{h}_{\bar{a}\bar{j}} - h_a f_{\bar{a}j} \bar{h}_{\bar{j}}) e^f \omega^n \\
&= \int_X (\mathrm{Ric}\ \omega_u(\partial h, \bar{\partial} \bar{h}) + h_{aj} \bar{h}_{\bar{a}\bar{j}} - i\partial\bar{\partial} f_u(\partial h, \bar{\partial} h)) e^f \omega^n \\
&= \int_X (\langle \partial h, \partial h \rangle + h_{aj} \bar{h}_{\bar{a}\bar{j}}) e^f \omega^n \\
&= \int_X \langle \partial h, \partial h \rangle e^f \omega^n + \int_X \langle \mathcal{L} h, \mathcal{L} h \rangle e^f \omega^n,
\end{aligned}$$

where in the last line we have used the identity $\mathrm{Ric}\ \omega_u - \omega_u = i\partial\bar{\partial} f_u$, and the expression of $\mathcal{L}$ in normal coordinates (see (5.12)). □

From (4.33) and Proposition 4.19 it follows that $\nabla^2 \mathcal{F}(u)(\cdot,\cdot)$ is indeed positive semi-definite. As an additional consequence we obtain that $\mathcal{B}$ is convex along ε-geodesics:

LEMMA 4.21. *Suppose $u_0, u_1 \in \mathcal{H}_\omega$ and $\varepsilon > 0$. Let $[0,1] \ni t \to u_t^\varepsilon \in \mathcal{H}_\omega$ be the smooth ε-geodesic joining u_0, u_1 (see (3.11)). Then $t \to \mathcal{B}(u_t^\varepsilon) := -\log\left(\int_X e^{-u_t^\varepsilon + f_0} \omega^n\right)$ is convex.*

PROOF. As it will not cause confusion, we will drop the reference to ε in our argument.

Using the L^2 Mabuchi structure of $\mathcal{H}_\omega$ we have the following formula:

$$\frac{d^2}{dt^2} \mathcal{B}(u_t) = \nabla^2 \mathcal{B}(u_t)(\dot{u}_t, \dot{u}_t) + d\mathcal{B}(u_t)(\nabla_{\dot{u}_t} \dot{u}_t).$$

From Proposition 4.20 it follows that $\nabla^2 \mathcal{B}(u_t)(\dot{u}_t, \dot{u}_t) \geq 0$. In addition to this, the equation of ε-geodesics (see (3.11)) gives $\nabla_{\dot{u}_t} \dot{u}_t > 0$, hence $d\mathcal{B}(u_t)(\nabla_{\dot{u}_t} \dot{u}_t) \geq 0$. Putting the last two facts together we get that $\frac{d^2}{dt^2} \mathcal{B}(u_t) \geq 0$. □

Finally, we argue that $\mathcal{F}$ is convex along the finite energy geodesics of $\mathcal{E}_1(X, \omega)$.

THEOREM 4.22. *Suppose $u_0, u_1 \in \mathcal{E}_1(X, \omega)$ and let $t \to u_t$ be the finite energy geodesic connecting u_0, u_1. Then $t \to \mathcal{F}(u_t)$ is convex and continuous on $[0, 1]$.*

PROOF. As $t \to I(u_t)$ is known to be affine (Proposition 3.42), we only need to argue that $t \to \mathcal{B}(u_t) := -\log\left(\int_X e^{-u_t + f_0} \omega^n\right)$ is also affine. As $u \to \mathcal{F}(u)$ and $u \to I(u)$ is d_1-continuous (see Proposition 3.40 and Theorem 4.17), we obtain that so is $u \to \mathcal{B}(u)$.

Let us assume first that $u_0, u_1 \in \mathcal{H}_\omega$. In this case $u_t^\varepsilon \nearrow u_t$ as $\varepsilon \to 0$, where $t \to u_t^\varepsilon$ is the ε–geodesic joining u_0, u_1 (see (3.19)). Consequently $d_1(u_t^\varepsilon, u_t) \to 0,\ t \in [0,1]$. Hence $\mathcal{B}(u_t^\varepsilon) \to \mathcal{B}(u_t),\ t \in [0,1]$. By Lemma 4.21 it follows that $t \to \mathcal{B}(u_t)$ is convex.

In the general case $u_0, u_1 \in \mathcal{E}_1(X,\omega)$, let $u_0^j, u_1^j \in \mathcal{H}_\omega$ be smooth decreasing approximants that exist by Theorem 2.1. Let $t \to u_t^j$ be the $C^{1,\bar{1}}$–geodesics joining u_0^j, u_1^j. By Proposition 3.15(i) it follows that $d_1(u_t^j, u_t) \to 0,\ t \in [0,1]$ and consequently $\mathcal{B}(u_t^j) \to \mathcal{B}(u_t),\ t \in [0,1]$, implying that $t \to \mathcal{B}(u_t)$ is convex, finishing the proof. □

For weak geodesics joining bounded potentials the above theorem was proved by Berndtsson in much more general context [**17**, Theorem 1.2]. The argument that we presented in this section follows more closely the simplified treatment in [**78**].

4.5. Uniqueness of KE metrics and reductivity of the automorphism group

In this section we will give the proof of an important theorem of Bando–Mabuchi according to which on a Fano manifold (X,ω) Kähler–Einstein metrics are unique up to pullback by an automorphism:

THEOREM 4.23 ([**2**]). *Suppose $u, v \in \mathcal{H}_\omega$ both solve (4.5), i.e., they are both KE potentials. Then there exists $g \in \mathrm{Aut}_0(X,J)$ such that $g^*\omega_u = \omega_v$.*

As we plan to use some of the machinery that we developed in previous parts, we will not follow the original proof of Bando–Mabuchi. Instead our proof will be a combination of the arguments of Berndtsson [**19**] and Berman–Berndtsson [**8**, Section 4], and bears similarities with the treatment in [**86**].

The proof will need a sequence of preliminary results about the automorphism group of Kähler–Einstein manifolds and will also use the classical theory of self adjoint elliptic differential operators on compact manifolds (see [**114**, Chapter 4]).

LEMMA 4.24. *Suppose X is a Fano manifold. Then $H^{0,q}(X,\mathbb{C})$ is trivial for $q \in \{1,\dots,n\}$.*

PROOF. We start with the observation $H^{0,q}(X,\mathbb{C}) \simeq H^{n,q}(X,-K_X)$. As $-K_X > 0$ by the Fano condition, the triviality of $H^{n,q}(X,-K_X)$ follows from the Kodaira vanishing theorem ([**61**, Theorem VII.3.3]). □

Note that by the Hodge decomposition we also have $H^1(X,\mathbb{C}) \simeq H^{1,0}(X,\mathbb{C}) \oplus H^{0,1}(X,\mathbb{C})$, hence this group is trivial as well. Though we will not use this, we mention that by an argument involving the Bonnet–Myers theorem and the Euler characteristic, we can further deduce that X is in fact simply connected.

Now we focus on the Lie algebra $\mathfrak{g}$ of $G := \mathrm{Aut}_0(X,J)$. Pick $U \in \mathfrak{g}$. As $U = U^{1,0} + \overline{U^{1,0}}$ is real holomorphic it follows that $\bar\partial(U^{1,0} \lrcorner \omega_u) = 0$ for all $u \in \mathcal{H}_\omega$. Indeed, this is immediate after one computes $\partial_{\bar l}(U_j^{1,0} {g^u}_{j\bar k}) = 0,\ l \in \{1,\dots,n\}$ in normal coordinates (here and below g^u is a local potential of ω_u).

Using the previous lemma it follows that there exists a unique $v_{\omega_u}^U \in C^\infty(X,\mathbb{C})$ with $\int_X v_{\omega_u}^U \omega_u^n = 0$ such that

$$U^{1,0} \lrcorner \omega_u = \bar\partial v_{\omega_u}^U.$$

Equivalently, using Hamiltonian formalism this can be written as $X^{\omega_u}_{1,0}(v^U_{\omega_u}) = U^{1,0}$, and we have the following identification for $\mathfrak{g}$ using ω_u:
(4.34)

$$\mathfrak{g} \simeq \mathfrak{g}_{\omega_u} := \{v \in C^\infty(X,\mathbb{C}) \; : \; X^{\omega_u}_{1,0}v \in T^{1,0}_{\mathbb{C}}X \text{ is holomorphic and } \int_X v\omega_u^n = 0\}.$$

Recall that the "complex" gradient $\nabla^{\omega_u}_{1,0}v = JX^{\omega_u}_{1,0}v$ (see (5.5)) is holomorphic precisely when $\mathcal{L}v = 0$, where $\mathcal{L}$ is the Lichnerowitz operator of the metric ω_u (see (5.12)).

When $\mathcal{H}_\omega$ contains a KE potential u then S_{ω_u} is trivially constant, hence by Proposition 5.3 we have that

$$\mathcal{L}^*\mathcal{L}(f) = \frac{1}{4}\Delta^{\omega_u}(\Delta^{\omega_u}f) + \langle Ric_{\omega_u}, i\partial\bar\partial f\rangle_{\omega_u} = \frac{1}{4}\Delta^{\omega_u}(\Delta^{\omega_u}f) + \frac{1}{2}\Delta^{\omega_u}f$$

is a real differential operator. As a result, $v \in \ker \mathcal{L}^*\mathcal{L} = \ker\mathcal{L}$ if and only if $\text{Re } v, \text{Im } v \in \ker\mathcal{L}$. Consequently, for KE Fano manifolds the above description of $\mathfrak{g}$ can be sharpened, to imply that $\text{Aut}_0(X,J)$ is reductive, which was one of the first known obstructions to existence of KE metrics:

PROPOSITION 4.25 ([**89**]). *Suppose (X,ω_u) is a Fano KE manifold. Introducing $\mathfrak{k}_{\omega_u} = \mathfrak{g}_{\omega_u} \cap C^\infty(X,\mathbb{R})$ we can write*

$$\mathfrak{g}_{\omega_u} = \mathfrak{k}_{\omega_u} \oplus i\mathfrak{k}_{\omega_u}.$$

In particular, $\text{Aut}_0(X,J)$ is the complexification of the compact connected Lie group $\text{Isom}_0(X,\omega_u,J)$, with Lie algebra $\mathfrak{k}_{\omega_u}$.

Here $\text{Isom}_0(X,\omega_u,J)$ is the identity component of the group of holomorphic isometries of the KE metric ω_u. As X is compact, the group of smooth isometries of (X,ω_u) is compact as well, hence so is its subgroup $\text{Isom}_0(X,\omega_u,J)$.

PROOF. The decomposition $\mathfrak{g}_{\omega_u} = \mathfrak{k}_{\omega_u} \oplus i\mathfrak{k}_{\omega_u}$ follows from the discussion preceding the proposition. We have to argue that the Lie algebra of $\text{Isom}_0(X,\omega_u,J)$ is exactly $\mathfrak{k}_{\omega_u}$.

Suppose $U \in \mathfrak{k}_{\omega_u}$. Trivially $v := v^U_{\omega_u} \in C^\infty(X,\mathbb{R})$, and since $U^{1,0} = X^{\omega_u}_{1,0}v$, by conjugation we obtain that in fact $U = X^{\omega_u}_v$. Consequently,

$$d(U \lrcorner \omega_u) = d(U^{1,0} \lrcorner \omega_u + U^{0,1} \lrcorner \omega_u) = d(\bar\partial h + \partial h) = ddh = 0,$$

hence U represents an infinitesimal symplectomorphism. Since U is holomorphic, by (5.2) U represents an infinitesimal isometry as well, i.e., $U \in \text{Lie}(\text{Isom}_0(X,\omega_u,J))$ as claimed.

Conversely, if $U \in \text{Lie}(\text{Isom}_0(X,\omega_u,J))$ then we have $0 = d(U \lrcorner \omega_u) = d(U^{1,0} + U^{0,1}) \lrcorner \omega_u = d(\bar\partial v^U_{\omega_u} + \partial \overline{v^U_{\omega_u}}) = 2i\partial\bar\partial \text{Im } v^U_{\omega_u}$. Consequently $v^U_{\omega_u} \in \mathfrak{g}_{\omega_u} \cap C^\infty(X,\mathbb{R}) = \mathfrak{k}_{\omega_u}$. □

According to the next result the action of a one parameter subgroup of automorphisms in the direction of $i\mathfrak{k}$ gives d_p–geodesic rays inside $\mathcal{H}_\omega$:

LEMMA 4.26. *Suppose $u \in \mathcal{H}_\omega \cap I^{-1}(0)$ is a KE potential. Let $U \in i\mathfrak{k}_{\omega_u}$ and $\mathbb{R} \ni t \to \rho_t \in \text{Aut}_0(X,J)$ be the associated one parameter subgroup. Then $t \to u_t := \rho_t.u$ is a smooth d_p–geodesic ray for any $p \geq 1$.*

PROOF. As $U \in i\mathfrak{k}_{\omega_u}$, it follows that $v^U_{\omega_u} = ih$ for some $h \in C^\infty(X,\mathbb{R})$ with $\int_X h\omega_u^n = 0$. Differentiating $\rho_t^*\omega_u = \omega_{u_t}$ we find that $i\partial\bar\partial \dot u_t = \rho_t^* d(U \lrcorner \omega_u) = \rho_t^* d(\bar\partial v^U_{\omega_u} + \partial \overline{v^U_{\omega_u}}) = 2i\rho_t^*\partial\bar\partial(\text{Im } v^U_{\omega_u}) = 2i\partial\bar\partial h \circ \rho_t$.

Since $I(u_t) = 0,\ t \geq 0$, Lemma 3.37 gives that $\int_X \dot u_t \omega^n_{u_t} = 0$. Also, $\int_X h \circ \rho_t \omega^n_{u_t} = \int_X h \circ \rho_t \rho_t^* \omega_u^n = \int_X h\omega_u^n = 0$, so we conclude that

$$\dot u_t = 2h \circ \rho_t. \tag{4.35}$$

Differentiating this identity we obtain

$$\begin{aligned}\ddot u_t = 2\rho_t^*(U \lrcorner dh) &= 2\rho_t^*\langle \nabla^{\omega_u} h, U\rangle_{\omega_u} = 2\rho_t^*\langle \nabla^{\omega_u} h, JX^{\omega_u} h\rangle_{\omega_u}\\ &= 2\langle \nabla^{\omega_{u_t}} h \circ \rho_t, \nabla^{\omega_{u_t}} h \circ \rho_t\rangle_{\omega_{u_t}} = \frac{1}{2}\langle \nabla^{\omega_{u_t}} \dot u_t, \nabla^{\omega_{u_t}} \dot u_t\rangle_{\omega_{u_t}},\end{aligned}$$

where we used (5.5) in the second to last equality, and (4.35) again in the last equality. The above arguments show that $t \to u_t$ satisfies (3.8), hence by Theorem 3.36 we obtain that $t \to u_t$ is a geodesic ray for any $p \geq 1$. □

The following result will play an important role in the proof of Theorem 4.23:

PROPOSITION 4.27. *Let (X,ω) be Fano. Suppose $u \in \mathcal{H}_\omega \cap I^{-1}(0)$ is a KE potential. Then the map $\text{Aut}_0(X,J) \ni h \to J_\omega(h.u) \in \mathbb{R}$ admits a minimizer $g \in \text{Aut}_0(X,J)$ that satisfies*

$$\int_X v\omega^n = 0 \quad \text{for all} \quad v \in \mathfrak{g}_{\omega_{g.u}}. \tag{4.36}$$

Recall the definition of the J functional from (3.68). To avoid the possibility of confusion with the complex structure, we denoted this functional with J_ω in the above proposition, and will continue to do so in the rest of this section.

PROOF. By Proposition 3.44 the functional J_ω has the same growth as the metric d_1. We turn to the group $\text{Aut}_0(X,J)$ which is reductive by Proposition 4.25, hence we can apply Proposition 5.14 in the appendix to deduce that the map $C : \text{Isom}_0(X,\omega_u,J) \oplus \mathfrak{k}_{\omega_u} \to \text{Aut}_0(X,J)$ given by $C(k,U) = k\exp_I(JU)$ is surjective.

For any $k \in \text{Isom}_0(X,\omega_u,J)$ and any $U \in \mathfrak{k}_{\omega_u}$ the previous proposition gives that $t \to k\exp_I(tJU).u = \exp_I(tJU).u =: u_t \in \mathcal{H}_\omega \cap I^{-1}(0)$ is a smooth d_1–geodesic. As the growth of J_ω is equivalent with the growth of the d_1 metric, it follows that the map

$$(K,\mathfrak{k}_{\omega_u}) \ni (k,U) \to \Theta(k,U) := J_\omega(C(k,U).u) = J_\omega(\exp_I(JU).u) = \Theta(I,U) \in \mathbb{R}$$

is proper (meaning that $\Theta(k_j,U_j) = \Theta(I,U_j) \to \infty$ if $|U_j| \to \infty$), hence it admits a minimizer. As C is surjective, it follows that $\text{Aut}_0(X,J) \ni h \to J_\omega(h.u) \in \mathbb{R}$ admits a minimizer $g \in \text{Aut}_0(X,J)$ as well.

Fix $v \in \mathfrak{k}_{\omega_{g.u}}$. We introduce the vector field $W = X^{\omega_{g.u}}v$, and by the previous proposition $[0,\infty) \ni t \to h_t := \exp(tJW).(g.u) \in \mathcal{H}_0 \cap I^{-1}(0)$ is a geodesic ray.

As the identity element minimizes $\text{Aut}_0(X,J) \ni h \to J_\omega(h.(g.u)) \in \mathbb{R}$, we can write

$$0 = \frac{d}{dt}\Big|_{t=0} J_\omega(\exp(tJW).(g.u)) = \frac{1}{V}\frac{d}{dt}\Big|_{t=0}\int_X h_t\omega^n = \frac{1}{V}\int_X \dot h_0\omega^n.$$

Using (4.35) we conclude that $\dot{h}_0 = 2v$, hence $\int_X v\omega^n = 0$. Finally, the decomposition formula of Proposition 4.25 implies that (4.36) in fact holds for any $v \in \mathfrak{g}_{\omega_{g.u}}$. □

For $u \in \mathcal{H}_\omega$, recall the self adjoint differential operator L^{f_u} from (4.30). Motivated by the explicit formula for the Hessian of $\mathcal{F}$ (see 4.27) we introduce the differential operator $\mathcal{D}^u : C^\infty(X, \mathbb{C}) \to \mathbb{R}$:

$$\mathcal{D}^u(h) = L^{f_u}(h) - h + \frac{1}{V}\int_X he^{f_u}\omega_u^n.$$

Integrating by parts in (4.27) we get the following formula, relating $\nabla^2\mathcal{F}(u)$ and $\mathcal{D}^u$:

$$\nabla^2\mathcal{F}(u)(\phi,\psi) = \frac{1}{V}\int_X \phi\overline{\Big(L^{f_u}\psi - \psi + \frac{1}{V}\int_X \psi e^{f_u}\omega_u^n\Big)}e^{f_u}\omega_u^n = \frac{1}{V}\int_X \phi\overline{\mathcal{D}^u(\psi)}e^{f_u}\omega_u^n.$$

$$= \frac{1}{V}\int_X \Big(L^{f_u}\phi - \phi + \frac{1}{V}\int_X \phi e^{f_u}\omega_u^n\Big)\overline{\psi}e^{f_u}\omega_u^n = \frac{1}{V}\int_X \mathcal{D}^u(\phi)\overline{\psi}e^{f_u}\omega_u^n. \tag{4.37}$$

This implies that $\mathcal{D}^u$ is a self–adjoint differential operator as well. Additionally, the kernel of $\mathcal{D}^u$ is exactly equal to the eigenspace of $\mathcal{L}^{f_u}$ corresponding to the eigenvalue $\lambda = 1$. In Proposition 4.20 (see especially the identity (4.32)) we gave an exact description of this space that we now recall:

$$\text{Ker } \mathcal{D}^u = \{v \in C^\infty(X,\mathbb{C}) \text{ s.t.} \int_X ve^{f_u}\omega_u^n = 0 \text{ and } \nabla^{\omega_u}_{1,0}v \in T^{1,0}_{\mathbb{C}}X \text{ is holomorphic}\}.$$

In case $u \in \mathcal{H}_\omega \cap I^{-1}(0)$ is a KE potential we trivially have $f_u = 0$, and comparing with (4.34) we get the following identification:

$$\text{Ker } \mathcal{D}^u = \mathfrak{g}_{\omega_u}. \tag{4.38}$$

As $\mathcal{D}^u$ is self–adjoint and elliptic, for any $h \in C^\infty(X,\mathbb{C})$ such that $h \perp \text{Ker } \mathcal{D}^u$ there exists $v \in C^\infty(X,\mathbb{C})$ such that $\mathcal{D}^u(v) = h$ (see [**114**, Theorem IV.4.11]). We note this fact in slightly more precise form in the following lemma:

LEMMA 4.28. *Let (X,ω_u) is a KE manifold. Suppose that $\int_X v\omega^n = 0$ for all $v \in \mathfrak{g}_{\omega_u}$. Then there exists $g \in C^\infty(X,\mathbb{R})$ such that*

$$\nabla^2\mathcal{F}(u)(f,g) = \frac{1}{V}\int_X f\overline{\mathcal{D}^u(g)}\omega_u^n = \frac{1}{V}\int_X f\omega^n, \quad \forall\, f \in C^\infty(X,\mathbb{C}).$$

PROOF. Denote $h := \omega^n/\omega_u^n \in C^\infty(X,\mathbb{R})$. By our assumption we have that $h \perp \mathfrak{g}_{\omega_u}$ with respect to the Hermitian product $\langle\alpha,\beta\rangle = \frac{1}{V}\int_X \alpha\overline{\beta}\omega_u^n$, hence by our above remarks there exists $g \in C^\infty(X,\mathbb{C})$ such that $\mathcal{D}^u(g) = h$. As u is a KE potential, we have $f_u = 0$, and as a result $\mathcal{D}^u$ is a real differential operator. Consequently, we can make sure that $g \in C^\infty(X,\mathbb{R})$, finishing the argument. □

PROOF OF THEOREM 4.23. Without loss of generality we can assume that our KE potentials u, v satisfy $u, v \in \mathcal{H}_\omega \cap I^{-1}(0)$. Also, by Proposition 4.27, after possibly pulling back u and v by an element of $\text{Aut}_0(X,J)$ we can assume that

$$\int_X h\omega^n = 0, \quad \forall \in h \in \mathfrak{g}_{\omega_u} \cup \mathfrak{g}_{\omega_v}.$$

Using this, by the previous lemma we can find $g_u, g_v \in C^\infty(X,\mathbb{R})$ such that

$$\nabla^2\mathcal{F}(u)(f,g_u) = \nabla^2\mathcal{F}(v)(f,g_v) = -\frac{1}{V}\int_X f\omega^n, \quad \forall\, f \in C^\infty(X,\mathbb{R}). \tag{4.39}$$

For the rest of the proof we will be working with the twisted Ding functional $\mathcal{F}_s : \mathcal{H}_\omega \to \mathbb{R}$, $s \geq 0$ given by the formula:

$$\mathcal{F}_s(h) = \mathcal{F}(h) + sJ_\omega(h).$$

For small enough $s > 0$ we can suppose that the potentials $u_0^s := u + sg_u$ and $u_1^s := v + sg_u$ satisfy $u_0^s, u_1^s \in \mathcal{H}_\omega$.

The differential of $\mathcal{F}_s$ is equal to $d\mathcal{F} + sdJ$. Choosing $w \in C^\infty(X, \mathbb{R})$ with $\int_X w\omega_u^n = 0$ a simple differentiation gives

$$\frac{d}{ds}\Big|_{s=0} d\mathcal{F}_s(u_0^s)(w) = \nabla^2 \mathcal{F}(u)(w, g_u) + d\mathcal{F}(u)(\nabla_{\frac{d}{ds}} w) + dJ_\omega(u)(w).$$

Since u minimizes $\mathcal{F}$ we have that $d\mathcal{F}(u)(\nabla_{\frac{d}{ds}} w) = 0$. Since $\int_X w\omega_u^n = 0$ we also have $dJ_\omega(u)(w) = \frac{1}{V}\int_X w\omega^n$, so we can continue the above identity and write:

$$\frac{d}{ds}\Big|_{s=0} d\mathcal{F}_s(u_0^s)(w) = \nabla^2 \mathcal{F}(u)(w, g_u) + \frac{1}{V}\int_X w\omega^n = 0,$$

where in the last identity we have used (4.39). Consequently $s \to d\mathcal{F}_s(u_0^s)(w) = O(s^2)$. Since $d\mathcal{F}_s(u_0^s)(w) = \frac{1}{V}\int_X w f_s \omega^n$ for some smooth curve $s \to f_s$, we can conclude that $f_s = O(s^2)$ and we have

$$\big|d\mathcal{F}_s(u_0^s)(w)\big| \leq Cs^2 \sup_X |w| \text{ for all } w \in C(X, \mathbb{R}). \tag{4.40}$$

A similar estimate holds for $\big|d\mathcal{F}_s(u_1^s)(w)\big|$ as well.

Let $[0,1] \ni t \to u_t^s \in \mathcal{H}_\omega^{1,\bar{1}}$ be the $C^{1,\bar{1}}$-geodesic connecting u_0^s and u_1^s. By convexity of $t \to \mathcal{F}(u_t^s)$ and $t \to J_\omega(u_t^s)$ it follows that

$$\begin{aligned} 0 \leq s\Big(\frac{d}{dt}\Big|_{t=1} J_\omega(u_t^s) - \frac{d}{dt}\Big|_{t=0} J_\omega(u_t^s)\Big) &\leq \frac{d}{dt}\Big|_{t=1} \mathcal{F}_s(u_t^s) - \frac{d}{dt}\Big|_{t=0} \mathcal{F}_s(u_t^s) \\ &= d\mathcal{F}_s(u_1^s)(\dot{u}_1^s) - d\mathcal{F}_s(u_0^s)(\dot{u}_0^s) \leq Cs^2, \end{aligned}$$

where the last inequality is a consequence of (4.40). Taking the limit $s \searrow 0$ in the above estimate, (by convexity) we obtain that $t \to J_\omega(u_t^0)$ is affine. By the lemma below, this implies that $u = u_0^0 = u_1^0 = v$, finishing the proof. □

LEMMA 4.29. *Suppose $u_0, u_1 \in \mathcal{H}_\omega \cap I^{-1}(0)$ and $t \to u_t$ is the $C^{1,\bar{1}}$-geodesic connecting u_0, u_1. If $t \to J_\omega(u_t)$ is affine then $u_0 = u_1$.*

PROOF. We know that $t \to I(u_t)$ is affine (Proposition 3.42). Using this, our assumption implies that $t \to \int_X u_t\omega^n$ is linear as well, hence $\frac{d}{d_t}\big|_{t=0} \int_X u_t\omega^n = \frac{d}{d_t}\big|_{t=1} \int_X u_t\omega^n$. This implies that $\int_X(\dot{u}_1 - \dot{u}_0)\omega^n = 0$. By convexity in the t variable we have $\dot{u}_1 \geq \dot{u}_0$, so we conclude that $\dot{u}_0 = \dot{u}_1$, hence $t \to u_t$ is affine, i.e.,

$$\dot{u}_0 = \dot{u}_1 = u_1 - u_0. \tag{4.41}$$

Since $t \to I(u_t)$ is affine too, we have $\frac{d}{dt}\big|_{t=0} I(u_t) = \frac{d}{dt}\big|_{t=1} I(u_t)$, and by (3.63) and (4.41) we get that $\int_X (u_1 - u_0)\omega_{u_0}^n = \int_X (u_1 - u_0)\omega_{u_0}^n$. Subtracting the right hand side from the left and integrating by parts we obtain

$$\sum_{j=0}^{n-1} \int_X i\partial(u_1 - u_0) \wedge \bar{\partial}(u_1 - u_0) \wedge \omega_{u_0}^j \wedge \omega_{u_1}^{n-1-j} = 0.$$

As all terms in the above sum are nonnegative, we get that $\int_X i\partial(u_1 - u_0) \wedge \bar\partial(u_1 - u_0) \wedge \omega_{u_0}^{n-1} = 0$. Consequently, $\langle \nabla^{\omega_{u_0}}(u_1 - u_0), \nabla^{\omega_{u_0}}(u_1 - u_0)\rangle_{\omega_{u_0}} = 0$, hence $u_1 = u_0 + c$. As $I(u_0) = I(u_1) = 0$, we have in fact $u_0 = u_1$. □

4.6. Regularity of weak minimizers of the Ding functional

In this short section we will show that minimizers of the extended $\mathcal{F}$ functional are actually smooth, with this verifying another important condition in the existence/properness principle described earlier (see Theorem 4.7):

THEOREM 4.30. *If $u \in \mathcal{E}_1(X,\omega)$ minimizes $\mathcal{F} : \mathcal{E}_1(X,\omega) \to \mathbb{R}$ then u is a smooth Kähler–Einstein potential.*

In case $u \in \mathcal{H}_\omega$ minimizes $\mathcal{F}$, then after computing the first order variation of $t \to \mathcal{F}(u + tv)$ for all $v \in C^\infty(X)$, Lemma 4.1 allows to conclude that:

$$\omega_u^n = \frac{V}{\int_X e^{-u+f_0}\omega^n} e^{-u+f_0}\omega^n, \tag{4.42}$$

hence u is indeed a KE potential. In case $u \in \mathcal{E}_1(X,\omega)$ we can't even guarantee that $t \to u + tv \in \mathcal{E}_1(X,\omega)$ for small t. Getting around this obstacle will represent one of main technical ingredients in the proof of Theorem 4.30.

Eventually we will be able to show that (4.42) holds for minimizers from $\mathcal{E}_1(X,\omega)$ as well. Notice that by Zeriahi's version of Skoda's theorem (Corollary 4.16) the right hand side of this equation does indeed makes sense for potentials of $\mathcal{E}_1(X,\omega)$. The proof is completed by appealing to work of Kolodziej and Tosatti–Székelyhidi on the apriori regularity theory of such equations:

THEOREM 4.31. *If $u \in \mathcal{E}_1(X,\omega)$ solves (4.42) then $u \in \mathcal{H}_\omega$, i.e., u is a smooth KE potential.*

SKETCH OF PROOF. As $u \in \mathcal{E}_1(X,\omega)$ Corollary 4.16 implies that $e^{-u+f_0} \in L^p(\omega^n)$ for all $p > 1$. By Kolodziej's theorem [**82**], since u solves (4.42), we obtain that u is bounded (for a full proof of this fact see [**75**, Theorem 14.1]). Using a result of Tosatti–Székelyhidi [**104**, Theorem 1.1] we obtain that u is actually a smooth KE potential (see also [**75**, Theorem 14.1]). □

PROOF OF THEOREM 4.30. Let $v \in C^\infty(X)$. Recall from (2.21) that $P(u + tv) \in \text{PSH}(X,\omega)$, $t \in \mathbb{R}$ is defined as follows:

$$P(u + tv) = \sup\{v \in \text{PSH}(X,\omega) \text{ s.t. } v \le u + tv\}.$$

Since $u - t\sup_X |v| \le P(u + tv)$, Corollary 2.8 implies that $P(u + tv) \in \mathcal{E}_1(X,\omega)$, and this allows to introduce the function

$$g(t) = -I(P(u + tv)) - \log \int_X e^{-u-tv+f_0}\omega^n + \log(V).$$

We claim that $g(t)$ is differentiable at $t = 0$ and the following formula holds:

$$\frac{d}{dt}\Big|_{t=0} g(t) = -\frac{1}{V}\int_X v\left(\omega_u^n - \frac{V}{\int_X e^{-u+f_0}\omega^n} e^{-u+f_0}\omega^n\right). \tag{4.43}$$

From the proposition below it follows that $\frac{d}{dt}\big|_{t=0} I(P(u + tv)) = \frac{1}{V}\int_X v\omega_u^n$. Consequently, to prove (4.43) it suffices to show that

$$\frac{d}{dt}\Big|_{t=0} \log \int_X e^{-u-tv+f_0}\omega^n = \frac{1}{\int_X e^{-u+f_0}\omega^n} e^{-u+f_0}\omega^n. \tag{4.44}$$

Using the elementary inequality $|e^x - e^y| \leq |x-y|(e^x + e^y)$ we can write that

$$\Big|\frac{e^{-u-tv+f_0} - e^{-u-lv+f_0}}{t-l}\Big| \leq |v|(e^{-u-tv+f_0} + e^{-u-lv+f_0}) \leq Ce^{-u}, \quad l,t \in (-1,1).$$

Corollary 4.16 implies that the right most quantity in this inequality is integrable, hence we can conclude (4.44) using the dominated convergence theorem.

Since $P(u+tv) \leq u+tv$, we notice that $g(0) = \mathcal{F}(u) \leq \mathcal{F}(P(u+tv)) \leq g(t)$, $t \in \mathbb{R}$. This implies that $\frac{d}{dt}\big|_{t=0} g(t) = 0$, and by (4.43) we can conclude that (4.42) holds for $u \in \mathcal{E}_1(X,\omega)$. Now using Theorem 4.31 we conclude that u is smooth, finishing the argument. □

PROPOSITION 4.32. [**9**, Lemma 3.10] *Suppose $u \in \mathcal{E}_1(X,\omega)$ and $v \in C^\infty(X)$. Then $P(u+tv) \in \mathcal{E}_1(X,\omega)$ for all $t \in \mathbb{R}$, and $t \to I(P(u+tv))$ is differentiable. More precisely,*

$$\frac{d}{dt}\Big|_{t=0} I(P(u+tv)) = \frac{1}{V}\int_X v\omega_u^n.$$

In our approach we will follow closely the simplified argument proposed by Lu and Nguyen [**87**].

PROOF. We want to show that

$$\frac{I(P(u+tv)) - I(u)}{t} \to \frac{1}{V}\int_X v\omega_u^n.$$

After changing v to $-v$, it suffices to consider $t > 0$ in the above limit. Using (3.58) we can write

$$\frac{I(P(u+tv)) - I(u)}{t} \leq \frac{1}{V}\int_X \frac{P(u+tv) - u}{t}\omega_u^n \leq \frac{1}{V}\int_X v\omega_u^n, \ t > 0.$$

By the same inequality we also have that

$$\frac{1}{V}\int_X \frac{P(u+tv) - u}{t}\omega_{P(u+tv)}^n \leq \frac{I(P(u+tv)) - I(u)}{t}.$$

By the lemma below $\omega_{P(u+tv)}^n$ is concentrated on the coincidence set $\{P(u+tv) = u+tv\}$, thus we have

$$\frac{1}{V}\int_X v\omega_{P(u+tv)}^n \leq \frac{I(P(u+tv)) - I(u)}{t}.$$

We also have $|P(u+tv) - u| \leq t\sup_X |v|$, and since the complex Monge–Ampère operator is continuous under uniform convergence we conclude that

$$\frac{1}{V}\int_X v\omega_u^n \leq \liminf_{t\to 0} \frac{I(P(u+tv)) - I(u)}{t},$$

finishing the argument. □

Finally, we provide the lemma promised in the proof of the above proposition:

LEMMA 4.33. *Suppose $u \in \mathcal{E}_1(X,\omega)$ and $v \in C^\infty(X)$. Then*

$$\int_{\{u+tv > P(u+tv)\}} \omega_{P(u+tv)}^n = 0.$$

PROOF. Using Theorem 2.1 we choose $u_k \in \mathcal{H}_\omega$ such that $u_k \searrow u$. By a classical Perron type argument it follows that $\int_{\{u_k+tv>P(u_k+tv)\}} \omega^n_{P(u_k+tv)} = 0$ (see [**3**, Corollary 9.2]). This is equivalent to $\int_X (u_k + tv - P(u_k + tv))\omega^n_{P(u_k+tv)} = 0$. As $u_k + t\sup_X |v| \geq P(u_k + tv)$, Proposition 2.11 allows to take the limit $k \to \infty$ and conclude that

$$\int_X (u + tv - P(u + tv))\omega^n_{P(u+tv)} = 0,$$

which is equivalent to $\int_{\{u+tv>P(u+tv)\}} \omega^n_{P(u+tv)} = 0$. □

4.7. Properness of the K–energy and existence of KE metrics

Given $u \in \mathcal{H}_\omega$, the average of the scalar curvature of the metric ω_u is independent of u, as by (5.9) and integration by parts yields

$$\bar{S} = \frac{1}{V}\int_X S_{\omega_u}\omega_u^n = \frac{n}{V}\int_X \text{Ric}\ \omega_u \wedge \omega_u^{n-1} = \frac{n}{V}\int_X \text{Ric}\ \omega \wedge \omega^{n-1}. \tag{4.45}$$

Next we introduce Mabuchi's K–energy functional $\mathcal{K} : \mathcal{H}_\omega \to \mathbb{R}$ [**88**]:

$$\mathcal{K}(u) = \frac{1}{V}\int_X [\log\Big(\frac{\omega_u^n}{\omega^n}\Big)\omega_u^n - u\sum_{j=0}^{n-1}\text{Ric}\ \omega \wedge \omega_u^j \wedge \omega^{n-j-1}] + \bar{S}I(u). \tag{4.46}$$

The reason behind this specific definition is the following variational formula, which shows that critical points of the K–energy are exactly the *constant scalar curvature Kähler* (csck) metrics:

PROPOSITION 4.34. *For a smooth curve* $(0,1) \ni t \to u_t \in \mathcal{H}_\omega$ *we have*

$$\frac{d}{dt}\mathcal{K}(u_t) = \frac{1}{V}\int_X \dot{u}_t(\bar{S} - S_{\omega_{u_t}})\omega^n_{u_t}.$$

PROOF. By straightforward calculations we arrive at the identities:

$$\frac{d}{dt}\Big[\log\Big(\frac{\omega^n_{u_t}}{\omega^n}\Big)\omega^n_{u_t}\Big] = \frac{1}{2}\Delta^{\omega_{u_t}}\dot{u}_t\omega^n_{u_t} + n\log\Big(\frac{\omega^n_{u_t}}{\omega^n}\Big)i\partial\bar{\partial}\dot{u}_t \wedge \omega^{n-1}_{u_t},$$

$$\frac{d}{dt}\Big[u_t\sum_{j=0}^{n-1}\text{Ric}\ \omega\wedge\omega^j_{u_t}\wedge\omega^{n-j-1}\Big] = \sum_{j=0}^{n-1}(\dot{u}_t\omega^j_{u_t} + ju_t\cdot i\partial\bar{\partial}\dot{u}_t\wedge\omega^{j-1}_{u_t})\wedge\text{Ric}\ \omega\wedge\omega^{n-j-1}.$$

Consequently, integration by parts and (5.9) gives:

$$\frac{d}{dt}\int_X \log\Big(\frac{\omega^n_{u_t}}{\omega^n}\Big)\omega^n_{u_t} = n\int_X \dot{u}_t\Big(\text{Ric}\ \omega - \text{Ric}\ \omega_u\Big)\wedge\omega^{n-1}_{u_t},$$

$$\frac{d}{dt}\int_X u_t\sum_{j=0}^{n-1}\text{Ric}\ \omega\wedge\omega^j_{u_t}\wedge\omega^{n-j-1} = n\int_X \dot{u}_t\text{Ric}\ \omega\wedge\omega^{n-1}_{u_t}.$$

The desired formula now follows after differentiating $t \to \mathcal{K}(u_t)$, and using the last two identities together with Lemma 3.37. □

Trivially, KE metrics are csck. By the next result, in case $c_1(X)$ is a multiple of $[\omega]$, the reverse is also true:

LEMMA 4.35. *Suppose* $c_1(X) = \lambda[\omega]$. *Then* $u \in \mathcal{H}_\omega$ *is a csck potential if and only if it is a KE potential.*

PROOF. Suppose u is a csck potential, i.e., $S_{\omega_u} = \bar{S}$. Since $c_1(X) = \lambda[\omega]$, by definition of the Ricci potential (see (4.4)) and invariance of $\bar{S}$ (see (4.45)) we have

$$\bar{S} = S_{\omega_u} = n\lambda + \frac{1}{2}\Delta^{\omega_u} f_u = \overline{S} + \frac{1}{2}\Delta^{\omega_u} f_u,$$

hence $\Delta^{\omega_u} f_u = 0$. The condition $\int_X e^{f_u}\omega^n = V$ gives $f_u = 0$, i.e., ω_u is a KE metric. □

By this last lemma, in case (X, ω) is Fano, the critical points of $\mathcal{K}$ are exactly the KE potentials, hence the K–energy plays a role similar to the $\mathcal{F}$ functional. Developing this analogy further, our main result in this section parallels Theorem 4.11, giving another characterization of existence of KE metrics, confirming a related conjecture of Tian (see [**107**, Conjecture 7.12],[**108**, p. 127]):

THEOREM 4.36. ([**58**, Theorem 2.4]) *Suppose (X, ω) is Fano with $c_1(X) = [\omega]$, and set $G := \text{Aut}(X, J)_0$. The following are equivalent:*
(i) there exists a KE metric in $\mathcal{H}$.
(ii) For some $C, D > 0$ the following holds:

$$\mathcal{K}(u) \geq Cd_{1,G}(G0, Gu) - D, \quad u \in \mathcal{H}_\omega \cap I^{-1}(0). \tag{4.47}$$

(iii) For some $C, D > 0$ the following holds:

$$\mathcal{K}(u) \geq CJ_G(Gu) - D, \quad u \in \mathcal{H}_\omega \cap I^{-1}(0). \tag{4.48}$$

For the resolution of other closely related conjectures we refer to [**58**]. It is possible to give a proof for this theorem by verifying the conditions of the existence/properness principle (Theorem 4.7) directly, the same way as we did with Theorem 4.11. Instead of doing this, we will rely on the special relationship between the K–energy and the $\mathcal{F}$ functional (see Proposition 4.41 below).

First we have to show that $\mathcal{K}$ is d_1–lsc, and (4.46) gives the d_1–lsc extension of this functional to $\mathcal{E}_1(X, \omega)$. This will be done in a series of lemmas and propositions:

LEMMA 4.37. [**58**, Lemma 5.23] *Suppose α is a smooth closed $(1,1)$-form on X. The functional $I_\alpha : \mathcal{H}_\omega \to \mathbb{R}$ given by*

$$I_\alpha(u) = \sum_{j=0}^{n-1} \int_X u\alpha \wedge \omega_u \wedge \omega^{n-1-j}$$

is d_1–continuous, and extends to a d_1–continuous functional $I_\alpha : \mathcal{E}_1(X, \omega) \to \mathbb{R}$. Additionally, I_α is bounded on d_1–bounded subsets of $\mathcal{E}_1(X, \omega)$.

PROOF. Let $u_k \in \mathcal{H}_\omega$ and $u \in \mathcal{E}_1(X, \omega)$ be such that $d_1(u_k, u) \to 0$. An argument similar to that yielding (3.58) shows that

$$I_\alpha(u_l) - I_\alpha(u_k) = \sum_{j=0}^{n-1} \int_X (u_l - u_k)\alpha \wedge \omega_{u_l}^j \wedge \omega_{u_k}^{n-1-j}.$$

For some $D > 0$ we have $-D\omega \leq \alpha \leq D\omega$. Thus, $\omega_{(u_l+u_k)/4} = \omega/2 + \omega_{u_l}/4 + \omega_{u_k}/4$ and for some $C > 0$ we can write

$$\big|I_\alpha(u_l) - I_\alpha(u_k)\big| \leq C \int_X |u_l - u_k|\omega^n_{(u_l+u_k)/4}. \tag{4.49}$$

Using Lemma 3.33 and the triangle inequality, $d_1(0,(u_l+u_k)/4)$ is bounded. By Corollary 3.50, we obtain that $\{I_\alpha(u_k)\}_k$ is a Cauchy sequence, showing that I_α is d_1–continuous and it that extends d_1–continuously to $\mathcal{E}_1(X,\omega)$.

To argue d_1–boundedness of I_α, we turn again to (4.49) (with $u_k = 0$). By this estimate and Corollary 3.50, it is enough to show that if $d_1(0,u)$ is bounded then so is $d_1(0,u/4)$. This is a consequence of Lemma 3.33. □

As we will see shortly, the entropy functional $u \to \int_X \log\big(\frac{\omega_u^n}{\omega^n}\big)\omega_u^n$ is only d_1–lsc. In fact, this functional is already lsc with respect to weak convergence of measures, as it follows from our discussion below and the next proposition.

Suppose ν,μ are Borel probability measures on X. If ν is absolutely continuous with respect to μ then the entropy of ν with respect to μ is $\text{Ent}(\mu,\nu) = \int_X \log\big(\frac{\nu}{\mu}\big)\frac{\nu}{\mu}\mu$, otherwise $\text{Ent}(\mu,\nu)=\infty$.

The next well known result follows from the classical Jensen inequality:

LEMMA 4.38. *Suppose ν,μ are Borel probability measures on X. Then we have $\text{Ent}(\mu,\nu)\geq 0$, and equality holds if and only if $\nu=\mu$.*

PROOF. $\text{Ent}(\mu,\nu)\geq 0$ follows from an application of Jensen's inequality to the convex weight $\rho(x) := x\log x$. As ρ is strictly convex on $\mathbb{R}^+$, the proof of Jensen's inequality implies that $\text{Ent}(\mu,\nu)\geq 0$ if and only if $\frac{\mu}{\nu}=1$ (see [**97**, Chapter 3, Theorem 3.3]). □

Let us recall the following classical formula for the entropy of two measures:

PROPOSITION 4.39. *Suppose μ,ν are probability Borel measures on X. Then the following holds:*

$$\text{Ent}(\mu,\nu) = \sup_{f\in B(X)}\Big(\int_X f\nu - \log\int_X e^f\mu\Big), \tag{4.50}$$

where $B(X)$ is the set of bounded Borel measurable functions on X.

PROOF. In case ν is not absolutely continuous with respect to μ then there exists a Borel set $M\subset X$ with $\mu(M)=0$ but $\nu(M)>0$. Then we trivially have that $\int_X c\mathbb{1}_M\nu - \log\int_X e^{c\mathbb{1}_M}\mu = \int_X c\mathbb{1}_M\nu - \log\int_X e^0\mu = c\nu(M)$ for any $c>0$ and consequently,

$$\sup_{f\in B(X)}\Big(\int_X f\nu - \log\int_X e^f\mu\Big) = \infty,$$

which is equal to $\text{Ent}(\mu,\nu)$ by definition.

We assume now that ν is absolutely continuous with respect to μ, i.e. $\nu = g\mu$ for some non–negative Borel measurable function g. To conclude (4.50) we need to show that

$$\int_X g\log g\mu = \sup_{f\in B(X)}\Big(\int_X f\mu - \log\int_X e^f\mu\Big). \tag{4.51}$$

By choosing $f_k=\log g_k := \log(\min(\max(g,1/k),k))$, $k\in\mathbb{N}$. We get that the right hand side of (4.51) is greater then $\int_X g\log g_k\mu - \log\int_X g_k\mu$. Letting $k\to\infty$, we conclude that the right hand side of (4.51) is greater then the left hand side.

For the other direction, we need to argue that $\int_X g \log g\mu \geq \int_X fg\mu - \log \int_X e^f \mu$ for any $f \in B(X)$. For this it is enough to invoke Jensen's inequality:

$$\log \int_X e^f \mu \geq \log \int_{\{g>0\}} \frac{e^f}{g} g\mu \geq \int_{\{g>0\}} (f - \log g) g\mu = \int_X (f - \log g) g\mu.$$

□

As the supremum of continuous functionals is lsc, it follows that $\nu \to \mathrm{Ent}(\mu, \nu)$ is lsc with respect to weak convergence of Borel measures.

Theorem 3.46(ii) implies that for any $u_k, u \in \mathcal{E}_1(X, \omega)$ we have that $d_1(u_k, u) \to 0$ implies $\omega^n_{u_k} \to \omega^n_u$ weakly. We arrive at the following important corollary:

COROLLARY 4.40. *The functional $\mathcal{E}_1(X, \omega) \ni u \to \mathrm{Ent}\big(\frac{1}{V}\omega^n, \frac{1}{V}\omega^n_u\big) \in \mathbb{R}$ is d_1–lsc.*

Comparing with (4.46), we observe that for $u \in \mathcal{H}_\omega$ we actually have

$$\mathcal{K}(u) = \mathrm{Ent}\Big(\frac{1}{V}\omega^n, \frac{1}{V}\omega^n_u\Big) - \frac{1}{V} I_{\mathrm{Ric}\,\omega}(u) + \bar{S} I(u). \tag{4.52}$$

This observation together with Proposition 3.40, Lemma 4.37 and Corollary 4.40 allows to conclude that $\mathcal{K}$ is d_1–lsc on $\mathcal{H}_\omega$ and it extends to $\mathcal{E}_1(X, \omega)$ in a d_1–lsc manner, using the formula of (4.52).

Lastly, before we prove Theorem 4.36, we provide a precise inequality between the K–energy and $\mathcal{F}$ functional, and we point out the relationship between the minimizers of these functionals:

PROPOSITION 4.41 ([**5**]). *Suppose (X, ω) is a Fano manifold with $c_1(X) = [\omega]$. For any $u \in \mathcal{E}_1(X, \omega)$ we have*

$$\mathcal{F}(u) \leq \mathcal{K}(u) - \frac{1}{V} \int_X f_0 \omega^n. \tag{4.53}$$

Moreover, for $u \in \mathcal{E}_1(X, \omega)$ the following are equivalent:
(i) $\mathcal{F}(u) = \mathcal{K}(u) - \frac{1}{V}\int_X f_0 \omega^n$.
(ii) u minimizes $\mathcal{F}$.
(iii) u minimizes $\mathcal{K}$.
(iv) u is a smooth KE potential.

Consequently, the minimizers of $\mathcal{K}$ and $\mathcal{F}$ on $\mathcal{E}_1(X, \omega)$ are the same and coincide with the set of smooth KE potentials.

PROOF. Let $u \in \mathcal{H}_\omega$. As both $\mathcal{K}$ and $\mathcal{F}$ are constant invariant, we can assume that $\int_X e^{-u+f_0}\omega^n = V$ and note the following identity:

$$\mathrm{Ent}\Big(\frac{1}{V} e^{-u+f_0}\omega^n, \frac{1}{V}\omega_u\Big) = \mathrm{Ent}\Big(\frac{1}{V} e^{f_0}\omega^n, \frac{1}{V}\omega_u\Big) + \frac{1}{V}\int_X u\omega^n_u.$$

By the formula of the next lemma, we can write that

$$\begin{aligned}\mathcal{K}(u) - \frac{1}{V}\int_X f_0\omega^n &= \mathrm{Ent}\Big(\frac{1}{V} e^{-u+f_0}\omega^n, \frac{1}{V}\omega_u\Big) - I(u)\\ &= \mathrm{Ent}\Big(\frac{1}{V} e^{-u+f_0}\omega^n, \frac{1}{V}\omega_u\Big) + \mathcal{F}(u).\end{aligned}$$

By Lemma 4.38 we have $\mathrm{Ent}\big(\frac{1}{V} e^{-u+f_0}\omega^n, \frac{1}{V}\omega_u\big) \geq 0$, hence (4.53) follows.

Moreover, by this same lemma, equality holds in (4.53) if and only if $\omega_u = e^{-u+f_0}\omega^n$, which is equivalent with u being a smooth KE potential (see Theorem 4.31). By Theorem 4.30 it immediately follows that (i), (ii) and (iv) are equivalent.

If (iv) holds, then u minimizes $\mathcal{F}$, and $\mathcal{F}(u) = \mathcal{K}(u) - \frac{1}{V}\int_X f_0\omega^n$, by (i). Consequently u minimizes $\mathcal{K}$ as well, hence (iii) holds.

Suppose (iii) holds. By Lemma 4.43(ii) below, there exists $v \in \text{PSH}(X,\omega)\cap L^\infty$ such that

$$\mathcal{K}(v) - \frac{1}{V}\int_X f_0\omega^n \leq \mathcal{F}(u) \leq \mathcal{K}(u) - \frac{1}{V}\int_X f_0\omega^n.$$

As u minimizes $\mathcal{K}$, it follows that the inequalities above are actually equalities, hence (i) holds. □

As pointed out in the above argument, we need a special expression for the K-energy:

LEMMA 4.42. *Suppose (X,ω) is a Fano manifold with $c_1(X) = [\omega]$. For $u \in \mathcal{H}_\omega$ we have*

$$\mathcal{K}(u) = \text{Ent}\Big(\frac{1}{V}e^{f_0}\omega^n, \frac{1}{V}\omega_u^n\Big) - I(u) + \frac{1}{V}\int_X u\omega_u^n + \frac{1}{V}\int_X f_0\omega^n.$$

PROOF. We start by deriving an alternative formula for $I_{\text{Ric}\,\omega}$:

$$\begin{aligned}
I_{\text{Ric}\,\omega}(u) &= \sum_{j=0}^{n-1}\int_X u\,\text{Ric}\,\omega\wedge\omega_u^j\wedge\omega^{n-1-j} = \sum_{j=0}^{n-1}\int_X u(\omega + i\partial\bar\partial f_0)\wedge\omega_u\wedge\omega^{n-1-j}\\
&= \sum_{j=0}^{n-1}\int_X u\omega_u^j\wedge\omega^{n-j} + \sum_{j=0}^{n-1}\int_X f_0 i\partial\bar\partial u\wedge\omega_u\wedge\omega^{n-1-j}\\
&= \sum_{j=0}^{n-1}\int_X u\omega_u\wedge\omega^{n-j} + \sum_{j=0}^{n-1}\int_X f_0\omega_u^{j+1}\wedge\omega^{n-1-j} - \sum_{j=0}^{n-1}\int_X f_0\omega_u^j\wedge\omega^{n-j}\\
&= \sum_{j=0}^{n-1}\int_X u\omega_u\wedge\omega^{n-j} + \int_X f_0\omega_u^n - \int_X f_0\omega^n.
\end{aligned}$$

Since $c_1(X) = [\omega]$, we have $\bar S = n$ (see (4.45)), so by the above we conclude that

$$\begin{aligned}
-\frac{1}{V}I_{\text{Ric}\,\omega}(u) + \bar S I(u) &= -\frac{1}{V}I_{\text{Ric}\,\omega}(u) + nI(u)\\
&= -I(u) + \frac{1}{V}\Big(\int_X u\omega_u^n - \int_X f_0\omega_u^n + \int_X f_0\omega^n\Big).
\end{aligned}$$

We note that $\text{Ent}(\frac{1}{V}\omega^n, \frac{1}{V}\omega_u^n) = \text{Ent}(\frac{1}{V}e^{f_0}\omega^n, \frac{1}{V}\omega_u^n) + \frac{1}{V}\int_X f_0\omega_u^n$. Adding this identity to the above, and comparing with (4.52) finishes the proof. □

In the next lemma, we will make us of the inverse Ricci operator, introduced in [**94**] in connection with the so-called Ricci iteration. Given a potential $u \in \mathcal{E}_1(X,\omega)$, Corollary 4.16 and Kolodziej's estimate [**82, 83**] give another potential $\text{Ric}^{-1}(u) \in \text{PSH}(X,\omega)\cap L^\infty$, unique up to a constant, such that

$$\omega^n_{\text{Ric}^{-1}(u)} = \frac{V}{\int_X e^{-u+f_0}\omega^n}e^{-u+f_0}\omega^n.$$

In case $u \in \mathcal{H}_\omega$, we notice that $\text{Ric}\,\omega_{\text{Ric}^{-1}u} = \omega_u$, motivating the terminology.

By the next lemma, the inverse Ricci operator decreases the $\mathcal{F}$ functional and sheds further light on the intimate relationship between $\mathcal{K}$ and $\mathcal{F}$:

LEMMA 4.43 ([**94**]). *Suppose $u \in \mathcal{E}_1(X,\omega)$. Then the following hold:*
(i) $\mathcal{F}(\mathrm{Ric}^{-1}(u)) \leq \mathcal{F}(u)$.
(ii) $\mathcal{K}(\mathrm{Ric}^{-1}(u)) - \frac{1}{V}\int_X f_0\omega^n \leq \mathcal{F}(u)$.
(iii) $\mathcal{K}(\mathrm{Ric}^{-1}(u)) \leq \mathcal{K}(u)$.

PROOF. First we argue (ii). We introduce $v := \mathrm{Ric}^{-1}(u)$. As both $\mathcal{F}$ and $\mathcal{K}$ are constant invariant, we can assume that $\int_X e^{-u+f_0}\omega^n = V$. By the previous lemma, notice that we have

$$\begin{aligned}\mathcal{K}(v) - \frac{1}{V}\int_X f_0\omega^n &= \mathrm{Ent}\Big(\frac{1}{V}e^{f_0}\omega^n, \frac{1}{V}\omega_v^n\Big) - I(v) + \frac{1}{V}\int_X v\omega_v^n \\ &= \mathrm{Ent}\Big(\frac{1}{V}e^{f_0}\omega^n, \frac{1}{V}e^{-u+f_0}\omega^n\Big) - I(v) + \frac{1}{V}\int_X v\omega_v^n \\ &= -I(v) + \frac{1}{V}\int_X (v-u)\omega_v^n \leq -I(u) = \mathcal{F}(u),\end{aligned}$$

where in the penultimate estimate we have used (3.59).

Lastly, (i) and (iii) follow from (ii) and (4.53). □

We now prove the main compactness theorem of the space $(\mathcal{E}_1(X,\omega), d_1)$:

THEOREM 4.44. *Let $u_j \in \mathcal{E}_1(X,\omega)$ be a d_1-bounded sequence and we assume that $\mathrm{Ent}\big(\frac{1}{V}\omega^n, \frac{1}{V}\omega_{u_j}^n\big)$ is also bounded. Then $\{u_j\}_j$ contains a d_1-convergent subsequence.*

PROOF. As a consequence of Corollary 4.16, for any $p > 0$ there exists $C(p) > 0$ such that that $\int_X e^{-pu_j}\omega^n \leq C$. Since $|\sup_X u_j|$ is bounded, we get that

$$\int_X e^{|pu_j|}\omega^n \leq C. \tag{4.54}$$

Consider $\phi, \psi : \mathbb{R} \to \mathbb{R}^+$ given by

$$\phi(|t|) = \frac{(|t|+1)\log(|t|+1) - |t|}{\log 2} \quad \text{and} \quad \psi(t) = \frac{e^{|t|} - |t| - 1}{e-1}.$$

An elementary calculation verifies that both of these functions are normalized Young weights and $\phi^* = \psi$ (in the sense of (1.1)).

Since $\mathrm{Ent}\big(\frac{1}{V}\omega^n, \frac{1}{V}\omega_{u_j}^n\big)$ is bounded, so is $\int_X \phi(\omega_{u_j}^n/\omega^n)\omega^n$. As ϕ is convex and $\phi(0) = 0$, for some $D \in (0,1)$ we get that

$$\int_X \phi\Big(D\frac{\omega_{u_j}^n}{\omega^n}\Big)\frac{1}{V}\omega^n \leq 1. \tag{4.55}$$

From d_1-boundedness we have that $|\sup_X u_j|$ and $I(u_j)$ are bounded. By Lemma 4.13, after possibly taking a subsequence, we can find $u \in \mathcal{E}_1(X,\omega)$ such that $\int_X |u_j - u|\omega^n \to 0$.

Recalling the definition of Orlicz norms from (1.3), we can use the Hölder inequality (1.6) to deduce that

(4.56)
$$\frac{1}{V}\int_X |u_j - u|\omega_{u_j}^n = \frac{1}{V}\int_X |u_j - u|\frac{\omega_{u_j}^n}{\omega^n}\omega^n \leq \Big\|\frac{1}{D}(u_j - u)\Big\|_{\psi,\frac{1}{V}\omega^n}\Big\|D\frac{\omega_{u_j}^n}{\omega^n}\Big\|_{\phi,\frac{1}{V}\omega^n}.$$

From (4.55) it follows that $\|D\omega^n_{u_j}/\omega^n\|_{\phi,\frac{1}{V}\omega^n} \leq 1$. Since $\psi(t) \leq t^2 e^{|t|}$, for any $r > 0$ we can write

$$\int_X \psi\Big(\frac{r}{D}(u_j - u)\Big)\omega^n \leq \int_X \frac{r^2}{D^2}|u_j - u|^2 e^{\frac{r}{D}|u_j-u|}\omega^n$$
$$\leq \frac{r^2}{D^2}\|(u_j - u)^2\|_{L^3(\omega^n)}\big\|e^{\frac{r}{D}|u|}\big\|_{L^3(\omega^n)}\big\|e^{\frac{r}{D}|u_j|}\big\|_{L^3(\omega^n)}.$$

The last two terms on right hand side are bounded by (4.54). As $\int_X |u_j - u|\omega^n \to 0$ it follows that $\int_X |u_j - u|^6\omega^n \to 0$, hence $\lim_j \|D^{-1}(u_j - u)\|_{\psi,\frac{1}{V}\omega^n} \leq r$.

Using (4.56), we conclude that $\int_X |u_j - u|\omega^n_{u_j} \to 0$. As a result, (3.59) (or rather the extension of this inequality to $\mathcal{E}_1(X,\omega)$) gives that $\liminf_j I(u_j) \geq I(u)$. Together with Corollary 4.14 we obtain that $\lim_j I(u_j) = I(u)$. Since additionally $\int_X |u_j - u|\omega^n \to 0$, Theorem 3.46(i) implies that $d_1(u_j, u) \to 0$, finishing the proof. □

Lastly, we prove our main theorem:

PROOF OF THEOREM 4.36. The equivalence between (ii) and (iii) follows from Lemma 4.4.

Suppose (i) holds. Then Proposition 4.41 and Theorem 4.11 implies properness of $\mathcal{K}$, giving (ii).

Now assume that (ii) holds. Then $\mathcal{K}$ is bounded from below and we can find $u_j \in \mathcal{E}_1(X,\omega)$ that is d_1–bounded and $\lim_j \mathcal{K}(u_j) = \inf_{v\in\mathcal{E}_1(X,\omega)} \mathcal{K}(v)$.

In particular, $\{\mathcal{K}(u_j)\}_j$ is bounded. Recalling (4.52), and the fact that both I and $I_{\mathrm{Ric}\,\omega}$ are bounded on d_1–bounded sets (see Proposition 3.40 and Lemma 4.37), it follows that $\mathrm{Ent}\big(\frac{1}{V}\omega^n, \frac{1}{V}\omega^n_{u_j}\big)$ is bounded as well.

By the previous compactness theorem, after possibly passing to a subsequence, we have $d_1(u_j, u) \to 0$ for some $u \in \mathcal{E}_1(X,\omega)$. Since $\mathcal{K}$ is d_1–lsc, we immediately obtain that $\mathcal{K}(u) = \inf_{v\in\mathcal{E}_1(X,\omega)} \mathcal{K}(v)$, i.e., u minimizes $\mathcal{K}$.

By the equivalence between (iii) and (iv) in Theorem 4.41, u is in fact a smooth KE potential. □

Brief historical remarks. Motivated by results in conformal geometry, the relationship between energy properness and existence of canonical metrics in Kähler geometry goes back to the work of Tian and collaborators in the nineties [**105**, **107**]. Numerous conjectures were proposed during this time, a number of which where adressed in the case of Fano manifolds without vectorfields [**65**, **106**, **111**]. For general Fano manifolds, all the remaining conjectures where addressed in [**58**], and we refer to this work for more details.

The sharpest form of the energy properness condition was identified in [**90**] and was later adopted in the literature, including in the present work.

Regarding general Kähler manifolds, in [**58**] the equivalence between energy properness and existence of consant scalar curvature (csck) metrics is linked to a regularity problem for fourth order PDE's. In case a csck metric exists, this regularity conjecture was confirmed in [**14**], showing that on such manifolds the K-energy is indeed proper, partially generalizing Theorem 4.36.

In this chapter we tackled problems related to energy properness directly via the existence/properness principle of [**58**] (Theorem 4.7). The use of geodesic convexity in this context was initially proposed by X.X. Chen, however he advocated for the use of the L^2 Mabuchi geometry instead [**35**].

One of the advantages of our method (that uses pluripotential theory predominantly) over previous approaches in the literature is its adaptability to Kähler structures with mild singularities [**50**, **63**]. For generalizations in other directions, as well as a more thourough overview of the vast related literature we refer to [**58**, **96**].

Part 5. Appendix

Appendix A. Basic formulas of Kähler geometry

In this section we recall the most important facts about Kähler manifolds. Our minimalist approach follows [**22**] closely, and we refer to [**61**],[**114**] and [**103**] for more exhaustive treatments.

Suppose (X, J) is compact connected complex manifold with holomorphic structure J. A Hermitian structure (X, h) is the complex analog of a Riemannian structure, i.e., a smooth choice of $J-$compatible Hermitian metrics on each fiber of TX. In local complex coordinates h can be expressed as

$$h = g_{j\bar{k}} dz_j \otimes d\bar{z}_k,$$

where we have used the Einstein summation. The real part of h induces a Riemannian structure:

$$\langle \cdot, \cdot \rangle = g_{j\bar{k}} dz_j \otimes d\bar{z}_k + g_{k\bar{j}} d\bar{z}_j \otimes dz_k.$$

This metric is compatible with J in the sense that $\langle \cdot, \cdot \rangle = \langle J(\cdot), J(\cdot) \rangle$.

By $T_{\mathbb{C}}X = TX \otimes \mathbb{C}$ we denote the complexification of TX and extend J and $\langle \cdot, \cdot \rangle$ to $T_{\mathbb{C}}X$ in a $\mathbb{C}$–linear way. In local coordinates $z_j = x_j + iy_j$ the vector fields $\partial/\partial x_j$, $\partial/\partial y_j$ span TX over $\mathbb{R}$. We also have $J\partial/\partial x_j = \partial/\partial y_j$, $J\partial/\partial y_j = -\partial/\partial x_j$ and the eigenbase of J composed of the vector fields

$$\Big\{\partial_j = \frac{\partial}{\partial z_j} = \frac{1}{2}\Big(\frac{\partial}{\partial x_j} - i\frac{\partial}{\partial y_j}\Big)\Big\}_{\{j=1,\dots,n\}}, \ \Big\{\partial_{\bar{j}} = \frac{\partial}{\partial \bar{z}_j} = \frac{1}{2}\Big(\frac{\partial}{\partial x_j} + i\frac{\partial}{\partial y_j}\Big)\Big\}_{\{j=1,\dots,n\}}$$

that span $T^{(1,0)}_{\mathbb{C}}X$ and $T^{(0,1)}_{\mathbb{C}}X$ respectively over $\mathbb{C}$. We also have the identities

$$J\partial_j = i\partial_j \ \text{ and } \ J\partial_{\bar{j}} = -i\partial_{\bar{j}}.$$

Compared to the literature, we chose to multiply by a factor of two in the definition of $\langle \cdot, \cdot \rangle$ so that the following formula holds:

$$h(\partial_j, \partial_{\bar{k}}) = \langle \partial_j, \partial_{\bar{k}} \rangle = g_{j\bar{k}}.$$

Consequently, for any real vector field $Y \in C^\infty(X, TX)$ we have that $Y = Y^j\partial_j + \overline{Y^j}\partial_{\bar{j}}$ and $\|Y\|^2 = \langle Y, Y \rangle = 2Y^j\overline{Y^k}g_{j\bar{k}}$.

We will be also interested in the imaginary part of h:

$$\omega = -2\mathrm{Im}\ h = ig_{j\bar{k}} dz_j \wedge d\bar{z}_k. \tag{5.1}$$

It is straightforward to see that

$$\omega(\cdot, \cdot) = \langle J(\cdot), \cdot \rangle. \tag{5.2}$$

We say that ω is a *Kähler form* if $d\omega = 0$. In this case (X, ω) is called a *Kähler manifold* and we fix such a manifold for the remainder of this section.

By the Poincaré lemma [**114**, Lemma II.2.15], the Kähler condition implies that for any $x \in X$ there exists an open neighborhood $U \ni x$ and a *local Kähler potential* $g \in C^\infty(\overline{U})$ satisfying

$$\omega|_U = i\partial\bar{\partial} g = i\frac{\partial^2 g}{\partial z_j \partial \bar{z}_k} dz_j \wedge d\bar{z}_k.$$

From this and (5.1) it follows that $g_{j\bar{k}} = \partial^2 g/\partial z_j \partial\bar{z}_k$, hence in Kähler geometry, when doing calculations in local coordinates, one can interchange indices of the metric with its partial derivatives. We will do this quite frequently in our study.

The Christoffel symbols. Given a hermitian structure (X, J, h), it is possible to derive the following identity relating ω, the Riemannian metric $\langle \cdot, \cdot \rangle$ and its Levi–Civita connection:

$$3d\omega(W, Y, Z) - 3d\omega(W, JY, JZ) = 2\langle (\nabla_W J)Y, Z\rangle,$$

where W, Y, Z are smooth vector fields on X. Using the Kähler condition, this identity gives that

$$\nabla_{(\cdot)} J = 0.$$

As a consequence of this we obtain the following formulas for Christoffel symbols of $\nabla_{(\cdot)}(\cdot)$ in holomorphic local coordinates:

$$\begin{aligned} \nabla_{\partial_j} \partial_{\bar{k}} &= \Gamma^{l}_{j\bar{k}} \partial_l + \Gamma^{\bar{l}}_{j\bar{k}} \partial_{\bar{l}} = 0, \ \nabla_{\partial_{\bar{j}}} \partial_k = \Gamma^{l}_{\bar{j}k} \partial_l + \Gamma^{\bar{l}}_{\bar{j}k} \partial_{\bar{l}} = 0, \\ \nabla_{\partial_j} \partial_k &= \Gamma^{l}_{jk} \partial_l = g^{l\bar{h}} \partial_j g_{k\bar{h}} \partial_l = g^{l\bar{h}} g_{jk\bar{h}} \partial_l, \\ \nabla_{\partial_{\bar{j}}} \partial_{\bar{k}} &= \Gamma^{\bar{l}}_{\bar{j}\bar{k}} \partial_{\bar{l}} = g^{h\bar{l}} \partial_{\bar{j}} g_{h\bar{k}} \partial_{\bar{l}} = g^{h\bar{l}} g_{h\bar{j}\bar{k}} \partial_{\bar{l}}. \end{aligned} \tag{5.3}$$

More concretely, to derive the above formulas we used that :
(i) the Levi–Civita connection is torsion free, i.e. $\Gamma^{j}_{l\bar{k}} = \Gamma^{j}_{\bar{k}l}$, $\Gamma^{j}_{lk} = \Gamma^{j}_{kl}$, etc.
(ii) the Levi–Civita connection is real, i.e., $\overline{\Gamma^{j}_{l\bar{k}}} = \Gamma^{\bar{j}}_{\bar{l}k}$, $\overline{\Gamma^{j}_{lk}} = \Gamma^{\bar{j}}_{\bar{l}\bar{k}}$, etc.
(iii) $\nabla_{(\cdot)} J = 0$, i.e., $i\nabla_{\partial_j} \partial_k = \nabla_{\partial_j} J\partial_k = J\nabla_{\partial_j} \partial_k$, $-i\nabla_{\partial_j} \partial_{\bar{k}} = \nabla_{\partial_j} J\partial_{\bar{k}} = J\nabla_{\partial_j} \partial_{\bar{k}}$.
(iv) the product rule, $g_{lj\bar{k}} = \partial_l \langle \partial_j, \partial_{\bar{k}} \rangle = \langle \nabla_{\partial_l} \partial_j, \partial_{\bar{k}} \rangle + \langle \partial_j, \nabla_{\partial_l} \partial_{\bar{k}} \rangle = \langle \nabla_{\partial_l} \partial_j, \partial_{\bar{k}} \rangle = \Gamma^{h}_{lj} g_{h\bar{k}}$.

The volume form, gradient, Laplacian and Hamiltonian. For Kähler geometers the volume form is equal to ω^n. This is a non–degenerate top form that is a scalar multiple of the usual volume form of Riemannian geometry.

For a smooth function $v \in C^\infty(X)$ the Kähler gradient and Laplacian are given as follows:
(5.4)

$$\nabla^\omega v = \nabla^\omega_{1,0} v + \nabla^\omega_{1,0} v = g^{j\bar{k}} v_{\bar{k}} \partial_j + g^{k\bar{j}} v_k \partial_{\bar{j}}, \quad \Delta^\omega v = 2g^{j\bar{k}} v_{j\bar{k}} = 2ni\partial\bar{\partial} v \wedge \omega^{n-1}/\omega^n.$$

The length squared of the gradient is $\langle \nabla^\omega v, \nabla^\omega v \rangle = 2ni\partial v \wedge \bar{\partial} v \wedge \omega^{n-1}/\omega^n = 2g^{j\bar{k}} v_j v_{\bar{k}}$. Also, given $u, v \in C^\infty(X)$, integration by parts gives the well known identity

$$\int_X u(\Delta^\omega v)\omega^n = -\int_X \langle \nabla^\omega u, \nabla^\omega v \rangle \omega^n.$$

Lastly, from (5.2) it follows that $dv = \langle \nabla^\omega v, \cdot \rangle = -\omega(J\nabla^\omega v, \cdot)$, hence we have the following formula between the gradient and Hamiltonian of v:

$$\nabla^\omega v = JX^\omega v. \tag{5.5}$$

The Ricci and scalar curvatures. We recall the definition of the curvature tensor. Suppose T, Y, Z, W are smooth sections of the complexified tangent bundle $TX^{\mathbb{C}}$. The Riemannian curvature tensor is introduced by the formulas:

$$R(T, Y)Z = \nabla_T \nabla_Y Z - \nabla_Y \nabla_T Z - \nabla_{[T,Y]} Z,$$

$$R(T, Y, Z, W) = \langle R(T, Y)Z, W \rangle.$$

The classical curvature identities are:

$$\begin{aligned} & R(T,Y,Z,W) = -R(Y,T,Z,W), \\ (5.6) \qquad & R(T,Y,Z,W) = R(Z,W,T,Y), \\ & R(T,Y,Z,W) + R(Z,T,Y,W) + R(Y,Z,T,W) = 0. \end{aligned}$$

From $\nabla_{(\cdot)} J = 0$ it follows that $R(T,Y)J = JR(T,Y)$, hence also

$$R(T,Y,Z,W) = R(T,Y,JZ,JW).$$

From the curvature identity $R(T,Y,Z,W) = R(Z,W,T,Y)$ we also obtain

$$R(JT,JY,Z,W) = R(T,Y,Z,W),$$

hence $R(JT,JY)Z = R(T,Y)Z$. We use this to compute the curvature in local coordinates:

$$\begin{aligned} & R(\partial_{\bar{j}}, \partial_{\bar{k}})\partial_l = R(\partial_{\bar{j}}, \partial_{\bar{k}})\partial_{\bar{l}} = R(\partial_j, \partial_k)\partial_l = R(\partial_j, \partial_k)\partial_{\bar{l}} = 0, \\ (5.7) \qquad & R(\partial_j, \partial_{\bar{k}})\partial_l = -\nabla_{\partial_{\bar{k}}} \nabla_{\partial j} \partial_l = -(\partial_{\bar{k}} \Gamma^h_{jl})\partial_h, \\ & R(\partial_j, \partial_{\bar{k}})\partial_{\bar{l}} = \nabla_{\partial_j} \nabla_{\partial_{\bar{k}}} \partial_{\bar{l}} = (\partial_j \Gamma^{\bar{h}}_{\bar{k}\bar{l}})\partial_{\bar{h}}. \end{aligned}$$

The Ricci curvature is the trace of the curvature tensor:

$$\text{Ric}(T,Y) = \text{Tr}\{Z \to R(Z,T)Y\}.$$

From the curvature identities (5.6) it follows that Ric is symmetric. Moreover, from (5.7) it also follows that $\text{Ric}(JT,JY) = \text{Ric}(T,Y)$ and in local coordinates we have:

$$\text{Ric}_{jk} = 0, \ \text{Ric}_{\bar{j}\bar{k}} = 0,$$

$$\text{Ric}_{j\bar{k}} = R(\partial_{\bar{l}}, \partial_j, \partial_{\bar{k}}, \partial_m) g^{m\bar{l}} = -\partial_j \Gamma^{\bar{l}}_{\bar{l}\bar{k}} = -\partial_j (g^{h\bar{l}} g_{\bar{k}h\bar{l}}) = -\partial_j \partial_{\bar{k}} \log(\det(g_{p\bar{q}})).$$

Similar to the contruction of the Kähler form (5.1) we consider the Ricci form:

$$(5.8) \qquad \text{Ric}\ \omega = -2\text{Im Ric} = i\text{Ric}_{j\bar{k}} dz_j \wedge d\bar{z}_k = -i\partial\bar{\partial}\log(\det(g_{p\bar{q}})).$$

This last formula tells us that the Ricci form is closed, and given two Kähler metrics ω, ω' on X we have

$$(5.9) \qquad \text{Ric}\ \omega - \text{Ric}\ \omega' = i\partial\bar{\partial}\log\Big(\frac{\omega'^n}{\omega^n}\Big).$$

The scalar curvature S_ω is the trace of the Ricci curvature. In local coordinates it can be expressed as

$$(5.10) \qquad S_\omega = g^{j\bar{k}} \text{Ric}_{j\bar{k}}.$$

We compute the variation of the Ricci and scalar curvatures in the next proposition:

PROPOSITION 5.1. *Suppose* $[0,1] \ni t \to \omega_t$ *is a smooth curve of Kähler metrics on* X. *Then the following holds for the variation of the Ricci and scalar curvatures:*

$$\frac{d}{dt}\text{Ric}\ \omega_t = -i\partial\bar{\partial}\text{Tr}_{\omega_t}\Big(\frac{d}{dt}\omega_t\Big),$$

$$\frac{d}{dt}S_{\omega_t} = -\frac{1}{2}\Delta_{\omega_t}\text{Tr}_{\omega_t}\Big(\frac{d}{dt}\omega_t\Big) - \Big\langle \text{Ric}\ \omega_t, \frac{d}{dt}\omega_t \Big\rangle_{\omega_t}.$$

PROOF. The proof of the formulas follow after we differentiate (5.8) and (5.10). □

Normal coordinates. One of the advantages of dealing with Kähler metrics is the fact that for local calculations one can choose coordinates in a convenient way:

PROPOSITION 5.2. *Suppose (X,ω) is a Kähler manifold. Given $x \in X$ there exist holomorphic local coordinates $z = (z_1, \dots, z_n)$ mapping a neighborhood of x to a neighborhood U of $0 \in \mathbb{C}^n$ such that in these coordinates ω can be expressed as:*

$$\omega|_U = idz_j \wedge d\bar{z}_j + ig_{j\bar{k}p\bar{q}} z_p \bar{z}_q dz_j \wedge d\bar{z}_k + O(|z^3|)$$

PROOF. First choose an arbitrary coordinate patch that maps x and a neighborhood of x to $0 \subset \mathbb{C}^n$ and $V \subset \mathbb{C}^n$ respectively. It is possible to apply a linear change of coordinates in $\mathbb{C}^n$ so that $\omega|_x = i\partial\bar{\partial} g(0) = idz_j \wedge d\bar{z}_j$. Now we deal with the first order coefficients by making the following local change of coordinates:

$$F_m(z) = z_m + c_{mjk} z_j z_k, \ m \in \{1, \dots, n\}.$$

An elementary calculation for $F^*\omega = i\partial\bar{\partial} g \circ F =: i\partial\bar{\partial}\tilde{g}$ yields

$$\tilde{g}_{j\bar{k}l}(0) = g_{j\bar{k}l}(0) + 2c_{kjl}.$$

Choosing $c_{kjl} = -g_{j\bar{k}l}(0)/2$, we obtain that after composing our local coordinate map with F the first order coefficients of ω vanish. Now we deal with the second order coefficients. For this we introduce a local diffeomorphism of the form

$$H_m(z) = z_m + b_{mjkl} z_j z_k z_l, \ m \in \{1, \dots, n\}.$$

Pulling back again by H, for $\tilde{g} = g \circ H$ we can write

$$\tilde{g}_{j\bar{k}ab}(0) = g_{j\bar{k}ab}(0) + 6b_{kjab}.$$

Choosing $b_{kjab} = -g_{j\bar{k}ab}(0)/6$ we get $\tilde{g}_{j\bar{k}ab}(0) = 0$. By taking conjugates and applying the Leibniz rule for derivatives we also get $\tilde{g}_{j\bar{k}\bar{a}\bar{b}}(0) = 0$, finishing the proof. ☐

It immediately follows that all Christoffel symbols of ω vanish at x after a choice of *normal coordinates* around this point, as described in the above proposition. Also, from (5.7),(5.8) and (5.10) we obtain convenient formulas for geometric quantities related to curvature:

$$R(\partial_j, \partial_{\bar{k}}, \partial_p, \partial_{\bar{q}})(x) = -g_{j\bar{k}p\bar{q}}, \quad \mathrm{Ric}_{j\bar{k}}(x) = -g_{j\bar{k}p\bar{p}}, \quad S_\omega(x) = -g_{j\bar{j}p\bar{p}}. \tag{5.11}$$

The Lichnerowicz operator. Given a complex valued function $u \in C^\infty(X)$, the Riemannian Hessian of u is computed by the well known formula $\nabla^2 u(X,Y) = Xdu(Y) - du(\nabla_X Y)$. We denote by $\mathcal{L}u$ the $(0,2)$ part of $\nabla^2 u$. In local coordinates this can be expressed as

$$\begin{aligned} \mathcal{L}u = (\mathcal{L}u)_{\bar{j}\bar{k}} d\bar{z}_j \otimes d\bar{z}_k &= (u_{\bar{j}\bar{k}} - \Gamma^{\bar{l}}_{\bar{j}\bar{k}} u_{\bar{l}}) d\bar{z}_j \otimes d\bar{z}_k \\ &= (u_{\bar{j}\bar{k}} - g^{h\bar{l}} g_{h\bar{j}\bar{k}} u_{\bar{l}}) d\bar{z}_j \otimes d\bar{z}_k. \end{aligned} \tag{5.12}$$

After switching to normal coordinates, one can easily see that $\mathcal{L}u = 0$ if and only if $(g^{j\bar{k}} u_{\bar{k}})_{\bar{l}} = 0$ for all $\bar{l} \in \{1, \dots, n\}$ which is equivalent to $\nabla^\omega_{1,0} u$ being a holomorphic vector field. $\mathcal{L}$ is the *Lichnerowicz operator* and the following formula for the self–adjoint complex operator $\mathcal{L}^*\mathcal{L}$ will be very useful for us:

PROPOSITION 5.3. $\mathcal{L}^*\mathcal{L}u = \frac{1}{4}\Delta^\omega(\Delta^\omega u) + \langle \mathrm{Ric}_\omega, i\partial\bar{\partial}u\rangle_{\omega_u} + \langle \partial S_\omega, \partial u\rangle, \ u \in C^\infty(X)$.

To clarify, in the above formula $\mathcal{L}^*$ is the Hermitian dual of $\mathcal{L}$, as will be detailed in the proof below. Also, $\langle \partial S_\omega, \partial u\rangle$ denotes the Hermitian inner product of ∂S_ω and ∂u (equaling $g^{j\bar{k}} S_{\omega j} u_{\bar{k}}$ in local coordinates).

PROOF. Suppose $x \in X$. One has the following local expression for $\langle \mathcal{L}u, \mathcal{L}v\rangle$:

$$\langle \mathcal{L}u, \mathcal{L}v\rangle = (\mathcal{L}u)_{\bar{j}\bar{k}} g^{p\bar{j}} g^{q\bar{k}} \overline{(\mathcal{L}v)_{\bar{p}\bar{q}}} = (u_{\bar{j}\bar{k}} - g^{h\bar{l}} g_{h\bar{j}\bar{k}} u_{\bar{l}}) g^{p\bar{j}} g^{q\bar{k}} (v_{pq} - g^{a\bar{b}} g_{pq\bar{b}} v_a)$$

We choose normal coordinates around x as in Proposition 5.2. Using integration by parts and (5.12) one can write

$$\mathcal{L}^* \mathcal{L}u = ((u_{\bar{j}\bar{k}} - g^{h\bar{l}} g_{h\bar{j}\bar{k}} u_{\bar{l}}) g^{p\bar{j}} g^{q\bar{k}})_{pq} - (((u_{\bar{j}\bar{k}} - g^{h\bar{l}} g_{h\bar{j}\bar{k}} u_{\bar{l}}) g^{p\bar{j}} g^{q\bar{k}}) g^{a\bar{b}} g_{pq\bar{b}})_a.$$

As we are working in normal coordinates, all terms of type $g_{abc\bar{d}}, g_{ab\bar{c}}$ vanish, hence the second term in the above sum is 0. By expanding the first term we arrive at:

$$\mathcal{L}^* \mathcal{L}u = u_{jk\bar{j}\bar{k}} - g_{ljk\bar{j}\bar{k}} u_{\bar{l}} - g_{lj\bar{j}\bar{k}} u_{k\bar{l}} - g_{lk\bar{j}\bar{k}} u_{j\bar{l}}.$$

By (5.4) and (5.11), in normal coordinates we have $4\Delta^\omega(\Delta^\omega u) = g^{a\bar{b}}(g^{c\bar{d}} u_{c\bar{d}})_{a\bar{b}} = u_{a\bar{a}c\bar{c}} - u_{c\bar{d}} g_{a\bar{a}c\bar{d}}$, $\langle \partial S_\omega, \partial u\rangle = g_{ljk\bar{j}\bar{k}} u_{\bar{l}}$ and $\langle \mathrm{Ric}_\omega, i\partial\bar{\partial}u\rangle = -g_{lj\bar{j}\bar{k}} u_{k\bar{l}}$ hence the desired identity follows. □

Appendix B. Approximation of ω–psh functions on Kähler manifolds

Given a compact Kähler manifold, we show that any ω–psh function can be approximated by a decreasing sequence of smooth Kähler potentials. Much stronger results have been derived by Demailly [**59, 60**], using sophisticated techniques. Here we will follow closely the arguments of Blocki–Kolodziej [**23**]. Our main result is the following:

THEOREM 5.4. *Let (X, ω) be a compact Kähler manifold. Given $u \in \mathrm{PSH}(X, \omega)$, there exists a decreasing sequence $\{u_k\}_k \subset \mathcal{H}_\omega$ such that $u_k \searrow u$.*

The main technical ingredient will be the following proposition, which provides an intermediate result:

PROPOSITION 5.5. *Let (X, ω) be a Kähler manifold (not necessarily compact). Let $X' \subset X$ be a relatively compact open set. If the Lelong numbers of φ are zero at any $x \in X$, then there exists $\varphi_k \in \mathrm{PSH}(X', (1 + \frac{1}{k})\omega) \cap C^\infty(X')$ such that $\varphi_k \searrow \varphi$.*

Based on this proposition we quickly give the proof of Theorem 5.4:

PROOF OF THEOREM 5.4. We can assume that $u \leq -1$, and consider the cut-offs $v_k := \max(u, -k) \in \mathrm{PSH}(X, \omega) \cap L^\infty$. An application of Proposition 5.5 to $\{v_k\}_k$ gives a sequence $w_k := \mathcal{H}_{(1+\frac{1}{k})\omega}$ such that $w_k \searrow u$ and $v_k \leq w_k \leq 0$. Introducing $u_k := \frac{1}{k} + (1 - \frac{1}{k}) w_k \in \mathcal{H}_\omega$, we see that $\{u_k\}_k$ is decreasing and $u_k \searrow u$. □

Returning to the local situation of a moment, let $v \in \mathrm{PSH}(U)$ for some open set $U \subset \mathbb{C}^n$. It is well known that the correspondence $r \to f(r) = \sup_{B(z_0, e^r)} v = \max_{\partial B(z_0, e^r)} v$ is convex for any $z_0 \in U$ (this follows from the comparison principle for the complex Monge–Ampère operator and the fact that $i\partial\bar{\partial}(\log|z - z_0|)^n = 0$ on $\mathbb{C}^n \setminus \{z_0\}$). Convexity of $r \to f(r)$ gives in particular that this map is continuous, hence so are the maps $v_\delta \in \mathrm{PSH}(U_\delta)$, defined by the formula

$$v_\delta(x) := \sup_{y \in B(x,\delta)} v(y), \tag{5.13}$$

where $U_\delta = \{x \in U,\ B(x, \delta) \subset U\}$. Lastly, as v is usc, we also obtain that $v_\delta \searrow v$.

Given $u \in \mathrm{PSH}(X,\omega)$ and $x_0 \in X$, recall the definition of the Lelong number $\mathcal{L}(u,x_0)$ from (2.19):

$$\mathcal{L}(u,x_0) := \sup\{r \geq 0 \mid u(x) \leq r\log|x| + C_r,\ \forall\, x \in U_r, \text{ for some } C_r > 0\},$$

where $U_r \subset \mathbb{C}^n$ is some coordinate neighborhood of x_0 (dependent on r) that identifies x_0 with $0 \in \mathbb{C}^n$. We now give an alternative description of $\mathcal{L}(u,x_0)$ that will be of great use in the proof of Proposition 5.5. Choose a coordinate neighborhood $U \subset \mathbb{C}^n$ of x_0, that identifies x_0 with $0 \in \mathbb{C}^n$, and a potential $f \in C^\infty(X)$ such that $i\partial\bar\partial f = \omega$ on U. Clearly $u + f \in \mathrm{PSH}(U)$, and the following is well known:

$$\mathcal{L}(u,x_0) = \lim_{r\to 0} \frac{\sup_{B(x_0,r)}(u+f)}{\log r}. \tag{5.14}$$

The limit on the right hand side is well defined as $\log r \to \sup_{B(x_0,r)}(u+f)$ is convex, hence the quotients involved are decreasing as $r \to 0$. Also, this limit is clearly independent of the choice of potential f. For an extensive treatment of Lelong numbers we refer to [**75**, Section 2.3].

Finally, we arrive at the following result which will allow to compare approximation via (5.13) using different coordinate charts:

LEMMA 5.6. *Let $U, V \subset \mathbb{C}^n$ and $F : U \to V$ a biholomorphic map. Suppose $u \in \mathrm{PSH}(U)$ has zero Lelong numbers. For $u^F_\delta := (u \circ F^{-1})_\delta \circ F$, the difference $u_\delta - u^F_\delta$ converges to zero uniformly on compact sets as $\delta \to 0$.*

PROOF. Fix $a > 1$, $r > 0$ and $z \in U$. As $\log\delta \to u_\delta(z)$ is convex, for small enough $\delta > 0$ we can write

$$0 \leq u_{a\delta}(z) - u_\delta(z) \leq \frac{\log a}{\log r - \log \delta}(u_r(z) - u_\delta(z)).$$

Consequently, as $\mathcal{L}(u,z) = 0$, (5.14) implies that $u_\delta - u_{a\delta} \to 0$ uniformly on compact sets as $\delta \to 0$.

Next we notice that

$$u^F_\delta(z) = \max_{F^{-1}(B(F(z),\delta))} u.$$

As F is a biholomorphism, for a fixed compact set $K \subset U$ it is possible to find $a > 1$ and $\delta_0 > 0$ such that for $\delta \in (0,\delta_0)$ and $z \in K$ we have

$$B(F(z),\delta) \subset F(B(z,a\delta)),\quad F(B(z,\delta)) \subset B(F(z),a\delta).$$

As a result of these containments, we get that $u^F_\delta \leq u_{a\delta}$ and $u_\delta \leq u^F_{a\delta}$ on K. Since $(u_\delta - u_{a\delta}) \to 0$ uniformly on compacts, the statement of the lemma follows. □

PROOF OF PROPOSITION 5.5. First we find $\tilde\varphi_j \in \mathrm{PSH}(X', (1+\frac{1}{j})\omega) \cap C(X')$ satisfying the requirements of the proposition.

Fix $\varepsilon > 0$. We can find a finite number of nested charts $V_\alpha \subset U_\alpha$ such that $\{V_\alpha\}_\alpha$ covers X', $\overline{V_\alpha} \subset U_\alpha$, and $i\partial\bar\partial f_\alpha = \omega$ on U_α, for some potentials $f_\alpha \in C^\infty(U_\alpha)$.

Then $\varphi_\alpha := \varphi + f_\alpha \in \mathrm{PSH}(U_\alpha)$ and by the previous lemma we have

$$\varphi_{\alpha,\delta} - \varphi_{\beta,\delta} = \varphi_{\alpha,\delta} - \varphi^F_{\alpha,\delta} + (\varphi_\alpha - \varphi_\beta)^F_\delta \to f_\alpha - f_\beta \tag{5.15}$$

locally uniformly on $U_\alpha \cap U_\beta$ as $\delta \to 0$, where F is the change of coordinates on the overlap $U_\alpha \cap U_\beta$. Let η_α be smooth on U_α such that $\eta_\alpha = 0$ on V_α and $\eta_\alpha = -1$ in a neighborhood of ∂U_α. We have $i\partial\bar\partial\eta_\alpha > -C\omega$ some $C > 0$.

We set

$$\tilde{\varphi}_\delta := \max_\alpha \Big(\varphi_{\alpha,\delta} - f_\alpha + \frac{\varepsilon\eta_\alpha}{C}\Big).$$

By (5.15), for $\delta > 0$ sufficiently small, the values of $\varphi_{\alpha,\delta} - f_\alpha + \varepsilon\eta_\alpha/C$ do not contribute to the maximum on the set $\{\eta_\alpha = -1\}$, thus $\tilde{\varphi}_\delta$ is continuous on X' and also $\tilde{\varphi}_\delta \in \mathrm{PSH}(X', (1+\varepsilon)\omega)$. It is also clear that $\tilde{\varphi}_\delta$ decreases to φ as $\delta \searrow 0$, allowing to construct $\tilde{\varphi}_j \in \mathrm{PSH}(X', (1+\frac{1}{j})\omega) \cap C(X')$ such that $\tilde{\varphi}_j \searrow \varphi$.

Next we argue that is also possible to approximate with smooth potentials. For this we only need to show that any ψ that is continuous and $(1+\varepsilon)\omega$–psh on an open neighborhood Y of $\overline{X'}$ can be approximated uniformly on X' by smooth $(1+2\varepsilon\omega)$–psh potentials.

This can be done by using classical Richberg approximation. Indeed, let $V_\alpha \subset U_\alpha$, f_α and η_α be as in the first part of the proof. Additionally, let $\rho \in C_c^\infty(B(0,1))$ be a smooth, non–negative, compactly supported and spherically invariant bump function with $\int_{\mathbb{C}^n} \rho = 1$.

We introduce the smooth mollifications $\psi_{\alpha,\delta} := (\psi + (1+\varepsilon) f_\alpha) * \rho_\delta \in \mathrm{PSH}(U_{\alpha,\delta})$ that converge uniformly to $\psi + (1+\varepsilon) f_\alpha$ on V_α, since ψ is continuous. As a result, we can apply the Richberg regularized maximum ([**61**, Corollary I.5.19 and Theorem II.5.21]) to get that

$$\psi_\delta := M_\alpha\Big(\psi_{\alpha,\delta} - (1+\varepsilon) f_\alpha + \frac{\varepsilon\eta_\alpha}{C}\Big)$$

is smooth on $\cup_\alpha V_\alpha$ (for small enough δ) and also $\psi_\delta \in \mathrm{PSH}(\cup_\alpha V_\alpha, (1+2\varepsilon)\omega)$. Consequently, $\psi_\delta \to \psi$ uniformly on $X' \subset \cup_\alpha V_\alpha$, finishing the proof. □

Appendix C. Regularity of envelopes of ω–psh functions

Suppose (X, ω) is a Kähler manifold and f is an usc function on X. Recall the definition of the envelope $P(f)$ from (2.21):

$$P(f) = \sup\{u \in \mathrm{PSH}(X,\omega) \text{ s.t. } u \le f\}. \tag{5.16}$$

As it was argued in Section 3.4, $P(f)$ is ω–psh and the purpose of this short section is to show the following regularity result:

THEOREM 5.7. *If $f \in C^\infty(X)$ then $\|P(f)\|_{C^{1,\bar{1}}} \le C(X, \omega, \|f\|_{C^{1,\bar{1}}})$.*

A bound on the $C^{1,\bar{1}}$ norm of $P(f)$ simply means a uniform bound on all mixed second order derivatives $\partial^2 P(f)/\partial z_j \bar{\partial} z_k$. Since $P(f)$ is ω-psh, this is equivalent to saying that $\Delta^\omega P(f)$ is bounded, and by the Calderon–Zygmund estimate [**76**, Chapter 9, Lemma 9.9], we automatically obtain that $P(f_1, f_2, \ldots, f_k) \in C^{1,\alpha}(X)$.

Theorem 5.7 was first proved by Berman–Demailly using methods from pluripotential theory[**15**]. Here we will follow an alternative path proposed by Berman, that uses more classical PDE techniques [**6**]. The point is to consider the following complex Monge–Ampère equation:

$$\omega_{u_\beta}^n = e^{\beta(u_\beta - f)}\omega^n. \tag{5.17}$$

As $f \in C^\infty(X)$, it follows from work of Aubin and Yau that this equation always has a unique smooth solution $u_\beta \in \mathcal{H}_\omega$ for any $\beta > 0$ [**1, 113**] (for a survey see [**75**, Theorem 14.1]). Theorem 5.7 will follow from the following regularity result:

PROPOSITION 5.8 ([6]). *The unique solutions* $\{u_\beta\}_{\beta>0}$ *of* (5.17) *satisfy the following:*
(i) $\|u_\beta - P(f)\|_{C^0} \to 0$ *as* $\beta \to \infty$.
(ii) there exists $\beta_0 > 0$ *and* $C > 0$ *such that* $\|\Delta^\omega u_\beta\|_{C^0} < C$ *for all* $\beta \geq \beta_0$.

To argue Proposition 5.8(i), first we prove the following comparison principle :

LEMMA 5.9. *Assume that* $u, v \in \mathrm{PSH}(X, \omega) \cap L^\infty$ *such that*

$$\omega_v^n \geq e^{\beta(v-f)}\omega^n \quad \text{and} \quad \omega_u^n \leq e^{\beta(u-f)}\omega^n.$$

Then $v \leq u$.

PROOF. According to the comparison principle $\int_{\{u<v\}} \omega_v^n \leq \int_{\{u<v\}} \omega_u^n$ (Proposition 2.3). Consequently, we can write that $\int_{\{u<v\}} e^{\beta(v-f)}\omega^n \leq \int_{\{u<v\}} e^{\beta(u-f)}\omega^n$. As a result, $v \leq u$ a.e. with respect to the Lebesque measure. Since u and v are ω–psh, we obtain that in fact $v \leq u$ globally on X. □

Now we argue Proposition 5.8(i):

LEMMA 5.10. *There exists* $C := C(f) > 0$ *such that the unique solutions* $\{u_\beta\}_{\beta>0}$ *of* (5.17) *satisfy the following estimate:*

$$\sup_X |u_\beta - P(f)| \leq C\frac{\log\beta}{\beta}, \ \beta > 2.$$

PROOF. Since u_β is smooth, at the point $x_0 \in X$ where the maximum of $u_\beta - f$ is attained we have $i\partial\bar\partial(u_\beta - f) \leq 0$. Consequently, (5.17) implies that at x_0 we have

$$u_\beta - f \leq \frac{1}{\beta}\log\Big(\frac{\omega_f^n}{\omega^n}\Big).$$

Denoting $C := \sup_X \log\big(\frac{\omega_f^n}{\omega^n}\big)$, we obtain that $u_\beta - \frac{C}{\beta} \leq f$, hence

$$u_\beta - P(f) \leq \frac{C}{\beta}. \tag{5.18}$$

Conversely, fix $v \in \mathcal{H}_\omega$ and $D > 0$ such that $\omega_v \geq D\omega$, and $v \leq f$. Also, choose $\delta, \varepsilon \in (0,1)$ such that

$$\delta^n D^n \geq e^{-\beta\varepsilon}. \tag{5.19}$$

As a result, $u_{\delta,\varepsilon} := (1-\delta)P(f) + \delta v - \varepsilon \leq f - \varepsilon$ and

$$\omega_{u_{\delta,\varepsilon}}^n \geq \delta^n D^n \omega^n \geq e^{-\beta\varepsilon}\omega^n \geq e^{\beta(u_{\delta,\varepsilon} - f)}\omega^n.$$

By Lemma 5.9 we obtain that $u_{\delta,\varepsilon} \leq u_\beta$. Also, for $\beta := \beta(D) > 2$ big enough, the choice $\delta := 1/\beta$ and $\varepsilon := 2n\log\beta/\beta$ satisfies (5.19), and we obtain that $(1 - \frac{1}{\beta})P(f) + \frac{1}{\beta}v - \frac{2n\log\beta}{\beta} \leq u_\beta$. Putting this together with (5.18), we arive at

$$u_\beta - \frac{C}{\beta} \leq P(f) \leq \frac{\beta}{\beta-1}u_\beta + \frac{2n\log\beta}{\beta-1} - \frac{1}{\beta-1}\inf_X v.$$

In particular, since $P(f)$ is bounded, this implies that u_β is uniformly bounded, and the estimate of the lemma follows. □

Lastly, we argue Proposition 5.8(ii):

LEMMA 5.11. *There exists $C > 0$ and $\beta_0 > 0$ such that the unique solutions $\{u_\beta\}_\beta$ of* (5.17) *satisfy the following estimate:*

$$-C \leq \Delta^\omega u_\beta \leq C, \quad \beta \geq \beta_0,$$

where C only depends on an upper bound for $\Delta^\omega f$.

PROOF. The lower bound on $\Delta^\omega u_\beta$ follows immediately from $\omega + i\partial\bar{\partial}u_\beta \geq 0$. To prove the upper bound, we recall a variant of a Laplacian estimate due to Aubin–Yau (provided by Siu in this context [**100**, page 99], for a survey we refer to [**26**, Proposition 4.1.2]): if $u \in \mathcal{H}_\omega$ satisfies $\omega_u^n = e^g\omega^n$, then

$$\Delta^{\omega_u} \log \operatorname{Tr}_\omega \omega_u \geq \frac{\Delta^\omega g}{\operatorname{Tr}_\omega \omega_u} - 2B\operatorname{Tr}_{\omega_u}\omega,$$

where $B > 0$ depends only on the magnitude of the holomorphic bisectional curvature of ω. Applying this estimate to (5.17) we obtain

$$2B\operatorname{Tr}_{\omega_{u_\beta}}\omega + \Delta^{\omega_{u_\beta}} \log \operatorname{Tr}_\omega \omega_{u_\beta} \geq \beta \frac{\Delta^\omega(u_\beta - f)}{\operatorname{Tr}_\omega \omega_{u_\beta}}.$$

Rearranging terms we conclude:

$$2nB + \Delta^{\omega_{u_\beta}}(\log \operatorname{Tr}_\omega \omega_{u_\beta} - Bu_\beta) \geq \beta \frac{\operatorname{Tr}_\omega \omega_{u_\beta} - 2n - \Delta^\omega f}{\operatorname{Tr}_\omega \omega_{u_\beta}}.$$

Denoting $C := \sup_X(2n + \Delta^\omega f) > 0$, and multiplying with $\operatorname{Tr}_\omega \omega_{u_\beta} e^{-Bu_\beta}$, we arrive at:

$$(C\beta + 2nB\operatorname{Tr}_\omega \omega_{u_\beta})e^{-Bu_\beta} + \Delta^{\omega_{u_\beta}}(\log \operatorname{Tr}_\omega \omega_{u_\beta} - Bu_\beta)\operatorname{Tr}_\omega \omega_{u_\beta} e^{-Bu_\beta} \geq \beta \operatorname{Tr}_\omega \omega_{u_\beta} e^{-Bu_\beta}.$$

Let $s := \sup_X \operatorname{Tr}_\omega \omega_{u_\beta} e^{-Bu_\beta}$ and suppose that this last supremum is realized at $x_0 \in X$. By the previous estimate we can write:

$$\beta s \leq 2nBs + C\beta e^{-Bu_\beta(x_0)}.$$

By the previous lemma, u_β is uniformly bounded, hence for $\beta \geq \beta_0 := 3nB$ we obtain an upper bound for s. Using again that u_β is uniformly bounded, we conclude that $\Delta^\omega u_\beta \leq C$ for any $\beta \geq \beta_0$. □

Notice that $g := \min(f_1, f_2, \ldots, f_k) = -\max(-f_1, -f_2, \ldots, -f_k)$. From this it follows that $\Delta^\omega g$ is uniformly bounded from above by a common upper bound for $\Delta^\omega f_j$, $j \in \{1, \ldots, k\}$. Using the fact that the constant C in the previous lemma only depends on an upper bound for the Laplacian of f, we can conclude the following more general result, proved in [**57**] using different methods:

THEOREM 5.12. *Let $f_1, ..., f_k \in C^\infty(X)$. Then $P(f_1, f_2, ..., f_k) \in C^{1,\alpha}(X)$, $\alpha \in (0, 1)$. More precisely, the following estimate holds:*

$$\|P(f_1, f_2, ..., f_k)\|_{C^{1,\bar{1}}} \leq C(X, \omega, \|f_1\|_{C^{1,\bar{1}}}, \|f_2\|_{C^{1,\bar{1}}}, \ldots, \|f_k\|_{C^{1,\bar{1}}}).$$

Appendix D. Cartan type decompositions of Lie groups

In the proof of the Bando–Mabuchi uniqueness theorem, some basic facts about Lie groups are needed. This section is based on [**58**, Section 6.1]. For an extensive study of such groups we refer to [**28**, **80**].

Recall the following form of the classical Cartan decomposition for complexifications of compact semisimple Lie groups (see [**28**, Proposition 32.1, Remark 31.1]):

THEOREM 5.13. *Let K be a compact connected semisimple Lie group. Denote by $(K^{\mathbb{C}}, J)$ the complexification of K, namely the unique connected complex Lie group whose Lie algebra is the complexification of $\mathfrak{k}$, the Lie algebra of K. Then the map $C : K \times \mathfrak{k} \to K^{\mathbb{C}}$ given by $C(k, X) := k \exp_I JX$ is a diffeomorphism.*

The following result is a partial extension of the above classical theorem to compact but not necessary semisimple Lie groups. We state the result in a form that will be most useful for our applications in Kähler geometry, albeit it is likely not optimal.

PROPOSITION 5.14. *Let K be a compact connected subgroup of a connected complex Lie group (G, J) and denote by $\mathfrak{k}$ and $\mathfrak{g}$ their Lie algebras. If $\mathfrak{g} = \mathfrak{k} \oplus J\mathfrak{k}$ then the map $C : K \times \mathfrak{k} \to G$ given by $C(k, X) = k \exp_I JX$ is surjective.*

PROOF. First we note that

$$\mathfrak{k} = \mathfrak{z}(\mathfrak{k}) \otimes [\mathfrak{k}, \mathfrak{k}], \tag{5.20}$$

where $\mathfrak{z}(\mathfrak{k})$ is the Lie algebra of $Z(K)$. This follows from [**80**, Proposition 6.6 (ii), p. 132], as K is compact. Next we note the following identity for the Lie algebra of the center $Z(G)$:

$$\mathfrak{z}(\mathfrak{g}) = \mathfrak{z}(\mathfrak{k}) \oplus J\mathfrak{z}(\mathfrak{k}). \tag{5.21}$$

Since $\mathfrak{z}(\mathfrak{g})$ is complex, we immediately obtain that $\mathfrak{z}(\mathfrak{g}) \supset \mathfrak{z}(\mathfrak{k}) \oplus J\mathfrak{z}(\mathfrak{k})$. For the reverse inclusion, we use that $\mathfrak{g} = \mathfrak{k} \oplus J\mathfrak{k}$. Consequently, for any $X \in \mathfrak{z}(\mathfrak{g})$ we have $X = X_1 + X_2$, with $X_2 \in \mathfrak{k} \cap \mathfrak{z}(\mathfrak{g}) = \mathfrak{z}(\mathfrak{k})$, and $X_2 \in J\mathfrak{k} \cap \mathfrak{z}(\mathfrak{g}) = J(\mathfrak{k} \cap \mathfrak{z}(\mathfrak{g})) = J\mathfrak{z}(\mathfrak{k})$, since $\mathfrak{z}(\mathfrak{g})$ is complex. This finishes the proof of (5.21).

Next we claim that the map

$$\Theta_1 : Z(K) \times \mathfrak{z}(\mathfrak{k}) \to Z(G)$$

given by $(z, X) \mapsto z \exp_I JX$ is surjective. Indeed, (5.21) implies that $\dim Z(K) + \dim \mathfrak{z}(\mathfrak{k}) = \dim Z(G)$ and the differential of Θ_1 at $(I, 0)$ is invertible (see [**80**, Proposition 1.6, p. 104]). As we are dealing with abelian groups it follows that Θ_1 is a Lie group homomorphism, thus it must be surjective as its image is a connected subgroup of the same dimension as that of $Z(G)$.

Let L denote the connected compact Lie subgroup of K whose Lie algebra is $[\mathfrak{k}, \mathfrak{k}]$ (since the Killing form is negative definite on $[\mathfrak{k}, \mathfrak{k}]$, L is indeed compact). By [**80**, Proposition 6.6 (i), p. 132], L is semisimple. By Theorem 5.13, the map

$$\Theta_2 : L \times [\mathfrak{k}, \mathfrak{k}] \to L^{\mathbb{C}}$$

given by $\Theta_2(l, X) = l \exp_I JX$, is a diffeomorphism, where $L^{\mathbb{C}}$ is the complexification of L inside G.

Next we note that the multiplication maps $Z(K) \times L \to K$, $Z(G) \times L^{\mathbb{C}} \to G$ are surjective. By (5.20) the multiplication map $Z(K) \times L \to K$ is a local isomorphism near (I, I) by dimension count. The map is also a group homomorphism since elements of $Z(K)$ commute with elements of L. Thus, it is surjective. The same argument works for the multiplication map $Z(G) \times L^{\mathbb{C}} \to G$, with dimension count provided by (5.21).

Now we put all the above ingredients together. Given $k \in K$ and $X \in \mathfrak{k}$, observe that

$$C(k, X) = zl \exp_I JX,$$

for some $z \in Z(K)$ and $l \in L$ such that $k = zl$ (these exist by the surjectivity of the multiplication map $Z(K) \times L \to K$). Now let X_1 and X_2 be the unique elements such that $X_1 \in \mathfrak{z}(\mathfrak{k})$, $X_2 \in [\mathfrak{k}, \mathfrak{k}]$, and $X = X_1 + X_2$, given by 5.20. Since $\exp_I JX_1 \in Z(G)$ we can write

$$C(k, X) = z \exp_I JX_1 l \exp_I JX_2 = \Theta_1(z, X_1)\Theta_2(l, X_2).$$

Since both Θ_2 and Θ_1 are surjective, as well as the multiplication map $Z(G) \times L^{\mathbb{C}} \to G$, it follows that C is surjective, concluding the proof. □

References

[1] Thierry Aubin, *Équations du type Monge-Ampère sur les variétés kählériennes compactes* (French, with English summary), Bull. Sci. Math. (2) **102** (1978), no. 1, 63–95. MR494932

[2] Shigetoshi Bando and Toshiki Mabuchi, *Uniqueness of Einstein Kähler metrics modulo connected group actions*, Algebraic geometry, Sendai, 1985, Adv. Stud. Pure Math., vol. 10, North-Holland, Amsterdam, 1987, pp. 11–40. MR946233

[3] Eric Bedford and B. A. Taylor, *A new capacity for plurisubharmonic functions*, Acta Math. **149** (1982), no. 1-2, 1–40, DOI 10.1007/BF02392348. MR674165

[4] Eric Bedford and B. A. Taylor, *Fine topology, Šilov boundary, and $(dd^c)^n$*, J. Funct. Anal. **72** (1987), no. 2, 225–251, DOI 10.1016/0022-1236(87)90087-5. MR886812

[5] Robert J. Berman, *A thermodynamical formalism for Monge-Ampère equations, Moser-Trudinger inequalities and Kähler-Einstein metrics*, Adv. Math. **248** (2013), 1254–1297, DOI 10.1016/j.aim.2013.08.024. MR3107540

[6] R.J. Berman, From Monge–Ampére equations to envelopes and geodesic rays in the zero temperature limit, arXiv:1307.3008.

[7] Robert J. Berman, *K-polystability of $\mathbb{Q}$-Fano varieties admitting Kähler-Einstein metrics*, Invent. Math. **203** (2016), no. 3, 973–1025, DOI 10.1007/s00222-015-0607-7. MR3461370

[8] Robert J. Berman and Bo Berndtsson, *Convexity of the K-energy on the space of Kähler metrics and uniqueness of extremal metrics*, J. Amer. Math. Soc. **30** (2017), no. 4, 1165–1196, DOI 10.1090/jams/880. MR3671939

[9] Robert Berman and Sébastien Boucksom, *Growth of balls of holomorphic sections and energy at equilibrium*, Invent. Math. **181** (2010), no. 2, 337–394, DOI 10.1007/s00222-010-0248-9. MR2657428

[10] R. Berman, S. Boucksom, M. Jonsson, A variational approach to the Yau–Tian–Donaldson conjecture, arXiv:1509.04561.

[11] Robert J. Berman, Sébastien Boucksom, Vincent Guedj, and Ahmed Zeriahi, *A variational approach to complex Monge-Ampère equations*, Publ. Math. Inst. Hautes Études Sci. **117** (2013), 179–245, DOI 10.1007/s10240-012-0046-6. MR3090260

[12] R. Berman, S. Boucksom, P. Eyssidieux, V. Guedj, A. Zeriahi, Kähler-Einstein metrics and the Kähler-Ricci flow on log Fano varieties, arXiv:1111.7158.

[13] Robert J. Berman, Tamás Darvas, and Chinh H. Lu, *Convexity of the extended K-energy and the large time behavior of the weak Calabi flow*, Geom. Topol. **21** (2017), no. 5, 2945–2988, DOI 10.2140/gt.2017.21.2945. MR3687111

[14] R.J. Berman, T. Darvas, C.H. Lu, Regularity of weak minimizers of the K-energy and applications to properness and K-stability, arXiv:1602.03114.

[15] Robert Berman and Jean-Pierre Demailly, *Regularity of plurisubharmonic upper envelopes in big cohomology classes*, Perspectives in analysis, geometry, and topology, Progr. Math., vol. 296, Birkhäuser/Springer, New York, 2012, pp. 39–66, DOI 10.1007/978-0-8176-8277-4_3. MR2884031

[16] Robert J. Berman and Chinh H. Lu, *From the Kähler-Ricci flow to moving free boundaries and shocks* (English, with English and French summaries), J. Éc. polytech. Math. **5** (2018), 519–563. MR3852260

[17] Bo Berndtsson, *Curvature of vector bundles associated to holomorphic fibrations*, Ann. of Math. (2) **169** (2009), no. 2, 531–560, DOI 10.4007/annals.2009.169.531. MR2480611

[18] B. Berndtsson, Probability measures related to geodesics in the space of Kähler metrics, arXiv:0907.1806.

[19] Bo Berndtsson, *A Brunn-Minkowski type inequality for Fano manifolds and some uniqueness theorems in Kähler geometry*, Invent. Math. **200** (2015), no. 1, 149–200, DOI 10.1007/s00222-014-0532-1. MR3323577

[20] Z. Błocki, The complex Monge–Ampère operator in pluripotential theory, lecture notes, `http://gamma.im.uj.edu.pl/~blocki/publ/ln/wykl.pdf`.

[21] Z. Blocki, On uniqueness of the complex Monge–Ampère equation on compact Kähler manifolds, Institut Mittag–Leffler, Report No. 1, 2007/2008.

[22] Zbigniew Błocki, *The complex Monge-Ampère equation in Kähler geometry*, Pluripotential theory, Lecture Notes in Math., vol. 2075, Springer, Heidelberg, 2013, pp. 95–141, DOI 10.1007/978-3-642-36421-1_2. MR3089069

[23] Zbigniew Błocki and Sławomir Kołodziej, *On regularization of plurisubharmonic functions on manifolds*, Proc. Amer. Math. Soc. **135** (2007), no. 7, 2089–2093, DOI 10.1090/S0002-9939-07-08858-2. MR2299485

[24] Thomas Bloom and Norman Levenberg, *Pluripotential energy*, Potential Anal. **36** (2012), no. 1, 155–176, DOI 10.1007/s11118-011-9224-2. MR2886457

[25] Sébastien Boucksom, *Monge-Ampère equations on complex manifolds with boundary*, Complex Monge-Ampère equations and geodesics in the space of Kähler metrics, Lecture Notes in Math., vol. 2038, Springer, Heidelberg, 2012, pp. 257–282, DOI 10.1007/978-3-642-23669-3_7. MR2932446

[26] Sébastien Boucksom, Philippe Eyssidieux, and Vincent Guedj, *Introduction*, An introduction to the Kähler-Ricci flow, Lecture Notes in Math., vol. 2086, Springer, Cham, 2013, pp. 1–6, DOI 10.1007/978-3-319-00819-6_1. MR3185331

[27] Sébastien Boucksom, Philippe Eyssidieux, Vincent Guedj, and Ahmed Zeriahi, *Monge-Ampère equations in big cohomology classes*, Acta Math. **205** (2010), no. 2, 199–262, DOI 10.1007/s11511-010-0054-7. MR2746347

[28] Daniel Bump, *Lie groups*, 2nd ed., Graduate Texts in Mathematics, vol. 225, Springer, New York, 2013. MR3136522

[29] E. Calabi, The variation of Kähler metrics, Bull. Amer. Math. Soc. 60 (1954), 167-168.

[30] E. Calabi and X. X. Chen, *The space of Kähler metrics. II*, J. Differential Geom. **61** (2002), no. 2, 173–193. MR1969662

[31] Simone Calamai, *The Calabi metric for the space of Kähler metrics*, Math. Ann. **353** (2012), no. 2, 373–402, DOI 10.1007/s00208-011-0690-z. MR2915541

[32] Simone Calamai and Kai Zheng, *The Dirichlet and the weighted metrics for the space of Kähler metrics*, Math. Ann. **363** (2015), no. 3-4, 817–856, DOI 10.1007/s00208-015-1188-x. MR3412344

[33] Urban Cegrell, *Pluricomplex energy*, Acta Math. **180** (1998), no. 2, 187–217, DOI 10.1007/BF02392899. MR1638768

[34] Xiuxiong Chen, *The space of Kähler metrics*, J. Differential Geom. **56** (2000), no. 2, 189–234. MR1863016

[35] X.X. Chen, Space of Kähler metrics (IV)–On the lower bound of the K–energy, arXiv:0809.4081.

[36] X.X. Chen, J. Cheng, On the constant scalar curvature Kähler metrics, existence results, arXiv:1801.00656.

[37] X.X. Chen, J. Cheng, On the constant scalar curvature Kähler metrics, general automorphism group, arXiv:1801.05907.

[38] Xiuxiong Chen, Simon Donaldson, and Song Sun, *Kähler-Einstein metrics on Fano manifolds. I: Approximation of metrics with cone singularities*, J. Amer. Math. Soc. **28** (2015), no. 1, 183–197, DOI 10.1090/S0894-0347-2014-00799-2. MR3264766

[39] Xiuxiong Chen, Simon Donaldson, and Song Sun, *Kähler-Einstein metrics on Fano manifolds. II: Limits with cone angle less than* 2π, J. Amer. Math. Soc. **28** (2015), no. 1, 199–234, DOI 10.1090/S0894-0347-2014-00800-6. MR3264767

[40] Xiuxiong Chen, Simon Donaldson, and Song Sun, *Kähler-Einstein metrics on Fano manifolds. III: Limits as cone angle approaches 2π and completion of the main proof*, J. Amer. Math. Soc. **28** (2015), no. 1, 235–278, DOI 10.1090/S0894-0347-2014-00801-8. MR3264768
[41] X. X. Chen and W. Y. He, *On the Calabi flow*, Amer. J. Math. **130** (2008), no. 2, 539–570, DOI 10.1353/ajm.2008.0018. MR2405167
[42] Xiuxiong Chen, Song Sun, and Bing Wang, *Kähler-Ricci flow, Kähler-Einstein metric, and K-stability*, Geom. Topol. **22** (2018), no. 6, 3145–3173, DOI 10.2140/gt.2018.22.3145. MR3858762
[43] Xiuxiong Chen and Song Sun, *Space of Kähler metrics (V)—Kähler quantization*, Metric and differential geometry, Progr. Math., vol. 297, Birkhäuser/Springer, Basel, 2012, pp. 19–41, DOI 10.1007/978-3-0348-0257-4_2. MR3220438
[44] Jianchun Chu, Valentino Tosatti, and Ben Weinkove, *On the $C^{1,1}$ regularity of geodesics in the space of Kähler metrics*, Ann. PDE **3** (2017), no. 2, Art. 15, 12, DOI 10.1007/s40818-017-0034-8. MR3695402
[45] J. Chu, B. Zhou, Optimal regularity of plurisubharmonic envelopes on compact Hermitian manifolds, arXiv:1702.05230.
[46] Brian Clarke and Yanir A. Rubinstein, *Ricci flow and the metric completion of the space of Kähler metrics*, Amer. J. Math. **135** (2013), no. 6, 1477–1505, DOI 10.1353/ajm.2013.0051. MR3145001
[47] Tamás Darvas, *Morse theory and geodesics in the space of Kähler metrics*, Proc. Amer. Math. Soc. **142** (2014), no. 8, 2775–2782, DOI 10.1090/S0002-9939-2014-12105-8. MR3209332
[48] Tamás Darvas, *The Mabuchi completion of the space of Kähler potentials*, Amer. J. Math. **139** (2017), no. 5, 1275–1313, DOI 10.1353/ajm.2017.0032. MR3702499
[49] Tamás Darvas, *The Mabuchi geometry of finite energy classes*, Adv. Math. **285** (2015), 182–219, DOI 10.1016/j.aim.2015.08.005. MR3406499
[50] Tamás Darvas, *Metric geometry of normal Kähler spaces, energy properness, and existence of canonical metrics*, Int. Math. Res. Not. IMRN **22** (2017), 6752–6777, DOI 10.1093/imrn/rnw203. MR3737320
[51] Tamás Darvas, *Comparison of the Calabi and Mabuchi geometries and applications to geometric flows*, Ann. Inst. H. Poincaré Anal. Non Linéaire **34** (2017), no. 5, 1131–1140, DOI 10.1016/j.anihpc.2016.09.002. MR3742517
[52] Tamás Darvas, Eleonora Di Nezza, and Chinh H. Lu, *Monotonicity of nonpluripolar products and complex Monge-Ampère equations with prescribed singularity*, Anal. PDE **11** (2018), no. 8, 2049–2087, DOI 10.2140/apde.2018.11.2049. MR3812864
[53] Tamás Darvas and Weiyong He, *Geodesic rays and Kähler-Ricci trajectories on Fano manifolds*, Trans. Amer. Math. Soc. **369** (2017), no. 7, 5069–5085, DOI 10.1090/tran/6878. MR3632560
[54] Tamás Darvas and László Lempert, *Weak geodesics in the space of Kähler metrics*, Math. Res. Lett. **19** (2012), no. 5, 1127–1135, DOI 10.4310/MRL.2012.v19.n5.a13. MR3039835
[55] T. Darvas, C.H. Lu. , Uniform convexity in L^p Mabuchi geometry, the space of rays, and geodesic stability, with C.H. Lu, arXiv:1810.04661.
[56] T. Darvas, C.H. Lu, Y.A. Rubinstein, Quantization in geometric pluripotential theory, arXiv:1806.03800.
[57] Tamás Darvas and Yanir A. Rubinstein, *Kiselman's principle, the Dirichlet problem for the Monge-Ampère equation, and rooftop obstacle problems*, J. Math. Soc. Japan **68** (2016), no. 2, 773–796, DOI 10.2969/jmsj/06820773. MR3488145
[58] Tamás Darvas and Yanir A. Rubinstein, *Tian's properness conjectures and Finsler geometry of the space of Kähler metrics*, J. Amer. Math. Soc. **30** (2017), no. 2, 347–387, DOI 10.1090/jams/873. MR3600039
[59] Jean-Pierre Demailly, *Regularization of closed positive currents and intersection theory*, J. Algebraic Geom. **1** (1992), no. 3, 361–409. MR1158622
[60] J.P. Demailly, Regularization of closed positive currents of type (1,1) by the flow of a Chern connection, Actes du Colloque en l'honneur de P. Dolbeault (Juin 1992), édité par H. Skoda et J.M. Trépreau, Aspects of Mathematics, Vol. E 26, Vieweg, (1994) 105–126.
[61] J.P. Demailly, Complex Analytic and Differential Geometry, `https://www-fourier.ujf-grenoble.fr/~demailly/manuscripts/agbook.pdf`.

[62] Amir Dembo and Ofer Zeitouni, *Large deviations techniques and applications*, 2nd ed., Applications of Mathematics (New York), vol. 38, Springer-Verlag, New York, 1998. MR1619036

[63] Eleonora Di Nezza and Vincent Guedj, *Geometry and topology of the space of Kähler metrics on singular varieties*, Compos. Math. **154** (2018), no. 8, 1593–1632, DOI 10.1112/s0010437x18007170. MR3830547

[64] Wei Yue Ding, *Remarks on the existence problem of positive Kähler-Einstein metrics*, Math. Ann. **282** (1988), no. 3, 463–471, DOI 10.1007/BF01460045. MR967024

[65] W.Y. Ding, G. Tian, The generalized Moser-Trudinger inequality, in: Nonlinear Analysis and Microlocal Analysis (K.-C. Chang et al., Eds.), World Scientific, 1992, pp. 57–70.

[66] Sławomir Dinew, *Uniqueness in $\mathcal{E}(X,\omega)$*, J. Funct. Anal. **256** (2009), no. 7, 2113–2122, DOI 10.1016/j.jfa.2009.01.019. MR2498760

[67] S. K. Donaldson, *Symmetric spaces, Kähler geometry and Hamiltonian dynamics*, Northern California Symplectic Geometry Seminar, Amer. Math. Soc. Transl. Ser. 2, vol. 196, Amer. Math. Soc., Providence, RI, 1999, pp. 13–33, DOI 10.1090/trans2/196/02. MR1736211

[68] S. K. Donaldson, *Scalar curvature and projective embeddings. I*, J. Differential Geom. **59** (2001), no. 3, 479–522. MR1916953

[69] Simon Donaldson, *Kähler-Einstein metrics and algebraic geometry*, Current developments in mathematics 2015, Int. Press, Somerville, MA, 2016, pp. 1–25. MR3642542

[70] Philippe Eyssidieux, Vincent Guedj, and Ahmed Zeriahi, *Singular Kähler-Einstein metrics*, J. Amer. Math. Soc. **22** (2009), no. 3, 607–639, DOI 10.1090/S0894-0347-09-00629-8. MR2505296

[71] Akito Futaki, *Kähler-Einstein metrics and integral invariants*, Lecture Notes in Mathematics, vol. 1314, Springer-Verlag, Berlin, 1988. MR947341

[72] V. Guedj, The metric completion of the Riemannian space of Kähler metrics, arXiv:1401.7857.

[73] Vincent Guedj and Ahmed Zeriahi, *Intrinsic capacities on compact Kähler manifolds*, J. Geom. Anal. **15** (2005), no. 4, 607–639, DOI 10.1007/BF02922247. MR2203165

[74] Vincent Guedj and Ahmed Zeriahi, *The weighted Monge-Ampère energy of quasiplurisubharmonic functions*, J. Funct. Anal. **250** (2007), no. 2, 442–482, DOI 10.1016/j.jfa.2007.04.018. MR2352488

[75] Vincent Guedj and Ahmed Zeriahi, *Degenerate complex Monge-Ampère equations*, EMS Tracts in Mathematics, vol. 26, European Mathematical Society (EMS), Zürich, 2017. MR3617346

[76] David Gilbarg and Neil S. Trudinger, *Elliptic partial differential equations of second order*, Classics in Mathematics, Springer-Verlag, Berlin, 2001. Reprint of the 1998 edition. MR1814364

[77] Weiyong He, *On the space of Kähler potentials*, Comm. Pure Appl. Math. **68** (2015), no. 2, 332–343, DOI 10.1002/cpa.21515. MR3298665

[78] Weiyong He, *Kähler-Ricci soliton and H-functional*, Asian J. Math. **20** (2016), no. 4, 645–663, DOI 10.4310/AJM.2016.v20.n4.a3. MR3570456

[79] W. He, J. Li, Geometrical pluripotential theory on Sasaki manifolds, arXiv:1803.00687.

[80] Sigurdur Helgason, *Differential geometry, Lie groups, and symmetric spaces*, Graduate Studies in Mathematics, vol. 34, American Mathematical Society, Providence, RI, 2001. Corrected reprint of the 1978 original. MR1834454

[81] Lars Hörmander, *Notions of convexity*, Modern Birkhäuser Classics, Birkhäuser Boston, Inc., Boston, MA, 2007. Reprint of the 1994 edition [of MR1301332]. MR2311920

[82] Sławomir Kołodziej, *The complex Monge-Ampère equation*, Acta Math. **180** (1998), no. 1, 69–117, DOI 10.1007/BF02392879. MR1618325

[83] Sławomir Kołodziej, *The complex Monge-Ampère equation and pluripotential theory*, Mem. Amer. Math. Soc. **178** (2005), no. 840, x+64, DOI 10.1090/memo/0840. MR2172891

[84] Sławomir Kołodziej, *The Monge-Ampère equation on compact Kähler manifolds*, Indiana Univ. Math. J. **52** (2003), no. 3, 667–686, DOI 10.1512/iumj.2003.52.2220. MR1986892

[85] László Lempert and Liz Vivas, *Geodesics in the space of Kähler metrics*, Duke Math. J. **162** (2013), no. 7, 1369–1381, DOI 10.1215/00127094-2142865. MR3079251

[86] Long Li, *On the spectrum of weighted Laplacian operator and its application to uniqueness of Kähler Einstein metrics*, Math. Ann. **362** (2015), no. 3-4, 1349–1378, DOI 10.1007/s00208-014-1152-1. MR3368103

[87] Chinh H. Lu and Van-Dong Nguyen, *Degenerate complex Hessian equations on compact Kähler manifolds*, Indiana Univ. Math. J. **64** (2015), no. 6, 1721–1745, DOI 10.1512/iumj.2015.64.5680. MR3436233

[88] Toshiki Mabuchi, *Some symplectic geometry on compact Kähler manifolds. I*, Osaka J. Math. **24** (1987), no. 2, 227–252. MR909015

[89] Yozô Matsushima, *Sur la structure du groupe d'homéomorphismes analytiques d'une certaine variété kählérienne* (French), Nagoya Math. J. **11** (1957), 145–150. MR0094478

[90] D. H. Phong, Jian Song, Jacob Sturm, and Ben Weinkove, *The Moser-Trudinger inequality on Kähler-Einstein manifolds*, Amer. J. Math. **130** (2008), no. 4, 1067–1085, DOI 10.1353/ajm.0.0013. MR2427008

[91] D. H. Phong and Jacob Sturm, *The Monge-Ampère operator and geodesics in the space of Kähler potentials*, Invent. Math. **166** (2006), no. 1, 125–149, DOI 10.1007/s00222-006-0512-1. MR2242635

[92] Duong H. Phong and Jacob Sturm, *Test configurations for K-stability and geodesic rays*, J. Symplectic Geom. **5** (2007), no. 2, 221–247. MR2377252

[93] D. H. Phong and Jacob Sturm, *Lectures on stability and constant scalar curvature*, Current developments in mathematics, 2007, Int. Press, Somerville, MA, 2009, pp. 101–176. MR2532997

[94] Yanir A. Rubinstein, *Some discretizations of geometric evolution equations and the Ricci iteration on the space of Kähler metrics*, Adv. Math. **218** (2008), no. 5, 1526–1565, DOI 10.1016/j.aim.2008.03.017. MR2419932

[95] Yanir A. Rubinstein, *Smooth and singular Kähler-Einstein metrics*, Geometric and spectral analysis, Contemp. Math., vol. 630, Amer. Math. Soc., Providence, RI, 2014, pp. 45–138, DOI 10.1090/conm/630/12665. MR3328541

[96] Y.A. Rubinstein, Tian's properness conjectures: an introduction to Kähler geometry, arxiv:1807.00928.

[97] Walter Rudin, *Real and complex analysis*, 3rd ed., McGraw-Hill Book Co., New York, 1987. MR924157

[98] M. M. Rao and Z. D. Ren, *Theory of Orlicz spaces*, Monographs and Textbooks in Pure and Applied Mathematics, vol. 146, Marcel Dekker, Inc., New York, 1991. MR1113700

[99] Stephen Semmes, *Complex Monge-Ampère and symplectic manifolds*, Amer. J. Math. **114** (1992), no. 3, 495–550, DOI 10.2307/2374768. MR1165352

[100] Yum Tong Siu, *Lectures on Hermitian-Einstein metrics for stable bundles and Kähler-Einstein metrics*, DMV Seminar, vol. 8, Birkhäuser Verlag, Basel, 1987. MR904673

[101] Jeffrey Streets, *Long time existence of minimizing movement solutions of Calabi flow*, Adv. Math. **259** (2014), 688–729, DOI 10.1016/j.aim.2014.03.027. MR3197669

[102] Jeffrey Streets, *The consistency and convergence of K-energy minimizing movements*, Trans. Amer. Math. Soc. **368** (2016), no. 7, 5075–5091, DOI 10.1090/tran/6508. MR3456172

[103] Gábor Székelyhidi, *An introduction to extremal Kähler metrics*, Graduate Studies in Mathematics, vol. 152, American Mathematical Society, Providence, RI, 2014. MR3186384

[104] Gábor Székelyhidi and Valentino Tosatti, *Regularity of weak solutions of a complex Monge-Ampère equation*, Anal. PDE **4** (2011), no. 3, 369–378, DOI 10.2140/apde.2011.4.369. MR2872120

[105] Gang Tian, *The K-energy on hypersurfaces and stability*, Comm. Anal. Geom. **2** (1994), no. 2, 239–265, DOI 10.4310/CAG.1994.v2.n2.a4. MR1312688

[106] Gang Tian, *Kähler-Einstein metrics with positive scalar curvature*, Invent. Math. **130** (1997), no. 1, 1–37, DOI 10.1007/s002220050176. MR1471884

[107] Gang Tian, *Canonical metrics in Kähler geometry*, Lectures in Mathematics ETH Zürich, Birkhäuser Verlag, Basel, 2000. Notes taken by Meike Akveld. MR1787650

[108] Gang Tian, *Existence of Einstein metrics on Fano manifolds*, Metric and differential geometry, Progr. Math., vol. 297, Birkhäuser/Springer, Basel, 2012, pp. 119–159, DOI 10.1007/978-3-0348-0257-4_5. MR3220441

[109] Gang Tian, *K-stability and Kähler-Einstein metrics*, Comm. Pure Appl. Math. **68** (2015), no. 7, 1085–1156, DOI 10.1002/cpa.21578. MR3352459

[110] Gang Tian, *Kähler-Einstein metrics on Fano manifolds*, Jpn. J. Math. **10** (2015), no. 1, 1–41, DOI 10.1007/s11537-014-1387-3. MR3320994

[111] Gang Tian and Xiaohua Zhu, *A nonlinear inequality of Moser-Trudinger type*, Calc. Var. Partial Differential Equations **10** (2000), no. 4, 349–354, DOI 10.1007/s005260010349. MR1767718

[112] Valentino Tosatti, *Regularity of envelopes in Kähler classes*, Math. Res. Lett. **25** (2018), no. 1, 281–289, DOI 10.4310/mrl.2018.v25.n1.a12. MR3818623

[113] Shing Tung Yau, *On the Ricci curvature of a compact Kähler manifold and the complex Monge-Ampère equation. I*, Comm. Pure Appl. Math. **31** (1978), no. 3, 339–411, DOI 10.1002/cpa.3160310304. MR480350

[114] Raymond O. Wells Jr., *Differential analysis on complex manifolds*, 3rd ed., Graduate Texts in Mathematics, vol. 65, Springer, New York, 2008. With a new appendix by Oscar Garcia-Prada. MR2359489

[115] Ahmed Zeriahi, *Volume and capacity of sublevel sets of a Lelong class of plurisubharmonic functions*, Indiana Univ. Math. J. **50** (2001), no. 1, 671–703, DOI 10.1512/iumj.2001.50.2062. MR1857051

Department of Mathematics, University of Maryland, 4176 Campus Drive, College Park, Maryland 20742

Email address: tdarvas@math.umd.edu

Contemporary Mathematics
Volume **735**, 2019
https://doi.org/10.1090/conm/735/14823

Local singularities of plurisubharmonic functions

Sławomir Dinew

ABSTRACT. We survey the recent advances in the theory of local regularity of plurisubharmonic functions and the complex Monge-Ampère equation.

CONTENTS

1. Introduction

Plurisubharmonic functions appear naturally in many branches of modern mathematics such as potential theory, complex analysis or complex geometry. Their importance in complex analysis stems for the fact that logarithms of absolute values of holomorphic mappings are plurisubharmonic. Another important property is the invariance modulo compositions with holomorphic maps. The extremely powerful machinery of the L^2 estimates for the $\bar{\partial}$- operator rely crucially on plurisubharmonic weights. Geometers in turn encounter plurisubharmonic functions as local potentials of Kähler forms. Yet another geometric notion tightly linked to plurisubharmonic functions are the weights of metrics on holomorphic (ample) line bundles. The non-linear potential theory of plurisuharmonic functions, known as pluripotential theory, has become one of the fundamental tools for studying rough regularity of the complex Monge-Ampère equation.

All these applications justify an interest towards plurisubharmonic functions on their own. In this survey we shall focus only on local issues and therefore we shall work in domains in $\mathbb{C}^n$. Nevertheless we hope that the geometrically oriented Reader will also find this survey useful. In particular the manuscript gives some insight what kind of degenerations of Kähler metrics can occur if one ignores global

2010 *Mathematics Subject Classification.* Primary 32U05.

The author was supported by the NCN grant 2013/08/A/ST1/00312.

©2019 American Mathematical Society

phenomena. It may be also useful for Readers working in a global non-compact setting — for example on complete complex manifolds.

Roughly speaking plurisubharmonic functions are the complex analogues of convex functions and hence it is natural to expect that their theory is tightly linked to convexity theory. This naive picture doesn't take into account many new phenomena arising in the complex setting. Thus it is justified to say that the world of plurisubharmonic functions *do share* similarities with convex functions' theory but is much more complicated and harder to study. Needless to say this also makes the associated pluripotential theory interesting and nontrvial. We refer to the classical book of Hörmander [**25**], where a fairly detailed convexity viewpoint of pluripotential theory is presented.

A starting point for us will be the observation that while every convex function is locally Lipschitz, the only regularity that a general plurisubharmonic function admits (besides upper-semicontinuity) is L^p_{loc} regularity for any $1 \leq p < \infty$ (see [**25**]). A very distinctive feature of plurisubharmonic functions is that they may allow "poles" i.e. sets where they have value $-\infty$. These sets are called *pluripolar* and, keeping in mind the examples of the type $u(z) = log||F(z)||$ for some holomorphic mapping F, generalize *complex analytic sets*. Yet another distinction from the convex case is that the sublevel sets of a plurisubharmonic function may well be disconnected as the planar function $v(z) = log|z^2 - 1|$ shows. In general these sublevel sets can have very complicated geometry.

Even if we restrict attention to *bounded* plurisubharmonic functions there are still very distinctive properties which do not have convex analogues. Such a restriction however allows one to apply the extremely powerful machinery of pluripotential theory invented by Bedford and Taylor in [**4, 5**]. In particular one can then define the *Monge-Ampère* operator of such a function which plays the role of the Laplacian in standard potential theory.

Having all this in mind it is reasonable to ask about the local regularity theory of plurisubharmonic functions. As any such function is locally a decreasing limit of smooth strictly plurisubharmonic functions the question can be rephrased in the following way: what could be the limit of a sequence of local smooth Kähler potentials if no global phenomena are taken into account?

In this survey we shall mainly investigate the regularity of plurisubharmonic functions that additionally satisfy the complex Monge-Ampère equation

$$(dd^c u)^n = f dV, \tag{1.1}$$

where $f \geq 0$ is a given function and dV stands for the Lebesgue measure. As elementary examples in the next sections show, fixing the Monge-Ampère measure is way too weak assumption to get any reasonable local regularity theory. Thus we shall investigate what could be said under *additional assumptions* on the solution. A typical goal we shall pursue is to get minimal possible conditions guaranteeing smooth solutions.

It has to be stressed that while *global* estimates for this equation are reasonably well understood (see [**14, 28**]) the *interior* estimates remain to a large extent a mystery. There are good resons for this. First of all the "symmetries" of the equation are the biholomorphic mappings- way too large class of functions in order to expect an easy *compactness* theory (in the case of real Monge-Ampère equation the symmetries are the affine mappings). Next the geometric restrictions on the boundary imposed by the equation yield that the set of natural domains associated to the equation are the (strictly) *pseudoconvex* ones. Again these form a much richer and complicated class compared to convex domains. Yet another reason between the discrepancies in the real and the complex case is that in the former one can use the powerful techniques of convex geometry which have no analog in the complex realm. Last but not least the singularities of bounded plurisubharmonic functions are poorly understood- there is no satisfactory theory describing these.

The purpose of this note is to gather the scattered local and interior regularity results dealing with the complex Monge-Ampère equation. There are no new results. The survey in particular covers some of the recent advances in the field made by the author and his collaborators [**10**, **17**–**19**] and other contributors [**9**, **15**, **29**, **36**, **38**]. It is interesting also to compare the existent results with the reasonably complete theory of the *real* Monge-Ampère equation (see [**23**]).

The note is organised as follows: in the Preliminaries we fix the notation and we gather all the classical tools from pluripotential theory and PDEs that shall be used later on. Section 3 is devoted to the local theory of the solutions of the Monge-Ampère operator. There we gather a collection of classical (counter)examples which we hope shed some light on what should be expected within the theory. Next section is devoted to interior a priori estimates. The final two sections contain the recent developments within the theory. In the first of these we study the question of minimal possible assumptions yielding smoothness of solutions to the complex Monge-Ampère equation. In the second one we describe some attempts to describe (special types of) singularities of bounded plurisubharmonic functions.

2. Preliminaries

Notation. Throughout the note Ω will always denote a bounded domain in $\mathbb{C}^n$. We shall denote, as customary, by C various numerical constants that depend only on the relevant quantities in question. In particular these may vary from line to line. If a distinction between these in an argument will be needed these will be enumerated by C_1, C_2 etc. By $B_z(r)$ we shall denote the ball of radius r centered at z. $< z, w >$ will stand for $\sum_{j=1}^n z_j\bar{w}_j$- the standard Hermitian inner product in $\mathbb{C}^n$. By $|z|$ we shall denote the length of the vector z (regardless of the dimension we work in). By $PSH(\Omega)$ we shall denote the space of plurisubharmonic functions in a given domain Ω. Also, by an abuse of notation we shall identify the density of the volume form $(dd^c u)^n$ and the volume form itself. Partial differentiations $\frac{\partial u}{\partial z_j}$, $\frac{\partial u}{\partial \bar{z}_k}$ will be denoted simply by u_j and $u_{\bar{k}}$, respectively. dV notes the Lebesgue measure. Occasionally we shall also use $d\lambda^{2n}$ for it as the letter V is (traditionally) used for the Siciak-Zahariuta extremal function.

2.1. Elements of pluripotential theory.

Pluripotential theory is the theory of weak solutions to the complex Monge-Ampère equation

$$\det\Big(\frac{\partial^2 u}{\partial z_j \partial \bar{z}_k}\Big)(z) = f(z).$$

This is a fully nonlinear PDE which is elliptic whenever restricted to the class of functions with pointwise nonnegative definite complex Hessian. For this reason it is not obvious how to properly define the operator for nonsmooth *plurisubharmonic* functions — the upper semicontinuous functions which are locally integrable and satisfy $dd^c u \geq 0$ in the distributional sense. Here d is the standard exterior differentiation operator, while $d^c = iC(n)(\bar{\partial} - \partial)$ for a numerical constant $C(n) > 0$ such that $(dd^c|z|^2)^n = 1$. In a different language $dd^c u$ is *a closed positive* $(1,1)$-*current*.

The basic idea of Bedford-Taylor theory ([**5**], see also [**28**]) is that whenever u is moreover locally bounded, the inductively constructed positive currents

$$(dd^c u)^{k+1} := dd^c(u(dd^c u)^k)$$

are well defined. Furthermore it can be proved that these currents have measure coefficients. Thus the Monge-Ampère measure $det(u_{i\bar{j}}) := (dd^c u)^n$ is a well defined Radon measure.

It has to be emphasized that the operator defined above is *not* continuous with respect to weak convergence of plurisubharmonic functions ([**28**]), which is in sharp contrast to the real Monge-Ampère operator — see [**23**]. The following monotone convergence theorem due to Bedford and Taylor [**5**] partially makes up for this:

THEOREM 2.1. *Let u_k, $k = 1, 2, \cdots$ be a decreasing sequence of locally bounded plurisubharmonic functions convergent to the locally bounded plurisubharmonic function u. Then*

$$(dd^c u_k)^n \to (dd^c u)^n$$

as $k \to \infty$ in the weak sense of measures.

The theorem also holds for unbounded plurisubharmonic functions as long as the Monge-Ampère operator is well defined (see [**28**],[**22**]). For our later purposes we recall that this is the case if the functions in question admit at most *loglog* poles.

A fundamental fact exhibiting the ellipticity of the complex Monge-Ampère operator is the following comparison principle due to Bedford and Taylor (see [**28**] for a proof):

THEOREM 2.2. *If u, w are two bounded plurisubharmonic functions in a domain $\Omega \subset \mathbb{C}^n$, such that $liminf_{z \to \partial\Omega}(u - w)(z) \geq 0$. If moreover*

$$(dd^c w)^n \geq (dd^c u)^n$$

as measures then $u \geq w$ in Ω.

Another useful inequality which in the smooth case follows from the concavity of the map $det(A)^{1/n}$ restricted to the positive definite matrices is the following fact due to Błocki ([**6**]):

THEOREM 2.3. *Assume that u, w are two bounded plurisubharmonic functions in a domain $\Omega \subset \mathbb{C}^n$ such that*

$$(dd^c u)^n \geq f, \ (dd^c w)^n \geq g$$

for some continuous functions $f, g \geq 0$, then

$$(dd^c u)^k \wedge (dd^c w)^{n-k} \geq f^{k/n} g^{(n-k)/n}.$$

In particular

$$(dd^c(u + w))^n \geq (f^{1/n} + g^{1/n})^n.$$

We remark that the above theorem still holds if $f, g \in L^1(\Omega)$ — see [**28**] for a proof.

We shall need also the notion of an *extremal function*: Denote by $\mathcal{L}(\mathbb{C}^n)$ the class of plurisubharmonic functions of logarithmic growth

$$\mathcal{L}(\mathbb{C}^n) := \{u \in PSH(\mathbb{C}^n) \mid u(z) \leq \log(1 + ||z||) + C_u\},$$

where the constant C_u depends on the function u but not on z.

Let K be a compact subset of $\mathbb{C}^n$. The Green function of K with pole at infinity, also known as the Siciak-Zahariuta extremal function is defined by

$$V_K(z) := \sup\{v(z) \mid v \in \mathcal{L}(\mathbb{C}^n), v|_K \leq 0\}.$$

This is a lower semicontinuous function in general and its upper semicontinuous regularization V_K^* is defined by

$$V_K^*(z) := \limsup_{w \to z} V_K(w).$$

We recall the following classical theorem — its proof can be found, for example, in [**28**]:

THEOREM 2.4. *Let K be a compact subset of $\mathbb{C}^n$. Then $V_K^* \equiv +\infty$ if and only if K is a pluripolar set. If $V_K^* \not\equiv +\infty$ then it is a plurisubharmonic function in the class $\mathcal{L}(\mathbb{C}^n)$. Furthermore it is equal to zero on K off a (possibly empty) pluripolar set, and it is maximal outside K in the sense that $(dd^c V_K^*)^n \equiv 0$ off K.*

Of course in complex dimension one the last property means that V_K^* is harmonic off K.

The *maximality* of Green functions outside the set K implies that they decay to zero as the argument approaches the boundary of K in the *slowest* possible fashion among all plurisubharmonic functions in the class $\mathcal{L}$, of course off the aforementioned pluripolar set.

DEFINITION 2.5. A compact set K is called regular if V_K is a continuous function.

In particular for regular sets $V_K^* = V_K$ and $K = \{V_K^* = 0\}$, so the pluripolar set in the above theorem is empty. In our applications more regularity of V_K^* will be needed:

DEFINITION 2.6. A regular compact set K is said to have **Hölder continuity property** of order α ($K \in HCP(\alpha)$) if the function $V_K = V_K^*$ is α-Hölder continuous.

A condition partially converse to (HCP) is the so-called Łojasiewicz-Siciak condition:

DEFINITION 2.7. A regular compact set K is said to satisfy the **Łojasiewicz-Siciak condition** of order α ($K \in LS(\alpha)$) if the function $V_K = V_K^*$ satisfies the inequality

$$V_K(z) \geq C dist(z, K)^\alpha, \text{ if } dist(z, K) \leq 1$$

for some positive constant C independent of the point z. The distance is with respect to the usual Euclidean metric.

For more details in pluripotential theory we refer to [**28**] and the recent book [**22**].

2.2. Quasiconformal mappings. We shall need the notion of a quasiconformal map for the construction of examples of singular sets later on. Recall that the notion of a quasiconformal mapping is a generalization of the classical conformal maps. We restirct our attention to domains in $\mathbb{C}$ in this subsection. Below we present one of the equivalent definitions of a quasiconformal map:

DEFINITION 2.8. Let $f : U \to \Omega$ be a homeomorphism between domains in the complex plane. f is said to be k-quasiconformal for some constant $k \geq 1$ if for any $z \in U$

$$\limsup_{r \to 0^+} \frac{max_{|h|=r}|f(z+h) - f(z)|}{min_{|h|=r}|f(z+h) - f(z)|} \leq k.$$

k-quasiconformal mappings for $k = 1$ are exactly the conformal ones. For $k > 1$ these mappings are much more flexible, yet they share some of the basic properties of conformal maps.

The following is a classical regularity theorem for such maps (see for example Theorem 5 in [**2**]):

THEOREM 2.9. *If $f : U \to \Omega$ is k-quasiconformal, then for any compact $E \Subset U$ $f|_E$ is a $1/k$-Hölder mapping with Hölder constant dependent only on $dist(E, \partial U)$.*

The following corollary of this result will be used later on:

COROLLARY 2.10. If f is a conformal mapping from a domain U onto a domain Ω which admits a k-quasiconformal extension to a domain U', such that $U \Subset U'$, then f is $1/k$-Hölder continuous up to the boundary of U.

The following proposition will be crucial in what follows. We provide a proof for the sake of completeness:

PROPOSITION 2.11. Let K be a connected compact set. Let also g be the Riemann conformal map from $\mathbb{C} \setminus \mathbb{D}$ to the unbounded component of $\mathbb{C} \setminus K$, sending the infinity to infinity. If g extends to the boundary as an α-Hölder continuous mapping, then K satisfies the Łojasiewicz-Siciak condition of order $1/\alpha$.

PROOF. The complement on the Riemann sphere of a connected compact set is simply connected and hence the Riemann mapping exists. Let $z \in \mathbb{C} \setminus K$ be a point satisfying $dist(z, K) \leq 1$. Let $w \in \mathbb{C} \setminus \mathbb{D}$ be the preimage of z under g. If w_0 is the closest point to w lying on the unit circle then by assumption we obtain

$$C(|w| - 1)^\alpha = C|w - w_0|^\alpha \geq |g(w) - g(w_0)| = |z - g(w_0)|.$$

If now g^{-1} denotes the inverse mapping of g we have

$$|g^{-1}(z)| - 1 \geq \left(\frac{|z - g(w_0)|}{C}\right)^{1/\alpha} \geq \left(\frac{dist(z, K)}{C}\right)^{1/\alpha}. \tag{2.1}$$

The proof is finished by noticing that $V_K(z) = \log|g^{-1}(z)|$ in this case. Indeed, we recall that $\log|g^{-1}(z)|$ is harmonic, equal to 0 on ∂K and has logarithmic growth at infinity, since g^{-1} has a simple pole at infinity. □

We refer to [**1**] for more details regarding quasiconformal mappings.

2.3. Elements of viscosity theory. Below we gather the basic notions in viscosity theory. We refer to [**16**] for a detailed introduction of the general viscosity theory and to [**40**] for the viscosity theory associated to the complex Monge-Ampère equation.

DEFINITION 2.12. A function q defined on some neighborhood U of a point z is called a (local) differential test from above at z for the upper-semicontinuous function φ defined on a domain $\Omega \subset \mathbb{C}^n$, also containing z, if it is $\mathcal{C}^2$ smooth on U, and

$$\varphi(z) - q(z) = \sup_{w \in u \cap \Omega} (\varphi(w) - q(w)).$$

Note that if $q \in \mathcal{C}^2(U)$, $q \geq \varphi$ on $U \cap \Omega$ and $\{w \in \Omega \cap U | q(w) = \varphi(w)\} \ni z$ then q is a differential test from above for φ at z.

DEFINITION 2.13. An upper-semicontinuous function φ on a domain Ω is said to allow a differential test from above at $z \in \Omega$ if there exists $U \ni z$ such that the set of differential tests from above for φ at z is non-empty.

Clearly φ allows a differential test from above at any point at which it is twice differentiable.

DEFINITION 2.14. The set of all points $z \in \Omega$ for which a given upper-semi-continuous function φ admits upper differential tests is called the **contact set** (of the function φ). Its complement in Ω is called the **non-contact set**.

The viscosity theory for elliptic equations intuitively boils down to exchanging the possibly singular function u by its smooth majorant thus working with genuinely smooth objects. The price to pay is that at points where u does not admit differential tests (the *non-contact set*) one has no control. It is thus desirable to prove that the non-contact set is small — ideally of zero Lebesgue measure.

2.4. A basic regularity result for the Dirichlet problem. The regularity theory of the complex Monge-Ampère equation investigates the regularity of the solution u provided the right-hand side is fixed and some other constraints are provided. A milestone in the whole theory is the following global regularity theorem for the complex Monge-Ampère equation by Caffarelli, Kohn, Nirenberg and Spruck [**14**].

THEOREM 2.15. *Let Ω be a bounded strictly pseudoconvex domain of class $C^{3,1}$. Let also $\varphi \in C^{3,1}(\partial\Omega)$. For a fixed number $c > 0$ and a function $f \in C^{1,1}(\overline{\Omega})$, $f \geq c$ the Dirichlet problem*

$$\begin{cases} u \in C(\overline{\Omega}), u - \text{plurisubharmonic}; \\ (dd^c u)^n = f; \\ u(z)|_{\partial\Omega} = \varphi(z) \end{cases} \tag{2.2}$$

admits a unique solution $u \in C^{1,1}(\overline{\Omega})$. Moreover the $C^{1,1}$ norm of u is controlled by $n, c, ||f||_{C^{1,1}}, ||\varphi||_{C^{3,1}(\partial\Omega)}$ and the geometry of Ω.

The above theorem is *optimal* in terms of regularity assumptions on the data. In particular smoother data yields smoother solutions.

PROPOSITION 2.16. Let Ω, u, φ, c and f be as above. If additionally f, $\varphi \in C^{k,\alpha}(\overline{\Omega})$ for some $k \geq 2$, $\alpha \in (0, 1)$ then $u \in C^{k+2,\alpha}(\Omega)$.

PROOF. By Theorem 2.15 u is $C^{1,1}$ smooth up to the boundary. If we differentiate in a fixed direction η the equation (2.2), we will obtain that almost everywhere $\frac{\partial u}{\partial \eta}$ satisfies a linear uniformly elliptic equation. Then the result follows from classical linear elliptic theory if $k \geq 3$ and from Evans-Krylov theory if $k = 2$ — see [**21**]. □

3. Local solutions of the complex Monge-Ampère equation — basic examples

In this section we begin studying the purely local properties of plurisubharmonic solutions to the complex Monge-Ampère equation.

Our first example shows that, contrary to the Laplace equation, the Monge-Ampère mass control alone is insufficient to yield even the local boundedness of the solution:

EXAMPLE 3.1. Let φ_m be a sequence of decreasing functions on the unit sphere in $\mathbb{C}^n$ defined by $\varphi_m(z', z_n) = max\{-log(-log|z'|), -m\}$. Let u_m be the (plurisubharmonic) solution to the Dirichlet problem

$$\begin{cases} (dd^c u_m)^n = 1 \text{ in } B_0(1); \\ u_m|_{\partial B_0(1)} = \varphi_m. \end{cases}$$

Let finally $u(z) = lim_{n\to\infty} u_n(z)$. Then u is a plurisubharmonic function such that $(dd^c u)^n = 1$, but $u|_{z'=0} = -\infty$.

Indeed, the sequence u_m is decreasing by Theorem 2.2, hence by Theorem 2.1 and the remark after it $(dd^c u)^n = 1$ unless $u = -\infty$ everywhere. But again by Theorem 2.2 $u_m \geq max\{-log(-log|z'|), -m\} + |z|^2 - 1$, and hence u is locally bounded away from $z' = 0$ and has at worst a loglog poles there. As $u_m|_{\{z'=0\}}$ is also plurisubharmonic, by the comparison principle once again (Theorem 2.2) it is bounded from above by $-m$ there. Thus $u|_{z'=0} = -\infty$.

Heuristically the strict positivity of the right hand side of the equation plays the role analogous to the *strict ellipiticity* in linear PDEs.

The following example by Gamelin and Sibony [**20**] shows that $C^{1,1}$ regularity is optimal when considering the degenerate complex Monge-Ampère equation (that is we drop the condition of strict positivity of the right hand side):

EXAMPLE 3.2. Let u be defined in the unit ball $B_0(1)$ in $\mathbb{C}^2$ by

$$\begin{cases} u \in C(\overline{B_0(1)}), u - \text{plurisubharmonic}; \\ (dd^c u)^n = 0; \\ u(z, w)|_{\partial B_0(1)} = \varphi(z, w), \end{cases} \tag{3.1}$$

where $\varphi(z, w) = max\{|z|^2 - 1/2, |w|^2 - 1/2\}$. Then φ and $\partial B_0(1)$ are smooth but $u(z, w) = max\{0, |z|^2 - 1/2, |w|^2 - 1/2\} \in C^{1,1}(\overline{B_0(1)}) \setminus C^2(B_0(1))$.

Our next example shows that even if the right hand side is smooth and strictly positive then bounded plurisubharmonic solutions may fail to be $C^{1,1}$ or even C^1 smooth. This example is due to Błocki [**7**] who relied on real examples of Pogorelov (see [**23, 30**]). The slight generalisation of Błocki's example is taken from [**17**].

EXAMPLE 3.3. Set $z = (z', z'')$ with $z' = (z_1, \cdots, z_{n-k})$, $z'' = (z_{n-k+1}, \cdots, z_n)$ then the plurisubharmonic function

$$u_k(z) := |z'|^{2-\frac{2k}{n}} \left(1 + |z''|^2\right)$$

has Monge-Ampère density equal to

$$\left(\frac{n-k}{n}\right)^2 \left(1 + |z''|^2\right)^{n-k-1},$$

which is smooth and strictly positive, but u_k is singular along the k-dimensional subspace $z' = 0$.

It is crucial in the example above that the singularities are *global* i.e. propagate past the boundary of any bounded domain containing them. Indeed no such example with *compact* singularities exists — its existence would have violated Yau's regularity theorem for the complex Monge-Ampère equation on compact Kähler manifolds [**39**]. Another observation which we shall investigate in detail in the last section is that the size of the singular set in the examples above decreases with the increase of the regularity both in the Sobolev and Hölder scale.

4. Interior estimates

In this section we shall discuss what kind of estimates one can obtain in the *interior* of a given domain where the complex Monge-Ampère equation is being considered. We shall also stress the dependence on the boundary data whenever it is crucial.

4.1. Uniform estimates. The Example 3.1 clearly shows that there is no hope to obtain purely local uniform estimates through the Monge-Ampère density. Assuming however some control on the boundary of the domain of definition (i.e considering a possibly degenerate Dirichlet problem) changes the situation dramatically. We have the following fundamental theorem of Kołodziej — see [**27**]:

THEOREM 4.1. *Let Ω be a strictly pseudoconvex domain in $\mathbb{C}^n$. Let $\varphi \in C(\partial\Omega)$ and $f \geq 0$, $f \in L^p(\Omega)$ be given, where $p > 1$. The Dirichlet problem*

$$\begin{cases} u \in C(\overline{\Omega}), u - \text{plurisubharmonic}; \\ (dd^c u)^n = f; \\ u(z)|_{\partial\Omega} = \varphi(z), \end{cases} \tag{4.1}$$

admits a unique solution. Furthermore there is a constant C dependent only on $osc_{\partial\Omega}\varphi, n, p$ and Ω such that

$$osc_{\Omega} u \leq C||f||_{L^p}^{1/n}.$$

The proof of this fact (see [**28**]) exploits the notion of a *relative capacity* which is beyond the scope of the current survey. Instead relaxing a bit the conditions on the right hand side there is a much simpler proof due to Z. Błocki (see [**9**]). Below we will prove the Kołodziej theorem under the assumption that $f \in L^2(\Omega)$.

PROOF. The uniqueness is a direct consequence of the comparison principle (Theorem 2.2).

Assume first that f and φ are smooth and f is strictly positive i.e. $f \geq C > 0$ for some constant C. Then the existence of u follows from Theorem 2.15. Now we focus on the uniform bound of u.

If $min_{\overline{\Omega}}u \geq min_{\partial\Omega}\varphi$ there is nothing to prove. Suppose then that there exists a constant c such that $\{u < c\}$ is relatively compact in Ω. It suffices thus to bound the supremum of $c - min_{\overline{\Omega}}u$ over all such c having this property.

Recall that the Alexandrov maximum principle (see [**21, 23**]) states that if u is a C^2 function vanishing on the boundary of a domain Ω in $\mathbb{C}^n$ regarded as $\mathbb{R}^{2n}$ then

$$sup_{\Omega}(-u) \leq \frac{diam\Omega}{\lambda_{2n}^{1/(2n)}}\Big(\int_{\Gamma} detD^2u\Big)^{1/(2n)}, \tag{4.2}$$

where λ_{2n} denotes the volume of the unit ball in $\mathbb{R}^{2n}$ and Γ is the set where $D^2u \geq 0$. The Alexandrov maximum principle can be extended to merely continuous functions (see [**23**]) provided Γ is taken to be the set where u agrees with its *convex envelope*. We refer to [**23**] for the details.

By standard computation if D^2u is nonnegative as a matrix then

$$(dd^cu)^n \geq C(n)\sqrt{\det D^2u}$$

for some numerical constant $C(n)$.

Coupling these in a connected component of the set $\{u < c\}$ one has

$$c - min_{\overline{\Omega}}u = sup(c-u) \leq C\Big(\int_{\Omega}\{(dd^cu)^n\}^2\Big)^{1/2n} = C\Big(\int_{\Omega}|f|^2\Big)^{1/2n}$$

and the latter quantity is bounded by assumption. This settles the proof in the case of smooth data. Note that we haven't used the positive lower bound for f.

In the general case of continuous φ and $f \in L^2$ consider a sequence (φ_n, f_n) of approximating data $\varphi_n \in C^\infty(\partial\Omega), 1/n < f_n \in L^2(\Omega) \cap C^\infty(\overline{\Omega})$ such that $osc_{\partial\Omega}|\varphi_n - \varphi| \to 0$ and $||f_n - f||_{L^2(\Omega)} \to 0$ as n tends to infinity. Let u_n be the solution of the Dirichlet problem with φ_n as boundary data and f_n as the right hand side. It suffices to show that the sequence of plurisubharmonic functions u_n converges uniformly to some limit, which by the Bedford-Taylor convergence theorem would imply that the limiting function u solves the initial Dirichlet problem with non-smooth data.

We will show that u_n's form a Cauchy sequence in the uniform topology. To this end fix $n, m \in \mathbb{N}$, $\epsilon > 0$ and consider the Dirichlet problem

$$\begin{cases} \rho \in C(\overline{\Omega}), \rho - \text{plurisubharmonic}; \\ (dd^c\rho)^n = \tilde{max}\{f_n - f_m, \epsilon\}; \\ \rho(z)|_{\partial\Omega} = \varphi_n(z) - \varphi_m(z), \end{cases} \tag{4.3}$$

where $\tilde{max}(x,y)$ is the regularization of the maximum function satisfying the conditions

(1) $\tilde{max}$ is smooth and $\tilde{max}(x,y) \geq max\{x,y\}$;
(2) if $|x-y| \geq \epsilon$ then $\tilde{max}(x,y) = max\{x,y\}$.

As f_n is a Cauchy sequence in L^2 and φ_n is a Cauchy sequence in $C(\overline{\Omega})$ it is easy to see that the proof in the smooth case yields that once n and m are sufficiently large $osc_{\Omega}\rho \leq C\epsilon$.

But $\rho + u_m$ has Monge-Ampère measure larger than the measure of u_n. Furthermore these two functions share the same boundary values. Hence by Theorem 2.2 we have $u_m - u_n \leq -\rho$. Interchanging the role of n nd m we obtain a bound on the oscillation of $u_n - u_m$ which is uniformly small for m, n large enough. This shows that u_n is a Cauchy sequence in the uniform topology.

Finally the uniform bound in the non-smooth case follows the same lines as in the smooth case once the existence is established. □

4.2. Błocki gradient estimate in convex domains. Again it has to be stressed that no purely interior gradient estimate could hold for the complex Monge-Ampère equation. If in turn one assumes sufficiently regular boundary data a global gradient bound follows easily by maximum principle argument (compare the proof of Caffarelli-Kohn-Nirenberg-Spruck theorem (Theorem 2.15 in [**14**])). On the other hand there is a powerful technique of Walsh [**35**] which is useful in obtaining Hölder estimates for Hölder data on the boundary. Our goal will be to establish an interior gradient bound under minimal possible assumptions.

As Examples 3.3 show it is necessary to assume some boundary condition. But if $\partial\Omega$ fails to be sufficiently smooth there is a problem how to define the regularity of the boundary data. We shall then assume that the plurisubharmonic function vanishes on the boundary of the considered domain Ω but we shall make no assumptions on the regularity of $\partial\Omega$. As gradient estimates require differentiation of the equation it is reasonable to assume some gradient control on the right hand side. It turns out that the natural quantity to control is $||f^{1/n}||_{C^1(\Omega)}$. Then a question arises:

QUESTION 4.2. Suppose that u solves the problem

$$\begin{cases} u \in C(\overline{\Omega}), u - \text{plurisubharmonic;} \\ (dd^c u)^n = f, f \geq 0; \\ u(z)|_{\partial\Omega} = 0, \end{cases} \tag{4.4}$$

with $||f^{1/n}||_{C^1}$ uniformly bounded. Is it possible to obtain an interior gradient bound on u i.e. for every relatively compact subset K of Ω there exists a constant C dependent on K, n and $||f^{1/n}||_{C^1}$ such that

$$||u||_{C^1(K)} \leq C?$$

This is one of the central open problems in the local regularity theory of the complex Monge-Ampère equation. It is worth pointing out that its real counterpart is trivial since any convex function is locally Lipschitz.

Below we present the only existent partial result in this direction. It is due to Z. Błocki (see [**8**]). The crucial additional assumption made is that Ω is assumed to be *convex*:

THEOREM 4.3. *Let u, f and Ω be as above. Assume additionally that Ω is convex. Then u is locally Lipschitz with the local Lipschitz constant dependent on the quantities listed in the question above.*

PROOF. We follow the argument in [**8**].

Fix a relatively compact subdomain $\Omega' \subset \Omega$ which we without loss of generality assume to be convex. We may also suppose, shifting the coordinates if necessary, that $0 \in \Omega'$. Let $d := dist(\Omega', \partial\Omega)$. The core of the proof hinges on the following geometric fact: for any two points $a, b \in \Omega'$ such that $|a-b| \leq d$ the linear mapping $T(z) := (1 - \frac{|a-b|}{d})[z-a] + b$ maps Ω into itself (note that we use the convexity of Ω' and Ω here!). Moreover it sends a to b and $|T(z) - z| \leq (1 + \frac{diam\Omega}{d})|a-b|$.

Consider now the function

$$v(z) := u\Big(T(z)\Big) + C\Big(|z-a|^2 - (diam\Omega)^2\Big)|a-b|.$$

If we can prove that $v \leq u$ for some sufficiently large constant C, then

$$0 \geq v(a) - u(a) = u(b) - u(a) - C|a-b|(diam\Omega)^2.$$

Reversing the roles of a and b would then finish the proof.

We wish to apply Theorem 2.2. To this end note that on $\partial\Omega$ $v(z) \leq u(T(z)) \leq 0$. It remains to prove the corresponding inequality of the Monge-Ampère measures. Recall that $\left(dd^c u \circ T\right)^n = |det\left(T'\right)|^2\left(dd^c u\right)^n \circ T$. By Theorem 2.3 and the Lipschitz bound on $f^{1/n}$ we have

$$\begin{aligned}(dd^c v)^n &\geq \left((1 - \frac{|a-b|}{d})^2 f^{1/n} \circ T(z) + C|a-b|\right)^n \\ &\geq ((1 - \frac{|a-b|}{d})^2 f^{1/n}(z) - (1 - \frac{|a-b|}{d})^2 \tilde{C}|T(z) - z| + C|a-b|)^n \\ &\geq (f^{1/n}(z) + (C - \hat{C})|a-b|)^n\end{aligned}$$

for some constants $\tilde{C}, \hat{C}$ dependent on the Lipschitz norm of $f^{1/n}, d$ and $diam\Omega$. Taking now $C = \hat{C}$ yields

$$(dd^c v)^n \geq f = (dd^c u)^n$$

and the needed inequality is proven. □

4.3. Bedford-Taylor $C^{1,1}$ estimates in a ball. The result, taken from the paper of Bedford and Taylor [**4**], provides an interior $C^{1,1}$ bound on the solution for suitably regular boundary data and right-hand side function. It has to be stressed that this bound hinges on a *complex argument* i.e. cannot be repeated for the real Monge-Ampère equation. The crucial fact that makes the argument work is the transitivity of the automorphism group of the ball.

Here is the formulation of the theorem in a version that will be used later on:

THEOREM 4.4. *Let u be a plurisubharmonic function in the ball $B_{z_0}(r)$ in $\mathbb{C}^n$ which is continuous up to the boundary. Suppose that u solves the problem*

$$\begin{cases}(dd^c u)^n = f \\ u|_{\partial B_{z_0}(r)} = \varphi.\end{cases} \tag{4.5}$$

Suppose that for some positive constants M and N

$$||f^{1/n}||_{C^{1,1}(B_{z_0}(r))} \leq Mr^{-2}, \quad ||\varphi||_{C^{1,1}(\partial B_{z_0}(r))} \leq Nr^{-2}.$$

Then there exists a numerical constant C dependent only on n, M and N such that

$$||u||_{C^{1,1}(B_{z_0}(r/2))} \leq Cr^{-2}.$$

PROOF. By shifting we can assume without loss of generality that $z_0 = 0$. Upon rescaling $v(z) := u(rz)$ it suffices to work in the unit ball $B_0(1)$. Similarly to [**6**] we consider the following holomorphic automorphism of $B_0(1)$: for an $a \in B_0(1)$ let

$$T_a(z) := \frac{a - P_a(z) - s_a Q_a(z)}{1 - \langle z, a \rangle}, \tag{4.6}$$

where $P_a(z) = \frac{\langle z,a \rangle}{|a|^2}a$, $Q_a(z) = z - P_a(z)$ and $s_a = \sqrt{1-|a|^2}$. This is the automorphism sending a to zero and zero to a (it can be checked that $T \circ T = Id$ i.e. T_a is an involution).

Fix a vector h in $\mathbb{C}^n$ of short length, say $|h| \leq 1/2$. Define now

$$L(a,h,z) := T_{a+h}^{-1}\Big(T_a(z)\Big). \tag{4.7}$$

By construction $L(a,h,\cdot)$ is an automorphism "close" to the shift $z \to z+h$. As $L(a,h,z)$ is C^2 in h as long as $|h| < 1/2$, $|a| \leq 1/2$ and $L(a,0,z) = z$, it can be computed that for some constant $c(a,z)$

$$|L(a,h,z) - z - c(a,z)h| \leq C_1(a,z)|h|^2, \ |det\Big(L_z(a,h,z)'\Big) - 1| \leq C_2(a,z)|h|^2, \tag{4.8}$$

where L_z' stands for the Jacobian of L in the z variables and the quantities $C_1(a,z)$, $C_2(a,z)$ are uniform as long as $a \in B_0(1/2)$. Contrary to the shift, however, $L(a,h,z)$ sends the unit ball into itself.

Next, we claim that the function

$$u^{(a)}(z) := [u\Big(L(a,h,z)\Big) + u\Big(L(a,-h,z)\Big)]/2 + C|h|^2(|z|^2 - 2)$$

is less or equal than u for some sufficiently large constant C (uniform in a as long as a belongs to the ball $B_0(1/2)$). Indeed, suppose this claim is true. Plugging then $a = z$ we obtain

$$(u(z+h) + u(z-h))/2 - u(z) \leq C|h|^2(2 - |z|^2)$$

for every sufficiently short vector h, which immediately yields the theorem.

In order to get our claim we shall apply the comparison principle. Indeed by Theorem 2.3

$$\begin{aligned}
(dd^c u^{(a)}(z))^n &\geq \Big(\frac{1}{2}\{dd^c[u\Big(L(a,h,z)\Big)]^n\}^{1/n} \\
&\quad + \frac{1}{2}\{dd^c[u\Big(L(a,-h,z)\Big)]^n\}^{1/n} + C|h|^2\Big)^n \\
&= \Big(\frac{1}{2}\det\Big(L_z(a,h,z)'\Big)^{2/n} f^{1/n} \circ L(a,h,z) \\
&\quad + \frac{1}{2}\det\Big(L_z(a,-h,z)'\Big)^{2/n} f^{1/n} \circ L(a,-h,z) + C|h|^2\Big)^n.
\end{aligned}$$

At this moment we make use of the assumption that $f^{1/n} \in C^{1,1}$. Using the Taylor expansion of $f^{1/n} \circ L(a,h,z)$ at the point z we have

$$f^{1/n} \circ L(a,h,z) = f^{1/n}(z) + D(f^{1/n})(z)\Big(L(a,h,z) - z\Big) + O(|L(a,h,z) - z|^2).$$

Exploiting further the estimates (4.8) we have the inequality

$$f^{1/n} \circ L(a,h,z) \geq f^{1/n}(z) + D(f^{1/n})(z)c(a,z)h - C_3(a,z)|h|^2$$

for some constant $C_3(a,z)$ dependent on $C_1(a,z)$ and on the $C^{1,1}$ norm of $f^{1/n}$.

Thus the string of inequalities yields

$$\begin{aligned}
(dd^c u^{(a)}(z))^n &\geq \Big(\{1 - C_2(a,z)|h|^2\}^{2/n}[f^{1/n}(z) + D(f^{1/n})(z)c(a,z)h - C_3(a,z)|h|^2]/2 \\
&\quad + [\{1 - C_2(a,z)|h|^2\}^{2/n}[f^{1/n}(z) - D(f^{1/n})(z)c(a,z)h - C_3(a,z)|h|^2]/2 + C|h|^2\Big)^n \\
&\geq \Big(\{1 - C_2(a,z)|h|^2\}^{2/n} f^{1/n}(z) - C_4(a,z)|h|^2 + C|h|^2\Big)^n \\
&\geq \Big(f^{1/n}(z) - C_5(a,z)|h|^2 + C|h|^2\Big)^n,
\end{aligned}$$

for some constants $C_4(a,z)$ and $C_5(a,z)$ dependent on the $C^{1,1}$ norm of $f^{1/n}$ which are uniform for $a \in B_0(1/2)$.

Thus finally

$$(dd^c u^{(a)}(z))^n \geq f = \left(dd^c u(z)\right)^n$$

if C is taken large enough. We can apply the comparison principle provided $u^{(a)}(z)$ is majorized by $u = \varphi$ on $\partial B_0(1)$. Observe however that on $\partial B_0(1)$ one has

$$u^{(a)}(z) \leq [\varphi\Big(L(a,h,z)\Big) + \varphi\Big(L(a,-h,z)\Big)]/2 - C|h|^2 \leq N|h|^2 - C|h|^2.$$

Thus taking C large enough finishes the proof. □

4.4. C^3 Calabi interior estimates. In this subsection we shall discuss the local Calabi C^3 estimate. It establishes a C^3 control on solutions to the complex Monge-Ampère equation for sufficiently regular data. The bound depends on the C^2 norm of the solution. It is worth mentioning that this is one of the very few purely local results in the whole theory. Heuristically the proof hinges on the fact that uniform C^2 bound on the solution implies uniform ellipticity of the equation.

We shall state a version of the theorem which was essentially proven by Riebesehl and Schulz [**31**]. As we shall need for later purposes the exact dependencies on the relevant constants we provide a more detailed version of the proof which is taken from [**29**]:

THEOREM 4.5. *Let u be a C^5 plurisubharmonic solution to the equation*

$$(dd^c u)^n = e^{\varphi}$$

in the r-ball $B_0(r)$ centered at zero. Suppose that for some positive constants λ, Λ, M and N:

$$i)\lambda I \leq (u_{k\bar{j}}) \leq \Lambda I;$$
$$ii) - Mr^{-2} I \leq (\varphi_{k\bar{j}}) \leq Mr^{-2} I;$$
$$iii)(\sum_{i,j,k} |\varphi_{i\bar{j}k}|^2)^{1/2} \leq Nr^{-3}.$$

Then in the ball $B_0(r/2)$ one has the bound

$$\sum_{i,j,k} |u_{i\bar{j}k}| \leq C(n)\lambda^{-1}\Lambda^3((1+M)^{-2}N^2 + \lambda^{-1}\Lambda(1+M))r^{-2}$$

for some constant $C(n)$ dependent only on the dimension n.

PROOF. Observe that our assumptions are made so that they scale in the right way under the rescaling $v(z) := \frac{u(rz)}{r^2}$. For this reason it is no loss of generality to assume $r = 1$ i.e. we can work in the unit ball.

In the proof rather than estimating $\sum_{i,j,k} |u_{i\bar{j}k}|^2$ we shall work with the equivalent quantity

$$S := \sum_{i,k,j,p,q,r} u^{i\bar{p}} u^{k\bar{q}} u^{j\bar{r}} u_{j\bar{p}\bar{q}} u_{ik\bar{r}}. \tag{4.9}$$

At this point we adopt the following handy notation, borrowed from [**29**] and [**31**]: by upper indices we shall indicate the following

$$v^j := \sum_k u^{j\bar{k}} v_{\bar{k}},\ v^{\bar{j}} := \sum_k u^{k\bar{j}} v_k,\ v^{jk} := \sum_{m,l} u^{j\bar{m}} u^{k\bar{l}} v_{\bar{m}\bar{l}},\ \text{etc.} \tag{4.10}$$

Also from now on we drop the summation indices as we shall use the Einstein summation convention.

Notice that in this new notation $S = u_j^{ik} u_{ik}^j$. Taking the logarithm of the Monge-Ampère equation and differentiating we obtain the formulas (see also (1,20) and (1.3) in [**31**])

$$u_{ik}^i = \varphi_k,\ u_{ik\bar{k}}^i = u_{jk}^i u_{i\bar{k}}^j + \varphi_{k\bar{k}}.$$

Thus if $L(v) := u^{i\bar{j}} v_{i\bar{j}} = v_i^i$ denotes the Laplacian operator with respect to the Kähler metric $dd^c u$ one computes

$$L(\Delta(u)) = u_{jk}^i u_{i\bar{k}}^j + \Delta\varphi.$$

The computation of $L(S)$ is more involved. The exact formula (compare [**29**], formula (6.5)) is

$$L(S) = T + \varphi_k^{lm} u_{lm}^k + \varphi_{lm}^k u_k^{lm} - \varphi_k^p (u_p^{lm} u_{lm}^k + 2u_{pm}^l u_l^{km}),$$

where

$$T = (u_{ki}^{lm} - u_{pi}^l u_k^{pm} - u_{pi}^m u_k^{lp})(u_{lm}^{ki} - u_l^{pi} u_{pm}^k - u_m^{pi} u_{lp}^k) + (u_{lmi}^k - u_{pi}^k u_{lm}^p)(u_k^{lmi} - u_k^{pi} u_p^{lm}).$$

It can be checked directly (for example by diagonalizing $dd^c u$ at a point) that $T \geq 0$, in fact both terms in T are positive.

The expression T controls the gradient of S in the following way. First by direct computation (see [**29**], formula (6.7))

$$S_i = u_{lm}^k (u_{ki}^{lm} - u_{pi}^l u_k^{pm} - u_{pi}^m u_k^{lp}) + u_k^{lm} (u_{lmi}^k - u_{pi}^k u_{lm}^p).$$

We also have (see pages 60-61 in [**31**]):

$$S^i = u_k^{lm} (u_{lm}^{ki} - u_l^{pi} u_{pm}^k - u_m^{pi} u_{lp}^k) + u_{lm}^k (u_k^{lmi} - u_k^{pi} u_p^{lm}).$$

and hence by Schwarz inequality

$$||\nabla S||_{dd^c u}^2 = 2u^{j\bar{k}} S_j S_{\bar{k}} = 2S_j S^j \leq 4ST. \tag{4.11}$$

Exploiting the assumptions $i), ii)$ and $iii)$ one obtains

$$L(\Delta(u)) \geq \lambda S - nM,\ L(S) \geq T - 2\lambda^{-3/2} N S^{1/2} - 3\lambda^{-1} MS. \tag{4.12}$$

Consider now the function

$$G(z) := (|z|^2 - 9/16)^4 S + A\lambda^{-1} \Delta u,$$

for a constant $A > 0$ to be determined. Note that we can assume that the maximum of G in the ball $B_0(3/4)$ occurs in the interior for otherwise we are through.

From the inequalities (4.12) we obtain

$$\begin{aligned} L(G) \geq \Big(A + L((|z|^2 - 9/16)^4) - 3\lambda^{-1} M (|z|^2 - 9/16)^4\Big) S + (|z|^2 - 9/16)^4 T \\ - (|z|^2 - 9/16)^4 2\lambda^{-3/2} N S^{1/2} - A\lambda^{-1} nM + 2Re([(|z|^2 - 9/16)^4]_i S^i). \end{aligned} \tag{4.13}$$

It is straightforward to check, using (4.11) that

$$|2Re([(|z|^2-9/16)^4]_i S^i)| \leq 2||\nabla(|z|^2-9/16)^4||_{dd^cu}(ST)^{1/2}$$

$$\leq (|z|^2-9/16)^{-4}||\nabla(|z|^2-9/16)^4||_{dd^cu}S + (|z|^2-9/16)^4 T.$$

Again by direct computation

$$(|z|^2-9/16)^{-4}||\nabla(|z|^2-9/16)^4||_{dd^cu} \leq 9\lambda^{-1},$$

$$L((|z|^2-9/16)^4) \geq -9n/8\lambda^{-1}.$$

Thus (4.13) reduces to

$$L(G) \geq (A - C_1(1+M)\lambda^{-1})S - C_2\lambda^{-3/2}NS^{1/2} - C_3A\lambda^{-1}M. \tag{4.14}$$

The choice $A = 2C_1(1+M)\lambda^{-1}$ yields

$$L(G) \geq C_1(1+M)\lambda^{-1}S - C_2\lambda^{-3/2}NS^{1/2} - C_4\lambda^{-2}(1+M)M.$$

Recall now that the operator L is strictly elliptic. Thus at the maximum point x_0 of G one has

$$0 \geq C_1(1+M)\lambda^{-1}S(x_0) - C_2\lambda^{-3/2}NS^{1/2}(x_0) - C_4\lambda^{-2}(1+M)M.$$

The above inequality easily implies

$$S(x_0) \leq C_5\lambda^{-1}((1+M)^{-2}N^2 + M).$$

This, coupled with our choice of A and the assumption (i) results in the bound

$$G(x_0) \leq C_6\lambda^{-1}((1+M)^{-2}N^2 + \lambda^{-1}\Lambda(1+M)).$$

Hence

$$max_{x\in B_0(1/2)}S(x) \leq max_{x\in B_0(1/2)}G(x_0)/(|x|^2-9/16)^4$$

$$\leq C_7\lambda^{-1}((1+M)^{-2}N^2 + \lambda^{-1}\Lambda(1+M)).$$

Now the claimed result follows from the trivial bound

$$\sum_{i,j,k}|u_{i\bar{j}k}|^2 \leq \Lambda^3 S.$$

□

5. Local regularity theory

In this section we shall deal with the purely local regularity theory associated to the complex Monge-Ampère equation. As we have seen in Example 3.3 even smooth strictly positive right hand side may fail to yield differentiability of the solution. For this reason the additional assumption we shall make is that the solution will have some given a priori regularity and we ask whether one can deduct better smoothness. The guiding principle for us will be that all functions like these in Example 3.3 live in spaces of low regularity and, above certain thresholds, solutions are smooth for smooth data.

5.1. Sobolev scale regularity. Example 3.3 for $k = 1$ belongs to $W^{2,p}$ for any $p < n(n-1)$. Below we shall show that this exponent is *sharp*: any solution that is more regular must be smooth:

THEOREM 5.1. *Assume that $n \geq 2$ and $p > n(n-1)$. Let $u \in W^{2,p}(\Omega) \cap PSH(\Omega)$, where Ω is a domain in $\mathbb{C}^n$, be a plurisubharmonic solution of*

$$(dd^c u)^n = \psi > 0.$$

Assume that $\psi \in C^{1,1}(\Omega)$. Then for $\Omega' \Subset \Omega$

$$\sup_{\Omega'} \Delta u \leq C,$$

where C depends only on n, p, $\operatorname{dist}(\Omega', \partial\Omega)$, $\inf_\Omega \psi$, $||\psi||_{C^{1,1}(\Omega)}$ and $||\Delta u||_{L^p(\Omega)}$.

REMARK 5.2. This fact coupled with Theorem 4.5 and Proposition 2.16 immediately implies that $W^{2,p}$ solutions for $p > n(n-1)$ are smooth once $\psi > 0$ is smooth.

PROOF. We follow the argument from [**10**]. The proof hinges on Theorem 4.1 by Kołodziej which yields uniform bounds for solutions to the complex Monge-Ampère equation with L^p right hand side [**27, 28**]. We also exploit the fact that the cofactor minors are themselves products of $n-1$ second order derivatives of u and hence $u \in W^{2,p}$ yields $L^{p/(n-1)}$ summability of these cofactors.

As the result is local it is no loss of generality to assume that $\Omega = B$ is the unit ball in $\mathbb{C}^n$ and that u is defined in some neighborhood of $\bar{B}$.

We will first prove the theorem assuming that u is in C^4. Taking the logarithm of the equation and differentiating with respect to z_p and $\bar{z}_p$ we will get

$$u^{i\bar{j}} u_{i\bar{j}p} = (\log \psi)_p$$

and

$$u^{i\bar{j}} u_{i\bar{j}p\bar{p}} = (\log \psi)_{p\bar{p}} + u^{i\bar{l}} u^{k\bar{j}} u_{k\bar{l}\bar{p}} u_{i\bar{j}p}.$$

Therefore

$$u^{i\bar{j}} (\Delta u)_{i\bar{j}} \geq \Delta(\log \psi). \tag{5.1}$$

We will now use an idea from [**32**]. For some $\alpha, \beta \geq 1$ to be determined later set

$$w := \eta(\Delta u)^\alpha,$$

where

$$\eta(z) := (1 - |z|^2)^\beta.$$

Then

$$w_i = \eta_i (\Delta u)^\alpha + \alpha\eta(\Delta u)^{\alpha-1}(\Delta u)_i$$

and

$$\begin{aligned} u^{i\bar{j}} w_{i\bar{j}} &= \alpha\eta(\Delta u)^{\alpha-1} u^{i\bar{j}} (\Delta u)_{i\bar{j}} + \alpha(\alpha-1)\eta(\Delta u)^{\alpha-2} u^{i\bar{j}} (\Delta u)_i (\Delta u)_{\bar{j}} \\ &\quad + 2\alpha(\Delta u)^{\alpha-1} \operatorname{Re}\big(u^{i\bar{j}} \eta_i (\Delta u)_{\bar{j}}\big) + (\Delta u)^\alpha u^{i\bar{j}} \eta_{i\bar{j}}. \end{aligned}$$

By (5.1) and the Schwarz inequality for $t > 0$

$$\begin{aligned} u^{i\bar{j}} w_{i\bar{j}} &\geq \alpha\eta(\Delta u)^{\alpha-1} \Delta(\log \psi) + \alpha(\alpha-1)\eta(\Delta u)^{\alpha-2} u^{i\bar{j}} (\Delta u)_i (\Delta u)_{\bar{j}} \\ &\quad - t\alpha(\Delta u)^{\alpha-1} u^{i\bar{j}} (\Delta u)_i (\Delta u)_{\bar{j}} - \frac{1}{t}\alpha(\Delta u)^{\alpha-1} u^{i\bar{j}} \eta_i \eta_{\bar{j}} + (\Delta u)^\alpha u^{i\bar{j}} \eta_{i\bar{j}}. \end{aligned}$$

Therefore with $t = (\alpha - 1)\eta/\Delta u$ we get

$$u^{i\bar{j}}w_{i\bar{j}} \geq \alpha\eta(\Delta u)^{\alpha-1}\Delta(\log\psi) + (\Delta u)^{\alpha}u^{i\bar{j}}\left(\eta_{i\bar{j}} - \frac{\alpha}{\alpha-1}\frac{\eta_i\eta_{\bar{j}}}{\eta}\right).$$

We now have

$$\eta_i = -\beta\bar{z}_i\eta^{1-1/\beta}$$
$$\eta_{i\bar{j}} = -\beta\delta_{i\bar{j}}\eta^{1-1/\beta} + \beta(\beta-1)\bar{z}_i z_j\eta^{1-2/\beta}$$

and thus

$$|\eta_{i\bar{j}}|,\ |\frac{\eta_i\eta_{\bar{j}}}{\eta}| \leq C(\beta)\eta^{1-2/\beta}.$$

We will get

$$u^{i\bar{j}}w_{i\bar{j}} \geq -C_1(\Delta u)^{\alpha-1} - C_2 w^{1-2/\beta}(\Delta u)^{2\alpha/\beta}\sum_{i,j}|u^{i\bar{j}}|.$$

Fix q with $1 < q < p/(n(n-1))$. Since $||\Delta u||_{L^p(B)}$ is bounded, it follows that $||u_{i\bar{j}}||_p$ and $||u^{i\bar{j}}||_{p/(n-1)}$ are as well. If now

$$\alpha = 1 + \frac{p}{qn},\quad \beta = 2(\frac{qn+p}{p-qn(n-1)})$$

(our assumptions imply that $\alpha > 1, \beta > 1$) we have

$$||(u^{i\bar{j}}w_{i\bar{j}})_-||_{qn} \leq C_3(1 + (\sup_B w)^{1-2/\beta}),$$

where $f_- := -\min(f, 0)$.

By Theorem 4.1 we can find continuous plurisubharmonic v vanishing on ∂B and such that

$$\det(v_{i\bar{j}}) = ((u^{i\bar{j}}w_{i\bar{j}})_-)^n$$

(weakly). By Theorem 2.3 we have in turn that

$$\begin{aligned} u^{i\bar{j}}v_{i\bar{j}} &\geq n(\det(u^{i\bar{j}}))^{1/n}(\det(v_{i\bar{j}}))^{1/n} \\ &= n\psi^{-1/n}(u^{i\bar{j}}w_{i\bar{j}})_- \\ &\geq -\frac{1}{C_4}u^{i\bar{j}}w_{i\bar{j}}. \end{aligned}$$

It follows that $w \leq -C_4 v$ and by Theorem 4.1

$$\begin{aligned} \sup_B w &\leq C_5||\det(v_{i\bar{j}})||_q^{1/n} \\ &= C_5||(u^{i\bar{j}}w_{i\bar{j}})_-||_{qn} \\ &\leq C_6(1 + (\sup_B w)^{1-2/\beta}). \end{aligned}$$

Therefore $w \leq C_7$ and the desired estimate follows if $u \in C^4$.

The idea for a general $W^{2,p}$ function u is to exchange Δu in the proof above by the following approximation: we will consider for a fixed $\varepsilon > 0$ the quantity

$$T = T^{(\varepsilon)}u = \frac{n+1}{\varepsilon^2}(u_\varepsilon - u),$$

where

$$u_\varepsilon(z) = \frac{1}{\lambda(B(z,\varepsilon))}\int_{B(z,\varepsilon)} u\,dV.$$

Since $T^{(\varepsilon)}u \to \Delta u$ weakly as $\varepsilon \to 0$, it is enough to show a uniform upper bound for T independent of ε.

By Theorem 2.3 we have

$$u^{i\bar{j}}u_{\varepsilon,i\bar{j}} \geq n\psi^{-1/n}(\det(u_{\varepsilon,i\bar{j}}))^{1/n} \geq n\psi^{-1/n}(\psi^{1/n})_\varepsilon.$$

Coupling this with $u^{i\bar{j}}u_{i\bar{j}} = n$, we obtain the following inequality:

$$u^{i\bar{j}}T_{i\bar{j}} \geq n\psi^{-1/n}\frac{n+1}{\varepsilon^2}\left((\psi^{1/n})_\varepsilon - \psi^{1/n}\right) \geq -C_8.$$

Changing the definition of w to ηT^α (since u is plurisubharmonic, T is nonnegative, hence T^α is well defined) and repeating the previous computations we will get

$$u^{i\bar{j}}w_{i\bar{j}} \geq C_9T^{\alpha-1} - C_{10}w^{1-2/\beta}T^{2\alpha/\beta}\sum_{i,j}|u^{i\bar{j}}|.$$

The rest of the proof is now the same as before. □

5.2. Hölder scale regularity. In the previous subsection we have dealt with the local regularity problem in the setting of Sobolev spaces. It is also natural to ask whether the Example 3.3 for $k = 1$ is *the worst one* in the Hölder scale. It is easy to see that this example belongs to $C^{1,1-2/n}_{loc}$. Hence the following question appears:

QUESTION 5.3. Assume that $n \geq 2$ and $1 \geq \alpha > 1 - 2/n$. Let $u \in C^{1,\alpha}(\Omega) \cap PSH(\Omega)$, where Ω is a domain in $\mathbb{C}^n$, be a plurisubharmonic solution of

$$(dd^cu)^n = \psi > 0.$$

Is it true that $\psi \in C^\infty$ implies that u is smooth?

A bit surprisingly, given the complete resolution in the Sobolev case, this question is still open despite numerous partial results (see [**19, 29, 36, 38**]). The major difference from the previous setting is that Hölder regularity gives no information whatsoever on the cofactor minors of the Hessian matrix.

The natural approach to this problem would be to differentiate it (or the logarithm of it) and analyze some derivatives of u multiplied by a cut-off function. Note however that *if u is C^3 differentiable* this can be done in a standard way by applying Schauder theory to the linearized equation. Thus the core of the problem is to pass through C^2 regularity threshold.

In this low-regularity realm (say we want to prove that $u \in C^{2,\alpha}$ for some $\alpha > 0$) it is however unreasonable to input strong differentiability properties on the right hand side. Indeed simple computation yields that $u \in C^{2,\alpha}$ results in $\psi \in C^\alpha$ and in general we could not expect smoother ψ. Hence the core of the problem boils down to solving the following question:

QUESTION 5.4. Assume that $u \in PSH(\Omega)$ solves the problem

$$(dd^cu)^n = \psi > 0.$$

If $\psi \in C^\alpha$ for some $\alpha \in (0, 1)$ what minimal condition on u implies $u \in C^{2,\alpha}$?

Clearly the expected condition should be that $u \in C^{1,\beta}$ for $\beta > 1 - 2/n$. As we already explained this problem remains open. Below we shall show some partial results implying in particular that $\beta > 1 - \frac{\alpha}{n(2+\alpha)-1}$ is sufficient.

We shall begin with the following theorem which shall serve as a first step towards the (currently) optimal result:

THEOREM 5.5. *Let Ω be a domain in $\mathbb{C}^n$ and $u \in PSH(\Omega) \cap C^{1,1}(\Omega)$ satisfy the Monge-Ampère equation*

$$det(u_{i\bar{j}}) = f.$$

Suppose additionally that $f \geq \lambda > 0$ in Ω for some constant λ and $f \in C^{\alpha}(\Omega)$ for some $\alpha \in (0,1)$. Then $u \in C^{2,\alpha}(\Omega)$. Furthermore the $C^{2,\alpha}$ norm of u in any relatively compact subset is estimable in terms of $n, \alpha, \lambda, ||f||_{C^{\alpha}(\Omega)}$ and the distance of the set to $\partial\Omega$.

PROOF. We follow the reasoning from [**19**]. Heuristically, as we cannot differentiate the equation, we shall localize the quadratic approximants of u in small balls and show that those converge. To this end we exploit Bedford-Taylor $C^{1,1}$ estimate (Theorem 4.4) and Calabi's C^3 estimates (Theorem 4.5).

Fix $\Omega'' \Subset \Omega' \Subset \Omega$ and let $d := dist(\Omega'', \partial\Omega')$. It is enough to show that $u \in C^{2,\alpha}(\Omega'')$.

For any fixed $x_0 \in \Omega''$ let us consider the system of balls $B_{x_0}(d\rho^k)$, $k = 0, 1, 2, \cdots$, where we put $\rho := 1/2$. Associated to any such system let $u(x; x_0, k)$ be the solution to the following Dirichlet problem

$$\begin{cases} u(x;x_0,k) \in PSH(B_{x_0}(d\rho^k)) \cap C(\overline{B_{x_0}(d\rho^k)}) \\ det(u(x;x_0,k)_{i\bar{j}}) = f(x_0) \text{ in } B_{x_0}(d\rho^k), \\ u(x;x_0,k) = u \text{ on } \partial B_{x_0}(d\rho^k). \end{cases} \tag{5.2}$$

For notational ease we denote $B_{x_0}(d\rho^k)$ by B_k and $u(x;x_0,k)$ by u_k. In case two different systems (with different centers) will appear, we shall mark them by $\widetilde{u}_k$ and $\widehat{u}_k$ to make a distinction.

By the Bedford-Taylor interior $C^{1,1}$ estimate applied to each ball B_k we obtain that

$$|| \, u_k \, ||_{C^{1,1}(B_{k+1})} \leq c_1, \tag{5.3}$$

where the constant c_1 depends merely on the supremum of f on B_k and $||u||_{C^{1,1}(\Omega')}$. Let us stress here that this estimate is independent of λ. Now, since $f \geq \lambda$ in Ω we obtain that on B_{k+1}

$$c_2(\lambda, ||u||_{C^{1,1}(\Omega')}) dd^c|z|^2 \leq dd^c \, u_k \leq c_3(||u||_{C^{1,1}(\Omega')}) dd^c|z|^2. \tag{5.4}$$

Observe that the argument above applies with no changes if instead of u one uses its mollification $u^{(\epsilon)}$ for ϵ small enough, so that everything is well defined in Ω'. As $||u^{(\epsilon)}||_{C^{1,1}} \to ||u||_{C^{1,1}}$ we can temporarily work with $u^{(\epsilon)}$ (and we suppress the indice ϵ for the sake of readability). By Theorem 2.15 and the remark after it the solutions u_k with the new boundary data coming from $u^{(\epsilon)}$ are smooth.

This allows one to apply the Calabi estimate (Theorem 4.5) to the problem. We obtain

$$sup_{B_{k+2}} \sum_{i,j,k} |u_{i\bar{j}k}|^2 \leq C/\rho^{2k}.$$

Thus, by standard interpolation inequalities (see [**21**]) for any $\gamma \in (0,1)$, we have

$$|| \, u_k \, ||_{C^{2,\gamma}(B_{k+2})} \leq c_4(c_2, c_3, d, n)/\rho^{k\gamma}. \tag{5.5}$$

Letting now $\epsilon \to 0^+$ we obtain that this estimate remains true for the original function u_k.

Note now that on B_{k+3} the following holds

$$0 = \log(f(x_0)) - \log(f(x_0)) = \log(\det(u_{k;i\bar{j}})) - \log(\det(u_{k+1;i\bar{j}})) =$$
$$\int_0^1 \frac{d}{dt} log(det((t\, u_k + (1-t)\, u_{k+1})_{i\bar{j}}))dt =$$
$$\sum_{i,j=1}^{n} \int_0^1 (t\, u_k + (1-t)\, u_{k+1})^{i\bar{j}} dt (u_k - u_{k+1})_{i\bar{j}},$$

where, as usual, $(a^{i\bar{j}})$ denotes the inverse transposed matrix to $(a_{i\bar{j}})$. Thus we obtain that the difference $v_k := u_k - u_{k+1}$ satisfies on B_{k+3} a linear elliptic equation

$$\sum_{i,j=1}^{n} b^{i\bar{j}}\, v_{k;i\bar{j}} = 0. \tag{5.6}$$

The coefficients $b^{i\bar{j}}$ satisfy, according to what we have proved so far, the estimates $c_5|\zeta|^2 \leq b^{i\bar{j}}\zeta_i\bar{\zeta}_j \leq c_6|\zeta|^2$ for any $\zeta \in \mathbb{C}^n$ and moreover $b^{i\bar{j}}$ are γ-Hölder continuous with $||b^{i\bar{j}}||_{C^\gamma(B_{k+3})} \leq c_7/\rho^{k\gamma}$.

This allows one to apply Schauder interior estimates to (5.6) (see [**21**], Theorem 6.2). So

$$\begin{aligned} &|| v_k ||_{C^0(B_{k+3})} + sup_{\{x\in B_{k+3}\}}(d_x|D\, v_k(x)|) + sup_{\{x\in B_{k+3}\}}(d_x^2|D^2\, v_k(x)|) + \\ &sup_{\{x,y\in B_{k+3}\}}(d_{x,y}^{2+\gamma}|D^2\, v_k(x) - D^2\, v_k(y)|/|x-y|^\gamma) \leq \\ &c_8\Big(c_5, c_6, (c_7/\rho^{k\gamma})(diam B_{k+3})^\gamma\Big)||v_k||_{C^0(B_{k+3})}. \end{aligned} \tag{5.7}$$

Here $d_z := dist(z, \partial B_{k+3})$, $d_{z,w} = min\{d_z,\ d_w\}$, and $D\, v_k$ (respectively $D^2\, v_k$) denotes any first order (resp. second order) partial derivative of v_k.

We wish to point out that while $||b^{i\bar{j}}||_{C^\gamma(B_{k+3})}$ may blow up as $k \to \infty$ the term $(diam B_{k+3})^\gamma$ compensates for this, so c_8 is a uniform constant independent of k.

The following estimates are straightforward consequences of (5.7):

$$|| v_k ||_{C^1(B_{k+4})} \leq (c_9/\rho^k)||v_k||_{C^0(B_{k+3})}; \tag{5.8}$$

$$|| v_k ||_{C^2(B_{k+4})} \leq (c_{10}/\rho^{2k})||v_k||_{C^0(B_{k+3})}; \tag{5.9}$$

$$|| v_k ||_{C^{2,\gamma}(B_{k+4})} \leq (c_{11}/\rho^{(2+\gamma)k})||v_k||_{C^0(B_{k+3})}. \tag{5.10}$$

Let now $\omega_f(r, x_0) = osc_{B(x_0,r)} f$ denote the modulus of continuity of f at x_0. Recall that u_k and u coincide on $\partial\, B_k$ and application of the comparison principle yields the inequalities

$$u_k(z) + \omega_f(d\rho^k, x_0)(|z|^2 - (d\rho^k)^2) \leq u(z) \leq u_k(z) - \omega_f(d\rho^k, x_0)(|z|^2 - (d\rho^k)^2)$$

for $z \in B_k$. Thus we get

$$|| u_k - u||_{C^0(B_k)} \leq (d\rho^k)^2\, \omega_f(d\rho^k, x_0). \tag{5.11}$$

Analogously

$$|| u_{k+1} - u||_{C^0(B_{k+1})} \leq (d\rho^{k+1})^2\, \omega_f(d\rho^{k+1}, x_0), \tag{5.12}$$

and coupling (5.11) and (5.12) we have

(5.13)

$$|| v_k ||_{C^0(B_{k+4})} \leq || u_k - u||_{C^0(B_{k+4})} + || u_{k+1} - u||_{C^0(B_{k+4})} \leq c_{12}\rho^{2k}\, \omega_f(d\rho^k, x_0).$$

Now, since we have assumed that $f \in C^\alpha(\Omega)$ we know that

$$\omega_f(d\rho^k, x_0) \sim \rho^{k\alpha} \tag{5.14}$$

uniformly for $x_0 \in \Omega''$, and thus (5.13) together with (5.8) and (5.9) result in the fact that all the sequences $\{u_k(x_0)\}_k$, $\{D\,u_k(x_0)\}_k, \{D^2\,u_k(x_0)\}_k$ are Cauchy sequences and hence are all convergent. Analyzing the rate of convergence one easily sees that $\lim_{k\to\infty} u_k(x_0) = u(x_0)$ and $\lim_{k\to\infty} D\,u_k(x_0) = Du(x_0)$ for any first order partial derivative. The same in fact holds also for any second order partial derivative provided it exists at x_0. Since $u \in C^{1,1}$ by assumption, it follows from Rademacher Theorem that this is the case almost everywhere. Note however that the limits $w_{st}(x) := \lim_{k\to\infty} \frac{\partial^2 u(x;x,k)}{\partial z_s \partial z_t}$, $w_{s\bar{t}}(x) := \lim_{k\to\infty} \frac{\partial^2 u(x;x,k)}{\partial z_s \partial \bar{z}_t}$ are defined in the whole Ω''.

Below we will show that all the limit functions w_{st}, $w_{s\bar{t}}$ are in fact α-Hölder continuous. Thus second derivatives of u exist almost everywhere and are equal to some α-Hölder continuous functions defined everywhere. By classical distribution theory it follows that $u \in C^{2,\alpha}$.

It is enough to prove Hölder continuity for any fixed w_{st} (any $w_{s\bar{t}}$ goes the same way). To this end we fix two points $x, y \in \Omega$" and consider two cases:

Case 1. Let $|x - y| \geq d/16$. Then

$$|w_{st}(x) - w_{st}(y)|/|x-y|^\alpha \leq (16/d)^\alpha \Big(|w_{st}(x)| + |w_{st}(y)|\Big).$$

If $\widetilde{u}_k$ and $\widehat{u}_k$ are the solutions of the Dirichlet problems related to systems of balls centered at x and y respectively, by (5.9) and (5.13) we obtain that

$$|w_{st}(x)| = |\widetilde{u}_{0;st}(x) - \sum_{k=0}^{\infty} \widetilde{v}_{k;st}(x)| \tag{5.15}$$

with the obvious meaning of $\widetilde{v}_k$. The last quantity can be estimated as follows

$$|\widetilde{u}_{0;st}(x) - \sum_{k=0}^{\infty} \widetilde{v}_{k;st}(x)| \leq |\widetilde{u}_{0;st}(x)| + c_{14} \sum_{k=0}^{\infty} \rho^{k\alpha} \leq c_{15} < \infty,$$

where we have used the bound (5.3) to control the first term.

Analogously on can bound $w_{st}(y)$ and thus in this case

$$|w_{st}(x) - w_{st}(y)|/|x-y|^\alpha \leq c_{16}. \tag{5.16}$$

Case 2. Let now $|x - y| < d/16$. The we fix a $k \geq 0$, $k \in \mathbb{N}$ such that $\rho^{k+5} d \leq |x - y| < \rho^{k+4} d$.

Then, we further estimate

$$\begin{aligned} |w_{st}(x) - w_{st}(y)| \leq &|w_{st}(x) - \widetilde{u}_{k;st}(x)| + \\ |w_{st}(y) - \widehat{u}_{k;st}(y)| + |\widehat{u}_{k;st}(y) &- \widetilde{u}_{k;st}(x)| =: I_1 + I_2 + I_3. \end{aligned}$$

The term I_1 can be easily handled in the following way:

$$I_1 = |\sum_{j=k}^{\infty} \widetilde{v}_{j;st}(x)| \leq c_{17} \sum_{j=k}^{\infty} \rho^{j\alpha} \leq c_{18}(\rho^{k+5} d)^\alpha \leq c_{19}|x-y|^\gamma,$$

where we have used (5.9), (5.13) and (5.14) in the first inequality.

I_2 is estimated completely the same way. To control I_3 observe that $y \in B_x(\rho^{k+4} d)$, so $\widetilde{u}_i$ is defined near y for $i = 0, \cdots, k$. Fix some $\gamma > \alpha$ (the bigger the difference $\gamma - \alpha$ the better). We have

$$I_3 \leq |\widehat{u}_{k;st}(y) - \widetilde{u}_{k;st}(y)| + |\widetilde{u}_{k;st}(y) - \widetilde{u}_{k;st}(x)| \leq$$

$$|\widehat{u}_{k;st}(y) - \widetilde{u}_{k;st}(y)| + |\widetilde{u}_{0;st}(y) - \widetilde{u}_{0;st}(x)| + \sum_{j=0}^{k-1} |\widetilde{v}_{j;st}(y) - \widetilde{v}_{j;st}(x)| \leq$$

$$|\widehat{u}_{k;st}(y) - \widetilde{u}_{k;st}(y)| + c_{20}|x-y|^{\gamma} \sum_{j=0}^{k-1} \rho^{(\alpha-\gamma)j},$$

where the last inequality holds because of (5.14). Since $\alpha < \gamma$ the last term is controlled by $|x-y|^{\gamma}\rho^{(\alpha-\gamma)k} \sim |x-y|^{\alpha}$. Meanwhile observe that $\widetilde{u}_k - \widehat{u}_k$ satisfies in the domain $B_x(\rho^{k+2}d) \cap B_y(\rho^{k+2}d)$ a linear elliptic equation of type $c^{i\bar{j}}(\widetilde{u}_k - \widehat{u}_k)_{i\bar{j}} = log(f(x)) - log(f(y))$ analogous to the equation for v_k (with x and y fixed here!). For the same reason as before we have uniform constants corresponding to c_5, c_6, c_7 (in fact if we have chosen at the beginning c_5, c_6, c_7 sufficiently big the same constants will do for this new equation). Thus Schauder interior estimates (with non homogeneous yet constant right hand side) give us the inequality

$$|| \widetilde{u}_k - \widehat{u}_k ||_{C^2(B(x,\rho^{k+3}d) \cap B(y,\rho^{k+3}d))} \leq$$

$$c_{21}\Big((1/\rho^{2k}) || \widetilde{u}_k - \widehat{u}_k ||_{C^0(B(x,\rho^{k+2}d) \cap B(y,\rho^{k+2}d))} + |log(f(x)) - log(f(y))|\Big).$$

Arguing as in (5.13) and using that $f \geq \lambda > 0$ the latter quantity is bounded by $c_{22}\rho^{k\alpha} \leq c_{23}|x-y|^{\alpha}$, and that finishes the estimation of I_3. Coupling all the obtained bounds we get the α-Hölder continuity also in this case.

□

The assumption that $u \in C^{1,1}$ seems too strong to be the optimal constraint. It has been relaxed by Y. Wang to $\Delta u \in L^{\infty}$ in [**36**]. Note that this is a weaker statement in the complex realm as plurisubharmonicity then yields a control merely on the *mixed* second derivatives of u, while for convex functions the control on the Laplacian yields uniform bounds on *every* second order partial derivative. In a subsequant paper [**38**] Y. Wang relaxed this assumption. It is sufficient to assume that Δu has bounded mean oscillation i.e.

$$sup_Q \frac{1}{V(Q)} \int_Q |\Delta u - (\Delta u)_Q| dV < \infty,$$

where supremum is taken over all Euclidean cubes Q and $(\Delta u)_Q$ is the average $\frac{1}{V(Q)} \int_Q \Delta u dV$.

Very recently Li, Li and Zhang [**29**] were able to weaken substantially this condition. Their result (up to date the strongest one) reads as follows:

THEOREM 5.6. *Let Ω be a domain in $\mathbb{C}^n$. Let also the plurisubharmonic function u satisfy the equation*

$$det(u_{i\bar{j}}) = f.$$

Suppose additionally that $f \geq \lambda > 0$ in Ω for some constant λ and $f \in C^{\alpha}(\Omega)$ for some $\alpha \in (0,1)$. If $u \in C^{1,\beta}(\Omega)$ for $\beta > 1 - \frac{\alpha}{n(2+\alpha)-1}$. Then $u \in C^{2,\alpha}(\Omega)$.

PROOF. The idea of the proof is to use an "improvement of regularity" lemma to get in the very end that $u \in C^{1,1}$ and then use Theorem 5.5.

Heuristically the previous argument cannot be repeated simply because u is (a priori) too singular to apply it as a boundary value for the Dirichlet problems in small balls. Instead once we fix a mollifier $\rho_\varepsilon(z) := \frac{\rho(|z|/\varepsilon)}{\varepsilon^n}$ for a fixed nonnegative rotationally invariant cut-off function $\rho \in C_0^\infty(B_0(1))$ satisfying $\int_{B_0(1)} \rho(z)dV(z) = 1$. Then $u_\varepsilon := \rho_\varepsilon * u$ is a smooth plurisubharmonic approximant of u which is well defined in any relatively compact subdomain of Ω as long as ε is small enough. As $u \in C^{1,\beta}$ one can even bound the possible blow-up rate of the norms of u_ε namely (see [**29**]):

$$u \leq u_\varepsilon \leq u + C\varepsilon^{1+\beta}, \tag{5.17}$$

$$||D^2 u_\varepsilon|| \leq C\varepsilon^{\beta-1}. \tag{5.18}$$

The hope is to use u_ε for boundary data in a ball centered at x_0 and small radius t in order to get strong enough estimates for the solution of the Dirichlet problem

$$\begin{cases} \varphi_{t,x_0} \in PSH(B_{x_0}(t)) \cap C(\overline{B_{x_0}(t)}) \\ det((\varphi_{t,x_0})_{i\bar{j}}) = f(x_0) \text{ in } B_{x_0}(t), \\ \varphi_{t,x_0} = u_\varepsilon \text{ on } \partial B_{x_0}(t). \end{cases}$$

(We assume, scaling if necessary, that the relatively compact subdomain we work in is at a fixed distance, say 1, from $\partial\Omega$). Of course ε has to depend on t and in what follows we assume that $\varepsilon = t^\mu$ for a constant $\mu > 0$ to be chosen later.

Let us analyze φ_{t,x_0}. From the comparison principle (Theorem 2.2) it is easy to see that in $B_{x_0}(t)$ one has

$$u_{t^\mu}(z) + Ct^\alpha(|z-x_0|^2 - t^2) \leq \varphi_{t,x_0}(z) \leq u(z) + Ct^{\mu(\beta+1)} - Ct^\alpha(|z-x_0|^2 - t^2) \tag{5.19}$$

(here we exploit the Hölder continuity of f).

From the Bedford-Taylor estimate Theorem 4.4 together with the bound (5.18) one has

$$\Delta\varphi_{t,x_0} \leq Ct^{\mu(\beta-1)} \text{ in } B_{x_0}(t/2). \tag{5.20}$$

Then, Calabi's third order estimate (Theorem 4.5) implies

$$\begin{aligned} [(\varphi_{t,x_0})_{i\bar{j}}]^*_{1,B_{x_0}(t/2)} &:= sup_{x\in B_{x_0}(t/2)} dist\Big(x, \partial B_{x_0}(t/2)\Big)|(D\varphi_{t,x_0})_{i\bar{j}}(x)| \\ &\leq Ct^{(n+1)\mu(\beta-1)}. \end{aligned} \tag{5.21}$$

By standard interpolation of Hölder norms (see [**21**]) the latter bound implies

$$\begin{aligned} &[\Delta\varphi_{t,x_0}]^*_{\gamma,B_{x_0}(t/2)} := \\ &sup_{x,y\in B_{x_0}(t/2)} min\{dist\Big(x, \partial B_{x_0}(t/2)\Big), dist\Big(y, \partial B_{x_0}(t/2)\Big)\}^\gamma \frac{|\Delta u(x) - \Delta u(y)|}{|x-y|^\gamma} \\ &\leq Ct^{(n+1)\mu(\beta-1)} \end{aligned} \tag{5.22}$$

for any $\gamma \in (0, 1]$.

All these bounds will be used to show the following: for a proper choice of μ we can get for some $\delta > \beta$

$$a) lim_{t\to 0^+} D\varphi_{t,x_0}(x_0) = Du(x_0),$$

$$b) |D\varphi_{t,x_0}(x_0) - Du(x_0)| \leq Ct^\delta$$

and

$$c)|D\varphi_{t,x_0}(x) - D\varphi_{t,x_0}(y)| \leq C|x-y|^\delta$$

for any two points $x, y \in B_{x_0}(t/4)$. These estimates strongly suggest that in fact $u \in C^{1,\delta}$ for $\delta > \beta$ and we would get an improvement of regularity.

In fact it is easy to see that these last three inequalities yield that $u \in C^{1,\delta}$ provided that we can also control the differences of derivatives of φ_{t,x_0} for different centers x_0. In fact what suffices is a bound

$$d)|D\varphi_{t,x}(x)(w) - D\varphi_{t,y}(y)(w)| \leq Ct^\delta$$

for any $w \in B_z(t/8)$ and here z is such that $B_z(t/4) \subset B_x(t/2) \cap B_y(t/2)$ (assuming that $|x-y| \leq t/8$).

Thus we are left with the proof of these four inequalities while keeping track how δ would depend on μ and β.

Just as in Theorem 5.5 we construct sequences of balls with radii $t_k := 2^{-k}t$ and consider the corresponding Dirichlet problems. Set

$$u_k := \varphi_{t_k,x_0}(x_0), \ v_k = u_{k-1} - u_k.$$

Inequalities (5.19) clearly yield

$$||u_{k-1} - u||_{C^0(B_{x_0}(t_{k+1}))}, ||u_k - u||_{C^0(B_{x_0}(t_{k+1}))} \leq C(t_k^{\mu(1+\beta)} + t_k^{2+\alpha}).$$

In turn the bound (5.22) implies

$$[\Delta u_{k-1}]^*_{\gamma,B_{x_0}(t_{k+1})} \leq Ct_k^{(n+1)\mu(\beta-1)},$$

$$[\Delta u_k]^*_{\gamma,B_{x_0}(t_{k+1})} \leq Ct_k^{(n+1)\mu(\beta-1)}.$$

Hence

$$||v_k||_{C^0(B_{x_0}(t_{k+1}))} \leq C(t_k^{\mu(1+\beta)} + t_k^{2+\alpha})$$

and

$$[\Delta v_k]^*_{\gamma,B_{x_0}(t_{k+1})} \leq Ct_k^{(n+1)\mu(\beta-1)}.$$

The interior Schauder estimates for the Laplacian (see [**21**]) then imply

$$[v_k]^*_{2,\gamma,B_{x_0}(t_{k+1})}$$

$$:= sup_{x,y\in B_{x_0}(t_{k+1})} min\{dist\Big(x, \partial B_{x_0}(t^{k+1})\Big), dist\Big(y, \partial B_{x_0}(t_{k+1})\Big)\}^\gamma$$

$$\times|\frac{D^2u(x) - D^2u(y)}{|x-y|^\gamma}| \leq Ct_k^{(n+1)\mu(\beta-1)+2}.$$

Interpolation inequalities in Hölder spaces yield for any $\epsilon > 0$

$$[v_k]^*_{1,\delta,B_{x_0}(t_{k+1})} \leq C\Big(\epsilon^{-(1+\delta)}(t_k^{\mu(1+\beta)} + t_k^{2+\alpha}) + \epsilon^{1+\gamma-\delta}t_k^{(n+1)\mu(\beta-1)+2}\Big). \tag{5.23}$$

The asymptotically optimal choice of ϵ making the right hand side as small as possible is

$$\epsilon^{2+\gamma} = (t_k^{\mu(1+\beta)} + t_k^{2+\alpha})(t_k^{(n+1)\mu(\beta-1)+2})^{-1}.$$

Such a choice yields a global $C^{1,\delta}$ bound in the smaller ball B_{k+2}:

$$[v_k]_{1,\delta,B_{x_0}(t_{k+2})} \leq C((t_k^{\frac{1+\gamma-\delta}{2+\gamma}\mu(1+\beta)} + t_k^{\frac{1+\gamma-\delta}{2+\gamma}(2+\alpha)})t_k^{\frac{1+\delta}{2+\gamma}(n+1)\mu(\beta-1)+2)-(1+\delta)}). \tag{5.24}$$

The $C^{1,\delta}$ norm of v_k will be small as long as both exponents

$$\frac{1+\gamma-\delta}{2+\gamma}\mu(1+\beta) + \frac{1+\delta}{2+\gamma}((n+1)\mu(\beta-1)+2)$$

and

$$\frac{1+\gamma-\delta}{2+\gamma}(2+\alpha)+\frac{1+\delta}{2+\gamma}((n+1)\mu(\beta-1)+2)$$

are strictly larger than $1+\delta$. It is easy to see that μ has to satisfy

$$\frac{\gamma(1+\delta)}{(1+\gamma-\beta)(1+\beta)-(1+\delta)(1-\beta)(n+1)}<\mu<\frac{(1+\gamma-\delta)(2+\alpha)-\gamma(1+\delta)}{(1+\delta)(1-\beta)(n+1)}.$$

Thus the necessary and sufficient condition is

$$\frac{\gamma(1+\delta)}{(1+\gamma-\beta)(1+\beta)-(1+\delta)(1-\beta)(n+1)}<\frac{(1+\gamma-\delta)(2+\alpha)-\gamma(1+\delta)}{(1+\delta)(1-\beta)(n+1)},$$

for then we take μ to be the arithmetic mean of the two quantities.

The previous inequality translates into

$$\beta>1-\frac{2((2+\alpha)(1+\gamma-\delta)-(1+\delta))}{(2+\alpha)(1+\delta)(n+1)+((2+\alpha)(1+\gamma-\delta)-(1+\delta)}.$$

At this stage we recall that up to now we haven't fixed the value of γ.

Note that once

$$\beta>1-\frac{2((2+\alpha)(2-\delta)-(1+\delta))}{(2+\alpha)(1+\delta)(n+1)+((2+\alpha)(2-\delta)-(1+\delta)}.$$

by choosing γ sufficiently close to one the claimed inequality will be satisifed.

Of course we have to show that $\delta>\beta$ so that we get an improvement in the regularity. It is however easy to see that once $\delta>1-\frac{\alpha}{n(2+\alpha)-1}$, then

$$\phi(\delta):=1-\frac{2((2+\alpha)(2-\delta)-(1+\delta))}{(2+\alpha)(1+\delta)(n+1)+((2+\alpha)(2-\delta)-(1+\delta)}<\delta.$$

The proof is finished by the following observations: first it is easy to see that the bound (5.24) just as in the proof of Theorem 5.5 yields the bounds a), b), c) and d). Thus $u\in C^{1,\delta}$.

Next we want to check how to apply this improvement of regularity to obtain in the end $C^{1,1}$ regularity. Note that the sequence $\{\delta_k\}_{k=0}^{\infty}$, $\delta_0=1, \delta_{i+1}=\phi(\delta_i)$ is decreasing and convergent to $1-\frac{\alpha}{n(2+\alpha)-1}$. Hence for any $\beta>1-\frac{\alpha}{n(2+\alpha)-1}$ there exists a smallest indice $k\in\mathbb{N}$, such that $\beta>\delta_k$. But then by our regularity improvement lemma $u\in C^{1,\delta_{k-1}}$ (in fact even $u\in C^{1,\delta_k'}$ for some $\delta_k'>\delta_k$) and by bootstrapping $u\in C^{1,\delta_0}=C^{1,1}$. □

6. Singular sets

In geometric analysis it is quite often the case that the solutions to a given equation may fail to be smooth. Nevertheless it is still of great importance to understand *to what extent the smoothness fails.* One of the natural approaches is to try to establish estimates for the size of the *singular set* i.e. the set where the solution fails to be smooth.

No such bounds have been established locally in the case of the complex Monge-Ampère equation even in the case of smooth strictly positive right hand side. This is one of the reasons for the lack of understanding of the local behavior of the solutions.

In this section we investigate two special types of sets associated to a solution of the Monge-Ampère equation

$$(dd^c u)^n = f$$

with f strictly positive. The first is the set where the *bounded* solution u reaches its minimum — the *minimum set*. Strictly speaking u may still be regular there but we will show that it can be a rich source of singularities just as in Examples 3.3.

The second type of singular sets are the *non-contact sets* i.e. the collections of points where u fails to admit a *local upper differential test*. These sets, derived from viscosity theory of elliptic PDEs, are important obstructions for regularity, since one has no viscosity information on this locus. Thus morally these sets should be *small*, othervise the viscosity approach could not be exploited without serious troubles.

In the first subsection we discuss the minimum sets, and the non-contact sets will be dealt with in the second subsection. Amazingly these seemingly very different types of sets share a lot of similarities which we shall discuss in detail.

6.1. Minimum sets. Let u be a bounded plurisubharmonic function. Adding a constant if necessary we can assume that $u \geq 0$. In this subsection we shall investigate the properties of the set $\{u = 0\}$.

In this generality not much can be said. For instance, as we are going to see below, in the planar case $n = 1$ any compact set which is regular in the sense of potential theory is such a minimum set. The situation changes dramatically if one assumes additionally that $u \in C^2$ and is *strictly plurisubharmonic*: a classical theorem of Harvey and Wells [**24**] states that the zero set of such a function is contained in a C^1 totally real hypersurface. In particular this implies that the Hausdorff dimension of the zero set is *small* compared to the dimension of the ambient space, and the zero set has *no analytic structure.*

We shall investigate below what could happen if we drop the smoothness assumption and instead of strict plurisubharmonicity we assume that $(dd^c u)^n \geq c > 0$.

Our motivation for the investigation of such minimum sets comes from the study of compactness properties of solutions to the complex Monge-Ampère equation. Analogous theory for the real Monge-Ampère equation was developed by Caffarelli [**11, 12**] and the analysis of the corresponding minimal sets is crucial there.

Heuristically the reason is as follows: suppose that the smooth plurisubharmonic functions u_j solve the equations $(dd^c u_j)^n = f_j$ with f_j uniformly bounded and strictly positive. Assuming that the supremums of their laplacians diverge to infinity and coupling this with the uniform bounds of the Monge-Ampère measures would yield that the global minimum of the least eigenvalue of $dd^c u_j$ converges to zero. If u is the limit of u_j then it is expected that the corresponding minimum eigenvectors of $dd^c u_j$ should converge to the kernel vectors of $dd^c u$ on a kernel set L. Such a kernel set is in turn an analytic subvariety for a sufficiently smooth u and it is expected that in the singular setting it will share many properties of analytic sets. Assuming the analyticity of the kernel one would get that u restricted to L is pluriharmonic. Using an averaging procedure this can be easily reduced to the case u is constant on L and attains local minimum there. Note that Examples 3.3 show exactly the behavior described above.

Below we shall try to answer the following two questions: how large the minimum sets can be and under what conditions the minimum sets contain no analytic subvarieties.

We start with the planar case $n = 1$. The trivial example $u(z) = Re^2(z)$ shows that the Hausdorff dimension of the minimum set can be equal to one (which is in line with the Harvey-Wells theorem).

As already noted in the introduction, every regular compact set K in the complex plane is the minimum set of a subharmonic function. In order to make K a minimum set of a *strictly subharmonic function* the basic idea is to perturb the function V_K^* suitably. Heuristically the function $u_K := (V_K^*)^2$ is "more subharmonic" with Laplace density equal to

$$\Delta u_K = 2V_K^* \Delta V_K^* + 2\left|\frac{\partial V_K^*}{\partial z}\right|^2, \tag{6.1}$$

with the the first term vanishing as K is a regular set since by Theorem 2.4 $\Delta V_K^* = 0$ off K, while on a regular set$K$ the function V_K^* vanishes according to Definition 2.5. The nontrivial issue is then to establish a lower bound on $|\frac{\partial V_K^*}{\partial z}|$ *up to the boundary* ∂K. What matters is the exact rate of convergence of V_K^* to zero as $z \to z_0 \in K$ (i.e. the exponent in the Łojasiewicz-Siciak condition from Definition 2.7). Also it is important to rule out clustering of vanishing points for $\frac{\partial V_K^*}{\partial z}$ to the boundary of K.

Below we prove a criterion which allows to decide whether a compact set K is a minimum set of a *strictly subharmonic function.*

THEOREM 6.1. *Let K be a compact set with empty interior satisfying the Łojasiewicz-Siciak exponent $\alpha < 2$. If K is further connected and does not disconnect the plane then it is a minimum set of a strictly subharmonic function.*

PROOF. Observe that $\hat{\mathbb{C}} \setminus K$ is simply connected and hence V_K^* is equal to the logarithm of the conformal map sending $\hat{\mathbb{C}} \setminus K$ onto the exterior of the unit disc in $\hat{\mathbb{C}}$ (recall Proposition 2.11). Then by the Koebe distortion theorem

$$\frac{\sinh V_K^*(w)}{4\left|\frac{\partial V_K^*}{\partial w}\right|} \leq dist(w, K) \leq \frac{\sinh V_K^*(w)}{\left|\frac{\partial V_K^*}{\partial w}\right|}. \tag{6.2}$$

In particular $\frac{\partial V_K^*}{\partial w}$ never vanishes on $\mathbb{C} \setminus K$.

Let $u_K := (V_K^*)^{2/\alpha}$. Then

$$\begin{aligned}\Delta u_K &= \frac{2}{\alpha}(V_K^*)^{2/\alpha-1}\Delta V_K^* + \frac{2}{\alpha}\left(\frac{2}{\alpha}-1\right)(V_K^*)^{2/\alpha-2}\left|\frac{\partial V_K^*}{\partial w}\right|^2 \\ &= \frac{2}{\alpha}\left(\frac{2}{\alpha}-1\right)(V_K^*)^{2/\alpha-2}\left|\frac{\partial V_K^*}{\partial w}\right|^2.\end{aligned}$$

By (6.2) this behaves like

$$\frac{2}{\alpha}\left(\frac{2}{\alpha}-1\right)(V_K^*)^{2/\alpha-2}\left|\frac{\sinh V_K^*}{dist(w, K)}\right|^2.$$

Note that this quantity close to the boundary is asymptotically equal to

$$\frac{2}{\alpha}\left(\frac{2}{\alpha}-1\right)\frac{(V_K^*)^{2/\alpha}}{|dist(w,K)|^2},$$

which is bounded below by the Łojasiewicz - Siciak condition.

□

This criterion is strong enough to produce minimum sets with Hausdorff dimension larger than one:

EXAMPLE 6.2. Let J_λ be the Julia set of the polynomial $f_\lambda(z) = z^2 + \lambda z$, $|\lambda| < 1$. Then for λ sufficiently close to zero J_λ is a minimum set of a strictly subharmonic function. The Hausdorff dimension of J_λ satisfies $dim_{\mathcal{H}} J_\lambda \geq 1 + 0.36|\lambda|^2$.

PROOF. We follow closely the argument in Theorem B from [**3**]. In particular it is well known that for small λ the Julia set is connected and its complement consists of two simply connected domains. As in [**3**] we note that the conformal map g_λ from the complement of the unit disc to the unbounded component U of $\mathbb{C} \setminus J_\lambda$ admits a k-quasiconformal extension (denoted by $\tilde{g}_\lambda$) to the whole of $\mathbb{C}$ for $\frac{k-1}{k+1} = |\lambda|$. In particular the conformal map g_λ is $1/k$-Hölder continuous up to the boundary by Theorem 2.9, and if $k < 2$ Proposition 2.11 implies that $V_{J_\lambda} = \log|g_\lambda^{-1}|$ satisfies $LS(\alpha)$ for $\alpha < 2$. Thus by Theorem 6.1 there is a perturbation $\tilde{V}_{J_\lambda}$ which is strictly subharmonic, nonnegative and vanishing continuously at the boundary.

In order to complete the proof we need to "fill in" the bounded component of $\mathbb{C} \setminus J_\lambda$. To this end note that if h_λ is the conformal map from the unit disc to this component (normalized by fixing zero) then the quasiconformal reflection

$$\tilde{h}_\lambda(z) = \begin{cases} h_\lambda(z) & \text{for } |z| \leq 1 \\ \tilde{g}_\lambda\left(1/\overline{\tilde{g}_\lambda^{-1}(h_\lambda(1/\overline{z}))}\right) & \text{for } |z| > 1 \end{cases}$$

is a k^2-quasiconformal mapping, hence it is $1/k^2$-Hölder continuous. Taking the Green function $G(z,0)$ with pole at zero we can apply the same reasoning (away from 0) for $-G$ as for the function V_K (note that $-G$ is still harmonic except at zero). Thus there is a strictly subharmonic function $\tilde{G}$ on the bounded component (with small neighborhood of the origin deleted) which vanishes continuously on the boundary.

Finally the function

$$H := \begin{cases} \tilde{V}_{J_\lambda}(z) \text{ if } z \in U \\ \tilde{G} \text{ if} z \in \mathbb{C} \setminus (J_\lambda \cup U \cup \{0\}) \\ 0 \text{ if } z \in J_\lambda \end{cases}$$

satisfies all the requirements. □

REMARK 6.3. Another type of examples- generalizations of Koch snowflakes with Hausdorff dimension larger than one was considered in [**18**].

Note that the Examples 3.3 show that the minimum set can have Hausdorff dimension much larger than n in $\mathbb{C}^n, n \geq 2$.

We now focus on the existence of analytic structures in the minimum set. We shall prove the following theorem which says that the Examples 3.3 are *optimal* in terms of Hölder regularity:

THEOREM 6.4. *Let $u \geq 0$ be a plurisubharmonic function satisfying $(dd^c u)^n \geq 1$. If additionally $u \in C^{1,\alpha}$ for $\alpha > 1 - \frac{2k}{n}$ if $2k \leq n$ or $u \in C^{0,\beta}$ for $\beta > 2 - \frac{2k}{n}$ if $2k > n$, then no analytic set of dimension $\geq k$ can be contained in $u^{-1}(0)$.*

PROOF. We follow the argument from [**17**]. We shall exploit an old idea of Urbas from [**34**] with suitable modifications.

We shall deal with both cases simultaneously writing $\beta = 1 + \alpha$ if necessary - this will not affect the argument. Suppose on contrary that X is a k dimensional analytic subset of $u^{-1}(0)$. Our goal will be to construct a barrier w on a thin domain close to a (modification of) X which will contradict the regularity that u has.

Pick a point x_0 in the regular part of X. Then there is a biholomorphic mapping $\pi : U \to W$ of an open ball U in $\mathbb{C}^n$ to a neighborhood W of x_0, such that

$$\pi^{-1}(X \cap W) = \{z \in U | z_1 = 0, \cdots, z_{n-k} = 0\},$$

with $(z_1, \cdots, z_n)$ being the coordinates in U centered at $0 = \pi^{-1}(x_0)$. We can also assume that the Jacobian of π at zero is equal to 1.

Consider now the function $\tilde{u}(z) := u(\pi(z))$. Then

$$(dd^c \tilde{u}(z))^n = (dd^c u)^n |_{\pi(z)} |Jac_\pi(z)|^2 \geq 1/2, \tag{6.3}$$

where Jac_π stands for the (complex) Jacobian of the mapping π and the last inequality follows by the smoothness of Jac_π (we can shrink U further if necessary). Denote by M the α-Hölder constant for $\nabla\tilde{u}$, which can be made as close to the Hölder constant of ∇u as necessary if U is further shrunk.

Let now $z' = (z_1, \cdots, z_{n-k}), z'' = (z_{n-k+1}, \cdots, z_n)$ Then

$$\tilde{u}(z', z'') \leq \tilde{u}(0, z'') + M|z'|^{1+\alpha} \leq A|z'|^2 + A^{-\gamma}C_0, \tag{6.4}$$

$$C_0 = M^{\frac{2}{1-\alpha}} \left(\left(\frac{1+\alpha}{2} \right)^{\frac{1+\alpha}{1-\alpha}} - \left(\frac{1+\alpha}{2} \right)^{\frac{2}{1-\alpha}} \right)$$

(recall $\alpha < 1$ in our convention).

Consider now the polydisc

$$P := \{z \in U | \ |z'| \leq \rho, |z_{n-k+1}| \leq \rho, \cdots, |z_n| \leq \rho\}.$$

If ρ is taken small enough, then $P \subset U$. Fix such ρ and consider the barrier function

$$w(z) := A|z'|^2 + A^{-\gamma}C_0 + \sum_{j=n-k+1}^{n} \frac{\varepsilon}{\rho}(n\rho - Re(z_j)) + B \sum_{j=n-k+1}^{n} (|z_j|^2 - \rho Re(z_j)),$$

with $0 < B \leq 1$ and $\varepsilon << 1$ to be chosen later on.

Note that if $|z'| = \rho$ then $w \geq A\rho^2 + A^{-\gamma}C_0 + k[(n-1)\varepsilon - B\frac{\rho^2}{4}]$, since $|z_j|^2 - \rho Re(z_j) = |z_j - \rho/2|^2 - \frac{\rho^2}{4}$. Thus if $A \geq \sup_U \tilde{u} + \frac{k}{4}$ we get $w \geq \tilde{u}$.

On the other hand if for some $n \geq j \geq n-k+1$ we have $|z_j| = \rho$ then

$$w(z', z'') \geq A|z'|^2 + A^{-\gamma}C_0 + k(n-1)\varepsilon - (k-1)B\frac{\rho^2}{4}. \tag{6.5}$$

Fixing $\varepsilon = \frac{k-1}{(n-1)k}B\frac{\rho^2}{4}$ if $k > 1$ and a small multiple of $B\frac{\rho^2}{4}$ if $k = 1$ (if ρ is small and $B \leq 1$ this quantity is clearly small) and exploiting (6.4) we again obtain $w \geq \tilde{u}$.

If one can prove that $(dd^c w)^n \leq \frac{1}{2} \leq (dd^c \tilde{u})^n$ then by comparison principle it would follow that $w \geq u$ over the whole polydisc. Note that $(dd^c w)^n = A^{n-k}B^k$, hence the choice $B = \left(\frac{1}{2A^{n-k}}\right)^{1/k}$ (if A is large enough this is clearly less than one) satisfies this requirement.

Under such a choice of constants we obtain

$$0 \leq \tilde{u}\left(0', \rho/2, \cdots, \rho/2\right) \leq w\left(0', \rho/2, \cdots, \rho/2\right) = A^{-\gamma} C_0 + k\left(n - \frac{1}{2}\right)\varepsilon - kB\frac{\rho^2}{4}.$$

We claim that the sum of the last two terms is negative. Indeed this is the case for $k = 1$ and for $k > 1$ we obtain

$$k\left(n - \frac{1}{2}\right)\varepsilon - kB\frac{\rho^2}{4} = \left(\frac{(k-1)(n-1/2)}{n-1} - k\right)B\frac{\rho^2}{4},$$

by our choice of ε, and the latter quantity is equal to $-\frac{2n-k-1}{2(n-1)}B\frac{\rho^2}{4}$. Comparing this with the first term above we end up with

$$0 \leq A^{-\gamma} C_0 - A^{-\frac{n-k}{k}} C_1 \frac{\rho^2}{4}$$

for some numerical constant C_1. This must hold (for fixed small ρ) for every sufficiently large constant A, thus implying $\frac{n-k}{k} \geq \gamma$. This in turn reads

$$\alpha \leq 1 - \frac{2k}{n},$$

which is a contradiction. □

We finish this subsection by showing that the Examples 3.3 are also optimal in terms of Sobolev regularity. This is a result due to C. Mooney and T. Collins [**15**]:

THEOREM 6.5. *Let $u \geq 0$ be a plurisubharmonic function satisfying $(dd^c u)^n \geq 1$. If $\Delta u \in L^p$ for $p \geq \frac{n}{k}(n-k)$, then no analytic set of dimension $\geq k$ can be contained in $u^{-1}(0)$.*

Before we start the proof we need the following fact proven by Collins and Mooney which is of independent interest:

LEMMA 6.6. *Let $p > \frac{n}{2}$ be fixed. Let u be a $W^{2,p}$ function (not necessarily plurisubharmonic) which is nonnegative in $B_0(2) \in \mathbb{R}^n$ and $u(0) = 0$. Then there exists c_0 dependent on n, p such that for all $\varepsilon \in (0,1)$ there exists a constant $\delta = \delta(\varepsilon, n, p)$ such that either*

$$\left(\int_{B_0(2)\setminus B_0(\varepsilon)} |D^2 u|^p\right)^{1/p} \geq \delta sup_{\partial B_0(1)} u$$

or

$$\left(\int_{B_0(2)} |D^2 u|^p\right)^{1/p} \geq c_0 \varepsilon^{\frac{n}{p}-2} sup_{\partial B_0(1)} u.$$

PROOF. Without loss of generality we assume that $sup_{\partial B_0(1)} u = 1$. The Sobolev-Poincare and Morrey inequalities yield (for some linear function l)

$$||u - l||_{C^0(B_0(2)\setminus B_0(\varepsilon))} \leq C(n,p,\varepsilon)\delta.$$

Take δ so small that the right hand side is at most $1/8$. Picking the point η where $u(\eta) = 1$ one has $l(\eta) \geq 7/8$. This coupled with the information about u yields $l(-2\eta) \geq -1/8$. These inequalities imply that

$$l(0) \geq 1/2.$$

Consider $\tilde{u} := (u - l)(\varepsilon x)$. Obviously $|\tilde{u}| \le 1/8$ in $B_2(0) \setminus B_1(0)$ and $\tilde{u}(0) \le -1/2$. But then

$$\Big(\int_{B_0(2)} |D^2\tilde{u}|^p\Big)^{1/p} \ge c_0(n,p),$$

which after the rescaling yields the lemma. □

Now we are ready to prove the theorem.

PROOF. We repeat the reasoning from [**15**]. Again as in the Hölder case we can assume, by contradiction, that u is defined in a unit polydisc and $u = 0$ on $\{z_1 = \cdots = z_{n-k} = 0\}$. By assumption and classical elliptic regularity we know that

$$||u||_{W^{2,\frac{n}{k}(n-k)}} \le C < \infty.$$

We begin with a preliminary claim about the growth of u in the first $(n-k)$-variables:

Claim There is a numerical constant $c(n)$, such that

$$sup_{|z''|\le 1/4} sup_{|(z_1,\cdots,z_{n-k})|=r} u(z_1,\cdots,z_{n-k},z'') \ge c(n) r^{2-2k/n}$$

for all $0 < r < 1$.

The claim can be proved by contradiction. To this end note that if it is violated for some $r_0 \in (0,1)$ then setting $h = c(n)r_0^{2-2k/n}$ one has

$$\{|z''| < 1/4\} \cap \{|(z_1,\cdots,z_{n-k})| < (c(n)^{-1}h)^{\frac{n}{2(n-k)}}\} \subset \{u < h\}.$$

Consider now the quadratic polynomial

$$Q_t(z_1,\cdots,z_{n-k},z'') = 2h(16|z''|^2 + (c(n)^{-1}h)^{\frac{-n}{n-k}}[\sum_{j=1}^{n-k}|z_j|^2]) + t.$$

Note that $(dd^cQ_t)^n = Cc(n)^n \le 1$ if $c(n)$ is small enough, while $Q_0 > u$ on $\partial(\{|z''| < 1/4\} \cap \{|(z_1,\cdots,z_{n-k})| < (c(n)^{-1}h)^{\frac{n}{2(n-k)}}\})$. Observe that $Q_0(0,0'') = 0$. Hence taking Q_t for a negative t close enough to zero, we still have $Q_t > u$ on $\partial(\{|z''| < 1/4\} \cap \{|(z_1,\cdots,z_{n-k})| < (c(n)^{-1}h)^{\frac{n}{2(n-k)}}\})$ but now $Q_t(0,0'') < u(0,0'')$. Thus the set $\{Q_t < u\}$ is relatively compact in $\{|z''| < 1/4\} \cap \{|(z_1,\cdots,z_{n-k})| < (c(n)^{-1}h)^{\frac{n}{2(n-k)}}\}$ which contradicts Theorem 2.2. This proves the claim.

Consider now the following scaling in the $z_1,\cdots z_{n-k}$ directions

$$u_r(z_1,\cdots,z_{n-k},z'') := \frac{1}{r^{2-2k/n}}u(rz_1,\cdots,rz_{n-k},z'') \tag{6.6}$$

for any $0 < r < 1/2$.

The proof hinges on the following *second claim* which we shall prove later on: *there exists $\varepsilon,\delta > 0$ and small dependent on n,k and the $W^{2,\frac{n}{k}(n-k)}$ norm of u but not on r such that*

$$\int_{|z''|<1,\ \varepsilon<|(z_1,\cdots,z_{n-k})|<2} |D_z^2 u|^{\frac{n}{k}(n-k)} > \delta, \tag{6.7}$$

with D_z^2 denoting the Hessian in the first $(n-k)$ variables $z_1,\cdots,z_{n-k}$.

Assuming the truth of this observation we proceed as follows. First by rescaling back one obtains

$$\int_{|z''|<1,\ \varepsilon r/2<|(z_1,\cdots,z_{n-k})|<r} |D_z^2 u_r|^{\frac{n}{k}(n-k)} > \delta.$$

Summing these inequalites over the annuli $|z''| < 1$, $(\varepsilon/2)^{m+1} < |(z_1, \cdots, z_{n-k}| < (\varepsilon/2)^m$, $m = 1, 2, \cdots$ we would have $||u||_{W^{2,\frac{n}{k}(n-k)}} = \infty$, a contradiction.

We are left with the proof of the second claim:

By the growth estimate given by the first claim there is a point $(z_1, \cdots z_{n-k}, z'')$ such that $|z''| < 1/4$, $|(z_1, \cdots z_{n-k})| = 1$ and

$$u_r(z_1, \cdots z_{n-k}, z'') \geq c(n).$$

Denote by $M(w)$ the function $u_r(z_1, \cdots, z_{n-k}, w)$ with the z coordinates being the coordinates of this chosen point. By the mean value property of subharmonic functions this implies that

$$\int_{|z"|<1} M(w) \geq c(n).$$

By Lemma 6.6 either

$$\left(\int_{\varepsilon<|(z_1,\cdots,z_{n-k})|<2} |D_z^2 u_r|^{\frac{n}{k}(n-k)}\right)^{\frac{k}{n(n-k)}} > \delta M(w)$$

or

$$\left(\int_{|(z_1,\cdots,z_{n-k})|<2} |D_z^2 u_r|^{\frac{n}{k}(n-k)}\right)^{\frac{k}{n(n-k)}} > c(n,k)\varepsilon^{-\frac{2(n-k)}{n}} M(w).$$

Let A_ε denote the set of all points w such that the first case holds and A_ε^c — its complement.

Then exploiting the boundendess of u in Sobolev norm we obtain

$$\begin{aligned} C \geq c(n,k) \int_{A_\varepsilon^c} \left(\int_{|(z_1,\cdots,z_{n-k})|<2} |D_z^2 u_r|^{\frac{n}{k}(n-k)}\right)^{\frac{k}{n(n-k)}} dV(w) \\ \geq c(n,k)\varepsilon^{-\frac{2(n-k)}{n}} \int_{A_\varepsilon^c} M(w) dV(w). \end{aligned}$$

Thus, for ε small enough the mass of $M(w)$ over A_ε^c is less than its mass on A_ε. Hence

$$\begin{aligned} \int_{|z''|<1,\ \varepsilon<|(z_1,\cdots,z_{n-k})|<2} |D_z^2 u_r|^{\frac{n}{k}(n-k)} \geq c(n,k)\delta \int_{A_\varepsilon} M(w) dV(w) \\ \geq c(n,k)\delta. \end{aligned}$$

□

6.2. Non-contact sets. In this subsection we will discuss the *upper contact sets*. Recall that these are the collections of points x_0 where a fixed plurisubharmonic function u admits a C^2 differential test from above i.e. a local C^2 smooth function q such that $q(x) \geq u(x)$ with equality at x_0.

The viscosity method was introduced to deal with Hamilton-Jacobi equations and initial developments focused on first-order equations. Soon it was realized that the methods apply also to second-order nonlinear PDE's, but still only in the real setting. Recently there has been much interest in applying these techniques in the complex setting (see [**37**, **40**]).

It is easy to see (see [**40**]) that the contact set is always dense. The measure theoretic properties of the non-contact sets are substantially subtler.

We prove that the upper non-contact set can have big Hausdorff dimension. Our examples are exactly mimimum sets of strictly plurisubharmonic functions i.e. the types of singular sets we discussed before.

Now we will prove that for small $|\lambda|$, the curve J_λ from Example 6.2 contains no points at which differential test from above exists for a certain subharmonic function. The crucial fact we shall use is that J_λ is a compact set satisfying the Łojasiewicz-Siciak condition with an exponent $\alpha < 2$.

Let

$$\tilde{V}_{J_\lambda}(z) = \begin{cases} V^*_{J_\lambda}(z), \text{ if } z \text{ lies in the unbounded component of } \mathbb{C} \setminus J_\lambda \\ 0, \text{ if } z \in J_\lambda \\ V^*_{\iota(J_\lambda)}(\frac{1}{z}), \text{ if } z \text{ lies in the bounded component of } \mathbb{C} \setminus J_\lambda \end{cases},$$

where $\iota(z) = \frac{1}{z}$ is the standard inversion. It is clear that $\tilde{V}_{J_\lambda}$ is subharmonic on $\mathbb{C} \setminus \{0\}$.

THEOREM 6.7. *For any λ, such that J_λ is a minimum set of a strictly subharmonic function, the set J_λ contains only points at which the function $\tilde{V}_{J_\lambda}$ allows no differential test from above. In particular such a set can have Hausdorff dimension greater than* 1.

PROOF. Fix $z_0 \in J_\lambda$ and suppose the contrary, that is, there exists a q which is a differential test from above in some small disc $B_{z_0}(r)$. First, since $\tilde{V}_{J_\lambda}(z_0) - q(z_0) = \sup_{w \in B_{z_0}(r)}(\tilde{V}_{J_\lambda}(w) - q(w))$, we can assume that $q(z_0) = 0$ by subtracting a constant. Now since $q \geq V_{J_\lambda} \geq 0$, q has a local minimum at z_0 and since $q \in C^2$, $\nabla q(z_0) = 0$. It follows by the Taylor expansion that $q(z) \leq c|z - z_0|^2 + o(|z - z_0|^2)$ for some constant $c \geq 0$, that is, $\tilde{V}_{J_\lambda}(z) \leq c|z - z_0|^2 + o(|z - z_0|^2)$ in $B_{z_0}(r)$.

We recall the estimate (6.2): If K is compact regular set such that $\hat{\mathbb{C}} \setminus K$ is simply connected and $K \in LS(\alpha)$, for $\alpha < 2$ then $\Delta V_K^{*\frac{2}{\alpha}}(z) \geq \frac{2}{\alpha}\left(\frac{2}{\alpha} - 1\right)\frac{V_K^{*\frac{2}{\alpha}}}{dist(z,K)^2} \geq c_1 > 0$, at every point where the Laplacian is well defined. This is satisfied both for $V^*_{J_\lambda}$ and $V^*_{\iota(J_\lambda)}$ and hence for $\tilde{V}_{J_\lambda}$. The Laplacian of $(\tilde{V}_{J_\lambda})^{\frac{2}{\alpha}}$ is singular only on J_λ which has planar Lebesgue measure 0. Hence the measure integral $\int_{B_{z_0}(r)} \Delta \tilde{V}_{J_\lambda}^{\frac{2}{\alpha}}$ is equal to the Lebesgue integral $\int_{B_{z_0}(r) \setminus J_\lambda} \Delta \tilde{V}_{J_\lambda}^{\frac{2}{\alpha}} d\lambda^2$.

The Jensen formula gives us

$$\frac{1}{2\pi}\int_0^{2\pi} \tilde{V}_{J_\lambda}^{\frac{2}{\alpha}}\left(z_0 + re^{i\theta}\right)d\theta = \frac{1}{2\pi}\int_0^{2\pi} \tilde{V}_{J_\lambda}^{\frac{2}{\alpha}}\left(z_0 + re^{i\theta}\right)d\theta - \tilde{V}_{J_\lambda}^{\frac{2}{\alpha}}(0)$$

$$= \frac{1}{2\pi}\int_0^r s^{-1}\int_{B(z_0,s)} \Delta \tilde{V}_{J_\lambda}^{\frac{2}{\alpha}}(z)\, d\lambda^2 ds \geq \frac{1}{2\pi}\int_0^r s^{-1}\pi c_1 s^2 ds = c_1 r^2/4.$$

On the other hand

$$\frac{1}{2\pi}\int_0^{2\pi} \tilde{V}_{J_\lambda}^{\frac{2}{\alpha}}\left(z_0 + re^{i\theta}\right)d\theta \leq \frac{1}{2\pi}\int_0^{2\pi} q(z_0 + re^{i\theta})^{\frac{2}{\alpha}} d\theta \leq \frac{1}{2\pi}\int_0^{2\pi} (cr^2 + o(r^2))^{\frac{2}{\alpha}} d\theta$$

$$= c_2 r^{\frac{4}{\alpha}} + o(r^{\frac{4}{\alpha}}).$$

Hence

$$c_1 r^2/4 \leq c_2 r^{\frac{4}{\alpha}} + o(r^{\frac{4}{\alpha}}).$$

Now since $\alpha < 2$ this is a contradiction for small enough r. □

This example easily generalizes to higher dimensions:

THEOREM 6.8. *The set $E \subset \Omega \subset \mathbb{C}^n$ of points, at which no differential test from above exists for a given $\varphi \in PSH(\Omega)$, can have Hausdorff dimension greater than $2n-1$.*

PROOF. Consider the function

$$\varphi(z_1, \cdots, z_n) := \max\{\log|z_1 \cdots z_{n-1}|, 0\} + \tilde{V}_{J_\lambda}(z_n).$$

Clearly $\varphi \in PSH(\mathbb{C}^{n-1} \times (\mathbb{C} \setminus \{0\})$. Fix a point in $w = (w_1, \cdots, w_n) \in \mathbb{D}^{n-1} \times J_\lambda$, where $\mathbb{D}^{n-1}$ is the unit polydisc of dimension $n-1$, and suppose that φ allows a differential test from above q at w. Then φ restricted to the line $l : \mathbb{C} \setminus \{0\} \ni \lambda \to (w_1, \cdots, w_{n-1}, \lambda)$ is subharmonic and $\varphi \circ l = \tilde{V}_{J_\lambda}$. Moreover, the $q \circ l$ is a differential test from above for $\varphi \circ l$ at w_n. The assertion now follows from Theorem 6.7 and the fact that the Hausdorff dimension of $\mathbb{D}^{n-1} \times J_\lambda$ is larger than $2n-1$. □

Our last theorem shows that the upper contact set is always of Lebesgue measure zero:

THEOREM 6.9. *Let $\Omega \subset \mathbb{C}^n$ be a domain and let $\varphi \in PSH(\Omega)$. Then almost everywhere in Ω (with respect to the Lebesgue measure) φ allows a differential test from above.*

PROOF. The proof is almost the same as in [**33**] with the necessary modifications. We follow the argument from [**18**].

First, we reduce the problem to the case when φ is bounded.

Take a general $\psi \in PSH(\Omega)$. The function e^ψ is still plurisubharmonic and bounded below. For any differential test from above q for e^ψ at the point $w \in \Omega$, $\log(q + |z-w|^2)$ is a differential test from above for ψ, except at points w where e^ψ vanishes. These points however constitute the set $\{\psi = -\infty\}$, which is thus pluripolar and hence of Lebesgue measure zero. Finally we can exhaust Ω with open sets $\Omega_n = \{z \in \Omega | e^{\psi(z)} < n\}$. Note that e^ψ is a bounded plurisubharmonic function in Ω_n. In particular the Monge-Ampère operator is well-defined.

Denote by E, $E \subset \Omega$ the set where φ allows a differential test from above. First we note that

$$(dd^c(r+\eta))^n - (dd^c r)^n \geq (dd^c \eta)^n \geq 0, \tag{6.8}$$

as measures, for any plurisubharmonic r and η in the domain of definition of the Monge-Ampère operator.

If φ is bounded on Ω then denote by $\omega_0 := \sup_{\tilde{w} \in \Omega} \varphi(\tilde{w}) - \inf_{w \in \Omega} \varphi(w)$ the oscillation of φ. For $\varepsilon > 0$ we define $\Omega_{\varepsilon\sqrt{2\omega_0}} = \{z \in \Omega | dist(z, \partial\Omega) > \varepsilon\sqrt{2\omega_0}\}$ and

$$\varphi_\varepsilon(z) = \sup_{w \in \Omega_{\varepsilon\sqrt{2\omega_0}}} \left(\varphi(w) - \frac{|w-z|^2}{2\varepsilon^2} \right).$$

The supremum is attained at some point $\tilde{z} \in \Omega$ such that $|z - \tilde{z}| < \varepsilon\sqrt{2\omega_0}$, assuming that $dist(z, \partial\Omega) > \varepsilon\sqrt{2\omega_0}$. Hence for $z \in \Omega_{\varepsilon\sqrt{2\omega_0}}$ the supremum can be taken over the whole Ω. It is a standard fact that the functions φ_ε decrease to φ as $\varepsilon \to 0^+$. Moreover φ_ε Lipschitz and $\varphi_\varepsilon(z) + k|z|^2$ is a convex function in $\Omega_{\varepsilon\sqrt{2\omega_0}}$ for any $k \geq \frac{1}{2\varepsilon^2}$. The latter property means that φ_ε is semi-convex and hence twice differentiable almost everywhere by the classical theorem of Alexandrov. Also φ_ε is plurisubharmonic in $\Omega_{\varepsilon\sqrt{2\omega_0}}$ (see [**40**]) and hence $(dd^c\varphi_\varepsilon)^n \geq 0$ as measure.

Fix a ball $B_z(R) \subset \Omega$ with $R < dist(z, \partial\Omega)$. For ε small enough we have $B_z(R) \subset \Omega_{\varepsilon\sqrt{2\omega_0}}$. We restrict further considerations to such small ε. For $k \in \mathbb{N}\backslash\{0\}$ let

$$\varphi_{k,\varepsilon}(w) = \varphi_\varepsilon(w) - k(|w-z|^2 - R^2).$$

For any k, $\varphi_{k,\varepsilon}$ agrees with φ_ε on $\partial B_z(R)$.

Let $E_{k,\varepsilon}$ be the upper contact set of $\varphi_{k,\varepsilon}$, that is the set of points, where the graph of $\varphi_{k,\varepsilon}$ lies below a supporting hyperplane in $\mathbb{R}^{2n+1}$,

$$E_{k,\varepsilon} = \{w \in B_z(R) | \varphi_{k,\varepsilon}(\tilde{w}) \leq \varphi_{k,\varepsilon}(w) + \langle p, \tilde{w} - w\rangle\},$$

for some $p \in \mathbb{R}^{2n}$, depending on w, and any $\tilde{w} \in B_z(R)$ (here $\langle\cdot,\cdot\rangle$ denotes the real inner product in $\mathbb{R}^{2n}$ and z and w are regarded as points in $\mathbb{R}^{2n} = \mathbb{R}^n + i\mathbb{R}^n \cong \mathbb{C}^n$). The set $E_{k,\varepsilon}$ is relatively closed in $B(z,r)$ and hence measurable. Note that by construction $-\varphi_{k,\varepsilon}$ would be convex on every open subset of $E_{k,\varepsilon}$. Thus $(dd^c(-\varphi_{k,\varepsilon}))^n \geq 0$ as measure on $E_{k,\varepsilon}$.

Likewise we introduce the upper contact set of $\varphi - k(|w-z|^2 - R^2)$:

$$E_k = \{w \in B(z,r) | \varphi(\tilde{w}) - k(|\tilde{w}-z|^2 - R^2) \leq \varphi(w) - k(|w-z|^2 - R^2) + \langle p, \tilde{w} - w\rangle\}.$$

Note that $E_k \subset E$, since by construction $q(\tilde{w}) = k(|\tilde{w}-z|^2 - R^2) + \langle p, \tilde{w} - z\rangle$ is a differential test from above at $w \in E_k$. Also the sets $E_{k,\varepsilon}$ approximate E_k as $\varepsilon \to 0^+$, hence

$$\limsup_{\varepsilon\to 0^+} \lambda^{2n}(E_{k,\varepsilon}) \leq \lambda^{2n}(E_k) \leq \lambda^{2n}(E \cap B(z,R)).$$

Using (6.8) we have

$$\begin{aligned}(dd^c(-\varphi_{k,\varepsilon}))^n &\leq (dd^c k(|w-z|^2 - R^2))^n - (dd^c\varphi_\varepsilon)^n \\ &\leq (dd^c k(|w-z|^2 - R^2))^n = 4^n n! k^n.\end{aligned}$$

Now by the Alexandrov maximum principle (recall inequality (4.2))

$$\sup_{B_z(R)} \varphi_{k,\varepsilon} \leq \sup_{\partial B_z(R)} \varphi_{k,\varepsilon} + C diam B_z(R) \left(\int_{E_{k,\varepsilon}} |\det D^2(\varphi_{k,\varepsilon})|\right)^{\frac{1}{2n}} \tag{6.9}$$

$$\leq \sup_{\partial B_z(R)} \varphi_\varepsilon + 2C_1 R \left(\int_{E_{k,\varepsilon}} f d\lambda^{2n}\right)^{\frac{1}{2n}} \leq \sup_{\partial B_z(R)} \varphi_\varepsilon + C_2 R k (\lambda^{2n}(E_{k,\varepsilon}))^{\frac{1}{2n}},$$

where $\sqrt{f} d\lambda^{2n} = (dd^c(-\varphi_{k,\varepsilon}))^n$. Note that we have used $D^2(-\varphi_{k,\varepsilon}) \geq 0$ on $E_{k,\varepsilon}$. The second inequality follows from the estimate (see [**9**]) $\det\left(\frac{\partial^2}{\partial z_i \partial \bar{z}_j} u\right) \geq 2^{-n}\sqrt{\det D^2 u} = 2^{-n}\sqrt{|\det D^2(-u)|}$ for any u at any point where it is twice differentiable and its real Hessian is positive definite.

If k is big enough (6.9) gives us

$$kR^2 + \varphi(z) = \sup_{B_z(R)} \varphi_{k,\varepsilon} \leq \sup_{\partial B_z(R)} \varphi_\varepsilon + C_2 R k (\lambda^{2n}(E_{k,\varepsilon}))^{\frac{1}{2n}}.$$

Hence passing with $\varepsilon \to 0^+$

$$R^2 \leq \frac{1}{k}\omega_0 + CR(\lambda^{2n}(E_k))^{1/(2n)}.$$

Finally,

$$\frac{\lambda^{2n}(E\cap B_z(R))}{\lambda^{2n}(B_z(R))}\geq \limsup_{k\to\infty}\frac{\lambda^{2n}(E_k)}{\lambda^{2n}(B_z(R))}\geq \limsup_{k\to\infty}\frac{(\frac{R}{C}-\frac{\omega_0}{kCR})^{2n}}{\frac{\pi^n}{n!}R^{2n}}=\frac{n!}{\pi^n C^{2n}}>0.$$

This means that the Lebesgue density at z of the set E, that is

$$\lim_{R\to 0+}\frac{\lambda^{2n}(E\cap B_z(R))}{\lambda^{2n}(B_z(R))}$$

is not 0 for any $z\in\Omega$. On the other hand it must be 0 almost everywhere in $\Omega\setminus E$ by the Lebesgue density theorem. Now it is clear that E is of full measure. □

Acknowledgments. I am deeply indebted to the referees whose remarks and suggestions improved the exposition significantly.

References

[1] Lars V. Ahlfors, *Lectures on quasiconformal mappings*, 2nd ed., University Lecture Series, vol. 38, American Mathematical Society, Providence, RI, 2006. With supplemental chapters by C. J. Earle, I. Kra, M. Shishikura and J. H. Hubbard. MR2241787

[2] Lars V. Ahlfors, *On quasiconformal mappings*, J. Analyse Math. **3** (1954), 1–58; correction, 207–208, DOI 10.1007/BF02803585. MR0064875

[3] J. Becker and Ch. Pommerenke, *On the Hausdorff dimension of quasicircles*, Ann. Acad. Sci. Fenn. Ser. A I Math. **12** (1987), no. 2, 329–333, DOI 10.5186/aasfm.1987.1206. MR951982

[4] Eric Bedford and B. A. Taylor, *The Dirichlet problem for a complex Monge-Ampère equation*, Invent. Math. **37** (1976), no. 1, 1–44, DOI 10.1007/BF01418826. MR0445006

[5] Eric Bedford and B. A. Taylor, *A new capacity for plurisubharmonic functions*, Acta Math. **149** (1982), no. 1-2, 1–40, DOI 10.1007/BF02392348. MR674165

[6] Zbigniew Błocki, *The complex Monge-Ampère operator in hyperconvex domains*, Ann. Scuola Norm. Sup. Pisa Cl. Sci. (4) **23** (1996), no. 4, 721–747 (1997). MR1469572

[7] Zbigniew Błocki, *On the regularity of the complex Monge-Ampère operator*, Complex geometric analysis in Pohang (1997), Contemp. Math., vol. 222, Amer. Math. Soc., Providence, RI, 1999, pp. 181–189, DOI 10.1090/conm/222/03161. MR1653050

[8] Zbigniew Błocki, *Interior regularity of the complex Monge-Ampère equation in convex domains*, Duke Math. J. **105** (2000), no. 1, 167–181, DOI 10.1215/S0012-7094-00-10518-2. MR1788046

[9] Zbigniew Błocki, *On uniform estimate in Calabi-Yau theorem*, Sci. China Ser. A **48** (2005), no. suppl., 244–247, DOI 10.1007/BF02884710. MR2156505

[10] Zbigniew Błocki and Sławomir Dinew, *A local regularity of the complex Monge-Ampère equation*, Math. Ann. **351** (2011), no. 2, 411–416, DOI 10.1007/s00208-010-0609-0. MR2836666

[11] L. A. Caffarelli, *A localization property of viscosity solutions to the Monge-Ampère equation and their strict convexity*, Ann. of Math. (2) **131** (1990), no. 1, 129–134, DOI 10.2307/1971509. MR1038359

[12] Luis A. Caffarelli, *Interior $W^{2,p}$ estimates for solutions of the Monge-Ampère equation*, Ann. of Math. (2) **131** (1990), no. 1, 135–150, DOI 10.2307/1971510. MR1038360

[13] Luis A. Caffarelli and Xavier Cabré, *Fully nonlinear elliptic equations*, American Mathematical Society Colloquium Publications, vol. 43, American Mathematical Society, Providence, RI, 1995. MR1351007

[14] L. Caffarelli, J. J. Kohn, L. Nirenberg, and J. Spruck, *The Dirichlet problem for nonlinear second-order elliptic equations. II. Complex Monge-Ampère, and uniformly elliptic, equations*, Comm. Pure Appl. Math. **38** (1985), no. 2, 209–252, DOI 10.1002/cpa.3160380206. MR780073

[15] Tristan C. Collins and Connor Mooney, *Dimension of the minimum set for the real and complex Monge-Ampère equations in critical Sobolev spaces*, Anal. PDE **10** (2017), no. 8, 2031–2041, DOI 10.2140/apde.2017.10.2031. MR3694014

[16] Michael G. Crandall, Hitoshi Ishii, and Pierre-Louis Lions, *User's guide to viscosity solutions of second order partial differential equations*, Bull. Amer. Math. Soc. (N.S.) **27** (1992), no. 1, 1–67, DOI 10.1090/S0273-0979-1992-00266-5. MR1118699

[17] Sławomir Dinew and Żywomir Dinew, *The minimum sets and free boundaries of strictly plurisubharmonic functions*, Calc. Var. Partial Differential Equations **55** (2016), no. 6, Art. 148, 17, DOI 10.1007/s00526-016-1069-5. MR3568053

[18] Sławomir Dinew and Żywomir Dinew, *Differential tests for plurisubharmonic functions and Koch curves*, Potential Anal. **50** (2019), no. 3, 381–400, DOI 10.1007/s11118-018-9686-6. MR3925586

[19] Sławomir Dinew, Xi Zhang, and Xiangwen Zhang, *The $C^{2,\alpha}$ estimate of complex Monge-Ampère equation*, Indiana Univ. Math. J. **60** (2011), no. 5, 1713–1722, DOI 10.1512/iumj.2011.60.4444. MR2997006

[20] T. W. Gamelin and N. Sibony, *Subharmonicity for uniform algebras*, J. Funct. Anal. **35** (1980), no. 1, 64–108, DOI 10.1016/0022-1236(80)90081-6. MR560218

[21] David Gilbarg and Neil S. Trudinger, *Elliptic partial differential equations of second order*, Classics in Mathematics, Springer-Verlag, Berlin, 2001. Reprint of the 1998 edition. MR1814364

[22] Vincent Guedj and Ahmed Zeriahi, *Degenerate complex Monge-Ampère equations*, EMS Tracts in Mathematics, vol. 26, European Mathematical Society (EMS), Zürich, 2017. MR3617346

[23] Cristian E. Gutiérrez, *The Monge-Ampère equation*, Progress in Nonlinear Differential Equations and their Applications, vol. 44, Birkhäuser Boston, Inc., Boston, MA, 2001. MR1829162

[24] F. Reese Harvey and R. O. Wells Jr., *Zero sets of non-negative strictly plurisubharmonic functions*, Math. Ann. **201** (1973), 165–170, DOI 10.1007/BF01359794. MR0330510

[25] Lars Hörmander, *Notions of convexity*, Progress in Mathematics, vol. 127, Birkhäuser Boston, Inc., Boston, MA, 1994. MR1301332

[26] Robert Jensen, *The maximum principle for viscosity solutions of fully nonlinear second order partial differential equations*, Arch. Rational Mech. Anal. **101** (1988), no. 1, 1–27, DOI 10.1007/BF00281780. MR920674

[27] Sławomir Kołodziej, *The complex Monge-Ampère equation*, Acta Math. **180** (1998), no. 1, 69–117, DOI 10.1007/BF02392879. MR1618325

[28] Sławomir Kołodziej, *The complex Monge-Ampère equation and pluripotential theory*, Mem. Amer. Math. Soc. **178** (2005), no. 840, x+64, DOI 10.1090/memo/0840. MR2172891

[29] Chao Li, Jiayu Li, and Xi Zhang, *A $\mathcal{C}^{2,\alpha}$ estimate of the complex Monge-Ampère equation*, J. Funct. Anal. **275** (2018), no. 1, 149–169, DOI 10.1016/j.jfa.2018.01.020. MR3799626

[30] A. V. Pogorelov, *A regular solution of the n-dimensional Minkowski problem*, Soviet Math. Dokl. **12** (1971), 1192–1196. MR0284956

[31] Dieter Riebesehl and Friedmar Schulz, *A priori estimates and a Liouville theorem for complex Monge-Ampère equations*, Math. Z. **186** (1984), no. 1, 57–66, DOI 10.1007/BF01215491. MR735051

[32] Neil S. Trudinger, *Local estimates for subsolutions and supersolutions of general second order elliptic quasilinear equations*, Invent. Math. **61** (1980), no. 1, 67–79, DOI 10.1007/BF01389895. MR587334

[33] Neil S. Trudinger, *On the twice differentiability of viscosity solutions of nonlinear elliptic equations*, Bull. Austral. Math. Soc. **39** (1989), no. 3, 443–447, DOI 10.1017/S0004972700003361. MR995142

[34] John I. E. Urbas, *Regularity of generalized solutions of Monge-Ampère equations*, Math. Z. **197** (1988), no. 3, 365–393, DOI 10.1007/BF01418336. MR926846

[35] J. B. Walsh, *Continuity of envelopes of plurisubharmonic functions*, J. Math. Mech. **18** (1968/1969), 143–148. MR0227465

[36] Yu Wang, *On the $C^{2,\alpha}$-regularity of the complex Monge-Ampère equation*, Math. Res. Lett. **19** (2012), no. 4, 939–946, DOI 10.4310/MRL.2012.v19.n4.a16. MR3008426

[37] Yu Wang, *A viscosity approach to the Dirichlet problem for complex Monge-Ampère equations*, Math. Z. **272** (2012), no. 1-2, 497–513, DOI 10.1007/s00209-011-0946-z. MR2968239

[38] Yu Wang, *Local regularity of the complex Monge-Ampère equation*, Analysis, complex geometry, and mathematical physics: in honor of Duong H. Phong, Contemp. Math., vol. 644, Amer. Math. Soc., Providence, RI, 2015, pp. 291–305, DOI 10.1090/conm/644/12769. MR3372473

[39] Shing Tung Yau, *On the Ricci curvature of a compact Kähler manifold and the complex Monge-Ampère equation. I*, Comm. Pure Appl. Math. **31** (1978), no. 3, 339–411, DOI 10.1002/cpa.3160310304. MR480350

[40] Ahmed Zeriahi, *A viscosity approach to degenerate complex Monge-Ampère equations* (English, with English and French summaries), Ann. Fac. Sci. Toulouse Math. (6) **22** (2013), no. 4, 843–913, DOI 10.5802/afst.1390. MR3137252

Institute of Mathematics, Jagiellonian University, ul Łojasiewicza 6, 30-348 Kraków, Poland

Email address: `slawomir.dinew@im.uj.edu.pl`

Contemporary Mathematics
Volume **735**, 2019
https://doi.org/10.1090/conm/735/14824

Pluriharmonics in general potential theories

F. Reese Harvey and H. Blaine Lawson, Jr.

Abstract. The general purpose of this paper is to investigate the notion of "pluriharmonics" for the general potential theory associated to a convex cone subequation $F \subset \mathrm{Sym}^2(\mathbb{R}^n)$. For such F there exists a maximal linear subspace $E \subset F$, called the *edge*, and F decomposes as $F = E \oplus F_0$. The *pluriharmonics* or *edge functions* are u's with $D^2u \in E$. Many subequations F have the same edge E, but there is a unique smallest such subequation. These are the focus of this investigation. Structural results are given. Many examples are described, and a classification of highly symmetric cases is given. Finally, the relevance of edge functions to the solutions of the Dirichlet problem is established.

Table of Contents

2010 *Mathematics Subject Classification.* Primary 35E20, 35E99.
Partially supported by the NSF.

©2019 American Mathematical Society

1. Introduction.

This paper is concerned with the edge E of a convex cone subequation $F \subset \mathrm{Sym}^2(\mathbb{R}^n)$, obtained from the decomposition

$$F \;=\; E \oplus F_0 \tag{1.1}$$

into a vector subspace $E \subset \mathrm{Sym}^2(\mathbb{R}^n)$ and a cone $F_0 \subset E^\perp$, called the *reduced constraint set*, which contains no lines. (See §2 for definition of *subequation.*) The interest in the edge E is that it gives us a notion of *pluriharmonics*, or *edge functions*, for the potential theory associated to the subequation F. These edge functions u are defined by $D^2u \in E$.

The edge E of a subequation is, in a sense, a crude invariant, since many subequations F have the same edge E. However, there is a canonical choice for F completely determined by E, namely $E + \mathcal{P}$, where $\mathcal{P} \equiv \{A : A \geq 0\}$. This is a subequation with edge E, and it must be contained in all other subequations with edge E (since by definition we always have $\mathcal{P} \subset F$). A large part of this paper is devoted to studying and classifying these subequations $E+\mathcal{P}$ for which the edge E is a determinative invariant. It is for these **minimal subequations** with edge E (Def. 5.2) that the pluriharmonics or edge functions play the most important role.

Let's look at some examples. The simplest is just $\mathcal{P}$ itself. Here the edge $E = \{0\}$ and the $\mathcal{P}$-subharmonics are the convex functions. The edge functions are those u with $D^2u \equiv 0$, that is, the affine functions.

The other orthogonally invariant edge is $E = \{A : \mathrm{tr}\, A = 0\}$, and we have $\mathcal{P} + E = \{A : \mathrm{tr}\, A \geq 0\} \equiv \Delta$. Here the subsolutions are the classical subharmonics, and the edge functions are just the harmonics.

The examples become more interesting if we look at $\mathrm{U}(n)$-invariant edges in $\mathbb{C}^n = (\mathbb{R}^{2n}, J)$. The reduction into irreducibles is

$$\mathrm{Sym}^2(\mathbb{R}^{2n}) \;=\; (\mathbb{R}\cdot \mathrm{Id}) \oplus \mathrm{Herm}_0^{\mathrm{sym}} \oplus \mathrm{Herm}^{\mathrm{skew}},$$

$\mathrm{Herm}_0^{\mathrm{sym}} = \{A : AJ = JA \text{ and } \mathrm{tr}\, A = 0\}$ and $\mathrm{Herm}^{\mathrm{skew}} = \{A : AJ = -JA\}$. Here there are two new edges.

The first is where we set $E = \mathrm{Herm}^{\mathrm{skew}}$. This gives the complex Monge-Ampère subequation:

$$\mathcal{P}_{\mathbb{C}} \;=\; E + \mathcal{P} \;=\; \{A : A_{\mathbb{C}} = A - JAJ \geq 0\}.$$

The subsolutions are the plurisubharmonics, and the edge functions (or pluriharmonics) are the classical pluriharmonic functions in complex analysis.

The other is where $E = \mathrm{Herm}_0^{\mathrm{sym}}$. This subequation is rather new.

$$\mathcal{P}(\mathrm{Lag}) \;=\; E + \mathcal{P} \;=\; \left\{A : \mathrm{tr}\left(A\big|_W\right) \geq 0 \;\text{ for all Lagrangian planes } W\right\}.$$

The subsolutions are the *Lagrangian plurisubharmonics* which were studied in [**10**]. In this case the edge functions are certain quadratic functions.

If one now looks for $\mathrm{Sp}(n)\cdot\mathrm{Sp}(1)$-invariant edges, there are many interesting examples. In fact, in Chapter 7 a wide variety of subequations are given. The reader might enjoy this section.

One interesting result concerning these minimal subequations is that they can be used to characterize the dual subharmonics (negatives of the superharmonics). (In this paper, degree-2 means degree ≤ 2.)

THEOREM 6.4. *Let $\mathcal{P}^+ = E + \mathcal{P}$ be a minimal subequation. The following conditions on an upper semi-continuous function u are equivalent.*

(1) *u is dually $\mathcal{P}^+$-subharmonic.*

(2) *u is "sub" the edge functions.*

(3) *u is locally "sub" the edge functions.*

(4) *u is locally "sub" the degree-2 polynomial edge functions.*

Here the notion of subharmonic is taken in the viscosity sense (see below). We note that u is "sub" a function v if for each compact set $\overline{\Omega}$, one has $u \le v$ on $\partial\Omega \Rightarrow u \le v$ on $\overline{\Omega}$.

For example, when $F = \mathcal{P}$ (the first example above), this theorem says that the $\widetilde{\mathcal{P}}$-subharmonic functions (the negatives of supersolutions for the real Monge-Ampère equation) are characterized by being "sub" the affine functions.

When $F = \mathcal{P}_{\mathbb{C}}$, the $\widetilde{\mathcal{P}}_{\mathbb{C}}$-subharmonic functions (the negatives of supersolutions for the complex Monge-Ampère equation) are characterized by being "sub" the standard pluriharmonics in complex analysis. In fact the degree-2 polynomial pluriharmonics will do.

When $F = \mathcal{P}(\mathrm{Lag})$, the $\widetilde{\mathcal{P}}(\mathrm{Lag})$-subharmonic functions (the negatives of supersolutions for the Lagrangian Monge-Ampère equation [**10**]) are characterized by being "sub" an explicit family of quadratic polynomials.

Chapter 2 of this paper lays down the fundamental notions of the edge E, the span S and the reduced constraint set F_0, which appears in (1.1) above. By definition S is the linear span of F_0, but it could also be defined as the orthogonal complement of E. This chapter then looks at geometrically defined equations, which give many interesting examples.

Chapter 3 introduces a further refinement of the structure of convex cone subequations. These equations are partitioned into two classes. The first consists of subequations which are *extremely degenerate* (see Definition 3.1). A basic example is the Laplacian on $\mathbb{R}^k$, considered as a subequation on $\mathbb{R}^n$ where $n > k$. The second class consists of those subequations which are *dimensionally complete* (see Definition 3.1*). This means that all the variables in $\mathbb{R}^n$ are required to define the subequation. These complementary concepts are determined entirely by properties of the edge, or equivalently the span, of F. For example,

$$F \text{ is complete} \quad \iff \quad E \cap \mathcal{P} = \{0\} \quad \iff \quad S \cap (\operatorname{Int}\mathcal{P}) \neq \emptyset.$$

There are several other equivalent criteria; see Propositions 3.5 and 3.10. A pair of orthogonal subspaces E and S of $\mathrm{Sym}^2(\mathbb{R}^n)$, which satisfy these criteria for E and S, is called an **edge-span pair**.

In Chapter 4 the structure of the subequation F is further illuminated by proving that there exists a unique subspace $W \subset \mathbb{R}^n$, called the **support of** F, with the property that

$$F = \mathrm{Sym}^2(W)^\perp \oplus F_1 \qquad \text{where} \quad F_1 \subset \mathrm{Sym}^2(W) \tag{1.2}$$

is a complete convex cone subequation in W. This F_1 is called the **the supporting subequation of** F.

Combining (1.1) and (1.2) give the decomposition

$$F = \mathrm{Sym}^2(W)^\perp \oplus E_1 \oplus F_0 \tag{1.3}$$

where the edge of F is $E = \mathrm{Sym}^2(W)^\perp \oplus E_1$ and E_1 is the edge of the supporting subequation F_1.

In Chapter 5 our minimal subequations are defined and discussed. We start with any basic edge $E \subset \mathrm{Sym}^2(\mathbb{R}^n)$, i.e, one which satisfies $E \cap \mathcal{P} = \{0\}$. Then (Lemma 5.1) the sum

$$\mathcal{P}^+ \equiv E + \mathcal{P} \quad \text{is a subequation, and it has edge } E.$$

(By the edge criteria $\mathcal{P}^+$ is then complete.) Such subequations are called **minimal** and are the main focus of this paper. These subequations have many special properties. Theorem 5.4 mentions eight of them, while Theorem 5.5 claims that any one of these eight properties implies the subequation is minimal.

In Chapter 7 many examples of minimal subequations are given.

In Chapter 8 we classify all the minimal subequations which are invariant under the compact group G where $G = \mathrm{O}_n, \mathrm{U}_n, \mathrm{Sp}_n \cdot \mathrm{Sp}_1, \mathrm{Sp}_n$ and $\mathrm{Sp}_n \cdot \mathrm{S}^1$ (all acting on their fundamental representation spaces).

We note that the general definition of F-plurisubharmonics is based on viscosity theory [**3**], [**2**]. The reader is referred to [**5**], [**7**] or [**8**] for definitions and properties.

In Chapter 6 the generalized pluriharmonics (or edge functions) are introduced from a viscosity point of view, and basic properties are discussed.

One might speculate, in light of Theorem 6.4 above, that for minimal subequations the Dirichlet Problem can be solved by replacing the standard Perron family with the subfamily consisting only of $\mathcal{P}^+$-pluriharmonics. For the two extreme subequations – the convexity and the Laplacian subequations – this is in fact true. In §8 we prove something close for all minimal subequations.

THEOREM 9.3. *Let $\mathcal{P}^+$ be a minimal subequation, and $\Omega \subset \mathbb{R}^n$ a domain with smooth strictly convex boundary. Then the standard solution to the Dirichlet problem for any $\varphi \in C(\partial\Omega)$ is the upper envelope of functions in the Perron subfamily of functions which can be written locally as the maximum of a finite number of pluriharmonics.*

2. Preliminaries.

In this section we review the basic properties of convex cone subequations and define many of the associated objects (cf. [**6**]).

We start with a closed convex cone $\mathcal{P}^+$ in $\mathrm{Sym}^2(\mathbb{R}^n)$, which we always assume is a non-empty proper subset. Using the natural inner product $\langle A, B\rangle = \mathrm{tr}(AB)$ we have

$$\textbf{(Polar Cone)} \qquad \mathcal{P}_+ \equiv \{A : \langle A, B\rangle \geq 0 \ \forall\, B \in \mathcal{P}^+\}. \tag{2.1}$$

If, in addition, $\mathcal{P}^+$ satisfies the following positivity condition, then $\mathcal{P}^+$ is referred to as a subequation. (We will frequently evoke the bipolar theorem, which says that the polar of the polar is the original convex cone.)

Definition 2.1. $\mathcal{P}^+$ is a **subequation** if satisfies the **positivity condition**

$$\mathcal{P}^+ + \mathcal{P} = \mathcal{P}^+, \qquad \text{i.e. } \mathcal{P} \subset \mathcal{P}^+ \tag{P}$$

or equivalently if $\mathcal{P}_+ \subset \mathcal{P}$. The equivalence follows since $\mathcal{P}$ is self-polar.

Subequations have the important **topological property**

$$\mathcal{P}^+ = \overline{\mathrm{Int}\,\mathcal{P}^+} \tag{T}$$

since $\mathcal{P}^+ + \epsilon I \subset \mathcal{P}^+ + \mathrm{Int}\,\mathcal{P} \subset \mathrm{Int}\,\mathcal{P}^+$ for all $\epsilon > 0$.

The following is an important class of examples.

Example 2.2. (The Geometric Case). Given a closed subset $\mathbb{G} \subset G(p, \mathbb{R}^n)$ of the Grassmannian of unoriented p-planes in $\mathbb{R}^n$, let

$$\mathcal{P}(\mathbb{G}) \equiv \{A : \langle A, P_W \rangle = \mathrm{tr}(A\big|_W) \geq 0 \ \forall W \in \mathbb{G}\}$$

where $W \in \mathbb{G}$ is identified with P_W, orthogonal projection onto W. Each such $\mathcal{P}^+ = \mathcal{P}(\mathbb{G})$ is a convex cone subequation with polar $\mathcal{P}_+ =$ Convex Cone Hull $(\mathbb{G})$ $\equiv CCH(\mathbb{G})$.

In addition to the polar cone $\mathcal{P}_+$ we associate two vector spaces E and S with $\mathcal{P}^+$ which form an orthogonal decomposition $\mathrm{Sym}^2(\mathbb{R}^n) = E \oplus S$.

The Edge and the Span

$$\textbf{(The Edge } E\textbf{)} \qquad E \equiv \mathcal{P}^+ \cap (-\mathcal{P}^+) = \{A : A + \mathcal{P}^+ = \mathcal{P}^+\} \tag{2.2}$$

$$\textbf{(The Dual Span } S\textbf{)} \qquad S \equiv \mathrm{span}\, \mathcal{P}_+ \tag{2.3}$$

Note that the edge E is the unique maximal vector space contained in $\mathcal{P}^+$. To verify the equality in (2.2), use the fact that $A + \mathcal{P}^+ = \mathcal{P}^+ \iff -A + \mathcal{P}^+ = \mathcal{P}^+$, along with the fact that $\mathcal{P}^+$ is a convex cone with vertex 0. Note that S is by definition a vector space, whereas E is obviously closed under addition and scalar multiplication.

Lemma 2.3. *The Edge E of a subequation enjoys the properties:*
(2.4) (Orthogonality) E and S are orthogonal compliments in $\mathrm{Sym}^2(\mathbb{R}^n)$.
(2.5) $E \cap (\mathrm{Int}\, \mathcal{P}^+) = \emptyset$. *In particular,* $E \cap (\mathrm{Int}\, \mathcal{P}) = \emptyset$.

Proof of (2.4). It is easy to see that $E \perp S$. Then since $\mathcal{P}_+ \subset S \Rightarrow S^\perp \subset \mathcal{P}^+$, and since $S^\perp$ is a vector subspace, this implies that $S^\perp \subset E$. Therefore, $\mathrm{Sym}^2(\mathbb{R}^n) = S^\perp + S \subset E + S$ thereby proving that $E + S = \mathrm{Sym}^2(\mathbb{R}^n)$ is an orthogonal decomposition. ∎

Proof of (2.5). If this does not hold, we can pick $A \in E \cap \mathrm{Int}\, \mathcal{P}^+$. Given $B \in \mathrm{Sym}^2(\mathbb{R}^n)$, we have $A + \epsilon B \in \mathcal{P}^+$ if $\epsilon > 0$ is sufficiently small. Therefore, $B = -\frac{1}{\epsilon}A + (\frac{1}{\epsilon}A + B) \in E + \mathcal{P}^+ = \mathcal{P}^+$. This contradicts the assumption that $\mathcal{P}^+$ is a proper subset of $\mathrm{Sym}^2(\mathbb{R}^n)$. ∎

$$\text{Let } \pi : \mathrm{Sym}^2(\mathbb{R}^n) \longrightarrow S \text{ denote orthogonal projection.} \tag{2.6}$$

As a constraint on the second derivative, the important part of $\mathcal{P}^+$ is

$$\textbf{(The Reduced Constraint Set)} \quad \mathcal{P}_0^+ \equiv \pi(\mathcal{P}^+). \tag{2.7}$$

$$\mathcal{P}^+ = E \oplus \mathcal{P}_0^+, \qquad \text{i.e.,} \qquad A \in \mathcal{P}^+ \iff \pi(A) \in \mathcal{P}_0^+. \tag{2.8a}$$

$$\mathrm{Int}\, \mathcal{P}^+ = E \oplus \mathrm{Int}\, \mathcal{P}_0^+, \quad \text{i.e.,} \quad A \in \mathrm{Int}\, \mathcal{P}^+ \iff \pi(A) \in \mathrm{Int}\, \mathcal{P}_0^+. \tag{2.8b}$$

Since $\pi(A)$ captures the important part of $A \equiv D^2u$, $\pi(D^2u)$ is called **the reduced hessian of u for $\mathcal{P}^+$.**

Note that

$$\mathcal{P}_0^+ \text{ is the polar of } \mathcal{P}_+ \text{ in its span } S. \tag{2.9}$$

The closed convex cone $\mathcal{P}_0^+ \subset S$ is not a subequation unless $S \equiv \mathrm{Sym}^2(\mathbb{R}^n)$, i.e., $E = \{0\}$, in which case $\mathcal{P}_0^+ = \mathcal{P}^+$.

We say that the subequation $\mathcal{P}^+$ is **self polar** if

$$\mathcal{P}_0^+ = \mathcal{P}_+ \tag{2.10}$$

3. Extremely Degenerate Versus Complete

Our subequations divide into two kinds. The first is that of extreme degeneracy. These subequations on n-dimensional euclidean space $\mathbb{R}^n$ are better understood as subequations on a lower dimensional subspace (see Prop. 3.4 below and the Support Theorem 4.3).

Definition 3.1*. A convex cone subequation $\mathcal{P}^+$ is said to be **extremely degenerate** if there exists a proper subspace $W \subset \mathbb{R}^n$ such that the following equivalent conditions are satisfied:

(1*) The reduced constraint set $\mathcal{P}_0^+ \subset \mathrm{Sym}^2(W)$,

(2*) The polar $\mathcal{P}_+ \subset \mathrm{Sym}^2(W)$, or equivalently the dual span $S \subset \mathrm{Sym}^2(W)$.

Proof that (1*) $\iff$ (2*). Note that: $\mathcal{P}_+ \subset \mathrm{Sym}^2(W) \iff S \equiv \mathrm{span}\, \mathcal{P}_+ \subset \mathrm{Sym}^2(W) \iff \mathcal{P}_0^+ \subset \mathrm{Sym}^2(W)$ by (2.10). ∎

Remark. In [**6**] we said for (2*) that "$\mathcal{P}_+$ only involves the variables in W", and for (1*) that "$\mathcal{P}^+$ can be defined using the variables in W".

The remaining subequations are defined by taking the negations of (1*) and (2*).

Definition 3.1. A subequation which is not extremely degenerate will be called **dimensionally complete**, or just **complete**. In other words

(1) $\mathcal{P}_0^+ \not\subset \mathrm{Sym}^2(W)$ for any proper subspace $W \subset \mathbb{R}^n$, or

(2) $\mathcal{P}_+ \not\subset \mathrm{Sym}^2(W)$ for any proper subspace $W \subset \mathbb{R}^n$.

See [**6**] for many interesting results for complete convex cone subequations. The purpose of this paper is to investigate a special class of such subequations described in Section 5.

Extreme Degeneracy

Let $A\big|_W$ denote the restriction of A to the subspace $W \subset \mathbb{R}^n$ as a quadratic form. In terms of the 2×2-blocking of $\mathrm{Sym}^2(\mathbb{R}^n)$ induced by $\mathbb{R}^n = W \oplus W^\perp$, $A\big|_W$ is the $(1,1)$-component of A.

The subequation $\mathcal{P}^+$ can be restricted to a subequation $\mathcal{P}_W^+$ on W by defining

$$\mathcal{P}_W^+ \equiv \left\{ A\big|_W : A \in \mathcal{P}^+ \right\}. \tag{3.1}$$

Note that $\mathcal{P}_W^+ \subset \mathrm{Sym}^2(W)$ satisfies positivity, since if $P \in \mathrm{Sym}^2(W)$, with $P \geq 0$, then $A\big|_W + P = (A+Q)\big|_W$, where $Q \in \mathrm{Sym}^2(\mathbb{R}^n)$ restricts to P on W and has all other components 0, and therefore $A + Q \in \mathcal{P}^+$. Thus $\mathcal{P}_W^+$ is a subequation on W.

We have proved the following. Let $\mathrm{Sym}^2(W)^\perp$ denote the orthogonal complement of $\mathrm{Sym}^2(W)$ in $\mathrm{Sym}^2(\mathbb{R}^n)$.

Proposition 3.2. *If $\mathcal{P}^+$ is extremely degenerate, i.e., if $\mathcal{P}_W^+$ and W satisfy the equivalent conditions (1*) and (2*) above, then*

$$\mathcal{P}^+ = \mathcal{P}_W^+ \oplus \mathrm{Sym}^2(W)^\perp. \tag{3.2}$$

Moreover, $\mathcal{P}^+$ and $\mathcal{P}_W^+$ have the same reduced constraint set $\mathcal{P}_0^+$ since $\mathrm{Sym}^2(W)^\perp \subset E$, i.e., $S \subset \mathrm{Sym}^2(W)$.

Definition 3.3. If (3.2) is satisfied, we say that $\mathcal{P}^+$ **reduces to** $\mathcal{P}_W^+$, and that $\mathcal{P}^+$ **is the trivial extension of $\mathcal{P}_W^+$ from W to $\mathbb{R}^n$.**

Proposition 3.4. [**5**, Thm. A.4]. *Suppose that $\mathcal{P}^+$ reduces to a subequation $\mathcal{P}_W^+$ on W. If $z = (x, y) \in W \oplus W^\perp = \mathbb{R}^n$ denotes coordinates, then $u(x, y)$ is*

$\mathcal{P}^+$-subharmonic if and only if for each y, $u(x,y)$ is $\mathcal{P}^+_W$-subharmonic in x, but otherwise u is just upper semi-continuous in (x,y), i.e., there is no constraint on u with respect to the y-variable.

The Edge Criteria

Extreme degeneracy and completeness can be described in a very simple way in terms of edges.

Proposition 3.5. (The Edge Criteria). *The following conditions on a convex cone subequation $\mathcal{P}^+$ are equivalent.*

(1)≅(2) $\mathcal{P}^+$ is complete.

(3) $E \cap \mathcal{P} = \{0\}$.

(4a) $P_e \notin E$ for all $|e| = 1$.

(4b) $-P_e \notin \mathcal{P}^+$ for all $|e| = 1$.

Stated as the edge criteria for extreme degeneracy, we have that the following are equivalent.

(1)≅(2*) $\mathcal{P}^+$ is extremely degenerate.*

(3) $E \cap \mathcal{P} \neq \{0\}$.*

*(4*a) $P_e \in E$ for some $|e| = 1$.*

*(4*b) $-P_e \in \mathcal{P}^+$ for some $|e| = 1$.*

Proof. We will prove the extreme degeneracy version. First we note that (4*a) and (4*b) are equivalent. One key to the proof is the following Lemma taken from [**6**].

Lemma 3.6. *Suppose that W is a hyperplane in $\mathbb{R}^n$ with unit normal e. Then*

$$P_e \in E \qquad \iff \qquad \mathcal{P}_+ \subset \mathrm{Sym}^2(W).$$

Proof.

Corollary 3.7. *We have that (2*) $\iff$ (4*a).*

Proof. The only thing to note is that if condition (2*), that $\mathcal{P}_+ \subset \mathrm{Sym}^2(W)$ for some proper subspace $W \subset \mathbb{R}^n$, holds, then $\mathcal{P}_+ \subset \mathrm{Sym}^2(W')$ for any hyperplane $W' \supset W$. ■

Since (4*a) implies (3*) is trivial, the only thing left to prove is that (3*) implies (4*a).

Lemma 3.8. *Suppose $P \geq 0$ has null space $N \subset \mathbb{R}^n$. Then*

$$P \in E \qquad \Rightarrow \qquad \mathrm{Sym}^2(N^\perp) \ \subset \ E \tag{3.3}$$

Proof. The proof is modeled on the proof of (2.5). It suffices to show that $\mathrm{Sym}^2(N^\perp) \subset \mathcal{P}^+$. Given $A \in \mathrm{Sym}^2(N^\perp)$, we write $A = -tP + (A + tP)$ and note that since $P \in E$, we have $-tP \in E$ for all $t \geq 0$. Now $A + tP \in \mathcal{P}$ if $t >> 0$ is sufficiently large since $P\big|_{N^\perp}$ is positive definite. This proves that $A \in E + \mathcal{P} \subset \mathcal{P}^+$. ■

Corollary 3.9. *We have that (3*) $\Rightarrow$ (4*a).*

Proof. If (3*) holds, choose $P \in E \cap \mathcal{P}$ with $P \neq 0$. Since $P \neq 0$, the subspace $N^\perp \neq \{0\}$. Pick $e \in N^\perp$ with $|e| = 1$. Then $P_e \in E$. ■

The Span Criteria

The span criteria for completeness also provides an easy check in examples for completeness, and has important consequences for the subequation.

Proposition 3.10. *The following conditions on a convex cone subequation $\mathcal{P}^+$ are equivalent.*

(1)≅(2) $\mathcal{P}^+$ *is complete.*

(5) $S \cap (\mathrm{Int}\, \mathcal{P}) \neq \emptyset$.

(6) $\mathrm{Int}_S\, \mathcal{P}_+ \subset \mathrm{Int}\, \mathcal{P}$.

Proof. First we show that for a pair of vector spaces E and S which are orthogonal complements in $\mathrm{Sym}^2(\mathbb{R}^n)$, (3) and (5) are equivalent.

Lemma 3.11. *Suppose subspaces E and S of* $\mathrm{Sym}^2(\mathbb{R}^n)$ *are orthogonal complements. Then*

E satisfies the Edge Criteria (3) $\iff$ *S satisfies the Span Criteria (5).*

Proof. (3) $\Rightarrow$ (5). If (5) is false, i.e., $S \cap (\mathrm{Int}\, \mathcal{P}) = \emptyset$, then by the Hahn-Banach Theorem there exists an open half-space U with $S \subset \partial U$ and $\mathrm{Int}\, \mathcal{P} \subset U$. Let $N \in U$ denote the unit normal to the hyperplane ∂U. Then $S \subset \partial U \;\Rightarrow\; N \in E = S^\perp$, while $\mathrm{Int}\, \mathcal{P} \subset U \;\Rightarrow\; \langle N, P\rangle > 0 \;\forall\, P > 0$, which implies $N \in \mathcal{P}$. Thus we have $N \in E \cap \mathcal{P}$, but $N \neq 0$ so that (3) is false.

(5) $\Rightarrow$ (3). By (5) we can pick $P \in S \cap (\mathrm{Int}\, \mathcal{P})$. If $A \in E \cap \mathcal{P}$, then $\langle A, P\rangle = 0$ since $A \in E$ and $P \in S$. However, since $A \geq 0$ and $P > 0$, this implies $A = 0$. ∎

Proof that (5) $\Rightarrow$ (6). By (5) we can choose $P \in S \cap (\mathrm{Int}\, \mathcal{P})$. Given $A \in \mathrm{Int}_S\, \mathcal{P}_+$, for $\epsilon > 0$ sufficiently small we have $A - \epsilon P \in \mathcal{P}_+$. Thus for all non-zero $Q \in \mathcal{P} \subset \mathcal{P}^+$ we have $0 \leq \langle A - \epsilon I, Q\rangle = \langle A, Q\rangle - \epsilon \langle P, Q\rangle$. Since $P > 0$, one has $\langle P, Q\rangle > 0$, which proves that $\langle A, Q\rangle > 0$ for all non-zero $Q \geq 0$. Thus $A > 0$, which proves (6). ∎

Proof that (6) $\Rightarrow$ (5). Now $\mathcal{P}_+$ is a closed convex cone in S. Hence $\mathrm{Int}_S\, \mathcal{P}_+ \neq \emptyset$ is equivalent to S equaling the span of $\mathcal{P}_+$, which it does by the definition of S. Now pick $P \in \mathrm{Int}_S\, \mathcal{P}_+$. Then $P \in S$ and by (6) we have $P > 0$, which proves (5). ∎

This completes the proof of Proposition 3.9. ∎

The edge and span criteria (3) and (5) for completeness motivates the following definition, which will be used in the next section.

Definition 3.12.

(a) A subspace $E \subset \mathrm{Sym}^2(\mathbb{R}^n)$ is called a **basic edge subspace** if

(3) $E \cap \mathcal{P} \;=\; \{0\}$.

(b) A subspace $S \subset \mathrm{Sym}^2(\mathbb{R}^n)$ is called a **basic span subspace** if

(5) $S \cap (\mathrm{Int}\, \mathcal{P}) \;\neq\; \emptyset$.

(c) If in addition E and S are orthogonal complements, then E, S well be referred to as a **basic edge-span pair**.

4. The Supporting Subequation

For each subequation there is a smallest subspace W of $\mathbb{R}^n$ to which the subequation reduces.

Definition 4.1. (Support). Given a convex cone subequation $\mathcal{P}^+$ we define the **support of** $\mathcal{P}^+$ to be the subspace $W \subset \mathbb{R}^n$ which is the intersection of all subspaces $W' \subset \mathbb{R}^n$ which that

$$\mathcal{P}^+ \;=\; \mathcal{P}^+_{W'} \oplus \mathrm{Sym}^2(W')^\perp. \tag{4.1}$$

Lemma 4.2. *The orthogonal complement of the support W of $\mathcal{P}^+$ equals:*

$$V \;\equiv\; \mathrm{span}\,\{e \in \mathbb{R}^n : P_e \in E, |e| = 1\}. \tag{4.2}$$

Proof. Note that (4.1) holds $\iff$ $\mathrm{Sym}^2(W')^\perp \subset E$ $\iff$ $S \subset \mathrm{Sym}^2(W')$ $\iff$ $\mathcal{P}_+ \subset \mathrm{Sym}^2(W')$ $\iff$ $P_e \in E$ for all $e \perp W'$ with $|e| = 1$. ∎

The support illuminates the structure of the subequation.

THEOREM 4.3. (Structure Theorem). *Suppose $\mathcal{P}^+ \subset \mathrm{Sym}^2(\mathbb{R}^n)$ is a convex cone subequation with support $W \subset \mathbb{R}^n$. Then*

$$\mathcal{P}^+ \;=\; \mathcal{P}^+_W \oplus \mathrm{Sym}^2(W)^\perp \qquad \text{and} \tag{4.3}$$

$$\mathcal{P}^+_W \;\subset\; \mathrm{Sym}^2(W) \ \text{ is a complete subequation.} \tag{4.4}$$

Proof. To be done later.

Definition 4.4. If W is the support of $\mathcal{P}^+$, the subequation $\mathcal{P}^+_W$ will be called the **supporting subequation of $\mathcal{P}^+$**, and its edge E_W will be called the **supporting edge of $\mathcal{P}^+$**

Note that the edge of $\mathcal{P}^+$,

$$E \;=\; E_W \oplus \mathrm{Sym}^2(W)^\perp, \tag{4.5}$$

is larger than its supporting edge E_W unless $\mathcal{P}^+$ is complete.

Note also that the original subequation $\mathcal{P}^+$ and the supporting subequation $\mathcal{P}^+_W$ have the same span S and the same reduced constraint set $\mathcal{P}^+_0$.

5. Minimal Subequations

These subequations are the focus of this paper. They are all constructed as follows, starting with a basic edge-span pair.

Lemma 5.1. *Suppose $E, S \subset \mathrm{Sym}^2(\mathbb{R}^n)$ are orthogonal complements with $E \cap \mathcal{P} = \{0\}$, or equivalently $S \cap (\mathrm{Int}\,\mathcal{P}) \neq \emptyset$. That is, E, S is a basic edge-span pair. Then*

$$\mathcal{P}^+ \;\equiv\; E + \mathcal{P} \ \text{ is a subequation, and it has edge } E \text{ and span } S. \tag{5.1}$$

Moreover, if Q^+ is any subequation with edge E, then $\mathcal{P}^+ \subset Q^+$.

Proof. Obviously $\mathcal{P}^+$ satisfies positivity. It remains to show that $\mathcal{P}^+ \equiv E + \mathcal{P}$ is closed. Let $\pi : \mathrm{Sym}^2(\mathbb{R}^n) \to S$ denote orthogonal projection as in (2.5). Since $E + \mathcal{P} = E \oplus \pi(\mathcal{P})$,

$$\mathcal{P}^+ \text{ is closed if and only if } \pi(\mathcal{P}) \text{ is closed.} \tag{5.2}$$

Now we prove that:

$$\pi(\mathcal{P}) \text{ is closed.}$$

Let $K \equiv \mathcal{P} \cap \{\mathrm{tr} = 1\}$, a compact base for $\mathcal{P}$. The image $\pi(K)$ is a compact subset of S. The basic edge condition $E \cap \mathcal{P} = \{0\}$ is equivalent to $0 \notin \pi(K)$. This is enough to conclude that the cone on the compact convex set $\pi(K)$ is closed. Thus, $\mathcal{P}^+ \equiv E + \mathcal{P}$ is a subequation. ∎

To prove that $\mathcal{P}^+$ has edge E we must show that

$$\mathcal{P}^+ \cap (-\mathcal{P}^+) \;=\; E \qquad \text{or equivalently} \qquad \pi(\mathcal{P}) \cap (-\pi(\mathcal{P})) \;=\; \{0\}.$$

Suppose $A \in \pi(\mathcal{P}) \cap (-\pi(\mathcal{P}))$, i.e., $A = \pi(P_1) = -\pi(P_2)$ with $P_1, P_2 \in \mathcal{P}$. Then $\pi(P_1 + P_2) = 0$, i.e., $P_1 + P_2 \in E$. Since $E \cap \mathcal{P} = \{0\}$, $P_1 + P_2 = 0$. But this implies $P_1 = P_2 = 0$ and hence $A = 0$. Since $\mathcal{P}^+$ has edge E, it has span $S = E^\perp$. Finally, $\mathcal{P}^+ \subset Q^+$, since $E \subset Q^+$ and positivity for Q^+ implies $\mathcal{P}^+ \equiv E + \mathcal{P} \subset Q^+$. ∎

Definition 5.2. The subequation $\mathcal{P}^+ = E + \mathcal{P}$ constructed in Lemma 5.1 will be referred to as a **minimal subequation**, or the **minimal subequation with edge** E.

Corollary 5.3. *Suppose $\mathcal{P}^+$ is a minimal subequation with edge-span E, S. Then*

$$(a)\ \ E \cap \mathcal{P} = \{0\}, \qquad (b)\ \ S \cap (\operatorname{Int}\mathcal{P}) \neq \emptyset, \qquad (c)\ \ \mathcal{P}^+ \text{ is complete.}$$

Proof. By definition of minimal we have $\mathcal{P}^+ = E' + \mathcal{P}$ where E' satisfies (a). By Lemma 5.1 the edge E of $\mathcal{P}^+$ equals E'. Lemma 3.10 says that (a) and)b) are equivalent. Either the edge criteria (a) $\Rightarrow$ (c), or the span criteria (b) $\Rightarrow$ (c), completes the proof. ■

There are many additional interesting properties of minimal subequations, besides the various completeness criteria in Section 3.

THEOREM 5.4. (Minimality Properties). *Suppose $\mathcal{P}^+ \equiv E + \mathcal{P}$ is the minimal subequation with edge E and span S. Then*
(1) $\mathcal{P}^+ = E + \mathcal{P}$, (1a) $\mathcal{P}_0^+ = \pi(\mathcal{P})$, (1b) $\mathcal{P}^+ = E \oplus \pi(\mathcal{P})$
(2) $\operatorname{Int}\mathcal{P}^+ = E + \operatorname{Int}\mathcal{P}$, (2a) $\operatorname{Int}\mathcal{P}_0^+ = \pi(\operatorname{Int}\mathcal{P})$, (2b) $\operatorname{Int}\mathcal{P}^+ = E \oplus \operatorname{Int}\pi(\mathcal{P})$
(3) $\mathcal{P}_+ = S \cap \mathcal{P}$, and (3*) $\operatorname{Int}_S \mathcal{P}_+ = S \cap (\operatorname{Int}\mathcal{P})$.

In fact, for complete subequations each of these eight properties characterizes minimality.

THEOREM 5.5. (Minimality Criteria). *Suppose $\mathcal{P}^+ \subset \operatorname{Sym}^2(\mathbb{R}^n)$ is a complete convex cone subequation, with edge E span S, reduced constraint set $\mathcal{P}_0$, and polar cone $\mathcal{P}_+$. Then $\mathcal{P}^+$ is the minimal subequation with edge E if and only if any one of the eight equivalent conditions in Theorem 5.4 hold.*

Proof of Theorem 5.4. Assertion (1) is by Definition 4.2. Next we show the following.

$$(1), (1a) \text{ and } (1b) \text{ are equivalent for any subequation } \mathcal{P}^+ \text{ with edge } E. \tag{5.3}$$

(1) $\Rightarrow$ (1a): By definition $\mathcal{P}_0^+ = \pi(\mathcal{P}^+)$. Since $\pi(E) = \{0\}$, (1) implies that $\pi(\mathcal{P}^+) = \pi(\mathcal{P})$.
(1a) $\Rightarrow$ (1b): This follows because $\mathcal{P}^+ = E \oplus \pi(\mathcal{P}^+)$.
(1b) $\Rightarrow$ (1): This is obvious.
Proof of (2). Obviously the open set $E + \operatorname{Int}\mathcal{P} \subset \operatorname{Int}\mathcal{P}^+$. If $A \in \operatorname{Int}\mathcal{P}^+$, then for small $\epsilon > 0$, $A - \epsilon I \in \operatorname{Int}\mathcal{P}^+ \subset \mathcal{P}^+$. Hence there exist $B_0 \in E$ and $P \geq 0$ such that $A - \epsilon I = B_0 + P$. Therefore, $A = B_0 + (P + \epsilon I) \in E + \operatorname{Int}\mathcal{P}$, proving that $\operatorname{Int}\mathcal{P}^+ = E + \operatorname{Int}\mathcal{P}$.

Just as in (5.3), we have

$$(2), (2a) \text{ and } (2b) \text{ are equivalent for any subequation } \mathcal{P}^+ \text{ with edge } E. \tag{5.4}$$

Proof of (3). Since $0 \in \mathcal{P}^+$ and $\mathcal{P}^+$ is $\mathcal{P}$-monotone, we have $\mathcal{P} \subset \mathcal{P}^+$. Since $\mathcal{P}$ is self polar, taking polars implies that $\mathcal{P}_+ \subset \mathcal{P}$ and therefore $\mathcal{P}_+ \subset S \cap \mathcal{P}$.

Suppose $B \in S \cap \mathcal{P}$. To show $B \in \mathcal{P}_+$ it suffices to show that $\langle A, B\rangle \geq 0$ for all $A \in \mathcal{P}^+$. By minimality, if $A \in \mathcal{P}^+$, then $A = A_0 + P$ with $A_0 \in E$ and $P \in \mathcal{P}$. Now $\langle A, B\rangle = \langle P, B\rangle \geq 0$ since $\langle A_0, B\rangle = 0$. ■
Proof of (3*). Note that $S \cap (\operatorname{Int}\mathcal{P})$ is an open set in S, and it is contained in $S \cap \mathcal{P}$, which is a subset of $\mathcal{P}_+$ by (3). Hence, $S \cap (\operatorname{Int}\mathcal{P}) \subset \operatorname{Int}_S \mathcal{P}_+$.

The only non-trivial part (and the most important part) of showing $\operatorname{Int}_S \mathcal{P}_+ = S \cap (\operatorname{Int}\mathcal{P})$ is to show that:

$$\operatorname{Int}_S \mathcal{P}_+ \ \subset\ \operatorname{Int}\mathcal{P}. \tag{5.5}$$

Since S is a basic span subspace, $S \cap (\operatorname{Int} \mathcal{P}) \neq 0$. Choose $P \in S \cap (\operatorname{Int} \mathcal{P})$. Given $A \in \operatorname{Int}_S \mathcal{P}_+$, for $\epsilon > 0$ sufficiently small we have $A - \epsilon P \in \mathcal{P}_+$. Thus for all non-zero $Q \in \mathcal{P} \subset \mathcal{P}^+$ we have $0 \leq \langle A - \epsilon I, Q \rangle = \langle A, Q \rangle - \epsilon \langle P, Q \rangle$. Since $P > 0$, one has $\langle P, Q \rangle > 0$, which proves that $\langle A, Q \rangle > 0$ for all non-zero $Q \geq 0$. Thus $A > 0$. ∎

Proof of Theorem 5.5. By Theorem 5.4, if $\mathcal{P}^+$ is minimal, then $\mathcal{P}^+$ satisfies each of the eight conditions. For the converses we use the hypothesis that $\mathcal{P}^+$ is complete. By the edge criteria, Proposition 3.5(3), the edge E of $\mathcal{P}^+$ satisfies $E \cap \mathcal{P} = \{0\}$. Therefore we can apply the construction in Lemma 4.1 to yield a minimal subequation $\mathcal{Q}^+ \equiv E + \mathcal{P}$ satisfying all the eight conditions. If $\mathcal{P}^+$ satisfies (1) then $\mathcal{P}^+ = \mathcal{Q}^+$ and so it is minimal. Similarly, if $\mathcal{P}^+$ satisfies (2), then $\operatorname{Int} \mathcal{P}^+ = \operatorname{Int} \mathcal{Q}^+$, so that $\mathcal{P} = \mathcal{Q}^+$ is minimal.

By (5.3) we have that (1), (1a) and (1b) are equivalent.

By (5.4) we have that (2), (2a) and (2b) are equivalent.

Finally, if $\mathcal{P}^+$ satisfies (3*), then since $\mathcal{Q}_+ = S \cap \mathcal{P}$ also, we have $\mathcal{P}_+ = \mathcal{Q}_+$ and hence $\mathcal{P}^+ = \mathcal{Q}^+$ is minimal. As noted above, (3) $\Rightarrow$ (3*). ∎

Remark 5.6. The property (5.5) is extremely important and useful. See [**6**] for more details of the following.

Given $A \geq 0$ define $\Delta_A u \equiv \langle D^2 u, A \rangle$, or equivalently, from the subequation point of view,

$$\Delta_A \equiv \{B \in \operatorname{Sym}^2(\mathbb{R}^n) : \langle B, A \rangle \geq 0\}.$$

Then u is $\mathcal{P}^+$-subharmonic if and only if u is Δ_A-subharmonic for all $A \in \operatorname{Int}_{\mathrm{rel}} \mathcal{P}_+$. If (5.5) is true, then each such operator Δ_A is just a linear coordinate change of the standard Laplacian on $\mathbb{R}^n$ (or said differently, it is the Laplacian on $\mathbb{R}^n$ with a different metric). Thus results of standard potential theory, such as $u \in L^1_{\mathrm{loc}}$, are valid for $\mathcal{P}^+$-subharmonic functions.

One final property of minimal subequation is the following.

Proposition 5.7. *Suppose $\mathcal{P}^+$ is a minimal subequation. Then $\mathcal{P}^+$ is contained in its dual subequation*

$$\widetilde{\mathcal{P}^+} \equiv \sim(-\operatorname{Int} \mathcal{P}^+) = -(\sim \operatorname{Int} \mathcal{P}^+). \tag{5.6}$$

Proof. Since $\mathcal{P}^+ = E + \mathcal{P}$ and $\widetilde{\mathcal{P}}^+ + \mathcal{P} = \widetilde{\mathcal{P}}^+$, it suffices to show that $E \equiv \mathcal{P}^+ \cap (-\mathcal{P}^+) \subset \widetilde{\mathcal{P}}^+$. Suppose $A \notin \widetilde{\mathcal{P}}^+$, i.e., $-A \in \operatorname{Int} \mathcal{P}^+$. Then by 5.4(2) we have $-A = B_1 + P$ with $B_1 \in E$ and $P > 0$. If $A \in \mathcal{P}^+$ also, then $A = B_2 + Q$ with $B_2 \in E$ and $Q \geq 0$. Therefore, $P + Q = -B_1 - B_2 \in E$. However, $P + Q > 0$ contradicting Corollary 5.3(a). ∎

6. Edge Functions – Pluriharmonics

Suppose as before that $\mathcal{P}^+$ is a complete convex cone subequation with edge E.

Definition 6.1. An **edge function**, or **$\mathcal{P}^+$-pluriharmonic function** is a function u such that both

$$u \text{ and } -u \text{ are } \mathcal{P}^+\text{-subharmonic.} \tag{6.1}$$

Thus, by definition, u is continuous.

Definition 6.2. An upper semi-continuous function u is **"sub"** the edge functions on an open set $X \subset \mathbb{R}^n$ if for all domains $\Omega \subset\subset \mathbb{R}^n$ and all edge functions h on Ω

which are continuous on $\overline{\Omega}$,

$$u \leq h \quad \text{on } \partial\Omega \qquad \Rightarrow \qquad u \leq h \quad \text{on } \overline{\Omega}. \tag{6.2}$$

Proposition 6.3. *If u is dually $\mathcal{P}^+$-subharmonic on X, i.e., u is $\widetilde{\mathcal{P}}^+$-subharmonic for the dual subequation $\widetilde{\mathcal{P}}^+$ (see (5.6)), then u is "sub" the edge functions on X.*

Proof. Suppose u is $\widetilde{\mathcal{P}}^+$-subharmonic and h is an edge function. Then $-h$ is $\mathcal{P}^+$-subharmonic and (6.2) follows from comparison (see Thm. 6.2 in [**9**]). ∎

Now if a subequation becomes smaller, it dual subequation becomes larger. Consequently, the only subequation $\mathcal{P}^+$, with a given edge E, for which Proposition 6.3 might have a converse is the minimal subequation with edge E (see Definition 5.2).

THEOREM 6.4. *Suppose that $E \subset \mathrm{Sym}^2(\mathbb{R}^n)$ is a basic vector subspace, so that $\mathcal{P}^+ \equiv E + \mathcal{P}$ is the minimal subequation with edge E. Then the following conditions on a function u are equivalent.*

(1) *u is dually $\mathcal{P}^+$-subharmonic.*

(2) *u is "sub" the edge functions.*

(3) *u is locally "sub" the edge functions.*

(4) *u is locally "sub" the degree-2 polynomial edge functions.*

Proof. Because of Proposition 6.3 we need only prove that if u is locally "sub" the degree-2 polynomial edge functions, then u is dually $\mathcal{P}^+$-subharmonic. For this suppose that u is not $\widetilde{\mathcal{P}}^+$-subharmonic on X. Then (see Lemma 2.4 in [**7**]) there exists $z_0 \in X$, a quadratic polynomial test function φ, and $\alpha > 0$ such that

$$u(z) \leq \varphi(z) - \alpha|z - z_0|^2 \quad \text{near } z_0 \text{ with equality at } z_0, \tag{6.3}$$

but

$$D^2_{z_0}\varphi \notin \widetilde{\mathcal{P}}^+, \quad \text{i.e.,} \quad -D^2_{z_0}\varphi \in \mathrm{Int}\,\mathcal{P}^+. \tag{6.4}$$

By Theorem 4.4(2) we have $\mathrm{Int}\,\mathcal{P}^+ = \mathrm{Int}\,\mathcal{P} + E$. Thus

$$-D^2_{z_0}\varphi = P + B \quad \text{with} \quad P > 0 \quad \text{and} \quad B \in E. \tag{6.5}$$

Consider the degree-2 edge polynomial

$$\begin{aligned} h(z) &\equiv \varphi(z_0) + \langle D_{z_0}\varphi, z - z_0\rangle - \tfrac{1}{2}\langle B(z - z_0), z - z_0\rangle \\ &= \varphi(z) - \tfrac{1}{2}\langle D^2_{z_0}\varphi, z - z_0\rangle - \tfrac{1}{2}\langle B(z - z_0), z - z_0\rangle \\ &= \varphi(z) + \tfrac{1}{2}\langle P(z - z_0), z - z_0\rangle. \end{aligned}$$

Since $P > 0$ by (6.3) this implies that

$$u(z) \leq h(z) - \alpha|z - z_0|^2 \tag{6.6}$$

near z_0 with equality at z_0. This implies that u is not sub the function h on any small ball about z_0. Hence, u is not locally "sub" the degree-2 edge polynomial h. ∎

7. Further Discussion of Examples

Before turning to the examples we define the **(compact) invariance group of $\mathcal{P}^+$** to be

$$\{g \in \mathrm{O}_n : g^*\mathcal{P}^+ \ = \ \mathcal{P}^+\}. \tag{7.1}$$

It is easy to see that for the minimal subequation $\mathcal{P}^+$ for a basic E,

$$g^*\mathcal{P}^+ = \mathcal{P}^+ \iff g^*S = S \iff g^*E = E \tag{7.2}$$

by using the conditions in Theorems 5.4 and 5.5, and this yields two equivalent definitions of this group.

Definition 7.1. (Self Duality). If the two convex cones $\mathcal{P}_0^+$ (the reduced constraint set) and $\mathcal{P}_+$ are polars of each other in the vector space S, then we say the subequation $\mathcal{P}^+$ is **polar self dual** (not to be confused with a subequation which equals its dual subequation in the sense of [**5**]).

Remark 7.2. Note that this can only happen for a minimal subequation $\mathcal{P}^+$. This is because if $\mathcal{P}_0^+ = \mathcal{P}_+$ (self duality), then $\mathcal{P}_0^+ = \mathcal{P}_+ \subset \mathcal{P}$, and hence $\mathcal{P}_0^+ = \pi(\mathcal{P}_0^+) \subset \pi(\mathcal{P})$. Note that $\mathcal{P} \subset \mathcal{P}^+$ so that $\pi(\mathcal{P}) \subset \mathcal{P}_0^+$ is always true. This proves $\mathcal{P}_0^+ = \pi(\mathcal{P})$, so by Theorems 5.5 and 5.4(1a), $\mathcal{P}^+$ is minimal.

Given a closed subset $\mathbb{G} \subset G(k, \mathbb{R}^n)$ consider the *subequation geometrically defined by* $\mathbb{G}$:

$$\mathcal{P}(\mathbb{G}) \ \equiv \ \left\{A \in \mathrm{Sym}^2(\mathbb{R}^n) : \langle A, P_W\rangle = \mathrm{tr}(A\big|_W) \geq 0 \ \ \forall W \in \mathbb{G}\right\}.$$

We shall use the following notations introduced in Example 2.2:

$$\mathcal{P}^+ \equiv \mathcal{P}(\mathbb{G}), \quad \mathcal{P}_+ = CCH(\mathbb{G}), \quad S = \mathrm{span}\,(\mathbb{G}), \quad E = S^\perp, \quad \text{and} \quad \mathcal{P}_0^+.$$

Note that the compact invariance group of the subequation $\mathcal{P}^+ = \mathcal{P}(\mathbb{G})$ can also be defined by

$$\{g \in \mathrm{O}(n) : g(\mathbb{G}) = \mathbb{G}\}. \tag{7.2$'$}$$

The O(n)-Invariance Group

For our first two examples of minimal subequations we focus on the O_n-**orthogonal decomposition**

$$\mathrm{Sym}^2(\mathbb{R}^n) \ = \ \mathbb{R}\cdot\mathrm{Id} \oplus \mathrm{Sym}_0^2(\mathbb{R}^n) \tag{7.3}$$

into irreducible components under O_n.

Example 7.1. (Real Monge-Ampère). The subequation is $\mathcal{P}^+ = \mathcal{P}$. Here the edge $E = \{0\}$ is as small as possible, and $S = \mathrm{Sym}^2(\mathbb{R}^n)$, $\mathcal{P}_+ = \mathcal{P}$, so the subequation is self-dual, and we have $\mathbb{G} = G(1, \mathbb{R}^n)$. Obviously E, S is a basic edge-span pair (Definition 3.12c). The conditions in Theorem 5.4 are obvious as well as the fact that $\mathcal{P} = \mathcal{P}^+ = \mathcal{P}_+ = \mathcal{P}_0^+$ is dimensionally complete. The invariance group is O_n, and the extreme rays are

$$\mathrm{Ext}(\mathcal{P}) \ = \ \{\mathrm{Ray}(P_e) : |e| = 1\}.$$

Each $A \in S$ can be put in canonical form $A = \sum_j \lambda_j P_{e_j}$ under the action of O_n, and $\det(A) = \prod_j \lambda_j$, provides a nonlinear operator for $\mathcal{P}^+ = \{\lambda_{\min} \geq 0\}$ (the standard real Monge-Ampère operator).

Example 7.2. (The Laplacian). Here $\mathcal{P}^+ = \Delta = \{A : \mathrm{tr}(A) \geq 0\}$ is a closed half space, and $\mathbb{G} = \{\mathrm{Id}\} = G(n, \mathbb{R}^n)$, $E = \mathrm{Sym}_0^2(\mathbb{R}^n)$, the traceless part of $\mathrm{Sym}^2(\mathbb{R}^n)$, $S = \mathbb{R}\cdot\mathrm{Id}$, and $\mathcal{P}_+ = \mathbb{R}_+\cdot\mathrm{Id}$ is a ray. The invariance group is O_n. The reduced constraint set is $\mathcal{P}_0^+ = \mathcal{P}_+$ so Δ is self dual. Now it is obvious that Δ is a minimal subequation.

The U(n)-Invariance Group

We now consider $\mathbb{C}^n$ and the following U(n)-**orthogonal decomposition** of real symmetric matrices into U_n-irreducible subspaces:

$$\mathrm{Sym}^2_{\mathbb{R}}(\mathbb{C}^n) \;=\; \mathbb{R}\cdot \mathrm{Id} \oplus \mathrm{Herm}_0^{\mathbb{C}-\mathrm{sym}}(\mathbb{C}^n) \oplus \mathrm{Herm}^{\mathbb{C}-\mathrm{skew}}(\mathbb{C}^n) \tag{7.4}$$

multiples of the identity, traceless complex hermitian symmetric, and complex hermitian skew components. Given $A \in \mathrm{Sym}^2_{\mathbb{R}}(\mathbb{C}^n)$, this decomposition can be written as

$$A \;=\; \frac{\mathrm{tr}(A)}{2n}\,\mathrm{Id} + A_0^{\mathbb{C}-\mathrm{sym}} + A^{\mathbb{C}-\mathrm{skew}} \tag{7.5}$$

where with respect to multiplication I by i:

$$A^{\mathbb{C}-\mathrm{sym}} \;=\; \tfrac{1}{2}(A - IAI) \qquad \text{and} \qquad A^{\mathbb{C}-\mathrm{skew}} \;=\; \tfrac{1}{2}(A + IAI).$$

Example 7.3. (Complex Plurisubharmonics). The subequation is $\mathcal{P}^+ = \mathcal{P}(\mathbb{G})$ where $\mathbb{G} = \mathbb{P}(\mathbb{C}^n) \subset G_{\mathbb{R}}(2, \mathbb{C}^n)$ is the Grassmannian of complex lines in $\mathbb{C}^n$. The edge is $E = \mathrm{Herm}^{\mathbb{C}-\mathrm{skew}}(\mathbb{C}^n)$ and the span is $S = \mathrm{Herm}^{\mathbb{C}-\mathrm{sym}}(\mathbb{C}^n)$. Also $\mathcal{P}_0^+ = \mathcal{P}_+$ is the convex cone on non-negative complex hermitian symmetric bilinear forms on $\mathbb{C}^n$, so this third example is self dual. Note that the projection of $2P_e$ onto S is $P_{\mathbb{C}e}$, (orthogonal projection onto the complex line through e) since $P_e - IP_eI = P_{\mathbb{C}e}$. The convex cone $\mathcal{P}_0^+ = \mathcal{P}_+$ has extreme rays generated by $\{P_{\mathbb{C}e} : |e| = 1\} = \mathbb{P}(\mathbb{C}^n) = \mathbb{G}$. The invariance group is U_n. Each $A \in S$ can be put into canonical form $A = \sum_{j=1}^n \lambda_j P_{\mathbb{C}e_j}$ under the action of this group, and $\mathcal{P}_0^+ = \{\lambda_{\min} \geq 0\}$. The complex Monge-Ampère operator $\det(A) = \lambda_1(A)\cdots\lambda_n(A)$ provides the nonlinear operator for $\mathcal{P}^+ = \mathcal{P}(\mathbb{P}(\mathbb{C}^n))$, in tight analogue with the real case $\mathcal{P}$.

Now we finally get to a new example, which is the subject of [**10**].

Example 7.4. (Lagrangian Plurisubharmonics). The subequation is $\mathcal{P}^+ = \mathcal{P}(\mathrm{LAG})$, where $\mathrm{LAG} \subset G_{\mathbb{R}}(n, \mathbb{C}^n)$ is the set of Lagrangian n-planes in $\mathbb{C}^n = \mathbb{R}^{2n}$. The edge E and span S are given by

$$E = \mathrm{Herm}_0^{\mathbb{C}-\mathrm{sym}}(\mathbb{C}^n) \qquad \text{and} \qquad S = \mathbb{R}\cdot\mathrm{Id} \oplus \mathrm{Herm}^{\mathbb{C}-\mathrm{skew}}(\mathbb{C}^n).$$

In [**10**] we prove that E, S is a basic edge-span pair, so that $\mathcal{P}^+ = E + \mathcal{P}$ and $\mathcal{P}_+ = S \cap \mathcal{P}$. The extreme rays in $\mathcal{P}_+$ are generated by the projections P_W with $W \in \mathrm{LAG}$ a Lagrangian n-plane. The extreme rays in $\mathcal{P}_0^+$ are generated by the images $\pi(P_e)$ of P_e where e is a unit vector. Note that

$$\pi(P_e) = \tfrac{1}{2n}\,\mathrm{Id} + \tfrac{1}{2}(P_e + IP_eI) = \tfrac{1}{2n}\,\mathrm{Id} + \tfrac{1}{2}(P_e - P_{Ie}),$$

and that $\frac{1}{2}(P_e - P_{Ie})$ is the $\mathbb{C}$-skew component of P_e. This example is *not* self dual. However, since each $A \in S$ can be put in canonical form

$$A \;=\; \frac{\mathrm{tr}(A)}{2n} + \frac{1}{2}\sum_{j=1}^n \lambda_j\left(P_{e_j} - P_{Ie_j}\right)$$

there is again a nonlinear operator for $\mathcal{P}^+ = \mathcal{P}(\mathrm{LAG})$ (see [**10**]). The invariance group is U_n.

The Sp(n)·Sp(1)-Invariance Group

Let $M_n(\mathbb{H})$ denote the space of $n\times n$ matrices with entries in $\mathbb{H}$, and let $A^* = \overline{A}^t$ if $A \in M_n(\mathbb{H})$. Consider the two subspaces

$$\begin{aligned} M_n^{\rm sym}(\mathbb{H}) &= \{A \in M_n(\mathbb{H}) : A^* = A\}, \qquad \text{and} \\ M_n^{\rm skew}(\mathbb{H}) &= \{A \in M_n(\mathbb{H}) : A^* = -A\}. \end{aligned}$$

We let the scalars $\mathbb{H}$ act on the right. Then by letting $M_n(\mathbb{H})$ act on $x = (x_1, ..., x_n)^t \in \mathbb{H}^n$ on the left, one can identify $M_n(\mathbb{H})$ with $\mathrm{End}_{\mathbb{H}}(\mathbb{H}^n)$, the vector space of $\mathbb{H}$-linear maps of $\mathbb{H}^n$. Let

$$\begin{aligned} \mathrm{Herm}^{\mathbb{H}-\mathrm{sym}}(\mathbb{H}^n) &= \{A \in \mathrm{End}_{\mathbb{H}}(\mathbb{H}^n) : A = A^*\}, \qquad \text{and} \\ \mathrm{Herm}^{\mathbb{H}-\mathrm{skew}}(\mathbb{H}^n) &= \{A \in \mathrm{End}_{\mathbb{H}}(\mathbb{H}^n) : A = -A^*\}. \end{aligned}$$

so that $M_n^{\rm sym}(\mathbb{H}) = \mathrm{Herm}^{\mathbb{H}-\mathrm{sym}}(\mathbb{H}^n)$ are identified (same for the skew parts).

Let $\epsilon(x, y) = \sum_{\ell=1}^n \overline{x}_\ell y_\ell$ denote the *standard quaternionic hermitian bilinear form on* $\mathbb{H}^n$. The **quaternionic unitary group** is

$$\mathrm{Sp}_n = \{A \in M_n(\mathbb{H}) : \epsilon(Ax, Ay) = \epsilon(x, y)\}.$$

For each scalar $u \in \mathbb{H}$ let $R_u x \equiv xu$ denote right multiplication, and set $I \equiv R_i$, $J \equiv R_j$, $K \equiv R_k$. Then the group of unit scalars $\mathrm{Sp}_1 \equiv S^3 = \{R_u : u \in \mathbb{H}, |u| = 1\}$ acts on $\mathbb{H}^n$ on the right and the **enhanced quaternionic unitary group** is the group

$$\mathrm{Sp}_n \cdot \mathrm{Sp}_1 = \mathrm{Sp}_n \times \mathrm{Sp}_1/\mathbb{Z}_2.$$

Since the standard euclidean inner product on $\mathbb{R}^{4n} = \mathbb{H}^n$ is $\langle x, y\rangle = \mathrm{Re}\,\epsilon(x, y)$,

$$M_n^{\rm sym}(\mathbb{H}) = \mathrm{Herm}^{\mathbb{H}-\mathrm{sym}}(\mathbb{H}^n) \ \text{ is a real subspace of } \mathrm{Sym}^2(\mathbb{R}^{4n})$$

and

$$M_n^{\rm skew}(\mathbb{H}) = \mathrm{Herm}^{\mathbb{H}-\mathrm{skew}}(\mathbb{H}^n) \ \text{ is a real subspace of } \mathrm{Skew}^2(\mathbb{R}^{4n})$$

where $\mathrm{End}_{\mathbb{R}}(\mathbb{R}^{4n}) = \mathrm{Sym}^2(\mathbb{R}^{4n}) \oplus \mathrm{Skew}^2(\mathbb{R}^{4n})$ is the usual decomposition. Note also that for each unit imaginary quaternion $u \in \mathrm{Im}\mathbb{H}$, we have $R_u \in \mathrm{Skew}^2(\mathbb{R}^{4n})$, and hence $R_u A = AR_u \in \mathrm{Sym}^2(\mathbb{R}^{4n})$ for all $A \in M_n^{\rm skew}(\mathbb{H}) = \mathrm{Herm}^{\mathbb{H}-\mathrm{skew}}(\mathbb{H}^n)$. This embeds

$$\mathrm{Im}\mathbb{H} \otimes \mathrm{Herm}^{\mathbb{H}-\mathrm{skew}}(\mathbb{H}^n) = \mathrm{Im}\mathbb{H} \otimes M_n^{\rm skew}(\mathbb{H}) \subset \mathrm{Sym}^2(\mathbb{R}^{4n}). \tag{7.6}$$

The $\mathrm{Sp}_n \cdot \mathrm{Sp}_1$-**orthogonal decomposition**

$$\mathrm{Sym}^2(\mathbb{R}^{4n}) = \mathbb{R}\cdot \mathrm{Id} \oplus \mathrm{Herm}_0^{\mathbb{H}-\mathrm{sym}}(\mathbb{H}^n) \oplus \left(\mathrm{Im}\mathbb{H} \otimes \mathrm{Herm}^{\mathbb{H}-\mathrm{skew}}(\mathbb{H}^n)\right) \tag{7.7}$$

into irreducible components plays a role in the next two examples, and a key role in classifying all the $\mathrm{Sp}_n \cdot \mathrm{Sp}_1$-invariant minimal subequations. Projection onto $\mathrm{Herm}^{\mathbb{H}-\mathrm{sym}}(\mathbb{H}^n) = \mathbb{R}\cdot\mathrm{Id} \oplus \mathrm{Herm}_0^{\mathbb{H}-\mathrm{sym}}(\mathbb{H}^n)$ and $\mathrm{Im}\mathbb{H} \otimes \mathrm{Herm}^{\mathbb{H}-\mathrm{skew}}(\mathbb{H}^n)$ are given by $A = A^{\mathbb{H}-\mathrm{sym}} + A^{\mathbb{H}-\mathrm{skew}}$ where

$$A^{\mathbb{H}-\mathrm{sym}} = \tfrac{1}{4}(A - IAI - JAJ - KAK), \qquad \text{and} \tag{7.8a}$$

$$A^{\mathbb{H}-\mathrm{skew}} = \tfrac{1}{4}(3A + IAI + JAJ + KAK). \tag{7.8b}$$

Example 7.5. (Quaternionic Plurisubharmonics). The subequation $\mathcal{P}^+ \equiv \mathcal{P}(\mathbb{P}(\mathbb{H}^n))$ is geometric with $\mathbf{G} = \mathbb{P}(\mathbb{H}^n) \subset G_{\mathbb{R}}(4, \mathbb{H}^n)$. The edge and span are given by

$$E = \mathrm{Im}\mathbb{H} \otimes \mathrm{Herm}^{\mathbb{H}-\mathrm{skew}}(\mathbb{H}^n), \qquad \text{and} \qquad S = \mathrm{Herm}^{\mathbb{H}-\mathrm{sym}}(\mathbb{H}^n). \tag{7.9}$$

The set $\mathcal{P}_0^+ = \mathcal{P}_+$ is the convex cone of non-negative quaternionic hermitian symmetric bilinear forms on $\mathbb{H}^n$ (see [**1**] or [**4**] for more details). Under the identification of $\mathrm{Herm}^{\mathbb{H}-\mathrm{sym}}(\mathbb{H}^n)$ with the set of quaternionic $n \times n$ matrices $M_n(\mathbb{H})$ satisfying $A^* \equiv \overline{A}^t = A$, we have

$$\mathcal{P}_0^+ \;=\; \{A \in M_n(\mathbb{H}) : \;A^* = A \;\text{ and }\; \overline{x}^t A x \geq 0 \;\forall\, x \in \mathbb{H}^n\}.$$

This is a minimal subequation and has compact invariance group $\mathrm{Sp}_n \cdot \mathrm{Sp}_1$. Note that by (7.7a) the projection of P_e ($|e| = 1$) onto $\mathrm{Herm}^{\mathbb{H}-\mathrm{sym}}(\mathbb{H}^n)$ is just $P_{\mathbb{H}e}$, orthogonal projection onto the quaternionic line $\mathbb{H}e$. Hence, this example is self dual, i.e., $\mathcal{P}_0^+ = \mathcal{P}_+$. Each $A \in M_n(\mathbb{H})$ with $A^* = A$ can be put in canonical form

$$A^{\mathbb{H}-\mathrm{sym}} \;=\; \sum_{j=1}^n \lambda_j P_{\mathbb{H}e_j}$$

under the action of $\mathrm{Sp}_n \cdot \mathrm{Sp}_1$. The quaternionic Monge-Ampère operator

$$\det_{\mathbb{H}}(A) \;\equiv\; \prod_{j=1}^n \lambda_j \left(A^{\mathbb{H}-\mathrm{sym}}\right)$$

provides the nonlinear operator for $\mathcal{P}^+ = \mathcal{P}(\mathbb{P}(\mathbb{H}^n))$.

Example 7.6a. Reversing the roles of $\mathrm{Im}\mathbb{H} \otimes \mathrm{Herm}^{\mathbb{H}-\mathrm{skew}}(\mathbb{H}^n)$ and $\mathrm{Herm}_0^{\mathbb{H}-\mathrm{sym}}(\mathbb{H}^n)$ in (7.9) above results in a second $\mathrm{Sp}_n \cdot \mathrm{Sp}_1$-invariant minimal subequation $\mathcal{P}^+ \equiv E + \mathcal{P}$ with $\mathcal{P}_+ = S \cap \mathcal{P}$, where

$$E \;\equiv\; \mathrm{Herm}_0^{\mathbb{H}-\mathrm{sym}}(\mathbb{H}^n) \;\text{ and }\; S \;\equiv\; \mathbb{R}\cdot \mathrm{Id} \oplus (\mathrm{Im}\mathbb{H} \otimes \mathrm{Herm}^{\mathbb{H}-\mathrm{skew}}(\mathbb{H}^n)). \tag{7.10}$$

Note that for each $|e| = 1$,

$$\pi(P_e) \;=\; \tfrac{1}{4n}\,\mathrm{Id} + \tfrac{1}{4}(3P_e - P_{Ie} - P_{Je} - P_{Ke}). \tag{7.11}$$

We leave as a question: Does $\pi(P_e)$ generate an exposed ray in $\mathcal{P}_0^+ = \pi(\mathcal{P})$?

This edge $E \equiv \mathrm{Herm}_0^{\mathbb{H}-\mathrm{sym}}(\mathbb{H}^n)$ is reminiscent of the edge in Example 7.4 in the complex case. We now pursue this analogy. We say that a real n-plane W in $\mathbb{H}^n$ is **$\mathbb{H}$-Lagrangian** if

$$W \oplus IW \oplus JW \oplus KW \;=\; \mathbb{H}^n \quad \text{(orthogonal direct sum)}, \tag{7.12}$$

and let $\mathbb{H}\,\mathrm{Lag}$ denote the set of all such n-planes.

Example 7.6b. (Quaternionic Lagrangian Plurisubharmonics). These are defined as the subharmonics for the geometrically defined subequation $\mathcal{P}(\mathbb{H}\,\mathrm{Lag})$. Note that $\mathbb{H}\,\mathrm{Lag}$ and hence $\mathcal{P}(\mathbb{H}\,\mathrm{Lag})$ has compact invariance group $\mathrm{Sp}_n \cdot \mathrm{Sp}_1$. Furthremore, given $A \in \mathrm{Sym}^2_{\mathbb{R}}(\mathbb{H}^n)$ one can show that

$$\mathrm{tr}\, A\big|_W \;=\; 0 \quad \forall\, W \in \mathbb{H}\,\mathrm{Lag} \qquad \iff \qquad A \in \mathrm{Herm}_0^{\mathbb{H}-\mathrm{sym}}(\mathbb{H}^n)). \tag{7.13}$$

Consequently, $\mathcal{P}(\mathbb{H}\,\mathrm{Lag})$ has edge-span given by (7.10). At the moment we do not know whether or not $\mathcal{P}(\mathbb{H}\,\mathrm{Lag})$ is the minimal subequation with this edge-span. Of course one has

$$E + \mathcal{P} \;\subset\; \mathcal{P}(\mathbb{H}\,\mathrm{Lag}) \qquad \text{and} \qquad \mathcal{P}_+(\mathbb{H}\,\mathrm{Lag}) \;\subset\; S \cap \mathcal{P}, \tag{7.14}$$

where $\mathcal{P}_+(\mathbb{H}\,\mathrm{Lag})$ is the convex cone hull of $\{P_W : W \in \mathbb{H}\,\mathrm{Lag}\}$.

The $\mathbf{Sp}_n \cdot \mathbf{S}^1$–Invariance group

If U_n is replaced by the smaller subgroup SU_n, the decomposition (7.4) of $\mathrm{Sym}^2_{\mathbb{R}}(\mathbb{C}^n)$ remains the same, and so SU_n is not a compact invariance group for a minimal subequation. However, the decomposition (7.7) does not remain the same if we replace $\mathrm{Sp}_n \cdot \mathrm{Sp}_1$ by Sp_n. The new decomposition can be written as

$$\mathrm{Sp}_n : \quad \mathrm{Sym}^2_{\mathbb{R}}(\mathbb{H}^n) \;=\; \mathbb{R}\cdot \mathrm{Id} \oplus \mathrm{Herm}_0^{\mathbb{H}-\mathrm{sym}}(\mathbb{H}^n) \bigoplus_{j=1}^{3} I_j \,\mathrm{Herm}^{\mathbb{H}-\mathrm{skew}}(\mathbb{H}^n) \qquad (7.15)$$

where I_j vary over I, J, K, or in fact over any orthonormal basis of $\mathrm{Im}\mathbb{H}$. Note that the representations $I_j\,\mathrm{Herm}^{\mathbb{H}-\mathrm{skew}}(\mathbb{H}^n)$ are all equivalent.

The next example is a minimal subequation which is new.

Example 7.7. **(I-Complex and J, K-Lagrangian Plurisubharmonics).** This is a geometrically defined subequation given by the set

$$\mathbb{G} \;=\; \mathbb{G}(I; J, K) \;\subset\; G_{\mathbb{R}}(2n, \mathbb{H}^n)$$

of real $2n$-planes which a simultaneously I-complex and both J and K Lagrangian. (Note that any two of these conditions implies the third.) The associated subequation is $\mathcal{P}(\mathbb{G}(I; J, K))$.

Now $\mathcal{P}(\mathrm{JLAG})$ has edge

$$\mathrm{Herm}_0^{J\mathbb{C}-\mathrm{sym}}(\mathbb{C}^{2n}) = \mathrm{Herm}_0^{\mathbb{H}-\mathrm{sym}}(\mathbb{H}^n) \oplus J\,\mathrm{Herm}^{\mathbb{H}-\mathrm{skew}}(\mathbb{H}^n)$$

and $\mathcal{P}(\mathrm{KLAG})$ has edge

$$\mathrm{Herm}_0^{K\mathbb{C}-\mathrm{sym}}(\mathbb{C}^{2n}) = \mathrm{Herm}_0^{\mathbb{H}-\mathrm{sym}}(\mathbb{H}^n) \oplus K\,\mathrm{Herm}^{\mathbb{H}-\mathrm{skew}}(\mathbb{H}^n).$$

Hence the sum

$$\mathrm{Herm}_0^{\mathbb{H}-\mathrm{sym}}(\mathbb{H}^n) \oplus J\,\mathrm{Herm}^{\mathbb{H}-\mathrm{skew}}(\mathbb{H}^n) \oplus K\,\mathrm{Herm}^{\mathbb{H}-\mathrm{skew}}(\mathbb{H}^n) \;\subset\; \mathrm{Edge}(\mathcal{P}(\mathbb{G})).$$

Each $W \in \mathbb{G}$ has a real basis of the form

$$e_1, Ie_1, ..., e_n, Ie_n \qquad \text{where } e_1, ..., e_n \text{ is an } \mathbb{H}\text{-basis for } \mathbb{H}^n.$$

Thus $P_W = P_V + P_{IV}$ where $V \equiv \mathrm{span}_{\mathbb{R}}\{e_1, ..., e_n\}$. Note that $W \in \mathbb{G} \Rightarrow W^\perp = JW = KW \in \mathbb{G}$. Hence, $\mathrm{Id} = P_W + P_{W^\perp} \in S \equiv \mathrm{span}\,(\mathcal{P}(\mathbb{G})) \equiv \mathrm{span}\,(\mathbb{G})$. Now we have

$$P_W - P_{W^\perp} = P_W - P_{IW} = P_W + IP_W I \;\in\; I\,\mathrm{Herm}^{\mathbb{H}-\mathrm{skew}}(\mathbb{H}^n).$$

One can show (direct proof and invariance proof) that

$$S \;=\; \mathbb{R}\cdot \mathrm{Id} \oplus I\,\mathrm{Herm}^{\mathbb{H}-\mathrm{skew}}(\mathbb{H}^n) \qquad (7.16)$$

and hence

$$E \;=\; \mathrm{Herm}_0^{\mathbb{H}-\mathrm{sym}}(\mathbb{H}^n) \oplus J\,\mathrm{Herm}^{\mathbb{H}-\mathrm{skew}}(\mathbb{H}^n) \oplus K\,\mathrm{Herm}^{\mathbb{H}-\mathrm{skew}}(\mathbb{H}^n). \qquad (7.17)$$

Lemma 7.8. *Each $A \in I\,\mathrm{Herm}^{\mathbb{H}-\mathrm{skew}}(\mathbb{H}^n)$ commutes with I and anti-commutes with J and K. If e is an eigenvector of A with eigenvalue λ, then Ie, Je, Ke are eigenvectors with eigenvalues $\lambda, -\lambda, -\lambda$. Hence, A can be put in the canonical form (where $e_1, ..., e_n$ is an $\mathbb{H}$-basis for $\mathbb{H}^n$):*

$$A \;\equiv\; \sum_{j=1}^{n} \lambda_j \left(P_{e_j} + P_{Ie_j} - P_{Je_j} - P_{Ke_j}\right).$$

Corollary 7.9. *The element* $B \equiv \frac{t}{4n}\,\mathrm{Id} + A \in S$ *is* ≥ 0 *if and only if each* $|\lambda_j| \leq \frac{t}{2n}$. *Hence, taking* $t = \mathrm{tr}(B) = 2n$, *the non-negativity condition becomes*

$$|\lambda_j| \ \leq\ \frac{1}{2}, \quad j = 1, ..., n.$$

This describes a cube in $\mathbb{R}^n$. The 2^n extreme points are $\epsilon = (\pm\frac{1}{2}, ..., \pm\frac{1}{2})$, which yields

$$B(\epsilon) \ \equiv\ \tfrac{1}{2}\,\mathrm{Id} + \sum_{j=1}^{n} \pm\tfrac{1}{2}\left(P_{e_j} + P_{Ie_j} - P_{Je_j} - P_{Ke_j}\right) \ =\ P_{W(\epsilon)}$$

where

$$W(\epsilon) \ =\ \mathrm{span}\left\{(e_1, Ie_1 \text{ if } \ \epsilon_1 = \tfrac{1}{2}) \ \text{ or } \ (Je_1, Ke_1 \text{ if } \ \epsilon_1 = -\tfrac{1}{2}), \ \ \dots \ \text{etc.}\right\}.$$

This proves

Proposition 7.10.

$$S \cap \mathcal{P} \ =\ CCH\{P_W : W \in \mathbb{G}\} \ \equiv\ \mathcal{P}_+(\mathbb{G}).$$

Corollary 7.11. *The subequation* $\mathcal{P}(\mathbb{G})$ *is* **the** *minimal subequation with span* S *and edge* E *given by (7.9) and (7.10).*

Example 7.12. Consider the edge $E_I \equiv I\,\mathrm{Herm}^{\mathbb{H}-\mathrm{skew}}(\mathbb{H}^n)$ and the minimal subequation $\mathcal{P}^+ \equiv E_I + \mathcal{P}$. The compact invariance group is $\mathrm{Sp}_n \cdot S^1$, as in Example 7.7.

Lemma 7.13. *One has*

$$\begin{aligned}\mathcal{P}^+ \ \equiv\ E_I + \mathcal{P} \ &\subset\ \mathcal{P}(I\,\mathrm{Lag}) \cap \mathcal{P}(\mathbb{P}_J(\mathbb{C}^{2n})) \cap \mathcal{P}(\mathbb{P}_K(\mathbb{C}^{2n})) \\ &=\ \mathcal{P}\left((I\,\mathrm{Lag}) \cup \mathbb{P}_J(\mathbb{C}^{2n}) \cup \mathbb{P}_K(\mathbb{C}^{2n})\right)\end{aligned}$$

which has edge E_I.

Proof. Suppose for all $W \in I\,\mathrm{Lag} \cup \mathbb{P}_J(\mathbb{C}^{2n}) \cup \mathbb{P}_K(\mathbb{C}^{2n})$ that $\langle A, P_W\rangle \geq 0$. Taking $W \in I\,\mathrm{Lag}$ proves that $A \in \mathcal{P}(I\,\mathrm{Lag})$; taking $W \in \mathbb{P}_J(\mathbb{C}^{2n})$ proves that $A \in \mathcal{P}(\mathbb{P}_J(\mathbb{C}^{2n}))$; and taking $W \in \mathbb{P}_K(\mathbb{C}^{2n})$ proves that $A \in \mathcal{P}(\mathbb{P}_K(\mathbb{C}^{2n}))$. Conversely, if A belongs to the intersection of the three geometric subequations in the Lemma, then $\mathrm{tr}\,A\big|_W \geq 0$ for all $W \in I\,\mathrm{Lag} \cup \mathbb{P}_J(\mathbb{C}^{2n}) \cup \mathbb{P}_K(\mathbb{C}^{2n})$. This proves the last equality in the Lemma.

Since $E_I \subset E_{0,I}$, by Example 7.4,

$$\mathcal{P}^+ \ \equiv\ E_I + \mathcal{P} \subset E_{0,I} + \mathcal{P} \ =\ \mathcal{P}(I\,\mathrm{Lag}).$$

Since $E_I \subset E_{I,K}$, by Example 7.3,

$$\mathcal{P}^+ \ \equiv\ E_I + \mathcal{P} \subset E_{I,K} + \mathcal{P} \ =\ \mathcal{P}(\mathbb{P}_J(\mathbb{C}^{2n})).$$

Since $E_I \subset E_{I,J}$, by Example 7.3,

$$\mathcal{P}^+ \ \equiv\ E_I + \mathcal{P} \subset E_{I,J} + \mathcal{P} \ =\ \mathcal{P}(\mathbb{P}_K(\mathbb{C}^{2n})).$$

Finally since $E_I = E_{0,I} \cap E_{I,K} \cap E_{I,J}$, this proves that $\mathcal{P}\left((I\,\mathrm{Lag}) \cup \mathbb{P}_J(\mathbb{C}^{2n}) \cup \mathbb{P}_K(\mathbb{C}^{2n})\right)$ has edge E_I. ∎

It remains an open question whether or not $\mathcal{P}\left((I\,\mathrm{Lag}) \cup \mathbb{P}_J(\mathbb{C}^{2n}) \cup \mathbb{P}_K(\mathbb{C}^{2n})\right)$ is the minimal subequation $\mathcal{P}^+ \equiv E_I + \mathcal{P}$ with edge E_I.

8. Classifying the Invariant Minimal Subequations

Given a compact subgroup $G \subset \mathrm{O}_N$, one could ask which (if any) subequations have G as their exact invariance group. Now the compact invariance group for a minimal subequation $\mathcal{P}^+ = E + \mathcal{P}$ is the same as for its edge E (see (7.2)). Therefore we need only classify the possible invariant edges E. This is easily done as follows. First decompose $\mathrm{Sym}^2_{\mathbb{R}}(\mathbb{R}^N)$ into irreducible pieces $\mathrm{Sym}^2_{\mathbb{R}}(\mathbb{R}^N) = \mathbb{R} \cdot \mathrm{Id} \oplus E_0 \oplus E_1 \oplus \cdots \oplus E_k$, and note that $E_0 \oplus \cdots \oplus E_k = \mathrm{Sym}^2_0(\mathbb{R}^N)$, the traceless part. Hence any space $E = E_{i_1} \oplus \cdots \oplus E_{i_\ell}$, $0 \leq i_1 < \cdot < \imath_\ell \leq k$ can be chosen as a basic (invariant) edge. Note that $E = \{0\}$ is also a basic invariant edge, and $E + \mathcal{P} = \mathcal{P}$, which has compact invariance group O_N.

The $\mathbf{O}_n$-Case. Here we have

$$\mathrm{Sym}^2(\mathbb{R}^n) \;=\; \mathbb{R} \cdot \mathrm{Id} \oplus E_0 \qquad \text{with} \qquad E_0 \;\equiv\; \mathrm{Sym}^2_0(\mathbb{R}^n).$$

There are two examples: $E = \{0\}$ and $E = E_0$ given by Examples 7.1 and 7.2.

The $\mathbf{U}_n$-Case. Here it is more complicated:

$$\mathrm{Sym}^2_{\mathbb{R}}(\mathbb{C}^n) \;=\; \mathbb{R} \cdot \mathrm{Id} \oplus E_0 \oplus E_1 \qquad \text{with}$$
$$E_0 \;\equiv\; \mathrm{Herm}_0^{\mathbb{C}-\mathrm{sym}}(\mathbb{C}^n) \qquad \text{and} \qquad E_1 \;\equiv\; \mathrm{Herm}^{\mathbb{C}-\mathrm{skew}}(\mathbb{C}^n),$$

which are Examples 7.3 and 7.4.

The $\mathbf{Sp}_n{\cdot}\mathbf{Sp}_1$-Case. Here we have

$$\mathrm{Sym}^2_{\mathbb{R}}(\mathbb{H}^n) \;=\; \mathbb{R} \cdot \mathrm{Id} \oplus E_0 \oplus E_1 \qquad \text{with}$$
$$E_0 \;\equiv\; \mathrm{Herm}_0^{\mathbb{H}-\mathrm{sym}}(\mathbb{H}^n) \qquad \text{and} \qquad E_1 \;\equiv\; \mathrm{Im}\mathbb{H} \otimes \mathrm{Herm}^{\mathbb{H}-\mathrm{skew}}(\mathbb{H}^n).$$

Hence again there are two new examples $E = E_0$ and $E = E_1$ which are Examples 7.5 and 7.6a.

The $\mathbf{Sp}_n$ and $\mathbf{Sp}_n{\cdot}\mathbf{S}^1$-Cases. Under Sp_n we have

$$\mathrm{Sym}^2_{\mathbb{R}}(\mathbb{H}^n) \;=\; \mathbb{R} \cdot \mathrm{Id} \oplus E_0 \oplus E_I \oplus E_J \oplus E_K \qquad \text{with}$$
$$E_0 \;\equiv\; \mathrm{Herm}_0^{\mathbb{H}-\mathrm{sym}}(\mathbb{H}^n), \quad E_I \;\equiv\; I\,\mathrm{Herm}^{\mathbb{H}-\mathrm{skew}}(\mathbb{H}^n),$$
$$E_J \;\equiv\; J\,\mathrm{Herm}^{\mathbb{H}-\mathrm{skew}}(\mathbb{H}^n), \quad E_K \;\equiv\; K\,\mathrm{Herm}^{\mathbb{H}-\mathrm{skew}}(\mathbb{H}^n),$$

(see (7.15)). Of the possible edges we can exclude most of them as coming from the previous cases. For example, $E_{0,I} \equiv E_0 \oplus E_I = \mathrm{Herm}_0^{\mathbb{C}-\mathrm{sym}}(\mathbb{C}^n)$ for the I-complex case (as well as $E_{0,J}$, $E_{0,K}$) come from Example 7.4. The case $E_{J,K} \equiv E_J \oplus E_K = \mathrm{Herm}^{\mathbb{C}-\mathrm{skew}}(\mathbb{C}^n)$ (for the complex structure I) can be excluded, since this is Example 7.3. Similarly we exclude $E_{I,K}$ and $E_{I,J}$. The case $E = E_0$ is just Example 7.6a, while the case $E \equiv E_{I,J,K} = E_I \oplus E_J \oplus E_K = \mathrm{Im}\mathbb{H} \otimes \mathrm{Herm}^{\mathbb{H}-\mathrm{skew}}(\mathbb{H}^n)$ is Example 7.5.

This leaves, up to permuting I, J, K, two examples: $E = E_I$, which is Example 7.12, and $E = E_{0,J,K} = E_0 \oplus E_J \oplus E_K$ as in (7.10), which is Example 7.7. These last two examples have compact invariance group $\mathrm{Sp}_n{\cdot}\mathrm{S}^1$. Note that this proves that there are no minimal subequations with compact invariance group Sp_n.

9. An Envelope Problem for Minimal Subequations.

Suppose that $\mathbf{F} \equiv \mathcal{P}^+$ is a minimal subequation. In this section we investigate the role played by the edge functions in solving the Dirichlet problem. The key fact about $\mathbf{F}$ that will be used below is the following from Theorem 5.4(2):

$$\mathrm{Int}\,\mathbf{F} \;=\; E + \mathrm{Int}\,\mathcal{P}. \tag{9.1}$$

We recall that existence and uniqueness for the (DP) on a bounded domain $\Omega \subset \mathbb{R}^n$ and arbitrary $\varphi \in C(\partial\Omega)$ was established in [**5**] if $\partial\Omega$ is smooth and strictly $\mathbf{F}$- and $\widetilde{\mathbf{F}}$-convex (for any subequation $\mathbf{F} \subset \mathrm{Sym}^2(\mathbb{R}^n)$). Moreover, the solution H equals the Perron function

$$H(x) \equiv \sup_{u \in \mathcal{F}_{\mathbf{F}}(\varphi)} u(x) \qquad \text{for } x \in \overline{\Omega} \tag{9.2}$$

for the Perron family of $\mathbf{F}$-subharmonics

$$\mathcal{F}_{\mathbf{F}}(\varphi) \equiv \left\{u \in \mathbf{F}(\overline{\Omega}) : u\big|_{\partial\Omega} \leq \varphi\right\}. \tag{9.3}$$

By definition $u \in \mathbf{F}(\overline{\Omega})$ if u is $[-\infty, \infty)$-valued and upper semi-continuous on $\overline{\Omega}$ and $u\big|_{\Omega} \in \mathbf{F}(\Omega)$.

The proof of our main result here follows (as closely as possible) the existence proof for the Dirichlet Problem given in [**7**].

To begin we consider the following analogues of the above. Let

$$E(\overline{\Omega}) \equiv \left\{u \in C(\overline{\Omega}) : u\big|_{\Omega} \in E(\Omega)\right\} \tag{9.4}$$

denote the space of edge functions on $\overline{\Omega}$, and consider the family of edge functions

$$\mathcal{F}_E(\varphi) \equiv \left\{h \in E(\overline{\Omega}) : h\big|_{\partial\Omega} \leq \varphi\right\}. \tag{9.5}$$

A natural question to ask is:

Question 1. When is the envelope

$$U_E(x) \equiv \sup_{h \in \mathcal{F}_E(\varphi)} h(x) \text{ equal to the solution } H \text{ defined by (9.2)?}$$

There are two interesting extreme cases where the answer is positive.

Example 9.1. **($\mathbf{F} \equiv \mathcal{P}$).** Here $E(\Omega) \equiv \mathrm{Aff}(\mathbb{R}^n)$, the space of affine functions on $\mathbb{R}^n$. In this case

$$U_{\mathrm{Aff}} = H_{\mathcal{P}}$$

because, by the Hahn-Banach Theorem, for each point $x_0 \in \Omega$, there exists an affine function h with $h \leq H_{\mathcal{P}}$ on $\overline{\Omega}$ and $h(x_0) = H_{\mathcal{P}}((x_0)$.

Example 9.2. **($\mathbf{F} \equiv \Delta$).** Here $E(\Omega) \equiv \{h \in C(\overline{\Omega}) : h\big|_{\Omega}$ is Δ-harmonic$\}$. Therefore, $H \in \mathcal{F}_E(\varphi)$, proving that

$$U_\Delta = H_\Delta.$$

For other cases Question 1 remains open, so it is appropriate to consider larger families than $\mathcal{F}_E(\varphi)$. First, set

$$E^{\max}(\Omega) \equiv \{M : M = \max\{h_1, ..., h_N\} \text{ with } h_1, ..., h_N \in E(\Omega)\} \tag{9.6}$$

and consider the family

$$\mathcal{F}_{E^{\max}}(\varphi) \equiv \{M \in E^{\max}(\overline{\Omega}) : M\big|_{\partial\Omega} \leq \varphi\} \tag{9.7}$$

where by definition $M \in E^{\max}(\overline{\Omega})$ if $M \in \mathrm{USC}(\overline{\Omega})$ and $M\big|_{\Omega} \in E^{\max}(\Omega)$. Since the conditions $M \equiv \max\{h_1, ..., h_N\} \in E^{\max}(\overline{\Omega})$ and $M\big|_{\partial\Omega} \leq \varphi$ imply that each $h_k \in \mathcal{F}_E(\varphi)$, we have

$$U_{E^{\max}} = \sup_{M \in \mathcal{F}_{E^{\max}}(\varphi)} M = \sup_{h \in \mathcal{F}_E(\varphi)} h = U_E. \tag{9.8}$$

In particular, $U_{E^{\max}} = H \iff U_E = H$.

Now we consider a localized version

$$\mathcal{F}_{E^{\text{loc-max}}}(\varphi) \;=\; \{u \in E^{\text{loc-max}}(\overline{\Omega}) : u\big|_{\partial\Omega} \le \varphi\} \tag{9.9}$$

where by definition $u \in E^{\text{loc-max}}(\overline{\Omega})$ if $u \in \mathrm{USC}(\overline{\Omega})$ and for each point $x_0 \in \Omega$, there exists a neighborhood $B_r(x_0) \subset \Omega$ such that

$$u\big|_{B_r(x_0)} \;\in\; E^{\max}(B_r(x_0)). \tag{9.10}$$

Question 2. When is the envelope

$$U \;\equiv\; U_{E^{\text{loc-max}}} \;=\; \sup_{u\in\mathcal{F}_{E^{\text{loc-max}}(\varphi)}} u \ \text{ equal to the solution } H \text{ in (9.2)?}$$

We can answer this question.

THEOREM 9.3. *If $\mathbf{F} = \mathcal{P}^+$ is a minimal subequation and $\partial\Omega$ is smooth and strictly $\mathbf{F}$-convex, then*

$$U \;=\; H.$$

Proof. Since $\mathbf{F} \subset \widetilde{\mathbf{F}}$ (Thm. 5.7), the strict $\widetilde{\mathbf{F}}$-convexity of the boundary is automatic.

In what follows we shall shorten $\mathcal{F}_{E^{\text{loc-max}}}(\varphi)$ to $\mathcal{F}(\varphi)$.

Note that $\mathcal{F}(\varphi) \subset \mathcal{F}_{\mathbf{F}}(\varphi) \Rightarrow U \le H \Rightarrow U^* \le H \Rightarrow$

$$U^*\big|_{\partial\Omega} \;\le\; \varphi, \qquad \text{and we also have} \tag{9.11a}$$

$$\varphi \;\le\; U_*\big|_{\partial\Omega} \qquad \text{proved at the end.} \tag{9.11b}$$

Note 9.4. If $\partial\Omega$ is strictly convex, then Example 9.1 shows that $\varphi = U_{\mathcal{P}}\big|_{\partial\Omega}$ and $U_{\text{Aff}} = U_{\mathcal{P}}$. Since $\mathcal{P} \subset \mathbf{F}$ and $\text{Aff} \subset E$, we have $U_{\mathcal{P}} \le U_E \le U$. Hence $U_{\mathcal{P}} \le U_*$, so that (9.11b) holds under strict $\mathcal{P}$-convexity of $\partial\Omega$.

These two properties imply the following.

$$\text{(Boundary Continuity)} \qquad U_*\big|_{\partial\Omega} \;=\; U\big|_{\partial\Omega} \;=\; U^*\big|_{\partial\Omega} \;=\; \varphi. \tag{9.11}$$

By the "families bounded above property" we have

$$U^* \in \mathbf{F}(\overline{\Omega}). \tag{9.12}$$

Note that H (or $\sup_{\partial\Omega}\varphi$ if you wish) provides an upper bound for $\mathcal{F}(\varphi)$.

Assume for the moment that:

$$-U_* \in \widetilde{\mathbf{F}}(\overline{\Omega}). \tag{9.13}$$

Then the proof is easily completed as follows. By (9.12) and (9.13), $U^* - U_* \in \widetilde{\mathcal{P}}(\overline{\Omega})$ is subaffine on Ω (see [**5**]). Moreover, it is ≥ 0 on $\overline{\Omega}$ and equal to zero on $\partial\Omega$. Hence, by the (MP) for $\widetilde{\mathcal{P}}$, $U^* - U_*$ vanishes on $\overline{\Omega}$. That is,

$$U_* \;=\; U \;=\; U^* \quad \text{on } \overline{\Omega}. \tag{9.14}$$

This proves that U is $\mathbf{F}$-harmonic on Ω and equal to φ on $\partial\Omega$. By uniqueness for the (DP) this proves that $U = H$ on $\overline{\Omega}$. Thus it remains to prove (9.11b) and the following.

Lemma 9.5. $\quad -U_*\big|_{\Omega} \in \widetilde{\mathbf{F}}(\Omega)$.

Proof. We follow that bump argument given in the proof of Lemma $\widetilde{F}$ in [**7**, p. 455] as closely as possible.

Suppose $-U_*\big|_\Omega \notin \widetilde{\mathbf{F}}(\Omega)$. Then there exists $x_0 \in \Omega$, $\epsilon > 0$ and ψ, a degree-2 polynomial, satisfying

$$\begin{aligned} &(a) \quad -U_* \ \le\ \psi - \epsilon|x-x_0|^2 \quad \text{near } x_0, \quad \text{and} \\ &(b) \quad -U_*(x_0) \ =\ \psi(x_0), \quad \text{and} \\ &(c) \quad D^2\psi \notin \widetilde{\mathbf{F}}. \end{aligned} \tag{9.15}$$

Rewrite (a) and (c) as

$$\begin{aligned} &(a)' \quad -\psi \ \le\ U_* - \epsilon|x-x_0|^2 \quad \text{near } x_0, \quad \text{and} \\ &(c)' \quad D^2(-\psi) \in \operatorname{Int}\mathbf{F}. \end{aligned}$$

By the key fact (9.1) above we have that

$$D^2(-\psi) \ =\ e + P \quad \text{with } e \in E \text{ and } P > 0. \tag{9.16}$$

Therefore

$$-\psi \ =\ h + \tfrac{1}{2}\langle P(x-x_0), x-x_0\rangle \tag{9.17}$$

with h a degree 2 polynomial satisfying

$$(i)\ D^2h \ =\ e \qquad \text{and} \qquad (ii)\ h(x_0) \ =\ U_*(x_0). \tag{9.18}$$

The first part is just the statement that

$$(i)' \quad h \text{ is an edge function on } \mathbb{R}^n. \tag{9.18)(i)'$$

Now by (9.17) the inequality (a)$'$ says

$$h + \tfrac{1}{2}\langle P(x-x_0), x-x_0\rangle \ \le\ U_* - \epsilon|x-x_0|^2 \quad \text{on } B_{r_2}(x_0). \tag{9.19}$$

Choose $0 < r_1 < r < r_2$. Then by (9.19)

$$h + \delta \ <\ U_* \quad \text{on } B_{r_2}(x_0) - B_{r_1}(x_0) \tag{9.20}$$

where $\delta \equiv \inf_{|x-x_0|=r_1} \frac{1}{2}\langle P(x-x_0), x-x_0\rangle$ (or $\delta = \epsilon r_1^2$ also works). For each point $y \in \partial B_r(x_0)$ we have $h(y) + \delta < U(y)$ by (9.20). Hence, by the definition of the $U = U_{\mathcal{F}_E^{\text{loc-max}}}$ given in Question 2, there exists $u_y \in \mathcal{F}(\varphi)$ with

$$h(y) + \delta \ <\ u_y(y), \tag{9.21}$$

and since h and u_y are continuous, this holds in a neighborhood of y. Therefore, by compactness, there exist $u_1, ..., u_N \in \mathcal{F}(\varphi)$ with

$$h + \delta \ <\ u \ \equiv\ \max\{u_1, ..., u_N\} \quad \text{in a neighborhood of } \partial B_r(x_0). \tag{9.22}$$

Since $\mathcal{F}(\varphi)$ is closed under taking the maximum of a finite number of elements, we have

$$h + \delta \ <\ u \text{ in a neighborhood of } \partial B_r(x_0) \text{ with } u \in \mathcal{F}(\varphi). \tag{9.23}$$

This implies that

$$u' \ \equiv\ \begin{cases} u & \text{on } \overline{\Omega} - B_r(x_0) \\ \max\{u, h+\delta\} & \text{on } \overline{B_r(x_0)} \end{cases} \tag{9.24}$$

is an element of $\mathcal{F}(\varphi)$. (Note that $h + \delta$ and hence u' is not necessarily an element of $\mathcal{F}_E^{\max}(\varphi)$.) Since $u' \in \mathcal{F}(\varphi)$, we have $u' \le U$ on $\overline{\Omega}$. In particular, $h + \delta \le U$ on $B_r(x_0)$. Since h is continuous, this implies $h + \delta \le U_*$, and hence

$$h(x_0) + \delta \ \le\ U_*(x_0), \tag{9.25}$$

which contradicts (9.18 b) that $h(x_0) = U_*(x_0)$.

It only remains to do the following.

Proof of (9.11b). We fix $x_0 \in \partial\Omega$, and let ρ be a smooth, strictly **F**-convex defining function for $\partial\Omega$ defined in a neighborhood of x_0. Then by (9.1) there exist $\epsilon > 0$ and $r > 0$ such that

$$D_x^2\rho - \epsilon I \ \in \ \mathrm{Int}\,\mathbf{F} \ = \ E + \mathrm{Int}\,\mathcal{P} \qquad \forall\, x \in B_r(x_0).$$

In particular,

$$D_{x_0}^2\rho - \epsilon I \ = \ A + P \qquad \text{for} \ \ A \in E \ \ \text{and} \ \ P > 0.$$

By adding a linear function to $\frac{1}{2}\langle Ax, x\rangle$, we get a quadratic ψ with $D_{x_0}^2\psi = A$ and $\psi(x_0) = 0$ so that

$$\rho(x) - \frac{\epsilon}{2}|x - x_0|^2 \ = \ \psi(x) + \tfrac{1}{2}\langle P(x - x_0), x - x_0\rangle + O(|x - x_0|^3).$$

Taking ϵ smaller, we can get a smaller $r > 0$ so that

$$\rho(x) - \frac{\epsilon}{2}|x - x_0|^2 \ > \ \psi(x) + \tfrac{1}{2}\langle P(x - x_0), x - x_0\rangle \qquad \text{for} \ \ x \in \overline{B_r(x_0)} - \{x_0\}.$$

Since $\rho \leq 0$ on $\overline{\Omega}$ we have

$$-\frac{\epsilon}{2}|x - x_0|^2 - \tfrac{1}{2}\langle P(x - x_0), x - x_0\rangle \ \geq \ \psi(x) \qquad \text{for} \ \ x \in \overline{B_r(x_0)} \cap \overline{\Omega}. \tag{9.26}$$

We now fix $\delta > 0$ and shrink $r > 0$ so that

$$\varphi(x_0) - \delta \ < \ \varphi \qquad \text{for} \ \ x \in \overline{B_r(x_0)} \cap \partial\Omega. \tag{9.27}$$

From (9.26) above we have that there exists η with

$$0 \ > \ \eta \ \geq \ \psi(x) \qquad \text{for} \ \ x \in \left(\overline{B_r(x_0)} - B_{r/2}(x_0)\right) \cap \overline{\Omega}. \tag{9.28}$$

We now consider the edge function

$$\Psi(x) \ \equiv \ \varphi(x_0) - \delta + C\psi(x). \tag{9.29}$$

By (9.28) we see that for $C >> 0$ we will have

$$\Psi(x) \ < \ \inf \varphi \qquad \text{on} \quad \left(\overline{B_r(x_0)} - B_{r/2}(x_0)\right) \cap \Omega$$

Therefore

$$\underline{u} \ \equiv \ \begin{cases} \inf_{\partial\Omega}\varphi & \text{on} \ \ \overline{\Omega} - B_{r/2}(x_0) \\ \max\{\Psi, \inf_{\partial\Omega}\varphi\} & \text{on} \ \ \overline{B_r(x_0)} \cap \overline{\Omega} \end{cases}$$

is a well defined function on $\overline{\Omega}$, and it is locally the maximum of edge functions. Furthermore, by (9.27) and (9.29) we see that $\underline{u} \leq \varphi$ on $\partial\Omega$. Hence, $\underline{u}$ is in our Perron family for the Dirichlet problem, and so we have $\underline{u} \leq U$, which implies that $\underline{u} \leq U_*$. In particular, $\underline{u}(x_0) = \varphi(x_0) - \delta \leq U_*(x_0)$. Taking $\delta \to 0$ shows that $\varphi(x_0) \leq U_*(x_0)$. This proves (9.11b) and therefore Theorem 9.3. ∎

References

[1] Robert Bryant and Reese Harvey, *Submanifolds in hyper-Kähler geometry*, J. Amer. Math. Soc. **2** (1989), no. 1, 1–31, DOI 10.2307/1990911. MR953169

[2] Michael G. Crandall, *Viscosity solutions: a primer*, Viscosity solutions and applications (Montecatini Terme, 1995), Lecture Notes in Math., vol. 1660, Springer, Berlin, 1997, pp. 1–43, DOI 10.1007/BFb0094294. MR1462699

[3] Michael G. Crandall, Hitoshi Ishii, and Pierre-Louis Lions, *User's guide to viscosity solutions of second order partial differential equations*, Bull. Amer. Math. Soc. (N.S.) **27** (1992), no. 1, 1–67, DOI 10.1090/S0273-0979-1992-00266-5. MR1118699

[4] F. Reese Harvey, *Spinors and calibrations*, Perspectives in Mathematics, vol. 9, Academic Press, Inc., Boston, MA, 1990. MR1045637

[5] F. Reese Harvey and H. Blaine Lawson Jr., *Dirichlet duality and the nonlinear Dirichlet problem*, Comm. Pure Appl. Math. **62** (2009), no. 3, 396–443, DOI 10.1002/cpa.20265. MR2487853

[6] F. Reese Harvey and H. Blaine Lawson Jr., *Plurisubharmonicity in a general geometric context*, Geometry and analysis. No. 1, Adv. Lect. Math. (ALM), vol. 17, Int. Press, Somerville, MA, 2011, pp. 363–402. MR2882430

[7] F. Reese Harvey and H. Blaine Lawson Jr., *Dirichlet duality and the nonlinear Dirichlet problem on Riemannian manifolds*, J. Differential Geom. **88** (2011), no. 3, 395–482. MR2844439

[8] S.-T. Yau (ed.), *Surveys in differential geometry: differential geometry inspired by string theory*, Surveys in Differential Geometry, vol. 5, International Press, Boston, MA, 1999. A supplement to the Journal of Differential Geometry. MR1772270

[9] ———, *The AE Theorem and Addition Theorems for quasi-convex functions*, ArXiv:1309:1770.

[10] F. Reese Harvey and H. Blaine Lawson Jr., *Lagrangian potential theory and a Lagrangian equation of Monge-Ampère type*, Surv. Differ. Geom., vol. 22, Int. Press, Somerville, MA, 2018, pp. 217–257. MR3838119

Department of Mathematics, Rice University, 6100 Main St., Houston, Texas 77005
Email address: harvey@rice.edu

Department of Mathematics, Stony Brook University, Stony Brook, New York 11790
Email address: blaine@math.sunysb.edu

Contemporary Mathematics
Volume **735**, 2019
https://doi.org/10.1090/conm/735/14825

Orbifold regularity of weak Kähler-Einstein metrics

Chi Li and Gang Tian

ABSTRACT. We show that weak Kähler-Einstein metrics on $\mathbb{Q}$-Fano varieties are smooth orbifold metrics away from analytic subsets of complex codimension at least three. Our proof uses orbifold resolution of singularities.

1. Introduction

It has been a fundamental problem to study how to compactify the moduli of Einstein metrics and what are degenerate metrics in the compactification. It has found many applications in Kähler geometry. For instance, in the resolution of the YTD conjecture on the existence of Kähler-Einstein metrics on Fano manifolds (see [**Tia4**] and also [**CDS**]), a crucial tool is a compactness result on Kähler-Einstein metrics. In its simplest form, this result says that the Gromov-Hausdorff limit of a sequence of smooth Kähler-Einstein manifolds $(X_i, \omega_{i,\mathrm{KE}})$ is a normal Fano variety $X := X_\infty$ with klt singularities and that there is a weak Kähler-Einstein metric $\omega_{\infty,\mathrm{KE}}$ on X_∞ which is smooth on X_∞^{reg}. The existence of a Gromov-Hausdorff limit follows from Gromov's compactness theorem. The problem is about the regularity of X_∞. It follows from Cheeger-Colding's theory and Cheeger-Colding-Tian's theory (see [**CCT**] and the reference therein) that X_∞ is smooth outside a closed subset S of Hausdorff codimension at least 4 and $\omega_{\infty,\mathrm{KE}}$ is a Kähler-Einstein metric. It was the second author ([**Tia1**], [**Tia2**], see also [**Li**]) who first pointed out the route to prove that X_∞ is an algebraic variety is to establish a so-called partial C^0-estimate. He demonstrated in [**Tia1**] how to achieve this when the complex dimension n is equal to 2 by showing that a sequence of Kähler-Einstein surfaces converges to a Fano orbifold with a smooth orbifold Kähler-Einstein metric. Note that when $n = 2$, klt singularities are nothing but isolated quotient singularities or orbifold singularities. Two key ingredients to prove the partial C^0-estimate in dimension 2 are orbifold compactness result of Kähler-Einstein 4-manifolds and Hörmander's L^2-estimates.

2010 *Mathematics Subject Classification.* Primary 53C25, 53C55; Secondary 32W20, 14E15.

Key words and phrases. Kähler-Einstein metrics, Fano varieties, complex Monge-Ampère equations, smooth orbifold metrics.

The first author was supported in part by NSF Grants DMS-1405936, DMS-1810867.

The second author was supported in part by NSF Grants DMS-1309359,1607091 and NSFC Grant No.11331001.

©2019 American Mathematical Society

Recently, Donaldson-Sun [**DS**], and the second author [**Tia3**] independently, generalized the partial C^0-estimate to higher dimensional Kähler-Einstein manifolds. Here they need to rely on compactness results on Kähler-Einstein metrics in higher dimensions developed by Cheeger-Colding and Cheeger-Colding-Tian. It follows from the partial C^0-estimate (see [**Tia2**], [**Li**] and [**DS**]) that X_∞ is a normal variety. Furthermore, the partial C^0-estimate implies that there is a uniform C^2-estimate of the potential of $\omega_{\infty,\mathrm{KE}}$ on the regular part X_∞^{reg} of X_∞. Then the Evans-Krylov theory or Calabi's 3rd derivative estimate allows one to show that $\omega_{\infty,\mathrm{KE}}$ is smooth on X_∞^{reg} (see [**Tia1**], [**DS**], [**Tia4**]). Alternatively, using Păun's Laplacian estimate in [**Pău**] and Evans-Krylov theory, Berman-Boucksom-Eyssidieux-Guedj-Zeriahi [**BBEGZ**] showed directly that any weak Kähler-Einstein metric $\omega^w_{\infty,\mathrm{KE}}$ (which is unique up to complex automorphism) on a klt Fano variety X_∞ is smooth on X_∞^{reg}. Hence, S is a subvariety of complex codimension at least 2 and $X_\infty^{\mathrm{reg}} = X_\infty \backslash S$.

It remains to study the structure of $(X_\infty, \omega_{\infty,\mathrm{KE}})$ around S. Compared to the complex dimension 2 case, the second author conjectured that $\omega_{\infty,\mathrm{KE}}$ is a smooth orbifold metric away from an analytic subvariety Z of complex codimension 3. In this short paper, we affirm this conjecture about the regularity of $\omega_{\infty,\mathrm{KE}}$ on the orbifold locus X_∞^{orb} of X_∞.

First, if $(X, -K_X)$ is a klt Fano variety, then we have the following result which says klt spaces have quotient singularities in codimension 2:

THEOREM 1.1 ([**GKKP**, Theorem 9.3]). *Let X be a variety with klt singularities. Then there exists a closed subset $Z \subset X$ with $\mathrm{codim}_X Z \geq 3$ such that $X\backslash Z$ has quotient singularities. More precisely every point $x \in X \setminus Z$ has an analytic neighborhood that is biholomorphic to an analytic neighborhood of the origin in a variety of the form $\mathbb{C}^n/G$ where G is a finite subgroup of $GL(n,\mathbb{C})$ that does not contain any quasi-reflections.*

By the above result we just need to show the following regularity result. For the definition of weak Kähler-Einstein metric, see Definition 2.1.

THEOREM 1.2. *Assume that ω^w_{KE} is a weak Kähler-Einstein metric on X_∞. Then ω^w_{KE} is a smooth orbifold metric on X_∞^{orb}.*

Our current proof uses the existence of an orbifold resolution, i.e., Theorem 3.3 which is proved by algebraic methods. Theorem 3.3 claims that there is a proper birational morphism $f^{\mathrm{par}} : X^{\mathrm{par}} \to X$ such that X^{par} only has quotient singularity and f^{par} is an isomorphic over X^{orb}. However, we believe that it is not necessary. There should be a purely differential geometric proof of Theorem 1.2 which does not rely on Theorem 3.3. In last section, we will discuss problems on analyzing further structures of singularities of higher codimension. We also believe that our analysis may be used to yield a complete understanding of the singularity for any 3-dimensional weak Kähler-Einstein metrics.

2. Regularity on the orbifold locus

From now on we will denote by X any $\mathbb{Q}$-Fano variety with klt singularities. Assume $\iota : X \to \mathbb{P}^N$ is an embedding given by the linear system $|-mK_X|$ for $m > 0 \in \mathbb{Z}$ sufficiently large and divisible. Let $h_0 = (\iota^* h_{FS})^{1/m}$ be the pull back of the Fubini-Study Hermitian metric h_{FS} on $\mathcal{O}_{\mathbb{P}^N}(1)$ normalized to be a Hermitian metric on $-K_X$. The Chern curvature form of h_0 is

$$\omega_0 = -\sqrt{-1}\partial\bar{\partial}\log h_0$$

which is a positive $(1,1)$-current on X. ω_0 is a smooth positive definite $(1,1)$-form on X^{reg}. However, on the singular locus X^{sing}, ω_0 in general is not canonically related to the local structure of X. Assume $p \in X^{\text{orb}}$ is a quotient singularity. By this, we mean that there exists a small neighborhood $\mathcal{U}_p$ which is isomorphic to a quotient of a smooth manifold by a finite group which acts freely outside a codimension 2 subset. In particular, there exists a branched covering map $\tilde{\mathcal{U}}_p \to \tilde{\mathcal{U}}_p/G \cong \mathcal{U}_p$. The lifting of metric ω_0 to the cover $\tilde{\mathcal{U}}_p$ in general is degenerate.

Now we define an adapted volume for on X by

$$\Omega = |v^*|_{h_0}^{2/m}(\sqrt{-1}^{mn^2} v\wedge\bar{v})^{1/m}.$$

Here v is any local generator of $\mathcal{O}(mK_X)$ and v^* is the dual generator of $\mathcal{O}(-mK_X)$. The Kähler-Einstein equation

$$Ric(\omega_\phi) = \omega_\phi. \tag{2.1}$$

can be transformed into a complex Monge-Ampère equation:

$$(\omega_0 + \sqrt{-1}\partial\bar{\partial}\phi)^n = e^{-\phi}\Omega. \tag{2.2}$$

DEFINITION 2.1. A weak solution to the (2.2) is a bounded function $\phi \in L^\infty(X) \cap PSH(X,\omega)$ satisfying (2.2) in the sense of pluripotential theory.

Let's first recall the method to prove the regularity of ϕ on X^{reg} following [**BBEGZ**]. One first chooses a resolution $\pi : \tilde{X} \to X$ with simple normal crossing exceptional divisor $E = \pi^{-1}(X^{\text{sing}})$ such that π is an isomorphism over X^{reg}. Then we can pull back the equation (2.2) to $\tilde{X}$ and get:

$$(\pi^*\omega_0 + \sqrt{-1}\partial\bar{\partial}\psi)^n = e^{-\psi}\pi^*\Omega. \tag{2.3}$$

On the other hand we can write:

$$K_{\tilde{X}} = \pi^*K_X + \sum_{i=1}^r a_iE_i - \sum_{j=1}^s b_jF_j,$$

such that $E = \cup_{i=1}^r E_i \bigcup \cup_{j=1}^s F_j$ and $a_i > 0$, $b_j > 0$. The klt property implies: $a_i > 0$, and $0 < b_j < 1$. Analytically, choosing a smooth Kähler metric η on $\tilde{X}$, there exists $f \in C^\infty(\tilde{X})$ such that:

$$\pi^*\Omega = e^f\frac{\prod_{i=1}^r |s_i|^{2a_i}}{\prod_{j=1}^s |\sigma_j|^{2b_j}}\eta^n.$$

where s_i and σ_j are defining sections of E_i for F_j respectively and $|s_i|^2$ and $|\sigma_j|^2$ are some fixed hermitian norms of them. So we have:

$$(\pi^*\omega_0 + \sqrt{-1}\partial\bar{\partial}\psi)^n = e^{-\psi+f+\sum_i a_i\log|s_i|^2 - \sum_j b_j\log|\sigma_j|^2}\eta^n = e^{\psi_+-\psi_-}\eta^n, \tag{2.4}$$

Here we have denoted

$$\psi_+ = f + \sum_i a_i\log|s_i|^2, \quad \psi_- = \psi + \sum_j b_j\log|\sigma_j|^2.$$

It's easy to see that they satisfy the quasi-plurisubharmonic condition:

$$\sqrt{-1}\partial\bar{\partial}\psi_+ \geq -C\eta, \sqrt{-1}\partial\bar{\partial}\psi_- \geq -C\eta, \tag{2.5}$$

for some uniform constant $C > 0$. To get Laplacian estimate of ψ away from E, we can first regularize (2.4) to

$$(\omega_\epsilon + \sqrt{-1}\partial\bar{\partial}\psi_\epsilon)^n = e^{\psi_{+,\epsilon}-\psi_{-,\epsilon}}\eta^n. \tag{2.6}$$

where $\omega_\epsilon = \pi^*\omega_0 - \epsilon\theta_E$ is a Kähler metric on $\tilde{X}$, and $\psi_{\pm,\epsilon} \in C^\infty(\tilde{X})$ converges to $\psi_\pm$ in $L^\infty(\tilde{X}\backslash E)$ such that exponential $e^{\psi_{\pm,\epsilon}}$ converges to $e^{\psi_\pm}$ in $L^p(\tilde{X})$ for some $p > 1$. This is possible thanks Demailly's regularization theorem ([**Dem**]). Then it follows from a result of Kołodziej (see Theorem 2.3) that ψ_ϵ converges to $\psi \in C^0(\tilde{X})$ which is a solution to the degenerate Monge-Ampère equation (2.3).

To get the higher regularity of ψ on $X_\infty^{\rm reg}$, Păun in the work [**Pău**] used the condition (2.5) and modified the Laplacian estimate of Aubin-Yau to prove the Laplacian estimate for the solutions ψ_ϵ away from E (see Theorem 2.4). More precisely, for any compact set $K \Subset \tilde{X}\backslash E$, there exists a constant $A = A(\|\psi\|_\infty, K)$, such that

$$\Delta_\eta\psi_\epsilon \le A(\|\psi\|_\infty, K)e^{-\psi_{-,\epsilon}}.$$

Because ω_ϵ is locally uniformly elliptic on $\tilde{X} \setminus E$, by the complex version of Evan-Krylov's theory (see [**Bło**]), we know that ψ_ϵ is locally uniformly $C^{2,\alpha}$ and hence by bootstrapping, $C^{k,\alpha}$ on $\tilde{X}\backslash E$. More precisely, for any compact set $K \Subset \tilde{X}\backslash E$, there exists a constant $C = C(K, \|\psi\|_\infty, k)$ such that:

$$\|\psi_\epsilon\|_{C^{k,\alpha}(K)} \le C. \tag{2.7}$$

As a consequence, ψ_ϵ converges to ψ in C^k norm locally uniformly away from E, and hence we know that ψ is smooth on $\tilde{X}\backslash E$.

One can also prove the regularity on $X^{\rm reg}$ with the help of Kähler-Ricci flow. Starting from the work in [**CTZ**], this idea has been used several times in the literature to prove the regularity of weak solutions to complex Monge-Ampère equations. Recall that the Kähler-Ricci flow is a solution to the following equation:

$$\begin{cases} \frac{\partial\omega_t}{\partial t} = -Ric(\omega_t) + \omega_t; \\ \omega(0) = \omega_{\phi_0}. \end{cases} \tag{2.8}$$

As in the elliptic case, this equation can be transformed into the following Monge-Ampère flow

$$\begin{cases} \frac{\partial\phi}{\partial t} = \log\frac{(\omega_0+\sqrt{-1}\partial\bar{\partial}\phi)^n}{\Omega} + \phi; \\ \phi(0,\cdot) = \phi_0. \end{cases} \tag{2.9}$$

To define a solution to this Monge-Ampère flow on the singular variety X, Song-Tian [**ST**] pulled up the flow equation in (2.9) to $\tilde{X}$ to get:

$$\begin{cases} \frac{\partial\tilde{\phi}}{\partial t} = \log\frac{(\pi^*\omega_0+\sqrt{-1}\partial\bar{\partial}\tilde{\phi})^n}{\pi^*\Omega} + \tilde{\phi}; \\ \\ \tilde{\phi}(0,\cdot) = \pi^*\phi_0. \end{cases} \tag{2.10}$$

THEOREM 2.2 ([**ST**]). *Let $\phi_0 \in PSH_p(X, \omega_0)$ for some $p > 1$. Then the Monge-Ampère flow (2.10) on $\tilde{X}\backslash E$ has a unique solution $\tilde{\phi} \in C^\infty((0, T_0) \times \tilde{X}\backslash E) \cap C^0([0, T_0) \times \tilde{X}\backslash E)$ such that for all $t \in [0, T_0)$, $\tilde{\phi}(t, \cdot) \in L^\infty(\tilde{X}) \cap PSH(\tilde{X}, \pi^*\omega_0)$.*

Since $\tilde{\phi}$ is constant along (connected) fibre of π, $\tilde{\phi}$ descends to a solution $\phi \in C^\infty((0, T_0) \times X^{\rm reg}) \cap C^0([0, T_0) \times X^{\rm reg})$ of the Monge-Ampère flow .

Now suppose $\omega_{\rm KE}^w = \omega_0 + \sqrt{-1}\partial\bar{\partial}\phi_{\rm KE}^w$ is a weak solution to the equation (2.2). If one can prove that the solution $\phi(t)$ to (2.9) with the initial condition $\phi(0) = \phi_{\rm KE}^w$ is stationary, then it follows from Theorem (2.2) that $\omega_{\rm KE}^{w}$ is smooth on $X_\infty^{\rm reg}$. The idea to prove stationarity in [**CTZ**] is to show that the energy functional is decreasing along the flow solution $\phi(t)$ and to use the uniqueness of weak Kähler-Einstein metrics. These are indeed true in the current case by the work of [**BBEGZ**].

To prove Theorem 1.2, the main observation is that the above arguments can be used to prove the regularity of $\omega_{\rm KE}^{w}$ on $X^{\rm orb}$ as long as one can find a partial resolution by orbifolds: $\pi^{\rm par}: X^{\rm par} \to X$. Indeed, by the next section, there exist orbifold (partial) resolutions. If $\pi^{\rm par}: X^{\rm par} \to X$ is an orbifold resolution, then we can write:

$$K_{X^{\rm par}} = (\pi^{\rm par})^* K_X + \sum_i^r a_i E_i - \sum_{j=1}^s b_j F_j,$$

where $E = \cup_{i=1}^r E_i \bigcup \cup_{j=1}^s F_j$ is now a simple normal crossing divisor within orbifold category (in the sense of Satake [**Sat1, Sat2**]). The centers of E_i and F_j on X are contained in the closed subvariety Z of codimension at least 3. The klt property of X again implies $a_i > 0$ and $0 < b_i < 1$. Indeed, a_i (resp. $-b_j$) are just discrepancies of E_i (resp. F_j) on X, which does not depend on the birational morphism $\pi^{\rm par}$.

Then the similar arguments as in the proof of regularity of $\omega_{\rm KE}^{w}$ on $X_\infty^{\rm reg}$ carry over to the orbifold setting to prove the orbifold regularity of $\omega_{\rm KE}^{w}$ on $X^{\rm orb}$. Indeed, the main arguments in proving the regularity on $X_\infty^{\rm reg}$ depend either on the maximum principle on compact Kähler manifolds or the local regularity theory. On the one hand, the maximum principle also works on compact orbifold Kähler manifolds. On the other hand, the local arguments remain true by working on local uniformizing charts. For convenience of the reader, we write down the orbifold version of Kołodziej's result:

THEOREM 2.3 (see [**Koł, EGZ**]). *Let M be a compact Kähler orbifold and $\omega_M > 0$ be a closed orbifold smooth $(1,1)$-form. Let $\{f_j\} \subset C^\infty(M)$ be a sequence of orbifold smooth functions on M, such that the following conditions are satisfied:*

(i) $\sup_j \|\exp(f_j)\|_{L^p(M)} < +\infty$ *for some* $p > 1$;

(ii) $\int_M e^{f_j}\omega_M^n = \int_M \omega_M^n$ *for* $j \geq 1$.

Then there exist a constant $C > 0$ such that for each solution φ_j of the equation

$$(\omega_M + \sqrt{-1}\partial\bar\partial\varphi_j)^n = e^{f_j}\omega_M^n \tag{2.11}$$

such that $\int_M \varphi_j \omega_M^n = 0$ we have $\sup_M |\varphi_j| \leq C$. Moreover, if $\exp(f_j) \to \exp(f_\infty)$ in L^p, then there exist a continuous function φ_∞ such that $\varphi_j \to \varphi_\infty$ in $C^0(M)$ and φ_∞ is a solution of the equation:

$$(\omega_M + \sqrt{-1}\partial\bar\partial\varphi_\infty)^n = e^{f_\infty}\omega_M^n. \tag{2.12}$$

Again one can prove the above result by following the same proof of Kołodziej, which depends on local pluripotential theory. For example the key lemma [**Koł**, Lemma 2.3.1] for plurisubharmonic functions on domains of $\mathbb{C}^n$ can be proved on domains of the form $U/G \subset \mathbb{C}^n/G$ by arguing on the local uniformization charts $U \to U/G$. Similar remarks apply to the following orbifold version of Păun's theorem:

THEOREM 2.4 (see [**Pău, BBEGZ**]). *Let M be an orbifold and $\omega_M \geq 0$ be a semi-positive orbifold smooth $(1,1)$-form. Let μ be a positive measure on an orbifold M of the form $\mu = e^{\psi^+ - \psi^-} dV$ with $\psi^\pm$ quasi-psh with respect to M and $e^{-\psi^-} \in L^p$ for some $p > 1$. Assume φ is a bounded ω_M-psh function such that*

$(\omega_M + \sqrt{-1}\partial\bar{\partial}\varphi)^n = \mu$. Then we have $\Delta\varphi = O(e^{-\psi^-})$ locally in the ample locus where $\omega_M > 0$.

Note that it was already observed in [**ST**, Section 4.3] (via the same reasoning as above) that if X has only orbifold singularities, then the Kähler-Ricci flow smooths out initial metric to become an *orbifold* smooth metric immediately when $t > 0$.

3. Orbifold partial resolution

The results in this section were communicated to us by Chenyang Xu.

LEMMA 3.1 (Resolution of Deligne-Mumford stacks). Let $\mathcal{X}$ be an integral Deligne-Mumford stack which is of finite type over $\mathbb{C}$. Then there exists a birational proper representable morphism $g^{\rm sm} : \mathcal{X}^{\rm sm} \to \mathcal{X}$ from a smooth Deligne-Mumford stack $\mathcal{X}^{\rm sm}$. Furthermore, we can assume that $g^{\rm sm}$ is isomorphic over the smooth locus of $\mathcal{X}$, and the exceptional locus of $g^{\rm sm}$ is a normal crossing divisorial closed substacks of $\mathcal{X}^{\rm sm}$.

PROOF. This follows from the functoriality property of resolution of singularities (see [**Wło**], [**Kol**], [**BM**],[**Tem**]). Indeed, following the argument by Temkin in [**Tem**, Theorem 5.1], we first cover $\mathcal{X}$ with chart $U_\alpha \to U_\alpha/G_\alpha = V_\alpha$. Then we can resolve the singularity of U_α by the work of Hironaka. This means that there exists a birational morphism $\pi_\alpha : \tilde{U}_\alpha \to U_\alpha$ which is a composition of a sequence of blow-ups with regular centers such that $\tilde{U}_\alpha$ is regular and π_α is an isomorphism over the regular locs $U_\alpha^{\rm reg}$ of U_α. It has been proved in [**BM**] that the resolution algorithm can be made functorial with respect to regular morphisms. This allows us to show that π_U is equivariant under the G_α action. Denote $\tilde{V}_\alpha = \tilde{U}_\alpha/G_\alpha$. Then the functoriality also implies $\tilde{V}_\alpha$ can be glued to become a smooth Deligne-Mumford stack $\mathcal{X}^{\rm sm}$. □

LEMMA 3.2 (Blow up the indeterminacy locus). Let X be a projective scheme. Let $\mathcal{X}$ be a normal Deligne-Mumford stack with a dense open set $i_U : U \hookrightarrow \mathcal{X}$, such that U admits a morphism $f_U : U \to X$. Then we can blow up an ideal $\mathcal{I} \subset \mathcal{O}_{\mathcal{X}}$ to obtain a Deligne-Mumford stack $\tilde{\mathcal{X}}$ such that $\tilde{\mathcal{X}} \to \mathcal{X}$ is isomorphic over U and f_U extends to a morphism $f : \tilde{\mathcal{X}} \to X$.

PROOF. By the above lemma, we can assume $\mathcal{X}$ to be smooth. Because X is projective, we can replace X by $\mathbb{P}^N$. Let $\{H_i\}_{i=1}^{N+1}$ be the hyperplane section of $\mathbb{P}^N$. Let $D_i \subset U$ be the pull back of H_i by f_U and let $\bar{D}_i$ be the closure of D_i in $\mathcal{X}$. Let $\mathcal{I}$ be the ideal sheaf of $\cap_{i=1}^{N+1} D_i$. Then $\mathcal{I}$ is supported on $\mathcal{X} \setminus \mathcal{X}^\circ$. By using the same proof for the schemes as in [**Har**, II.7.17.3], ${\rm Bl}_{\mathcal{I}}\mathcal{X}$ satisfies the property in the statement. □

THEOREM 3.3. Let X be a quasi-projective normal variety. Let $X^{\rm orb}$ be the locus where X only has orbifold singularity. Then there exists $f^{\rm par} : X^{\rm par} \to X$ a proper birational morphism, such that $X^{\rm par}$ only has quotient singularity and $f^{\rm par}$ is an isomorphic over $X^{\rm orb}$.

PROOF. After taking the closure of $X \subset \mathbb{P}^N$, we can assume X is projective.

By [**Vis**, 2.8], we know there is a smooth Deligne-Mumford stack $\mathcal{X}^0$ whose coarse moduli space is $X^{\rm orb}$. It follows from [**Kre**, Theorem 4.4] that $\mathcal{X}^0 = [Z/G]$ for some quasi-projective scheme Z and linear algebraic group G. Actually, Z can be taken as the frame bundle of $X^{\rm orb}$ and $G = GL_n(\mathbb{C})$. Then by [**Kre**, Theorem

5.3], there is a proper Deligne-Mumford stack $\mathcal{X}$, such that $\mathcal{X}^0 \subset \mathcal{X}$ is a dense open set. We explain this by following the proof of [**Kre**]: for some N there is a linear action of G on $\mathbb{P}^N$ and an equivariant embedding $Z \to \mathbb{P}^N$ such that the embedding factors through the stable locus $(\mathbb{P}^N)^s$. By Kirwan's blow-up construction, there is a birational morphism $\mu : W \to \mathbb{P}^N$ which is a composition of blow-ups of nonsingular G-invariant subvarieties, such that μ is isomorphism over $(\mathbb{P}^N)^s$ and $W^s = W^{ss}$. Let $V \subset W$ be the closure of the strict transform of Z under μ. Then $\mathcal{X} = [V/G]$ satisfies the condition.

Consider the rational map $f : \mathcal{X} \dashrightarrow X$, by Lemma 3.2 we know that there is a blow up $\mathcal{Y} \to \mathcal{X}$ along the indeterminacy locus of f, such that there is a morphism $g : \mathcal{Y} \to X$. Moreover, by the construction, we know over X^{orb},

$$\mathcal{Y}^0 := g^{-1}(X^{\mathrm{orb}}) \cong \mathcal{X}^0.$$

By Lemma 3.1, we know that there is a smooth Deligne-Mumford stack $h : \mathcal{Y}^{\mathrm{sm}} \to \mathcal{Y}$, where h is a representable proper birational morphism which is isomorphic over the smooth locus of $\mathcal{Y}$. In particular, h is isomorphic over $\mathcal{Y}^0$.

As $\mathcal{X}$ has finite stabilizer and $\mathcal{Y}^{\mathrm{sm}} \to \mathcal{Y} \to \mathcal{X}$ is proper, we know that $\mathcal{Y}^{\mathrm{sm}}$ has also finite stabilizer. Thus it follows from [**KeM**] that $\mathcal{Y}^{\mathrm{sm}}$ admits a coarse moduli space, which we denote by X^{par}. It has a morphism $f^{\mathrm{par}} : X^{\mathrm{par}} \to X$ by the universal property. We can then easily check that they satisfy all the properties. □

4. Further discussions

In this section, we discuss some possible extensions of our theorem and open questions. Assume that (X_i, ω_i) is a sequence of Kähler-Einstein metrics satisfying:
(1) $Ric(\omega_i) = \lambda\omega_i$, where $\lambda = -1$,0 or 1;
(2) There is a $v > 0$ such that $Vol(B_1(x_i, \omega_i)) \geq v$, where $x_i \in X_i$.

By the Gromov-Hausdorff compactness, by taking a subsequence if necessary, we may assume that the sequence of pointed spaces (X_i, ω_i, x_i) converges to a length space $(X_\infty, d_\infty, x_\infty)$ in the Gromov-Hausdorff topology. By Cheeger-Colding's theory and Cheeger-Colding-Tian's theory, X_∞ is smooth outside a closed subset S of codimension at least 4 and d_∞ is given by a Kähler-Einstein metric ω_∞ on $X_\infty \backslash S$. Moreover, (X_i, ω_i, x_i) converges to $(X_\infty, \omega_\infty, x_\infty)$ in the Cheeger-Gromov topology, and in particular, in the C^∞-topology outside S. We expect that X_∞ is a Kähler variety whose singular set S is a subvariety of complex codimension at least 2 and $(X_\infty, \omega_\infty)$ is a Kähler-Einstein orbifold outside a subvariety $Z \subset S$ of complex codimension at least 3. Theorem 1.2 confirms this in the case of Fano manifolds with Kähler-Einstein metrics. More generally, we have:

THEOREM 4.1. With the same assumption as above, if $(X_\infty, \omega_\infty)$ has finite diameter and $m\,[\omega_i] \in H^2(X_i, \mathbb{Z})$ is uniformly bounded for some fixed $m \in \mathbb{Z}$, then X_∞ is a normal projective variety and ω_∞ is a positive closed $(1,1)$-current which is a smooth Kähler-Einstein orbifold metric outside a subvariety Z of complex codimension at least 3.

PROOF. As mentioned in the introduction, the first statement follows from partial C^0-estimate (see [**Tia2, Tia3**], [**Li**] and [**DS**]). The second statement is obtained from the same proof as that of Theorem 1.2. □

The proof of this theorem as well as Theorem 1.2, relies on a global partial resolution $X^{\rm par}$ of X_∞ which resolves all the non-quotient singularities. It is desirable to have a proof which does not use such a partial resolution and works locally. More precisely, let U be an open subset in an affine subvariety $Y \subset \mathbb{C}^N$ of the form $\tilde{U}/\Gamma$, where $\tilde{U}$ is open subset in $\mathbb{C}^n$ and Γ is a finite group acting on $\mathbb{C}^n$ holomorphically. Assume that ω is a weak Kähler-Einstein metric on U, that is, it satisfies: (1) ω is a smooth Kähler-Einstein metric on the regular part of U; (2) Near each $x \in U$, it can be written as $\sqrt{-1}\,\partial\bar{\partial}\varphi_x$ in the sense of currents for some bounded function φ_x; (3) $\omega \geq \omega_0|_U$ for some smooth Kähler metric on $\mathbb{C}^N$. Then we expect that $\pi^*\omega$ extends to be a smooth metric on $\tilde{U}$, where $\pi : \tilde{U} \mapsto U$ is the natural projection. If Γ is trivial, it is true as explained in the introduction: The condition (3) implies that there is a uniform C^2-estimate of the potential of ω on any compact subset of U; next, the Evans-Krylov theory or Calabi's 3rd derivative estimate allows one to show that ω is smooth on U.

It should be possible to understand singularities of weak Kähler-Einstein metrics in higher codimensions in an inductive way. It is proved in [**Tia1**] that the only singularities for 2-dimensional weak Kähler-Einstein metrics are quotient singularities. The next case is dimension 3. One should be able to have a complete understanding of singularities for weak Kähler-Einstein metrics in dimension 3. Here is a heuristic reasoning: Let $(X_\infty, \omega_\infty)$ be a limit of 3-dimensional compact Kähler-Einstein metrics (X_i, ω_i) in the Cheeger-Gromov topology. Then any tangent cone C_x at $x \in X_\infty$ is a complex cone over some complex 2-dimensional Fano variety Z (see [**DS**]) and Z is a Kähler-Einstein orbifold. It is believed that the structure of X_∞ near x should be modeled on this tangent cone, so we may be able to analyze the structure of X_∞ near x. Moreover, if $x_i \in X_i$ converge to x, then we may also decode information on X_i near x_i.

Another possible extension of Theorem 1.2 is about conic Kähler-Einstein metrics. For simplicity, we consider only the following situation. The general cases are similar. Assume that M is a Fano manifold, $D \subset M$ is a pluri-anti-canonical divisor and ω_i are conic Kähler-Einstein metrics on M with angles $2\pi\beta$ along D, where $\beta \in (0,1)$. We further assume that $(M_\infty, \omega_\infty)$ is the limit of (M, ω_i) in the Cheeger-Gromov topology and D_∞ is the limit of D. In view of [**Tia4**] and [**CDS**], we know that M_∞ is a normal variety and D_∞ is a divisor in M_∞. Furthermore, if we write

$$D_\infty = \sum_{i=1}^{k} m_i\, D_{\infty,i},$$

where each $D_{\infty,i}$ is irreducible, then ω_∞ is a weak conic Kähler-Einstein metric with conic angles $2\pi\beta_i$ along each $D_{\infty,i}$. Here β_i is given by the equality

$$m_i\,(1-\beta) \;=\; (1-\beta_i).$$

If $k=1$ and $m_1 = 1$, then it is not hard to show that ω_∞ is a conic Kähler-Einstein metric near any point of $D_\infty^{\rm reg} \cap M_\infty^{\rm reg}$, where $D_\infty^{\rm reg}$ denotes the regular part of D_∞ and $M_\infty^{\rm reg}$ denotes the regular part of M_∞. In particular, if M_∞ is smooth and D_∞ is a smooth divisor, ω_∞ is a genuine conic metric. In general, we expect a similar structure. More precisely, we conjecture that there is a subvariety Z of complex codimension at least 2 such that the restriction of ω_∞ to $M_\infty \backslash Z$ is a smooth conic Kähler-Einstein metric with angle $2\pi\beta_i$ along each $D_{\infty,i}$. It is possible that our arguments for proving Theorem 1.2 can be adapted to prove this conjecture by using the conic Kähler-Ricci flow.

Acknowledgments. We would like to thank Chenyang Xu for communicating the results in section 3 to us. The first author would like to thank Professor J. Starr for discussions about stacks.

References

[BBEGZ] R.J. Berman, S. Boucksom, P. Eyssidieux, V. Guedj, and A. Zeriahi, *Kähler-Einstein metrics and the Kähler-Ricci flow on log Fano varieties*, to appear in J. reine angew. Math., arXiv:1111.7158.

[BM] Edward Bierstone and Pierre D. Milman, *Functoriality in resolution of singularities*, Publ. Res. Inst. Math. Sci. **44** (2008), no. 2, 609–639, DOI 10.2977/prims/1210167338. MR2426359

[Bło] Zbigniew Błocki, *Interior regularity of the complex Monge-Ampère equation in convex domains*, Duke Math. J. **105** (2000), no. 1, 167–181, DOI 10.1215/S0012-7094-00-10518-2. MR1788046

[CCT] J. Cheeger, T. H. Colding, and G. Tian, *On the singularities of spaces with bounded Ricci curvature*, Geom. Funct. Anal. **12** (2002), no. 5, 873–914, DOI 10.1007/PL00012649. MR1937830

[CDS] Xiuxiong Chen, Simon Donaldson, and Song Sun, *Kähler-Einstein metrics on Fano manifolds. II: Limits with cone angle less than 2π*, J. Amer. Math. Soc. **28** (2015), no. 1, 199–234, DOI 10.1090/S0894-0347-2014-00800-6. MR3264767

[CTZ] X. X. Chen, G. Tian, and Z. Zhang, *On the weak Kähler-Ricci flow*, Trans. Amer. Math. Soc. **363** (2011), no. 6, 2849–2863, DOI 10.1090/S0002-9947-2011-05015-4. MR2775789

[Dem] Jean-Pierre Demailly, *Regularization of closed positive currents and intersection theory*, J. Algebraic Geom. **1** (1992), no. 3, 361–409. MR1158622

[DS] Simon Donaldson and Song Sun, *Gromov-Hausdorff limits of Kähler manifolds and algebraic geometry*, Acta Math. **213** (2014), no. 1, 63–106, DOI 10.1007/s11511-014-0116-3. MR3261011

[EGZ] Philippe Eyssidieux, Vincent Guedj, and Ahmed Zeriahi, *Singular Kähler-Einstein metrics*, J. Amer. Math. Soc. **22** (2009), no. 3, 607–639, DOI 10.1090/S0894-0347-09-00629-8. MR2505296

[GKKP] Daniel Greb, Stefan Kebekus, Sándor J. Kovács, and Thomas Peternell, *Differential forms on log canonical spaces*, Publ. Math. Inst. Hautes Études Sci. **114** (2011), 87–169, DOI 10.1007/s10240-011-0036-0. MR2854859

[Har] Robin Hartshorne, *Algebraic geometry*, Springer-Verlag, New York-Heidelberg, 1977. Graduate Texts in Mathematics, No. 52. MR0463157

[Hir] Heisuke Hironaka, *Resolution of singularities of an algebraic variety over a field of characteristic zero. I, II*, Ann. of Math. (2) **79** (1964), 109–203; ibid. (2) **79** (1964), 205–326. MR0199184

[KeM] Seán Keel and Shigefumi Mori, *Quotients by groupoids*, Ann. of Math. (2) **145** (1997), no. 1, 193–213, DOI 10.2307/2951828. MR1432041

[Kol] János Kollár, *Lectures on resolution of singularities*, Annals of Mathematics Studies, vol. 166, Princeton University Press, Princeton, NJ, 2007. MR2289519

[KoM] János Kollár and Shigefumi Mori, *Birational geometry of algebraic varieties*, Cambridge Tracts in Mathematics, vol. 134, Cambridge University Press, Cambridge, 1998. With the collaboration of C. H. Clemens and A. Corti; Translated from the 1998 Japanese original. MR1658959

[Koł] Sławomir Kołodziej, *The complex Monge-Ampère equation*, Acta Math. **180** (1998), no. 1, 69–117, DOI 10.1007/BF02392879. MR1618325

[Kre] Andrew Kresch, *On the geometry of Deligne-Mumford stacks*, Algebraic geometry—Seattle 2005. Part 1, Proc. Sympos. Pure Math., vol. 80, Amer. Math. Soc., Providence, RI, 2009, pp. 259–271, DOI 10.1090/pspum/080.1/2483938. MR2483938

[Li] Chi Li, *Kahler-Einstein metrics and K-stability*, ProQuest LLC, Ann Arbor, MI, 2012. Thesis (Ph.D.)–Princeton University. MR3078441

[Pău] Mihai Păun, *Regularity properties of the degenerate Monge-Ampère equations on compact Kähler manifolds*, Chin. Ann. Math. Ser. B **29** (2008), no. 6, 623–630, DOI 10.1007/s11401-007-0457-8. MR2470619

[Sat1] I. Satake, *On a generalization of the notion of manifold*, Proc. Nat. Acad. Sci. U.S.A. **42** (1956), 359–363, DOI 10.1073/pnas.42.6.359. MR0079769

[Sat2] Ichirô Satake, *The Gauss-Bonnet theorem for V-manifolds*, J. Math. Soc. Japan **9** (1957), 464–492, DOI 10.2969/jmsj/00940464. MR0095520

[ST] Jian Song and Gang Tian, *The Kähler-Ricci flow through singularities*, Invent. Math. **207** (2017), no. 2, 519–595, DOI 10.1007/s00222-016-0674-4. MR3595934

[Tia1] G. Tian, *On Calabi's conjecture for complex surfaces with positive first Chern class*, Invent. Math. **101** (1990), no. 1, 101–172, DOI 10.1007/BF01231499. MR1055713

[Tia2] G. Tian, *Einstein metrics on Fano manifolds. Metric and Differential Geomtry*, Proceeding of the 2008 conference celebrating J. Cheeger's 65th birthday, edited by Dai et al., Progress in Mathematics, volume 239. Birkhauser, 2012.

[Tia2] Gang Tian, *Existence of Einstein metrics on Fano manifolds*, Metric and differential geometry, Progr. Math., vol. 297, Birkhäuser/Springer, Basel, 2012, pp. 119–159, DOI 10.1007/978-3-0348-0257-4_5. MR3220441

[Tia3] Gang Tian, *Partial C^0-estimate for Kähler-Einstein metrics*, Commun. Math. Stat. **1** (2013), no. 2, 105–113, DOI 10.1007/s40304-013-0011-9. MR3197855

[Tia4] Gang Tian, *K-stability and Kähler-Einstein metrics*, Comm. Pure Appl. Math. **68** (2015), no. 7, 1085–1156, DOI 10.1002/cpa.21578. MR3352459

[Tem] Michael Temkin, *Functorial desingularization of quasi-excellent schemes in characteristic zero: the nonembedded case*, Duke Math. J. **161** (2012), no. 11, 2207–2254, DOI 10.1215/00127094-1699539. MR2957701

[Vis] Angelo Vistoli, *Intersection theory on algebraic stacks and on their moduli spaces*, Invent. Math. **97** (1989), no. 3, 613–670, DOI 10.1007/BF01388892. MR1005008

[Wło] Jarosław Włodarczyk, *Simple Hironaka resolution in characteristic zero*, J. Amer. Math. Soc. **18** (2005), no. 4, 779–822, DOI 10.1090/S0894-0347-05-00493-5. MR2163383

Department of Mathematics, Stony Brook University, Stony Brook, New York 11794

Current address: Department of Mathematics, Purdue University, West Lafayette, Indiana 47907

Email address: `li2285@purdue.edu`

School of Mathematical Sciences and BICMR, Peking University, Beijing 100871, China

Email address: `tian@math.princeton.edu`

Contemporary Mathematics
Volume **735**, 2019
https://doi.org/10.1090/conm/735/14826

Analysis of the Laplacian on the moduli space of polarized Calabi-Yau manifolds

Zhiqin Lu and Hang Xu

ABSTRACT. In this paper, we generalize the spectrum relation in the paper *On the spectrum of the Laplacian, Math. Ann., 359(1-2):211–238, 2014* (by Nelia Charalambous and Zhiqin Lu) to any Hermitian manifolds. We also prove that the closure of Laplace operator $\square = \delta d$ on the moduli space of polarized Calabi-Yau manifolds is self-adjoint.

1. Introduction

Let (M, g) be a Hermitian manifold with a holomorphic vector bundle (E, h). Suppose $\square$ is the Hodge Laplacian on smooth E-valued (p, q) forms. Though $\square$ in general is only symmetric but not self-adjoint, one can consider self-adjoint extensions of the Hodge Laplacian. One well-known self-adjoint extension is the so-called Gaffney extension $\square_G$ ([**5**]). In this note, we generalize the spectrum relations in [**2**] to the Gaffney extension on incomplete manifolds. One key ingredient for the spectrum relations is a generalized version of the Weyl's criterion.

Another well-know extension of $\square$ is the Friedrichs extension $\square_F$. $\square_G$ and $\square_F$ are in general different on incomplete manifolds. In the special case of the moduli space of polarized Calabi-Yau manifolds $\mathcal{M}$ with the Weil-Petersson metric ω_{WF}, we prove the Cauchy boundary of $\mathcal{M}$ has zero capacity, and therefore $\square_G = \square_F$ on functions. Furthermore, we also show that the Hodge Laplacian on functions with certain $Dom\square$ is essentially self-adjoint, which is a generalization of the results in [**6**] and [**10**].

Using the spectrum results we obtain on different self-adjoint extensions of the Laplacians, we study the L^2-estimates on incomplete manifolds. The L^2-estimate played one of the most crucial roles in several complex variables and complex geometry. The method allows us to construct a lot of holomorphic functions and holomorphic sections in various function spaces.

One of the most important applications of the L^2-estimate is the proof of Kodaira's embedding theorem. Let L be a positive line bundle over a compact complex manifold X. Then there exists a positive integer k such that the line bundle $L^k = L \otimes \cdots \otimes L$ has a lot of (ample) holomorphic sections.

2010 *Mathematics Subject Classification.* Primary 32Q15, 32Q25; Secondary 53C55.

The first author is partially supported by the NSF grant DMS-1547878.

Part of the paper is a refinement of the thesis of the second author.

©2019 American Mathematical Society

In this paper, we study the case when X is *not* a complete complex manifold. As it is well known, on a incomplete manifold, the extension of the Laplacian as a self-adjoint operator is not unique. So we need to specify the extension. Secondly, the L^2 estimates heavily depends on the spectrum gap on the bundle-valued $(0,1)$ forms. Therefore, it is useful to generalize the results in [**2**] to the incomplete case.

The main result of this paper is in §5, where we re-prove the results of Masamune [**10, 11**]. We found a gap in his proof and we showed this gap by a counter-example in §8.

In §7 and §8, we apply our results to the moduli space of Calabi-Yau manifolds.

Acknowledgments. We are very thankful to the anonymous referees for many helpful comments.

2. Two Self-Adjoint Extensions of Hodge Laplacian

In this section, we assume (M,g) is a Hermitian manifold with a holomorphic Hermitian vector bundle (E,h). Consider Hodge Laplacian on E-valued (p,q) forms with compact support. As the Hodge Laplacian is symmetric but not self-adjoint, we consider the self-adjoint extensions of the Hodge Laplacian via the corresponding closed quadratic forms. By endowing the quadratic form with different domain of definition, we will get two important self-adjoint extensions, which are respectively Gaffney extension and Friedrichs extension. For more details about this section, we recommend references [**9, 15**].

We begin with the d-bar differential operator

$$\bar{\partial}_{p,q} : L^2(M, \Lambda^{p,q}(E)) \to L^2(M, \Lambda^{p,q+1}(E)),$$

with

$$\begin{aligned} &Dom(\bar{\partial}_{p,q}) \\ &\quad = \{\varphi \in L^2(M, \Lambda^{p,q}(E)) : \text{the distributional derivative } \bar{\partial}\varphi \in L^2(M, \Lambda^{p,q+1}(E))\}. \end{aligned}$$

With the above domain of definition, the operator $\bar{\partial}_{p,q}$ is a densely defined closed operator. We denote the L^2 inner product on $L^2(M, \Lambda^{p,q}(E))$ as $(\cdot,\cdot)_{p,q}$. With respect to the L^2 inner product on $L^2(M, \Lambda^{p,q}(E))$ and $L^2(M, \Lambda^{p,q+1}(E))$, we have the adjoint operator of $\bar{\partial}_{p,q}$ as

$$\bar{\partial}^*_{p,q+1} : L^2(M, \Lambda^{p,q+1}(E)) \to L^2(M, \Lambda^{p,q}(E)),$$

with

$$\begin{aligned} Dom(\bar{\partial}^*_{p,q+1}) = \{\phi \in L^2(M, \Lambda^{p,q+1}(E)) : \exists\, \varphi \in L^2(M, \Lambda^{p,q}(E)) \text{ such that} \\ (\bar{\partial}u, \phi)_{p,q+1} = (u, \varphi)_{p,q} \text{ for any } u \in Dom(\bar{\partial}_{p,q})\}. \end{aligned}$$

And in the above notation, $\bar{\partial}^*\phi$ is defined to be φ.

In the following, we will suppress the indices p, q in the operators and inner product for simplicity when there is no confusion from context.

Now let us recall Hodge Laplacian and the associated quadratic form. We use the notation $\mathcal{D}(M, \Lambda^{p,q}(E))$ to denote the set of all smooth E-valued (p,q) forms with compact support.

DEFINITION 2.1. i) Let $\square : \mathcal{D}(M, \Lambda^{p,q}(E)) \to \mathcal{D}(M, \Lambda^{p,q}(E))$ be the Hodge Laplacian defined as

$$\square = \bar{\partial}\bar{\partial}^* + \bar{\partial}^*\bar{\partial}.$$

ii) Let $Q : \mathcal{D}(M, \Lambda^{p,q}(E)) \times \mathcal{D}(M, \Lambda^{p,q}(E)) \to \mathbb{C}$ be the quadratic form associated to $\square$ defined as

$$Q(\varphi, \phi) = (\bar{\partial}\varphi, \bar{\partial}\phi) + (\bar{\partial}^*\varphi, \bar{\partial}^*\phi) \text{ for any } \varphi, \phi \in \mathcal{D}(M, \Lambda^{p,q}(E)).$$

Since $\bar{\partial}, \bar{\partial}^*$ are closed operators, if we endow quadratic form Q with $Dom(Q) = Dom(\bar{\partial}) \cap Dom(\bar{\partial}^*)$, then Q is closed. That means, for any sequence $\varphi_n \in Dom(Q)$, $\varphi_n \xrightarrow{L^2} \varphi$ and $Q(\varphi_m - \varphi_n, \varphi_m - \varphi_n) \to 0$ as $m, n \to \infty$, then $\varphi \in Dom(Q)$ and $Q(\varphi_n - \varphi, \varphi_n - \varphi) \to 0$.

We cite the following theorem from [**14**] in Chapter VIII.6.

THEOREM 2.1 ([**14**]). *If Q is a closed semibounded quadratic form, then Q is the quadratic form of a unique self-adjoint operator.*

By applying this theorem to our quadratic form Q with $Dom(Q) = Dom(\bar{\partial}) \cap Dom(\bar{\partial}^*) \subset L^2(M, \Lambda^{p,q}(E))$, we get a self-adjoint extension of $\square$, which is called Gaffney extension and denoted as $\square_G$. The domain of $\square_G$ is

$$\begin{aligned} Dom(\square_G) = \{\varphi \in Dom(\bar{\partial}) \cap Dom(\bar{\partial}^*) : \exists \eta \in L^2(M, \Lambda^{p,q}(E)) \text{ such that} \\ Q(\varphi, \phi) = (\eta, \phi) \text{ for any } \phi \in Dom(\bar{\partial}) \cap Dom(\bar{\partial}^*)\}. \end{aligned} \tag{2.1}$$

And in the same notation as above, $\square_G\varphi$ is defined to be η.

The following Gaffney's Theorem from [**5**] (See also chapter 3 in [**9**]) tells us that Gaffney extension can be viewed as the composition of $\bar{\partial}$ and $\bar{\partial}^*$ as follows.

THEOREM 2.2 (Gaffney).

$$Dom(\square_G) = \{\varphi \in Dom(\bar{\partial}) \cap Dom(\bar{\partial}^*) : \bar{\partial}\varphi \in Dom(\bar{\partial}^*) \text{ and } \bar{\partial}^*\varphi \in Dom(\bar{\partial})\}. \tag{2.2}$$

And for any $\varphi \in Dom(\square_G)$, we have

$$\square_G\varphi = \bar{\partial}\bar{\partial}^*\varphi + \bar{\partial}^*\bar{\partial}\varphi.$$

Similarly, we will introduce Friedriechs extension by endowing Q with a different domain of definition. Let's first recall the following Sobolev spaces. We denote $Q_1(\cdot,\cdot) = Q(\cdot,\cdot) + (\cdot,\cdot)$. It is not hard to see Q_1 is an inner product on $\mathcal{D}(M, \Lambda^{p,q}(E))$.

DEFINITION 2.2 (Sobolev Spaces).

$$\begin{aligned} W_0^1(M, \Lambda^{p,q}(E) = &\text{Completion of } \mathcal{D}(M, \Lambda^{p,q}(E)) \\ &\text{with respect to } Q_1 \text{ inner product,} \end{aligned} \tag{2.3}$$

$$\begin{aligned} W^1(M, \Lambda^{p,q}(E)) = &\text{Completion of } \{\varphi \in \mathcal{C}^\infty(M, \Lambda^{p,q}(E)) : Q_1(\varphi, \varphi) < \infty\} \\ &\text{with respect to } Q_1 \text{ inner product.} \end{aligned} \tag{2.4}$$

REMARK 2.3. Note that φ is not necessarily in $Dom(\bar{\partial}^*_{p,q})$ when $\varphi \in \mathcal{C}^\infty(M, \Lambda^{p,q}(E))$. So in the definition of $W^1(M, \Lambda^{p,q}(E))$, to be precise, $Q_1(\varphi, \varphi) < \infty$ means $\varphi \in L^2(M, \Lambda^{p,q}(E))$ and the point-wise differentials $\bar{\partial}\varphi, \bar{\partial}^*\varphi$ belong to $L^2(M, \Lambda^{p,q+1}(E))$ and $L^2(M, \Lambda^{p,q-1}(E))$ respectively. And one can prove $\varphi \in$

$W^1(M, \Lambda^{p,q}(E))$ if and only if $\varphi \in L^2(M, \Lambda^{p,q}(E))$ and the distributional differentials $\overline{\partial}\varphi, \bar{\partial}^*\varphi$ belong to $L^2(M, \Lambda^{p,q+1}(E))$ and $L^2(M, \Lambda^{p,q-1}(E))$ respectively.

REMARK 2.4. Note that $W_0^1 \subset Dom(\overline{\partial}) \cap Dom(\bar{\partial}^*) \subset W^1$. But they are generally not equal to each other.

If we endow Q with $Dom(Q) = W_0^1$, then it becomes a closed quadratic form. By applying Theorem 2.1 again, we will get a different self-adjoint extension of Hodge Laplacian $\Box$, which is called Friedrichs extension and denoted as $\Box_F$. Note that $\Box_F$ is generally different from $\Box_G$ by Remark 2.4.

EXAMPLE 2.5. Take the Hermitian manifold $M = \Omega \subset \mathbb{C}^n$ be a bounded open set with smooth boundary. Let Hermitian vector bundle E be the trivial line bundle. Assume $u \in \mathcal{C}^\infty(\overline{\Omega}, \Lambda^{p,q})$. Let us investigate the boundary conditions induced from $\Box_G$ and $\Box_F$ in this case.

If $u \in Dom(\bar{\partial}^*)$, then

$$(\overline{\partial}\varphi, u) = (\varphi, \bar{\partial}^* u) \text{ for any } \varphi \in \mathcal{C}^\infty(\overline{\Omega}, \Lambda^{p,q-1}).$$

Note

$$(\overline{\partial}\varphi, u) = \int_\Omega \overline{\partial}\varphi \wedge *\overline{u} = \int_{\partial\Omega} \varphi \wedge *\overline{u} + (-1)^{p+q} \int_\Omega \varphi \wedge \overline{\partial} * \overline{u} = \int_{\partial\Omega} \varphi \wedge *\overline{u} + (\varphi, \bar{\partial}^* u).$$

Here $*$ is the Hodge star operator. The second equality follows from Stokes Theorem and the last one is based on the identity $\bar{\partial}^* = - * \partial *$. Therefore we have

$$\int_{\partial\Omega} \varphi \wedge *\overline{u} = 0 \text{ for any } \varphi \in \mathcal{C}^\infty(\overline{\Omega}, \Lambda^{p,q-1}).$$

It implies $*u|_{\partial\Omega} = 0$(the restriction of $*u$ to $\partial\Omega$). So by Theorem 2.2, $u \in Dom(\Box_G)$ implies the boundary condition $*u|_{\partial\Omega} = 0$ and $*\overline{\partial}u|_{\partial\Omega} = 0$.

For the Friedrichs extension, $u \in Dom(\Box_F)$ implies $u \in W_0^1$. Then there exists a sequence $u_j \in \mathcal{D}(M, \Lambda^{p,q})$ such that $u_n \to u$ in W_0^1. By Weitzenböck formula, we have

$$\Box u_j = -\sum_{i=1}^n \frac{\partial}{\partial z^i}\frac{\partial}{\partial z^{\bar{i}}} u_j = -\sum_{i=1}^n \frac{\partial}{\partial z^{\bar{i}}}\frac{\partial}{\partial z^i} u_j. \tag{2.5}$$

Therefore by taking the inner product with u_j,

$$Q(u_j, u_j) = \sum_{i=1}^n (\frac{\partial}{\partial z^{\bar{i}}} u_j, \frac{\partial}{\partial z^{\bar{i}}} u_j) = \sum_{i=1}^n (\frac{\partial}{\partial z^i} u_j, \frac{\partial}{\partial z^i} u_j). \tag{2.6}$$

Then we have

$$\frac{\partial}{\partial z^{\bar{i}}} u_j \to \frac{\partial}{\partial z^{\bar{i}}} u \quad \text{and} \quad \frac{\partial}{\partial z^i} u_j \to \frac{\partial}{\partial z^i} u \quad \text{in } L^2 \text{ norms for each } i. \tag{2.7}$$

If we write $u = u_{I\overline{J}} dz^I \wedge d\overline{z^J}$, then each function $u_{I\overline{J}}$ is in the standard Sobolev space $H_0^1(\Omega)$, which implies $u_{I\overline{J}}|_{\partial\Omega} = 0$ for each multi-index I, J.

3. Spectrums of Gaffney Extension

The main goal of this section is to prove the following spectrum relations of Gaffney extension.

THEOREM 3.1. *Let (M, g) be a Hermitian manifold with a holomorphic Hermitian vector bundle (E, h). Consider Gaffney extension of Hodge Laplacian, $\Box_{p,q}$: $L^2(M, \Lambda^{p,q}(E)) \to L^2(M, \Lambda^{p,q}(E))$. We have the following spectrum relations.*

$$Spec(\Box_{p,q}) \cup \{0\} = Spec(\overline{\partial}\bar{\partial}^*_{p,q+1}) \cup Spec(\bar{\partial}^*\overline{\partial}_{p,q-1}) \cup \{0\}. \tag{3.1}$$

$$Spec(\Box_{p,q}) \cup \{0\} = Spec(\overline{\partial}\bar{\partial}^*_{p,q}) \cup Spec(\bar{\partial}^*\overline{\partial}_{p,q}) \cup \{0\}. \tag{3.2}$$

REMARK 3.1. The above notation $\overline{\partial}\bar{\partial}^*_{p,q}$ means $\overline{\partial}_{p,q-1}\bar{\partial}^*_{p,q}$ and $\bar{\partial}^*\overline{\partial}_{p,q}$ means $\bar{\partial}^*_{p,q+1}\overline{\partial}_{p,q}$. Note that $\overline{\partial}\bar{\partial}^*_{p,q}$ and $\bar{\partial}^*\overline{\partial}_{p,q}$ are self-adjoint operators by Von Neumann's Theorem (see Chapter X in [**13**]) since both $\overline{\partial}_{p,q}$ and $\bar{\partial}^*_{p,q}$ are densely defined closed operators. In the following we will omit the sub-indices p, q when there is not confusion from context.

This is a generalization of results in [**2**], where similar spectrum relations were proved for complete Riemannian manifolds. One main tool we are going to use is the generalized Weyl criterion from [**2**]. The advantage of this generalized Weyl criterion is that we do not necessarily pick the test sequence from the domain of an unbounded operator. After proving it, we will mention a well known relation between Gaffney extension and L^2 estimates, which serves a preparation for later sections.

We will split the proof of Theorem 3.1 into to several Lemmas. First, we prove one containment relation of (3.1).

LEMMA 3.2. *Under the same assumption as Theorem* 3.1, *we have*

$$Spec(\Box_{p,q}) \subset Spec(\overline{\partial}\bar{\partial}^*_{p,q+1}) \cup Spec(\bar{\partial}^*\overline{\partial}_{p,q-1}) \cup \{0\}. \tag{3.3}$$

PROOF. In this proof, we will use $\Box$ to represent $\Box_{p,q}$ for simplicity. Take $\lambda_0 \in$ Spec$(\Box)$ and $\lambda_0 > 0$. By Weyl's criterion, there exists a sequence $u_j \in Dom(\Box)$ with $(u_j, u_j) = 1$ such that

$$(\Box - \lambda_0)u_j \to 0 \text{ as } j \to \infty.$$

Since $\Box$ is non-negative and self-adjoint, $(1+\Box)^{-1} : L^2(M, \Lambda^{p,q}(E)) \to Dom(\Box_{p,q}) \subset L^2(M, \Lambda^{p,q}(E))$ is a bounded operator. By identity (2.1), we have

$$Q((1+\Box)^{-2}u_j, (1+\Box)^{-2}u_j) = (\Box(1+\Box)^{-2}u_j, (1+\Box)^{-2}u_j). \tag{3.4}$$

Let $\{P_\lambda\}$ be the Projection Valued Measure of $\Box$. Then

$$(\Box(1+\Box)^{-2}u_j, (1+\Box)^{-2}u_j) = \int_0^\infty \frac{\lambda}{(1+\lambda)^4} d(P_\lambda u_j, u_j). \tag{3.5}$$

Take $C(\lambda_0) = \min_{\lambda \in [\frac{\lambda_0}{2}, \frac{3\lambda_0}{2}]} \frac{\lambda}{(1+\lambda)^4} > 0$. Then

(3.6)

$$\int_0^\infty \frac{\lambda}{(1+\lambda)^4} d(P_\lambda u_j, u_j) \geq C(\lambda_0) \int_{\frac{1}{2}\lambda_0}^{\frac{3}{2}\lambda_0} d(P_\lambda u_j, u_j) \geq C(\lambda_0) \|P_{(\frac{1}{2}\lambda_0, \frac{3}{2}\lambda_0)} u_j\|^2.$$

We denote $u_j^{(1)} = P_{(\frac{1}{2}\lambda_0, \frac{3}{2}\lambda_0)} u_j$ and $u_j^{(2)} = u_j - u_j^{(1)}$. By using the Projection Valued Measure again, we have

$$((\Box - \lambda_0)u_j, (\Box - \lambda_0)u_j) = \int_0^\infty (\lambda - \lambda_0)^2 d(P_\lambda u_j, u_j) \geq \frac{\lambda_0^2}{4} \|u_j^{(2)}\|^2.$$

Since we know $(\Box - \lambda_0)u_j \to 0$ as j goes to infinity, we have

$$\|u_j^{(2)}\| \to 0 \text{ as } j \to \infty,$$

whence

$$\|u_j^{(1)}\| \to 1 \text{ as } j \to \infty.$$

Together with (3.4), (3.5) and (3.6), we have for sufficiently large j

$$\|\bar{\partial}(1+\Box)^{-2}u_j\|^2 + \|\bar{\partial}^*(1+\Box)^{-2}u_j\|^2 \geq \frac{C(\lambda_0)}{2} > 0. \tag{3.7}$$

On the other hand, we have

$$\begin{aligned}
&\|(\bar{\partial}\bar{\partial}^* - \lambda_0)\bar{\partial}(1+\Box)^{-2}u_j\|^2 + \|(\bar{\partial}^*\bar{\partial} - \lambda_0)\bar{\partial}(1+\Box)^{-2}u_j\|^2\\
=&\|\bar{\partial}(\Box - \lambda_0)(1+\Box)^{-2}u_j\|^2 + \|\bar{\partial}^*(\Box - \lambda_0)(1+\Box)^{-2}u_j\|^2\\
=&\big(\Box(1+\Box)^{-2}(\Box-\lambda_0)u_j, (1+\Box)^{-2}(\Box-\lambda_0)u_j\big)\\
\leq&\|(\Box-\lambda_0)u_j\|^2.
\end{aligned}$$

The first equality is because $\bar{\partial}\circ\bar{\partial} = 0$ on $Dom(\bar{\partial})$ and $\bar{\partial}^*\circ\bar{\partial}^* = 0$ on $Dom(\bar{\partial}^*)$. The second one follows from (2.1) and the commutativity of $\Box$ and $(1+\Box)^{-1}$. And the last inequality follows from $\|(1+\Box)^{-1}\|_{L^2\to L^2} \leq 1$ and $\|\Box(1+\Box)^{-1}\|_{L^2\to L^2} \leq 1$. Therefore

$$\|(\bar{\partial}\bar{\partial}^* - \lambda_0)\bar{\partial}(1+\Box)^{-2}u_j\|^2 + \|(\bar{\partial}^*\bar{\partial} - \lambda_0)\bar{\partial}(1+\Box)^{-2}u_j\|^2 \to 0. \tag{3.8}$$

Combining (3.7) and (3.8), we have $\lambda_0 \in \mathrm{Spec}(\bar{\partial}\bar{\partial}^*_{p,q+1}) \cup \mathrm{Spec}(\bar{\partial}^*\bar{\partial}_{p,q-1})$ by Weyl criterion. So the result follows. $\Box$

Now we prove the other containment of (3.1).

LEMMA 3.3. *Under the same assumption as Theorem 3.1, we have*

$$Spec(\bar{\partial}\bar{\partial}^*_{p,q+1}) \cup Spec(\bar{\partial}^*\bar{\partial}_{p,q-1}) \subset Spec(\Box_{p,q}) \cup \{0\}.$$

In order to prove this lemma, we will use one generalized Weyl criterion from [**2**].

THEOREM 3.2 (Charalambous-Lu). *Let H be a non-negative self-adjoint operator on Hilbert space $\mathcal{H}$. A positive real number λ_0 is contained in $Spec(H)$ if there exists a sequence $u_j \in \mathcal{H}$ such that*

(1) *For any j, $\|u_j\| = 1$.*
(2) *$((H-\lambda_0)(1+H)^{-m}u_j, u_j) \to 0$ for $m = 1, 2$.*

Note that compared to the classical Weyl criterion, the above theorem does not require $u_j \in Dom(H)$. We give a proof of this theorem here for the completeness.

PROOF. Note that

$$(H-\lambda_0)^2(1+H)^{-2} = (H-\lambda_0)(1+H)^{-1} - (\lambda_0+1)(H-\lambda_0)(1+H)^{-2}.$$

The assumptions imply that

$$\big((H-\lambda_0)^2(1+H)^{-2}u_j, u_j\big) \to 0. \tag{3.9}$$

Let $\{P_\lambda\}$ be the Projection Valued Measure of H. Then

$$\big((H-\lambda_0)^2(1+H)^{-2}u_j, u_j\big) = \int_0^\infty \frac{(\lambda-\lambda_0)^2}{(1+\lambda)^2} d(P_\lambda u_j, u_j). \tag{3.10}$$

Define $u_j^{(1)} = P_{(\lambda_0-\varepsilon_j,\lambda_0+\varepsilon_j)}u_j$ and $u_j^{(2)} = u_j - u_j^{(1)}$. The constants $\varepsilon_j \in (0, \frac{\lambda_0}{2})$ are to be selected later. Note the integrand $\frac{(\lambda-\lambda_0)^2}{(1+\lambda)^2}$ in (3.10) has the following lower bound for $\lambda \notin (\lambda_0 - \varepsilon_j, \lambda_0 + \varepsilon_j)$.

$$\frac{(\lambda - \lambda_0)^2}{(1+\lambda)^2} \geq \min\left(\frac{\varepsilon_j^2}{(1+\lambda_0-\varepsilon_j)^2}, \frac{\varepsilon_j^2}{(1+\lambda_0+\varepsilon_j)^2}\right) \geq \frac{\varepsilon_j^2}{(1+\frac{3}{2}\lambda_0)^2}. \tag{3.11}$$

Therefore

$$\left((H-\lambda_0)^2(1+H)^{-2}u_j, u_j\right) \geq \frac{\varepsilon_j^2}{(1+\frac{3}{2}\lambda_0)^2}\|u_j^{(2)}\|^2. \tag{3.12}$$

Choose a sequence $\varepsilon_j \in (0, \frac{\lambda_0}{2})$ such that

i) $\varepsilon_j \to 0$.
ii) $\left((H-\lambda_0)^2(1+H)^{-2}u_j, u_j\right)/\varepsilon_j^2 \to 0$.

For example, we can take $\varepsilon_j = \left((H-\lambda_0)^2(1+H)^{-2}u_j, u_j\right)^{\frac{1}{3}}$. Therefore (3.12) implies

$$\|u_j^{(2)}\| \to 0 \text{ as } j \to \infty,$$

whence

$$\|u_j^{(1)}\| \to 1 \text{ as } j \to \infty. \tag{3.13}$$

On the other hand, as

$$\int_0^\infty \lambda^2 d(P_\lambda u_j^{(1)}, u_j^{(1)}) \leq (\lambda_0 + \varepsilon_j)^2\|u_j\|^2 < \infty,$$

the sequence $u_j^{(1)} \in Dom(H)$. So we can apply the classical Weyl Criterion to the sequence $u_j^{(1)}$. By Projection Valued Measure again,

$$\|(H-\lambda_0)u_j^{(1)}\|^2 = \int_0^\infty (\lambda-\lambda_0)^2 d(P_\lambda u_j^{(1)}, u_j^{(1)}) \leq \varepsilon_j^2 \to 0, \tag{3.14}$$

which implies $\lambda_0 \in \mathrm{Spec}(H)$. So the result follows. □

REMARK 3.4. Note that the condition (2) in the theorem can be weaken to $\left((H-\lambda_0)^2(1+H)^{-2}u_j, u_j\right) \to 0$ by the proof.

REMARK 3.5. The above theorem also holds for $\lambda_0 = 0$. And in fact we can also prove conditions (1) and (2) are not only sufficient but also necessary for $\lambda_0 \in \mathrm{Spec}(H)$. More details can be found in [**2**].

With the generalized Weyl criterion 3.2, we are ready to prove Lemma 3.3.

PROOF. Here we prove $\mathrm{Spec}(\bar\partial\bar\partial^*_{p,q+1}) \subset \mathrm{Spec}(\Box_{p,q}) \cup \{0\}$. The other containment $\mathrm{Spec}(\bar\partial^*\bar\partial_{p,q-1}) \subset \mathrm{Spec}(\Box_{p,q}) \cup \{0\}$ can be proved similarly.

Take $\lambda_0 \in \mathrm{Spec}(\bar\partial\bar\partial^*)$ and $\lambda_0 > 0$. By classical Weyl criterion, there exists a sequence $u_j \in Dom(\bar\partial\bar\partial^*)$ with $(u_j, u_j) = 1$ such that

$$((\bar\partial\bar\partial^* - \lambda_0)u_j, (\bar\partial\bar\partial^* - \lambda_0)u_j) \to 0. \tag{3.15}$$

We will verify that the sequence $\bar{\partial}^* u_j$ satisfies conditions in Theorem 3.2. For $m = 1, 2$,

$$\begin{aligned}
&\left((\square - \lambda_0)(1+\square)^{-m}\bar{\partial}^* u_j, \bar{\partial}^* u_j\right) \\
&= \left((\square - \lambda_0)(1+\square)^{-m} u_j, \overline{\partial}\bar{\partial}^* u_j\right) \\
&= \left((\overline{\partial}\bar{\partial}^* - \lambda_0)(1+\square)^{-m} u_j, \overline{\partial}\bar{\partial}^* u_j\right) \\
&= \left(\overline{\partial}\bar{\partial}^*(1+\square)^{-m} u_j, (\overline{\partial}\bar{\partial}^* - \lambda_0) u_j\right).
\end{aligned}$$

The first equality is because $(1+\square)^{-1}\bar{\partial}^* = \bar{\partial}^*(1+\square)^{-1}$ on $Dom(\bar{\partial}^*)$, which follows from Theorem 2.2. The second one follows from $\overline{\partial} \circ \overline{\partial} = 0$ on $Dom(\overline{\partial})$. The third one comes from the self-adjointness of $\overline{\partial}\bar{\partial}^*$ and straightforward calculations. Since

$$\|\overline{\partial}\bar{\partial}^*(1+\square)^{-m} u_j\| \leq \|\square(1+\square)^{-m} u_j\| \leq \|u_j\| = 1, \tag{3.16}$$

(3.15) implies

$$\left((\square - \lambda_0)(1+\square)^{-m}\bar{\partial}^* u_j, \bar{\partial}^* u_j\right) \to 0 \text{ for } m = 1, 2. \tag{3.17}$$

The other thing we need to verify is that $\|\bar{\partial}^* u_j\|$ has a positive lower bound uniformly for all j. This is from the following calculations:

$$(\bar{\partial}^* u_j, \bar{\partial}^* u_j) = ((\overline{\partial}\bar{\partial}^* - \lambda_0) u_j, u_j) + \lambda_0 \to \lambda_0 > 0. \tag{3.18}$$

Since $\|\bar{\partial}^* u_j\|$ has a uniform lower bound, we can apply Theorem 3.2 to the scaled sequence $\bar{\partial}^* u_j / \|\bar{\partial}^* u_j\|$ and the result follows immediately. $\square$

Now we are going to finish the proof of Theorem 3.1 in next lemma.

LEMMA 3.6. *Under the same assumption as Theorem* 3.1, *we have*

$$Spec(\square_{p,q}) \subset Spec(\overline{\partial}\bar{\partial}^*_{p,q}) \cup Spec(\bar{\partial}^*\overline{\partial}_{p,q}) \cup \{0\}.$$

PROOF. Take $\lambda_0 \in \mathrm{Spec}(\square)$ and $\lambda_0 > 0$. Then by classical Weyl criterion, there exists a sequence $u_j \in Dom(\square)$ with $\|u_j\| = 1$ such that

$$(\square - \lambda_0) u_j \to 0. \tag{3.19}$$

We will use $\overline{\partial}\bar{\partial}^*(1+\square)^{-2} u_j$ and $\bar{\partial}^*\overline{\partial}(1+\square)^{-2} u_j$ as the test sequences. By the fact that $\overline{\partial} \circ \overline{\partial} = 0$ on $Dom(\overline{\partial})$ and $(1+\square)^{-1}\square = \square(1+\square)^{-1}$ on $Dom(\square)$, we have

$$(\overline{\partial}\bar{\partial}^* - \lambda_0)\overline{\partial}\bar{\partial}^*(1+\square)^{-2} u_j = \overline{\partial}\bar{\partial}^*(1+\square)^{-2}(\square - \lambda_0) u_j. \tag{3.20}$$

Since $\|\overline{\partial}\bar{\partial}^*(1+\square)^{-2}\|_{L^2 \to L^2} \leq 1$, it implies

$$\|(\overline{\partial}\bar{\partial}^* - \lambda_0)\overline{\partial}\bar{\partial}^*(1+\square)^{-2} u_j\| \leq \|(\square - \lambda_0) u_j\| \to 0. \tag{3.21}$$

Similarly, we also have

$$\|(\bar{\partial}^*\overline{\partial} - \lambda_0)\bar{\partial}^*\overline{\partial}(1+\square)^{-2} u_j\| \leq \|(\square - \lambda_0) u_j\| \to 0. \tag{3.22}$$

Now we need to check either $\|\overline{\partial}\bar{\partial}^*(1+\square)^{-2} u_j\|$ or $\|\bar{\partial}^*\overline{\partial}(1+\square)^{-2} u_j\|$ has a positive lower bound. Note

$$\|\overline{\partial}\bar{\partial}^*(1+\square)^{-2} u_j\|^2 + \|\bar{\partial}^*\overline{\partial}(1+\square)^{-2} u_j\|^2 = \|\square(1+\square)^{-2} u_j\|^2. \tag{3.23}$$

Let $\{P_\lambda\}$ be the Projection Valued Measure of $\square$. Then

$$\|\square(1+\square)^{-2} u_j\|^2 = \int_0^\infty \frac{\lambda^2}{(1+\lambda)^4} d(P_\lambda u_j, u_j) \geq C(\lambda_0)\|P_{(\frac{1}{2}\lambda_0, \frac{3}{2}\lambda_0)} u_j\|^2. \tag{3.24}$$

Note $(\Box - \lambda_0)u_j \to 0$ implies

$$\|P_{(\frac{1}{2}\lambda_0, \frac{3}{2}\lambda_0)}u_j\| \to 1. \tag{3.25}$$

Therefore for sufficiently large j,

$$\|\bar{\partial}\bar{\partial}^*(1+\Box)^{-2}u_j\|^2 + \|\bar{\partial}^*\bar{\partial}(1+\Box)^{-2}u_j\|^2 \geq \frac{C(\lambda_0)}{2} > 0. \tag{3.26}$$

So $\lambda_0 \in \mathrm{Spec}(\bar{\partial}\bar{\partial}^*_{p,q}) \cup \mathrm{Spec}(\bar{\partial}^*\bar{\partial}_{p,q})$ by classical Weyl criterion and the result follows. $\Box$

One direct corollary from Theorem 3.1 is the following spectrum relations of Gaffney extensions.

COROLLARY 3.7. *Under the same assumption as Theorem* 3.1, *we have*

$$Spec(\Box_{p,q}) \subset Spec(\Box_{p,q+1}) \cup Spec(\Box_{p,q-1}) \cup \{0\}. \tag{3.27}$$

At the end of this section, let us recall the well known relation between the spectrum of Gaffney extension and L^2 estimates.

THEOREM 3.3. *Let (M, g) be a Hermitian manifold with a holomorphic Hermitian vector bundle (E, h). Assume the Gaffney extension of Hodge Laplacian $\Box_{p,q+1} : L^2(M, \Lambda^{p,q+1}(E)) \to L^2(M, \Lambda^{p,q+1}(E))$ satisfies $Spec(\Box_{p,q+1}) \subset [a, \infty)$ for some positive number a. Then for any $f \in \ker \bar{\partial}_{p,q+1} \subset L^2(M, \Lambda^{p,q+1}(E))$, there exists $u \in L^2(M, \Lambda^{p,q}(E))$ such that $\bar{\partial}u = f$ with the following estimate*

$$(u, u) \leq \frac{1}{a}(f, f). \tag{3.28}$$

PROOF. In the proof, we will use $\Box$ to represent $\Box_{p,q+1}$ for simplicity. By the condition $\mathrm{Spec}\Box \subset [a, \infty)$, we have $\Box^{-1} : L^2(M, \Lambda^{p,q+1}(E)) \to Dom(\Box) \subset L^2(M, \Lambda^{p,q+1}(E))$ is a bounded operator with

$$\|\Box^{-1}\|_{L^2 \to L^2} \leq \frac{1}{a}. \tag{3.29}$$

Take $u = \bar{\partial}^*\Box^{-1}f$ and we will verify u satisfies all the conclusions. First, since the Gaffney extension satisfies $\Box = \bar{\partial}\bar{\partial}^* + \bar{\partial}^*\bar{\partial}$ by Theorem 2.2, we have

$$\bar{\partial}u = \bar{\partial}\bar{\partial}^*\Box^{-1}f = f - \bar{\partial}^*\bar{\partial}\Box^{-1}f. \tag{3.30}$$

Therefore $f \in \ker \bar{\partial}$ implies $\bar{\partial}^*\bar{\partial}\Box^{-1}f \in \ker \bar{\partial}$. By taking the following inner product

$$0 = \left(\bar{\partial}\bar{\partial}^*\bar{\partial}\Box^{-1}f, \bar{\partial}\Box^{-1}f\right) = \left(\bar{\partial}^*\bar{\partial}\Box^{-1}f, \bar{\partial}^*\bar{\partial}\Box^{-1}f\right), \tag{3.31}$$

we have

$$\bar{\partial}^*\bar{\partial}\Box^{-1}f = 0. \tag{3.32}$$

Again by taking the following inner product with $\Box^{-1}f$

$$0 = \left(\bar{\partial}^*\bar{\partial}\Box^{-1}f, \Box^{-1}f\right) = \left(\bar{\partial}\Box^{-1}f, \bar{\partial}\Box^{-1}f\right), \tag{3.33}$$

we have

$$\bar{\partial}\Box^{-1}f = 0. \tag{3.34}$$

Together with (3.30), we have

$$\bar{\partial}u = f.$$

Second, we will verify the estimate (3.28). By (3.34) and straightforward calculations, we have

$$(u,u) = (\bar{\partial}^*\Box^{-1}f, \bar{\partial}^*\Box^{-1}f) = (\bar{\partial}\bar{\partial}^*\Box^{-1}f, \Box^{-1}f) = (f, \Box^{-1}f).$$

Therefore (3.29) implies the result. □

REMARK 3.8. Note we cannot directly use $\overline{\partial}_{p,q+2}\Box^{-1}_{p,q+1}f = \Box^{-1}_{p,q+2}\overline{\partial}_{p.q+1}f = 0$ in the proof as we do not know the existence of $\Box^{-1}_{p,q+2}$.

4. Spectrums of Friedrichs Extension

In this section, we assume (M,ω) is a Kähler manifold with a holomorphic Hermitian line bundle (L,h) . Since $Dom(\Box_F) \subset W^1_0$, we will not get any boundary term when doing integration by parts for sections in $Dom(\Box_F)$. By using the Weitzenböck formula, we will prove the following spectrum lower bound for Friedrichs extension under certain curvature conditions.

THEOREM 4.1. *Let (M,ω) be a Kähler manifold with a holomorphic Hermitian line bundle (L,h). Consider Friedrichs extension of Hodge Laplacian, $\Box_{0,q}$: $L^2(M,\Lambda^{0,q}(L)) \to L^2(M,\Lambda^{0,q}(L))$. If $\mathrm{Ric}(TM) + \mathrm{Ric}(L) \geq a\omega$ for some positive number a, then*

$$Spec\Box_{0,q} \subset [aq,\infty). \tag{4.1}$$

REMARK 4.1. In this section, $\Box_{0,q}$ always represents the Friedrichs extension and we will omit the subindex $\{0,q\}$ when there is no ambiguity.

PROOF. Take $\varphi \in Dom(\Box_{0,q})$. As $Dom(\Box_{0,q}) \subset W^1_0(M,\Lambda^{0,q}(L))$, there exists a sequence $\varphi_n \in \mathcal{D}(M,\Lambda^{0,q}(L))$ such that $\varphi_n \to \varphi$ in W^1_0. By the Weitzenböck formula $\Box_{0,q} = -\overline{\nabla}\nabla + q\,\mathrm{Ric}(TM) + q\,\mathrm{Ric}(L)$, we have

$$Q(\varphi_n,\varphi_n) = (\nabla\varphi_n,\nabla\varphi_n) + (q(\mathrm{Ric}(TM)+\mathrm{Ric}(L))\varphi_n,\varphi_n) \geq aq(\varphi_n,\varphi_n). \tag{4.2}$$

Letting $n\to\infty$, we have

$$Q(\varphi,\varphi) \geq aq(\varphi,\varphi). \tag{4.3}$$

As $Q(\varphi,\varphi) = (\Box_{0,q}\varphi,\varphi)$, the result follows. □

REMARK 4.2. Let $n = \dim M$. As the Weitzenböck formula for L-valued (n,q) form is $\Box_{n,q} = -\overline{\nabla}\nabla + q\,\mathrm{Ric}(L)$. If $\mathrm{Ric}(L) \geq a\omega$ for some positive constant a, then the Friedrichs extension $\Box_{n,q}$ satisfies $\mathrm{Spec}\Box_{n,q} \subset [aq,\infty)$.

5. Manifolds with Almost Polar Boundary

Let (M,g) be a Riemannian manifold. Similar as the Definition 2.2, we can define the Sobolev space for functions by taking the quadratic form $Q_1(\cdot,\cdot) = (\cdot,\cdot) + (d\cdot,d\cdot)$.

DEFINITION 5.1.

$$W^1_0(M) = \text{Completion of } \mathcal{D}(M) \text{ with respect to } Q_1 \text{ inner product}, \tag{5.1}$$

$$W^1(M) = \text{Completion of } \{\varphi\in\mathcal{C}^\infty(M) : Q_1(\varphi,\varphi)<\infty\} \text{ with respect to } Q_1 \text{ inner product.} \tag{5.2}$$

Generally we know $W^1(M) = W^1_0(M)$ for complete Riemannian manifolds. In [**10, 11**], Masamune proved $W^1(M) = W^1_0(M)$ for Riemannian manifolds with almost polar boundary. We will repeat the proof here for the sake of completeness and because there is a gap in Masamune's proof.

We first introduce the definition and notations. Let d be the distance function induced by the length of piecewise curves on M. Then (M, d) is a metric space. We use $(\overline{M}_c, d)$ to denote the Cauchy completion of (M, d). We define the Cauchy boundary $\partial_c M = \overline{M}_c - M$.

DEFINITION 5.2. We define the capacity of an open set $O \subset \overline{M}_c$ by

$$\operatorname{cap}(O) = \inf\{Q_1(u,u) : u \in W^1(M), 0 \le u \le 1 \text{ and } u|_{O\cap M} = 1\}. \tag{5.3}$$

We also define the capacity of an arbitrary set $\Sigma \subset \overline{M}_c$ by

$$\operatorname{cap}(\Sigma) = \inf\{\operatorname{cap}(O), \Sigma \subset O, O \subset \overline{M}_c \text{ is open}\}. \tag{5.4}$$

A set Σ is said to be almost polar if $\operatorname{cap}(\Sigma) = 0$.

REMARK 5.3. For any open set $O \subset \overline{M}_c$, $e \in W^1(M)$ is called the equilibrium potential of O if it satisfies

1. $Q_1(e, e) = \operatorname{cap}(O)$.
2. $e|_O = 1$.
3. $0 \le e \le 1$.

It is know that the equilibrium potential exists for any open set $O \subset \overline{M}_c$. See [**3**] for more details.

Here is the main theorem we are going to prove.

THEOREM 5.1. *Let (M, g) be a Riemannian manifold. If $cap(\partial_c M) = 0$, then*

$$W^1(M) = W^1_0(M). \tag{5.5}$$

Before going to the proof, let's explain the main idea. First we show that $L^\infty(M)\cap W^1(M) \subset W^1(M)$ is dense. Then it is sufficient to consider $f \in L^\infty(M)\cap W^1(M)$. Choosing a sequence of open sets $\{V_n\}$ decreasing to $\partial_c M$, by using the equilibrium potential of V_n, say e_n, we can approximate f by $(1 - e_n)f$ whose support is contained in $M - V_n$. In the last, we want to modify the function $(1 - e_n)f$ to be compactly supported. As $(\overline{M}_c, d)$ is only a complete metric space, the closed metric ball excluding an open set containing $\partial_c M$ might not be a compact set even if $\operatorname{cap}(\partial_c M) = 0$(See Section 8 for more details). So we will refer to the intrinsic distance and verify that the intrinsic distance induces the same topology as d on $M - V_n$. As the closed metric ball with respect to the intrinsic distance is compact by Hopf-Rinow-Cohn-Vossen Theorem (see Theorem 2.5.28 in [**1**]). And we will use some cut-off function to finish the modification on support.

We begin the proof with the following lemma described above.

LEMMA 5.4. *For any Riemannian manifold (M, g), $L^\infty(M)\cap W^1(M)$ is dense in $W^1(M)$.*

PROOF. Take $f \in W^1(M)$. Define a cut-off function $\rho \in \mathcal{C}^\infty(\mathbb{R})$ such that

$$\rho(x) = \begin{cases} 1 & x \le 1 \\ 0 & x \ge 2 \end{cases},$$

and

$$0 \leq \rho \leq 1, \quad -C \leq \rho' \leq 0.$$

We define $\rho_m(x) = \rho(\frac{x}{m})$ and $f_m = \rho_m(|f|)f$. Note $f_m \in L^\infty(M) \cap W^1(M)$ and we will prove $f_m \to f$ in $W^1(M)$. By dominated convergence theorem, we directly get $f_m \to f$ in $L^2(M)$.

As to df_m, we have

$$df_m - df = \big(\rho(\frac{|f|}{m}) - 1\big)df + \frac{1}{m}\rho'(\frac{|f|}{m})f \cdot d|f|. \tag{5.6}$$

The first term on the right hand side converges to 0 in $L^2(M, \Lambda^1)$ as $|\rho(\frac{|f|}{m}) - 1| \leq \chi_{\{|f| \geq m\}}$. For the second term, since

$$|\frac{1}{m}\rho'(\frac{|f|}{m})f \cdot d|f|| \leq 2C\chi_{\{m \leq |f| \leq 2m\}}|df|, \tag{5.7}$$

it follows that $\frac{1}{m}\rho'(\frac{|f|}{m})f \cdot d|f| \to 0$ in $L^2(M, \Lambda^1)$. So we have $f_m \to f$ in $W^1(M)$ and the result follows. □

In next two lemmas, we will construct open sets containing $\partial_c M$ with smooth boundary.

LEMMA 5.5. *$\partial_c M \subset \overline{M}_c$ is a closed subset.*

PROOF. Since M is the complement of $\partial_c M$ in $\overline{M}_c$, it is equivalent to check that $M \subset \overline{M}_c$ is an open subset. For any $x \in M$, let i_x be the injectivity radius of (M, g) at x. Then for any $r \in (0, i_x)$, by considering the exponential map at x, we know $\overline{B_M(x,r)} = \{y \in M, d(x,y) \leq r\}$ is compact, whence complete. Therefore $B_M(x,r) = B_{\overline{M}_c}(x,r) = \{y \in \overline{M}_c, d(x,y) < r\}$ since we will not add any new point to $B_M(x,r)$ during the Cauchy completion of M. So $B_{\overline{M}_c}(x,r) \subset M$ and the result follows. □

LEMMA 5.6. *For any open set $U \subset \overline{M}_c$ containing $\partial_c M$, there exists an open set $V \subset \overline{M}_c$ such that $\partial_c M \subset V \subset \overline{V} \subset U$ and $\partial(\overline{M}_c \setminus V) \subset M$ is a smooth submanifold of codimension 1.*

PROOF. Let U^C be the complement of U in $\overline{M}_c$. Since $\partial_c M$ and U^C are both closed in $(\overline{M}_c, d)$. By Urysohn's Lemma, there exists a function $f \in \mathcal{C}(\overline{M}_c)$ such that $0 \leq f \leq 1$, $f^{-1}(\{0\}) = \partial_c M$ and $f^{-1}(\{1\}) = U^C$. Take $S = f^{-1}([0, \frac{1}{2}))$. Then S is an open subset of $\overline{M}_c$ such that $\partial_c M \subset S \subset \overline{S} \subset U$.

Note that $\overline{S} \setminus \partial_c M = \overline{S} \cap M$ and U^C are both closed in M. By the Smooth Urysohn's Lemma in [**12**], there exists a function $g \in \mathcal{C}^\infty(M)$ such that $0 \leq g \leq 1$, $g^{-1}(\{0\}) = \overline{S} \setminus \partial_c M$ and $g^{-1}(\{1\}) = U^C$. By Sard's Theorem, without loss of generality, we can assume $\frac{1}{2}$ is a regular value of g. Take $V = g^{-1}([0, \frac{1}{2})) \cup \partial_c M \subset \overline{M}_c$. Then it's easy to see $V = g^{-1}([0, \frac{1}{2})) \cup S$. Therefore V is open in $\overline{M}_c$ such that $\partial_c M \subset V \subset \overline{V} = \overline{g^{-1}([0, \frac{1}{2}))} \cup \overline{S} \subset U$. The remaining part of the lemma follows from $\partial(\overline{M}_c \setminus V) = g^{-1}(\{\frac{1}{2}\})$ and $\frac{1}{2}$ is a regular value of g. □

Let V be an open subset satisfying the conclusion in the above lemma. Denote $V^C = \overline{M}_c \setminus V$ as the complement of V in $\overline{M}_c$. Then $V^C = \cup_{\lambda \in \Lambda} A_\lambda$, where each A_λ is a connected component of V^C and Λ is the index set. Since V^C is locally path connected, each A_λ is both open and closed in V^C. Define the intrinsic distance function d_{A_λ} on A_λ as

DEFINITION 5.7. Define the intrinsic distance on A_λ as $d_{A_\lambda} : A_\lambda \times A_\lambda \to [0, \infty)$,

$$d_{A_\lambda}(x, y) = \inf_{l \in L_{A_\lambda}} \|l\| \tag{5.8}$$

where $L_{A_\lambda} = \{$all piecewise smooth curves contained in A_λ from x to $y\}$ and $\|l\|$ denotes the length of curve l.

REMARK 5.8. $d(x, y) \leq d_{A_\lambda}(x, y)$ for any $x, y \in A_\lambda$ as d is the infimum over a larger set.

In general, d and d_{A_λ} are not globally equivalent to each other on A_λ. The next lemma shows that they are locally equivalent on A_λ.

LEMMA 5.9. *For any $x \in A_\lambda$, there exists $r = r(x) > 0$ such that*

$$d_{A_\lambda}(x, y) \leq 4d(x, y) \quad \text{for any } y \in B_{A_\lambda}(x, r). \tag{5.9}$$

where $B_{A_\lambda}(x, r) = \{y \in A_\lambda, d(x, y) < r\}$.

PROOF. For any $x \in A_\lambda \subset V^C \subset M$, either x is in the interior of V^C or $x \in \partial V^C$. In the first case, take $r < i_x$ (i_x denotes the injectivity radius at x) small enough such that $B_M(x, r) \subset A_\lambda$. Then for any $y \in B_M(x, r)$, there existed a minimizing geodesic $l \subset B_M(x, r)$ such that $\|l\| = d(x, y)$. Therefore $d_{A_\lambda}(x, y) = d(x, y)$ for any $y \in B_{A_\lambda}(x, r) = B_M(x, r)$.

In the second case, i.e. $x \in \partial V^C$, take $r < i_x$. We can identify $B_{\mathbb{R}^m}(o, r)$ (w.r.t the Euclidean metric g_x) with $B_M(x, r)$ by the exponential map Exp_x at x. By shrinking r, we can assume the Riemannian metric on $B_M(x, r)$ is equivalent to the metric at x, say $\frac{1}{2}g_x \leq g \leq 2g_x$. Let $\{e_i\}_{i=1}^m$ be the standard orthonormal basis of $\mathbb{R}^m$. Up to an orthonormal linear transformation, we can assume $\{e_i\}_{i=1}^{m-1} \subset T_x(\partial V^C)$ and e_m is the normal direction of ∂V^C at x. By Lemma 5.6, possibly shrinking r again, we can assume $\partial V^C = \{(x_1, x_2, \cdots, x_m) \in B(o, r), x_m = h(x_1, x_2 \cdots, x_{m-1})\}$ where $h \in \mathcal{C}^\infty(\mathbb{R}^{m-1})$ and $h(0, \cdots, 0) = 0$. Since $\{e_i\}_{i=1}^{m-1}$ are tangent vectors of ∂V^C at x, $\nabla h(0, \cdots, 0) = 0$. By shrinking r again, we can assume $|\nabla h| \leq 1$ in $B_{\mathbb{R}^{m-1}}(o, r)$.

For any point $y \in B_{\mathbb{R}^m}(o, r)$, consider the curve $l_1 = (ty_1, ty_2, \cdots, ty_{m-1}, h(ty_1, \cdots, ty_{m-1}))$ for $t \in [0, 1]$ and $l_2 = (y_1, y_2, \cdots, y_{m-1}, ty_m + (1-t)h(y_1, y_2, \cdots, y_{m-1}))$ for $t \in [0, 1]$. Then the concatenation $l_1 \cup l_2 \subset V^C$ is from x to y. The Euclidean length of l_1, l_2 are respectively

$$\begin{aligned}
\|l_1\|_{\mathbb{R}^m} =& \int_0^1 \sqrt{y_1^2 + y_2^2 + \cdots + y_{m-1}^2 + |\nabla h(ty_1, ty_2, \cdots, ty_{m-1}) \cdot (y_1, y_2, \cdots, y_{m-1})|^2} dt \\
\leq& 2\sqrt{y_1^2 + y_2^2 + \cdots + y_{m-1}^2}, \\
\|l_2\|_{\mathbb{R}^m} =& |y_m - h(y_1, y_2, \cdots, y_{m-1})| \\
\leq& |y_m| + \sqrt{y_1^2 + y_2^2 + \cdots + y_{m-1}^2}.
\end{aligned}$$

Therefore

$$\begin{aligned}
d_{A_\lambda}(x, y) \leq& \|l_1\| + \|l_2\| \\
\leq& 2\|l_1\|_{\mathbb{R}^m} + 2\|l_2\|_{\mathbb{R}^m} \\
\leq& 4\sqrt{y_1^2 + y_2^2 + \cdots + y_{m-1}^2 + y_m^2} \\
=& 4d(x, y).
\end{aligned}$$

The second inequality is because $\frac{1}{2}g_x \leq g \leq 2g_x$. So the result follows. □

Base on Remark 5.8 and Lemma 5.9, we have the following properties on $(A_\lambda, d_{A_\lambda})$.

PROPOSITION 5.10. *$(A_\lambda, d_{A_\lambda})$ satisfies the following property.*

(a). *$(A_\lambda, d_{A_\lambda})$ and (A_λ, d) have the same topology.*
(b). *$(A_\lambda, d_{A_\lambda})$ is locally compact.*
(c). *$(A_\lambda, d_{A_\lambda})$ is complete.*

PROOF. Part (a) directly follows from Remark 5.8 and Lemma 5.9.

Now we prove part (b). Since V^C is a closed subset of (M, d) and (M, d) is locally compact, (V^C, d) is locally compact. And we know A_λ is a closed subset of (V^C, d), therefore (A_λ, d) is locally compact. The result follows by part (a).

Last we prove part (c). Let $\{x_n\}_{n=1}^\infty$ be a Cauchy sequence in $(A_\lambda, d_{A_\lambda})$. By Remark 5.8, $\{x_n\}_{n=1}^\infty$ is also a Cauchy sequence in (A_λ, d). Since A_λ is closed in (V^C, d) and V^C is closed in the complete space $(\overline{M}_c, d)$, (A_λ, d) is complete. Then there exists some $x \in A_\lambda$ such that $\lim d(x, x_n) = 0$. By Lemma 5.9, $\lim d_{A_\lambda}(x_n, x) = 0$ and therefore the result follows. □

For any $x_0 \in A_\lambda$, define the function $r_{x_0} : A_\lambda \to [0, \infty)$ as $r_{x_0}(x) = d_{A_\lambda}(x_0, x)$. Then r_{x_0} has the following property.

PROPOSITION 5.11. *For the function r_{x_0} defined as above, we have*

$$|\nabla r|_g \leq 4. \tag{5.10}$$

PROOF. Since $|r(x) - r(y)| \leq d_{A_\lambda}(x, y)$, the result follows from Lemma 5.9. □

The closed metric ball induced by d_{A_λ} is compact though it is not the case for the closed metric ball induced by d. The following lemma is essentially Hopf-Rinow-Cohn-Vossen Theorem. See Theorem 2.5.28 in [**1**] for more details.

LEMMA 5.12. *For any $x \in A_\lambda, r > 0$, $\overline{B_{(A_\lambda, d_{A_\lambda})}(x, r)}$ is compact. Here $B_{(A_\lambda, d_{A_\lambda})}(x, r)$ denotes the set $\{y \in A_\lambda, d_{A_\lambda}(x, y) < r\}$.*

REMARK 5.13. By part (a) in Proposition 5.10, the closures of $B_{(A_\lambda, d_{A_\lambda})}(x, r)$ in (A_λ, d) and in (A_λ, d_λ) are the same. The compactness in (A_λ, d) and that in (A_λ, d_λ) are also the same. So there is no ambiguity in the above lemma.

PROOF. By part (b) in Proposition 5.10, the set

$$\{r > 0, \overline{B_{(A_\lambda, d_{A_\lambda})}(x, r)} \text{ is compact}\}$$

is nonempty. So we can define $r_0 = \sup\{r > 0, \overline{B_{(A_\lambda, d_{A_\lambda})}(x, r)} \text{ is compact}\}$. Now it suffices to prove $r_0 = \infty$. Assume not. Then $r_0 \in (0, \infty)$.

First, we prove that $\overline{B_{(A_\lambda, d_{A_\lambda})}(x, r_0)}$ is compact. Take an arbitrary $\varepsilon > 0$. For any $y \in \overline{B_{(A_\lambda, d_{A_\lambda})}(x, r_0)}$, since $d_{A_\lambda}(x, y) \leq r_0$, there exists a piecewise smooth curve $l \subset A_\lambda$ from x to y such that $\|l\| < r_0 + \varepsilon$. Reparametrize the curve l by arc length. Then the restriction $l|_{[r_0 - \varepsilon, \|l\|]}$ is a piecewise smooth curve from a point in $\overline{B_{(A_\lambda, d_{A_\lambda})}(x, r_0 - \varepsilon)}$ to y. Since

$$\|l|_{[r_0 - \varepsilon, \|l\|]}\| < 2\varepsilon, y \in B_{(A_\lambda, d_{A_\lambda})}(\overline{B_{(A_\lambda, d_{A_\lambda})}(x, r_0 - \varepsilon)}, 2\varepsilon).$$

Therefore

$$\overline{B_{(A_\lambda,d_{A_\lambda})}(x,r_0)} \subset B_{(A_\lambda,d_{A_\lambda})}(\overline{B_{(A_\lambda,d_{A_\lambda})}(x,r_0-\varepsilon)},2\varepsilon).$$

Since $\overline{B_{(A_\lambda,d_{A_\lambda})}(x,r_0-\varepsilon)}$ is compact by the definition of r_0, $\overline{B_{(A_\lambda,d_{A_\lambda})}(x,r_0)}$ is totally bounded in $(A_\lambda, d_{A_\lambda})$. Therefore $\overline{B_{(A_\lambda,d_{A_\lambda})}(x,r_0)}$ is compact by part (c) in Proposition 5.10.

Second, we prove that $\overline{B_{(A_\lambda,d_{A_\lambda})}(x,r_0+\delta)}$ is compact for some $\delta > 0$, which contradicts the definition of r_0 and therefore we get the result. Since $\overline{B_{(A_\lambda,d_{A_\lambda})}(x,r_0)}$ is also compact, together with part (b) in Proposition 5.10, $\overline{B_{(A_\lambda,d_{A_\lambda})}(x,r_0)}$ has a finite cover $\{B_{(A_\lambda,d_{A_\lambda})}(y_i,\delta_i)\}_{i=1}^N$, such that $y_i \in \overline{B_{(A_\lambda,d_{A_\lambda})}(x,r_0)}$, $\delta_i > 0$ and $\overline{B_{(A_\lambda,d_{A_\lambda})}(y_i,2\delta_i)}$ is compact for each i. Take $\delta = \min_{1\le i\le N}\delta_i$. Then

$$\overline{B_{(A_\lambda,d_{A_\lambda})}(x,r_0+\delta)} \subset \cup_{i=1}^N \overline{B_{(A_\lambda,d_{A_\lambda})}(y_i,2\delta_i)}$$

is compact. □

Now we are ready to prove the Theorem 5.1.

PROOF. Since $\operatorname{cap}(\partial_c M) = 0$, there exists a sequence of open sets $\{U_n\}_{n=1}^\infty$ such that $\partial_c M \subset U_n$ and $\lim \operatorname{cap}(U_n) = 0$. For U_1, by Lemma 5.6, there exists an open set V_1 such that $\partial_c M \subset V_1 \subset \overline{V_1} \subset U_1$ and $\partial(V_1^C)$ is a smooth submanifold. Then for $V_1 \cap U_2$, by Lemma 5.6, there exists an open set V_2 such that $\partial_c M \subset V_2 \subset \overline{V_2} \subset V_1 \cap U_2$ and $\partial(V_2^C)$ is a smooth submanifold. Inductively, we construct V_{i+1} by applying Lemma 5.6 to $V_i \cap U_{i+1}$. So we get a sequence of decreasing open sets $\{V_n\}_{n=1}^\infty$ such that $\partial_c M \subset V_n \subset \overline{V_n} \subset V_{n-1} \cap U_n$ and $\partial(V_n^C)$ is a smooth submanifold. Since in particular $V_n \subset U_n$, we have $\lim \operatorname{cap}(V_n) = 0$.

Take $f \in W^1(M) \cap L^\infty$. It suffices to prove $f \in W_0^1(M)$.

First, we approximate f by functions with support in some V_n^C. Let e_n be the equilibrium potential (see Remark 5.3) of V_n, i.e. e_n satisfies

- $e_n \in W$ and $Q_1(e_n, e_n) = \operatorname{cap}(V_n)$.
- $e_n|_{V_n} = 1$.
- $0 \le e_n \le 1$.

Since $\|e_n\|_W = \operatorname{cap}(V_n) \to 0$, we can assume $e_n \to 0$ a.e. by passing to a subsequence. Let $f_n = (1 - e_{n-1})f$. Then $f_n \to f$ in $W^1(M)$ and $\operatorname{supp}(f_n) \subset V_{n-1}^C \subset \overline{V_n}^C \subset \operatorname{interior}(V_n^C)$.

Secondly, we approximate each f_n with $\operatorname{supp}(f_n) \subset \operatorname{interior}(V_n^C)$ by functions with compact support. From now on, we fix f_n and V_n^C. For economy we suppress the index n. Write V^C into the disjoint union of connected component, $V^C = \cup_{\lambda\in\Lambda} A_\lambda$. Since $f \in W^1(M)$ and $\{A_\lambda\}_{\lambda\in\Lambda}$ is pairly disjoint, f vanishes on all but countably many A_λ, say $\{A_{\lambda_j}\}_{j=1}^\infty$. Denote $g_j = f\chi_{A_{\lambda_j}}$ where $\chi_{A_{\lambda_j}}$ is the characteristic function of A_{λ_j}. Note $g_j \in W^1(M)$ and $\nabla g_j = (\nabla f)\chi_{A_{\lambda_j}}$ by the fact that $\partial A_{\lambda_j} \subset \partial(V^C)$ and f vanishes close to $\partial(V^C)$ as $\operatorname{supp} f \subset \operatorname{interior}(V^C)$. Then $f = \sum_{j=1}^\infty g_j$ and $\|f\|_W^2 = \sum_{j=1}^\infty \|g_j\|_W^2$. Therefore for any $\varepsilon > 0$, there exists $N > 0$ such that $\|f - \sum_{j=1}^N g_j\|_W < \varepsilon$.

Now it suffices to approximate each g_j be compact supported function. Take $x_j \in A_{\lambda_j}$ and define $r_j : A_{\lambda_j} \to [0,\infty)$ as $r_j(x) = d_{A_{\lambda_j}}(x_j, x)$. Then $|\nabla r_j|_g \le 4$ by Proposition 5.11. Let $\varphi \in \mathcal{C}^\infty(R)$ satisfy the following conditions:

- φ is a decreasing function and $0 \le \varphi \le 1$.

- $\varphi|_{(-\infty,0]} = 1$ and $\varphi_{[1,\infty)} = 0$.
- $|\varphi'| \le C$ and C is a fixed constant.

Define $\varphi_k(x) = \varphi(\frac{x}{k})$. Then $\varphi_k \circ r_j \to 1$ a.e. on A_{λ_j} as $k \to \infty$ and $|\nabla(\varphi_k \circ r_j)|_g \le \frac{4C}{k}$. Therefore we have $(\varphi_k \circ r_j)g_j \to g_j$ in $W^1(M)$. And $\mathrm{supp}((\varphi_k \circ r_j)g_j) \subset \mathrm{supp}(\varphi_k \circ r_j) \subset \overline{B_{(A_{\lambda_j}, d_{A_{\lambda_j}})}(x_j, 2k)}$, which is compact by Lemma 5.12. So the result follows. □

6. Moduli Space of Polarized Calabi-Yau Manifolds

Let (M, L) be a Calabi-Yau manifold polarized by a positive line bundle L. That is, M is a compact Kähler manifold with a Ricci flat Kähler metric ω and the metric ω is contained in the first Chern class of L. Let $\mathcal{M}$ be the moduli space of Calabi-Yau manifolds polarized by a fixed positive line bundle L. In [**16**], Viehweg proved the moduli space $\mathcal{M}$ is a quasi-projective variety. Take $\overline{\mathcal{M}}$ as the compactification of $\mathcal{M}$. With the classical result of Hironaka, by resolution of singularities, we can choose $\overline{\mathcal{M}}$ in such a way that the divisor $Y = \overline{\mathcal{M}} \setminus \mathcal{M}$ is a divisor of normal crossings. After passing to a finite cover, we may assume $\mathcal{M}$ and $\overline{\mathcal{M}}$ are smooth manifolds (see Lemma 4.1 in [**8**]). From now on, we will work on this quasi-projective Kähler manifold $(\mathcal{M}, \omega_{WP})$ with the compactification $\overline{\mathcal{M}}$ as a compact Kähler manifold.

Here is the main theorem we are going to prove in this section.

THEOREM 6.1. *The moduli space of polarized Calabi-Yau manifolds $(\mathcal{M}, \omega_{WP})$ has almost polar Cauchy boundary, i.e. $cap(\partial_c\mathcal{M}) = 0$.*

REMARK 6.1. In general, the Cauchy completion $\overline{\mathcal{M}_c}$ is not necessarily identical to the compactification $\overline{\mathcal{M}}$.

It is well-known that there is a complete Kähler metric on $\mathcal{M}$ such that it is asymptotical to the Poincaré metric near infinity. We call it Poincaré metric and denote it by ω_P(See Lemme 3.1 in [**8**]). The key ingredient to prove Theorem 6.1 is the following lemma in [**7**].

LEMMA 6.2. *For any $\varepsilon > 0$ small enough, there is a smooth real valued function $\rho_\varepsilon \in \mathcal{D}(\mathcal{M})$ such that*

(a). *$0 \le \rho_\epsilon \le 1$;*

(b). *There is a constant C, independent of ϵ, such that $-C\omega_P \le \sqrt{-1}\partial\overline{\partial}\rho_\epsilon \le C\omega_P$;*

(c). *In a neighborhood of Y, $\rho_\epsilon = 0$ and $\rho_\epsilon(x) = 1$ if the Euclidean distance of $x \in M$ to Y is greater than 2ε.*

PROOF. As $Y \subset \overline{\mathcal{M}}$ is a divisor of normal crossings, by [**8**] (see Lemma 4.1), we can find a finite cover $\{U_\alpha\}_{\alpha=1}^t$ of $\overline{\mathcal{M}}$ such that $Y \subset \cup_{\alpha=1}^s U_\alpha$ and $U_{s+1} \cup \cdots \cup U_t) \cap Y = \emptyset$. Furthermore, we can assume that $U_\alpha - Y = (\Delta^*)^{a_\alpha} \times (\Delta)^{b_\alpha}$ with the coordinates $(s_1^\alpha, \cdots, s_{a_\alpha}^\alpha, w_1^\alpha, \cdots, w_{b_\alpha}^\alpha)$ for any $1 \le \alpha \le s$, where Δ^* and Δ are respectively the punctured unit disk and the unit disk in $\mathbb{C}$. Let $\eta : \mathbb{R} \to \mathbb{R}$ be a smooth decreasing function such that $0 \le \eta \le 1$ and

$$\eta = \begin{cases} 1 & x \le 0 \\ 0 & x \ge 1 \end{cases}.$$

Let

$$\eta_\varepsilon(z) = \begin{cases} 1 & |z| \le e^{-\frac{1}{\varepsilon}} \\ \eta\left(\frac{(\log\frac{1}{|z|})^{-1}-\varepsilon}{\varepsilon}\right) & e^{-\frac{1}{\varepsilon}} \le |z| \le e^{-\frac{1}{2\varepsilon}} \\ 0 & |z| \ge e^{-\frac{1}{2\varepsilon}} \end{cases}.$$

And let

$$\eta_\varepsilon^\alpha(s_1^\alpha, \cdots, s_{a_\alpha}^\alpha) = \prod_{j=1}^{a_\alpha}(1-\eta_\varepsilon(s_j^\alpha)).$$

Then define the function

$$\rho_\varepsilon = \sum_{\alpha=1}^{s} \psi_\alpha \eta_\varepsilon^\alpha + \sum_{\alpha=s+1}^{t} \psi_\alpha,$$

where $\{\psi_\alpha\}$ is a partition of unity subordinated to $\{U_\alpha\}$.

Then $0 \le \rho_\varepsilon \le 1$. By a straightforward calculation, we have

$$\begin{aligned} \bar\partial \eta_\varepsilon &= \frac{1}{2\varepsilon}\eta' \frac{d\bar z}{\bar z(\log\frac{1}{|z|})^2}, \\ \partial\bar\partial \eta_\varepsilon &= \frac{1}{4\varepsilon^2}\eta'' \frac{dz\wedge d\bar z}{|z|^2(\log\frac{1}{|z|})^4} + \frac{1}{2\varepsilon}\eta' \frac{dz\wedge d\bar z}{|z|^2(\log\frac{1}{|z|})^3}. \end{aligned}$$

Note that $\eta' = 0$ unless $\varepsilon \le (\log\frac{1}{|z|})^{-1} \le 2\varepsilon$. Therefore

$$|\bar\partial\eta_\varepsilon| \le C\left|\frac{d\bar z}{|z|\log\frac{1}{|z|}}\right|, \quad |\partial\bar\partial\eta_\varepsilon| \le C\left|\frac{dz\wedge d\bar z}{|z|^2(\log\frac{1}{|z|})^2}\right|,$$

where C is a constant independent of ϵ. Therefore we obtain part (b) as ψ_α are fixed smooth functions on $\overline{\mathcal{M}}$.

Let $x \in \mathcal{M}$. When x is sufficiently close to Y, $\psi_\alpha = 0$ for any $\alpha \ge s+1$ and $\eta_\varepsilon^\alpha = 0$ for any $\alpha \le s$. Therefore $\rho_\varepsilon = 0$ in a neighborhood of Y. If the distance of x to Y is at least 2ε, then there is a constant $C > 0$ such that $|s_j^\alpha| \ge C\varepsilon$ for any $1 \le j \le a_\alpha$ and $1 \le \alpha \le s$. Since $\varepsilon e^{\frac{1}{2\varepsilon}} \to \infty$ as $\varepsilon \to 0$, when ε is small enough we have $\rho_\varepsilon(x) = \sum \psi_\alpha = 1$. □

Now we are ready to prove Theorem 6.1.

PROOF. Take the function ρ_ε constructed in Lemma 6.2. As $\rho_\varepsilon \in \mathcal{D}(\mathcal{M})$ and $0 \le \rho_\varepsilon \le 1$, we have

$$\text{(6.1)} \qquad \operatorname{cap}(\partial_c\mathcal{M}) \le \int_{\mathcal{M}} |1-\rho_\varepsilon|^2 \frac{\omega_{WP}^n}{n!} + \int_{\mathcal{M}} |d(1-\rho_\varepsilon)|^2 \frac{\omega_{WP}^n}{n!}, \text{ for any } \varepsilon > 0.$$

Since $\rho_\varepsilon \to 1$ pointwise on $\mathcal{M}$ and the volume of Weil-Petersson metric is finite by Theorem 1.1 in [**8**],

$$\text{(6.2)} \qquad \lim_{\varepsilon\to 0}\int_{\mathcal{M}} |1-\rho_\varepsilon|^2 \frac{\omega_{WP}^n}{n!} = 0.$$

It suffices to prove that $\int_{\mathcal{M}} |d\rho|^2 \to 0$. Note

$$\int_{\mathcal{M}} |d\rho_\varepsilon|^2 \omega_{WP}^n = 2\int_{\mathcal{M}} |\overline{\partial}\rho_\varepsilon|^2 \omega_{WP}^n = 2n\int_{\mathcal{M}} \sqrt{-1}\partial\rho_\varepsilon \wedge \overline{\partial}\rho_\varepsilon \wedge \omega_{WP}^{n-1}$$
$$= -2n\int_{\mathcal{M}} \sqrt{-1}\rho_\varepsilon \partial\overline{\partial}\rho_\varepsilon \wedge \omega_{WP}^{n-1}.$$

Since $-C\omega_P \le \sqrt{-1}\partial\overline{\partial}\rho_\varepsilon \le C\omega_P$ and $\omega_{WP} \le C\omega_P$(see Proposition 3.1 in [**8**]), we have

$$\int_{\mathcal{M}} |\overline{\partial}\rho_\varepsilon|^2 \omega_{WP}^n \le C\int_{\mathrm{supp}(\overline{\partial}\rho_\varepsilon)} \omega_P^n.$$

Use the same cover of $\{U_\alpha\}_{\alpha=1}^t$ of $\overline{\mathcal{M}}$ as in Lemma 6.2. Then $Y \subset \cup_{\alpha=1}^s U_\alpha$, $U_{s+1} \cup \cdots \cup U_t) \cap Y = \emptyset$ and $U_\alpha - Y = (\Delta^*)^{a_\alpha} \times (\Delta)^{b_\alpha}$ with the coordinates $(s_1^\alpha, \cdots, s_{a_\alpha}^\alpha, w_1^\alpha, \cdots, w_{b_\alpha}^\alpha)$ for any $1 \le \alpha \le s$. When ε is small enough, we can assume that $\mathrm{supp}(\overline{\partial}\rho_\varepsilon) \cap U_\alpha \subset \{|s_j^\alpha| \le \frac{1}{2}, |w_j^\alpha| \le \frac{1}{2}\}$ for any $1 \le \alpha \le s$. Since in $U_\alpha - Y$ for any $1 \le \alpha \le s$, the Poincaré metric ω_P is asymptotic to

$$\frac{\sqrt{-1}}{2}\left(\sum_{j=1}^{a_\alpha} \frac{ds_j^\alpha \wedge d\bar{s}_j^\alpha}{|s_j^\alpha|^2 (\log \frac{1}{|s_j^\alpha|})^2} + \sum_{j=1}^{b_\alpha} dw_j^\alpha \wedge d\bar{w}_j^\alpha\right), \tag{6.3}$$

we have

(6.4)
$$\int_{\mathrm{supp}(\overline{\partial}\rho_\varepsilon)} \omega_P^n \le C\sum_{\alpha=1}^{s}\prod_{j=1}^{a_\alpha}\int_{e^{-\frac{1}{2\varepsilon}}}^{e^{-\frac{1}{\varepsilon}}} \frac{1}{|s_j^\alpha|(\log \frac{1}{|s_j^\alpha|})^2} d|s_j^\alpha| \prod_{j=1}^{b_\alpha}\int_0^{\frac{1}{2}} |w_j^\alpha| d|w_j^\alpha| \le C\varepsilon.$$

So we have

$$\lim_{\varepsilon\to 0}\int_{\mathcal{M}} |d\rho_\varepsilon|^2 \omega_{WP}^n = 0 \tag{6.5}$$

and the result follows. □

7. Self-Adjointness of the Laplacian on Moduli Space

In this section, we will consider the self-adjointness of Laplacian on $(\mathcal{M}, \omega_{WP})$. Let us consider the differential operators d and δ defined on C^1 functions and C^1 forms on $\mathcal{M}$ respectively. We define the domain $Dom(d)$ of d to be the set of C^1 functions f defined on $\mathcal{M}$ such that both f and df are in L^2. Similarly, we define the domain $Dom(\delta)$ of δ to be the set of C^1 1-forms w such that both w and δw are in L^2. We then define the Laplacian Δ with respect to ω_{WP} by Δ with $Dom(\Delta)$ given by the set of C^2 functions f such that $f \in Dom(d)$ and $df \in Dom(\delta)$. In this section, we will prove the closure $\overline{\Delta}$ of Δ is self-adjoint.

THEOREM 7.1. *On $(\mathcal{M}, \omega_{WP})$, the closure $\overline{\Delta}$ of Laplacian on functions is self-adjoint.*

It is proved in [**6**] that $\overline{\Delta}$ is self-adjoint on $M \setminus \Sigma_M$ when M is an algebraic variety with the induced Fubini-Study metric and Σ_M is the singular set at least of real codimension 2. Here our result is different as we are considering the Weil-Petersson metric.

PROOF. By the theorem of Gaffney in [4], in order to show $\overline{\Delta}$ is self-adjoint, it is sufficient to prove

$$(df, w) = (f, \delta w) \tag{7.1}$$

for any $f \in Dom(d)$ and $w \in Dom(\delta)$. By Theorem 6.1 and 5.1, we have $W^1(\mathcal{M}) = W_0^1(\mathcal{M})$. Since $Dom(d) \subset W^1(\mathcal{M})$, there exists a sequence $f_n \in \mathcal{D}(\mathcal{M})$ such that $f_n \to f$ in $W^1(\mathcal{M})$. As each f_n has compact support, by integration by parts, we have

$$(df_n, w) = (f_n, \delta w). \tag{7.2}$$

The result follows by taking $n \to \infty$. □

8. An Example

Let (M, g) be a Riemannian manifold. A closed metric ball in $(\overline{M}_c, d)$ excluding an open set containing $\partial_c M$ might not be compact even if $\mathrm{cap}(\partial_c M) = 0$. In this section, we will give a concrete example.

Consider the Riemannian manifold (M, g) as follows. $M = \mathbb{R}^3$ and in terms of the cylindrical coordinates (r, θ, z),

$$g = e^{2z}(dr^2 + f^2(r)d\theta^2 + dz^2). \tag{8.1}$$

Here the function $f \in \mathcal{C}^\infty([0, \infty))$ satisfies the following properties:

- $f(r) = r$ for $r \in [0, \frac{1}{2}]$.
- f is increasing on $[0, 1]$ and $f(1) = 1$.
- f is decreasing on $[1, \infty)$.
- $f(r) = e^{-r}$ for $f \in [2, \infty)$.

For any piecewise smooth curve $l : [a, b] \to M$, we denote the length of l by $\|l\|$, i.e.

$$\|l\| = \int_a^b e^{z(t)} \sqrt{\dot{r}^2(t) + f^2(r(t))\dot{\theta}^2(t) + \dot{z}^2(t)dt} \tag{8.2}$$

And define the distance function d as

$$d(p, q) = \inf_{l \in L} \|l\|,$$

where $L = \{$all piecewise smooth curves from p to $q\}$. Then we know (M, d) is a metric space.

LEMMA 8.1. *For any $P_1, P_2 \in M$, denote the coordinate of P_i as (r_i, θ_i, z_i) for $i = 1, 2$. Then*

$$d(P_1, P_2) \le e^{z_1} + e^{z_2}. \tag{8.3}$$

PROOF. For any $t_0 < \min(z_1, z_2)$. Define the following three smooth curves.

- $l_1 : (r_1, \theta_1, t)$ for $t \in [t_0, z_1]$ oriented form z_1 to t_0.
- $l_2 : (r_1 + (r_2 - r_1)t, \theta_1 + (\theta_2 - \theta_1)t, t_0)$ for $t \in [0, 1]$.
- $l_3 : (r_2, \theta_2, t)$ for $t \in [t_0, z_2]$.

Then $l_1 \cup l_2 \cup l_3$ is a piecewise smooth curve connecting P_1 and P_2. We can calculate the length of these curves straightforwardly.

$$\begin{aligned}\|l_1\| &= \int_{t_0}^{z_1} e^t dt = e^{z_1} - e^{t_0},\\ \|l_3\| &= \int_{t_0}^{z_2} e^t dt = e^{z_2} - e^{t_0},\\ \|l_2\| &= \int_0^1 e^{t_0}\sqrt{(r_2-r_1)^2 + (\theta_2-\theta_1)^2 f^2(r_1+(r_2-r_1)t)}dt\\ &\le e^{t_0}\sqrt{(r_2-r_1)^2+(\theta_2-\theta_1)^2}.\end{aligned}$$

Therefore

$$d(P_1,P_2) \le e^{z_1} + e^{z_2} - 2e^{t_0} + e^{t_0}\sqrt{(r_2-r_1)^2+(\theta_2-\theta_1)^2}.$$

Taking $t_0 \to -\infty$, the result follows. □

Define $H_I = \mathbb{R}^2 \times I = \{(r,\theta,z) : z \in I\}$ for any $I \subset \mathbb{R}$. And we will use diam S to denote the diameter of set $S \subset M$.

COROLLARY 8.2. *diam* $H_{(-\infty,0]} \le 2$.

PROOF. For any $P_1, P_2 \in H_{(-\infty,0]}$, we have $d(P_1,P_2) \le e^{z_1} + e^{z_2} \le 2$. □

LEMMA 8.3. *For any* $P_1, P_2 \in M$,

$$d(P_1,P_2) \ge |e^{z_1} - e^{z_2}|. \tag{8.4}$$

PROOF. For any piecewise smooth curve $l : [0,1] \to M$ from P_1 to P_2, we have

$$\begin{aligned}\|l\| &= \int_0^1 e^{z(t)}\sqrt{\dot r^2(t) + f^2(r(t))\dot\theta^2(t) + \dot z^2(t)}dt\\ &\ge \int_0^1 e^z|\dot z(t)|dt\\ &\ge |e^{z_1} - e^{z_2}|.\end{aligned}$$

□

Note that the metric space (M,d) is not complete. $\{(0,0,-n)\}_{n=1}^\infty$ is a Cauchy sequence since $d((0,0,-m),(0,0,-n)) \le e^{-m} + e^{-n}$. But it is not convergent in M.

THEOREM 8.1. *Let* $\overline{M}$ *be the completion of* M *with respect to metric* d. *Then* $\overline{M} = M \cup \{\infty\}$ *where* $\{\infty\}$ *is defined as the Cauchy sequence* $\{(0,0,-n)\}_{n=1}^\infty$.

We want show that for any Cauchy sequence $\{P_n\}_{n=1}^\infty$, either it is convergent in M or it is equivalent to the Cauchy sequence $\{(0,0,-n)\}_{n=1}^\infty$. We split the proof into following lemmas.

LEMMA 8.4. *Let* $\{P_n\}_{n=1}^\infty$ *be a Cauchy sequence in* M *and denote* $P_n = (r_n, \theta_n, z_n)$. *Then* $\{z_n\}_{n=1}^\infty$ *is either convergent in* $\mathbb{R}$ *or* $\lim_{n\to\infty} z_n = -\infty$.

PROOF. By inequality (8.4), we have $d(P_m,P_n) \ge |e^{z_m} - e^{z_n}|$. Therefore $\{e^{z_n}\}$ is a Cauchy sequence in $\mathbb{R}$. So the result follows. □

LEMMA 8.5. *Let* $\{P_n\}_{n=1}^\infty$ *be a Cauchy sequence in* M *and denote* $P_n = (r_n, \theta_n, z_n)$. *If* $\{z_n\}_{n=1}^\infty$ *is a Cauchy sequence in* $\mathbb{R}$, *then* $\{r_n\}_{n=1}^\infty$ *is a Cauchy sequence in* $\mathbb{R}$.

PROOF. Let $z_0 = \lim z_n$. By dropping finitely many beginning terms, we can assume $z_n \in [z_0 - 1, z_0 + 1]$. Let $\delta = \delta(z_0) = e^{z_0-1} - e^{z_0-2}$. Since $\{P_n\}$ is Cauchy, by dropping more beginning terms, we can assume further that $d(P_m, P_n) < \frac{\delta}{3}$ for any $m, n \in \mathbb{Z}^+$. By the definition of metric d, there exists a piecewise smooth curve $l_{mn} : [0, 1] \to M$ from P_m to P_n such that $\|l_{mn}\| \leq \frac{3}{2}d(P_m, P_n)$. We claim

$$\min_{t\in[0,1]} z(t) > z_0 - 2. \tag{8.5}$$

Assume not. Take $t = t_0 \in [0, 1]$ be the first time such that $z(t) = z_0 - 2$, which implies that $z(t) \geq z_0 - 2$ for $t \in [0, t_0]$. Then

$$\|l_{mn}\| \geq \int_0^{t_0} e^{z(t)}|\dot{z}(t)|dt \geq e^{z_m} - e^{z(t_0)} \geq e^{z_0-1} - e^{z_0-2} = \delta.$$

However, according to our assumption on l_{mn}, we have

$$\|l_{mn}\| \leq \frac{3}{2}d(P_m, P_n) < \frac{\delta}{2}, \tag{8.6}$$

which is a contradiction and therefore the claim follows. Thus we have

$$\frac{3}{2}d(P_m, P_n) \geq l_{mn} \geq \int_0^1 e^{z(t)}|\dot{r}(t)|dt \geq e^{z_0-2}|r_m - r_n|.$$

Therefore $\{r_n\}$ is a Cauchy sequence in $\mathbb{R}$. □

LEMMA 8.6. *Let $\{P_n\}_{n=1}^{\infty}$ be a Cauchy sequence in M and denote $P_n = (r_n, \theta_n, z_n)$. If $\{z_n\}_{n=1}^{\infty}$ is a Cauchy sequence in $\mathbb{R}$ and $\lim r_n > 0$, then $\{\theta_n\}_{n=1}^{\infty}$ is a Cauchy sequence in $\mathbb{R}$.*

PROOF. Let $z_0 = \lim z_n$ and $r_0 = \lim r_n$. By dropping finitely many beginning terms, we can assume that $z_n \in [z_0 - 1, z_0 + 1]$ and $r_n \in [\frac{1}{2}r_0, \frac{3}{2}r_0]$ for any $n \in \mathbb{Z}$. Define $\delta(z_0) = e^{z_0-1} - e^{z_0-2}$ and $\delta(r_0, z_0) = \frac{1}{4}r_0e^{z_0-2}$. And take $\delta = \min\{\delta(z_0), \delta(r_0, z_0)\}$. By dropping more beginning terms, we can assume further $d(P_m, P_n) < \frac{\delta}{3}$ for any $m, n \in \mathbb{Z}$. Again we take a piecewise smooth curve $l_{mn} : [0, 1] \to M$ from P_m to P_n such that $\|l_{mn}\| \leq \frac{3}{2}d(P_m, P_n)$. By the proof in Lemma 8.5, we have $\min z(t) \geq z_0 - 2$. Here we claim

$$r(t) \in [\frac{1}{4}r_0, \frac{7}{4}r_0] \text{ for any } t \in [0, 1]. \tag{8.7}$$

Assume not. Then let $t = t_0$ be the first time such that $r(t_0) = \frac{1}{4}r_0$ or $\frac{7}{4}r_0$. Then

$$\|l_{mn}\| \geq \int_0^{t_0} e^{z(t)}|\dot{r}(t)|dt \geq e^{z_0-2}|r(t_0) - r_m| \geq \frac{1}{4}r_0e^{z_0-2} = \delta(r_0, z_0).$$

But we also have

$$\|l_{mn}\| \leq \frac{3}{2}d(P_m, P_n) < \frac{\delta}{2},$$

which is a contradiction. So the claim follows. Therefore

$$\begin{aligned}\frac{3}{2}d(P_m, P_n) \geq \|l_{mn}\| &\geq \int_0^1 e^{z(t)}f(r(t))|\dot{\theta}(t)|dt \\ &\geq e^{z_0-2}\min\{f(\frac{1}{4}r_0), f(\frac{7}{4}r_0)\}|\theta_m - \theta_n|.\end{aligned}$$

It follows that $\{\theta_n\}$ is a Cauchy sequence. □

LEMMA 8.7. *Let $P_n = (r_n, \theta_n, z_n)$ be a sequence in M. If $r_n \to r_0, \theta_n \to r_0, z_n \to z_0$ in $\mathbb{R}$, then P_n converges to $P_0 = (r_0, \theta_0, z_0)$ with respect to metric d.*

PROOF. Since $z_n \to z_0$ in $\mathbb{R}$. By dropping finitely many beginning terms, we can assume $z_n \in [z_0 - 1, z_0 + 1]$. Define a smooth curve from P_0 to P_n as $l(t) = (r_0 + (r_n - r_0)t, \theta_0 + (\theta_n - \theta_0)t, z_0 + (z_n - z_0)t)$. Then

$$\begin{aligned} d(P_0, P_n) \leq \|l\| =& int_0^1 e^{z(t)} \sqrt{(r_n - r_0)^2 + f^2(r_0 + (r_n - r_0)t)(\theta_n - \theta_0)^2 + (z_n - z_0)^2} dt \\ \leq& e^{z_0+1} \sqrt{(r_n - r_0)^2 + (\theta_n - \theta_0)^2 + (z_n - z_0)^2} \end{aligned}$$

So the result follows. □

Now we are ready to prove prove Theorem 8.1.

PROOF. Let $\{P_n\}_{n=1}^{\infty}$ be a Cauchy sequence in M. By Lemma 8.4, we have either $\lim z_n = -\infty$ or $\lim z_n = z_0$ for some $z_0 \in \mathbb{R}$. In the first case, we have

$$d(P_n, (0, 0, -n)) \leq e^{z_n} + e^{-n} \to 0.$$

Therefore Cauchy sequence $\{P_n\}$ and $\{(0, 0, -n)\}$ are equivalent to each other.

In the second case that $z_0 = \lim z_n \in \mathbb{R}$, we can assume $z_n \in [z_0 - 1, z_0 + 1]$ for any $n \in \mathbb{Z}^+$. By Lemma 8.5, we know that $\{r_n\}$ is a Cauchy sequence in $\mathbb{R}$. Let $r_0 = \lim r_n$. We have two sub-cases, either $r_0 = 0$ or $r_0 > 0$. When $r_0 = 0$, take a smooth curve l from $(0, 0, z_0)$ to P_n as $l(t) = (r_n t, \theta_n t, z_0 + (z_n - z_0)t)$. Then

$$\begin{aligned} d((0, 0, z_0), P_n) \leq \|l\| =& \int_0^1 e^{z(t)} \sqrt{r_n^2 + f^2(r_n t)\theta_n^2 + (z_n - z_0)^2} dt \\ \leq& e^{z_0+1} \int_0^1 \sqrt{r_n^2 + 4\pi^2 f^2(r_n t) + (z_n - z_0)^2} dt \\ \to& 0 \quad \text{as } n \to \infty. \end{aligned}$$

Therefore $P_n \to (0, 0, z_0)$ in M.

In the second sub-case that $r_0 > 0$, by Lemma 8.6, we have that $\lim \theta_n = \theta_0$ for some $\theta_0 \in \mathbb{R}$. Then by Lemma 8.7, we have that P_n converges to $P_0 = (r_0, \theta_0, z_0)$ in M. So the result follows. □

THEOREM 8.2. *The capacity of $\partial_c M = \{\infty\} \subset \overline{M}_c$ is zero.*

PROOF. Define a decreasing function $\varphi \in C^\infty(\mathbb{R})$ such that

$$\varphi(z) = \begin{cases} 1 & z \leq 0 \\ 0 & z \geq 1. \end{cases}$$

For any $a \in \mathbb{R}$, define $\varphi_a \in C^\infty(M)$ as $\varphi_a(P) = \varphi(z - a)$ for any $P = (r, \theta, z) \in M$. Then $\varphi = 1$ on $H_{(-\infty,a)} = B(\infty, e^a)$ and $\varphi = 0$ outside $H_{(-\infty,a+1)} =$

$B(\infty, e^{a+1})$. Then

$$\begin{aligned}\int_M \varphi_a^2 dV_g &\le \int_{H_{(-\infty,a+1)}} dV_g \\ &= \int_{-\infty}^{a+1}\int_0^{2\pi}\int_0^\infty e^{3z} f(r)drd\theta dz \\ &= 2\pi e^{3a+3}\int_0^\infty f(r)dr \\ &\to 0, \quad \text{as } a \to -\infty.\end{aligned}$$

On the other hand,

$$\begin{aligned}\int_M |\nabla\varphi_a|_g^2 dV_g &= \int_{H_{(a,a+1)}} |\varphi'(z-a)|^2 e^{-2z} dV_g \\ &= \int_a^{a+1}\int_0^{2\pi}\int_0^\infty |\varphi'(z-a)|^2 e^z f(r)drd\theta dz \\ &\le 2\pi(e^{a+1}-e^a)\sup_{\mathbb{R}}|\varphi'|\int_0^\infty f(r)dr \\ &\to 0, \quad \text{as } a\to -\infty.\end{aligned}$$

Therefore the result follows. □

PROPOSITION 8.8. *Let $o = (0,0,0)$. Then $\overline{B(o,2)} \setminus B(\infty, e^{-1})$ is not compact in M.*

PROOF. By Corollary 8.2, we have

$$\overline{B(o,2)} - B(\infty, e^{-1}) \supset H_{(-\infty,0]} - H_{(-\infty,-1)} = H_{(-1,0]}.$$

Consider the sequence $P_n = (n,0,0)$ in $H_{(-1,0]}$. We claim

$$d(P_m,P_n) \ge \min(e^{-1}, 1-e^{-1}) \quad \text{for any } m \ne n. \tag{8.8}$$

Let $l : [0,1] \to M$ be an arbitrary smooth curve from P_m to P_n. Then either $l \subset H_{(-1,+\infty)}$ or l will hit the plane $z = -1$. In the first case, we have

$$\|l\| \ge \int_0^1 e^{z(t)}|\dot r(t)|dt \ge e^{-1}|r_m - r_n| \ge e^{-1}$$

In the second case, take $t = t_0$ be the first time l hit the plane $z = -1$. Then

$$\|l\| \ge \int_0^{t_0} e^{z(t)}|\dot z(t)|dt \ge e^{z(0)} - e^{z(t_0)} = 1 - e^{-1}.$$

Combining these two cases, we have $\|l\| \ge \min(e^{-1}, 1-e^{-1})$ for any piecewise smooth curve from P_m to P_n. So the claim follows. Therefore, there is no convergent subsequence of $\{P_n\}$ and thus the result follows. □

References

[1] Dmitri Burago, Yuri Burago, and Sergei Ivanov, *A course in metric geometry*, Graduate Studies in Mathematics, vol. 33, American Mathematical Society, Providence, RI, 2001. MR1835418

[2] Nelia Charalambous and Zhiqin Lu, *On the spectrum of the Laplacian*, Math. Ann. **359** (2014), no. 1-2, 211–238, DOI 10.1007/s00208-013-1000-8. MR3201899

[3] Masatoshi Fukushima, Yoichi Oshima, and Masayoshi Takeda, *Dirichlet forms and symmetric Markov processes*, Second revised and extended edition, De Gruyter Studies in Mathematics, vol. 19, Walter de Gruyter & Co., Berlin, 2011. MR2778606

[4] Matthew P. Gaffney, *The harmonic operator for exterior differential forms*, Proc. Nat. Acad. Sci. U. S. A. **37** (1951), 48–50, DOI 10.1073/pnas.37.1.48. MR0048138

[5] Matthew P. Gaffney, *Hilbert space methods in the theory of harmonic integrals*, Trans. Amer. Math. Soc. **78** (1955), 426–444, DOI 10.2307/1993072. MR0068888

[6] Peter Li and Gang Tian, *On the heat kernel of the Bergmann metric on algebraic varieties*, J. Amer. Math. Soc. **8** (1995), no. 4, 857–877, DOI 10.2307/2152831. MR1320155

[7] Zhiqin Lu and Michael R. Douglas, *Gauss-Bonnet-Chern theorem on moduli space*, Math. Ann. **357** (2013), no. 2, 469–511, DOI 10.1007/s00208-013-0907-4. MR3096515

[8] Zhiqin Lu and Xiaofeng Sun, *On the Weil-Petersson volume and the first Chern class of the moduli space of Calabi-Yau manifolds*, Comm. Math. Phys. **261** (2006), no. 2, 297–322, DOI 10.1007/s00220-005-1441-3. MR2191883

[9] Xiaonan Ma and George Marinescu, *Holomorphic Morse inequalities and Bergman kernels*, Progress in Mathematics, vol. 254, Birkhäuser Verlag, Basel, 2007. MR2339952

[10] Jun Masamune, *Essential self-adjointness of Laplacians on Riemannian manifolds with fractal boundary*, Comm. Partial Differential Equations **24** (1999), no. 3-4, 749–757, DOI 10.1080/03605309908821442. MR1683058

[11] Jun Masamune, *Analysis of the Laplacian of an incomplete manifold with almost polar boundary*, Rend. Mat. Appl. (7) **25** (2005), no. 1, 109–126. MR2142127

[12] Peter Petersen, *Manifold theory*, Available online at `http://www.math.ucla.edu/~petersen/manifolds.pdf`.

[13] Michael Reed and Barry Simon, *Methods of modern mathematical physics. II. Fourier analysis, self-adjointness*, Academic Press [Harcourt Brace Jovanovich, Publishers], New York-London, 1975. MR0493420

[14] Michael Reed and Barry Simon, *Methods of modern mathematical physics. I*, 2nd ed., Academic Press, Inc. [Harcourt Brace Jovanovich, Publishers], New York, 1980. Functional analysis. MR751959

[15] Emil J. Straube, *Lectures on the $\mathcal{L}^2$-Sobolev theory of the $\overline{\partial}$-Neumann problem*, ESI Lectures in Mathematics and Physics, European Mathematical Society (EMS), Zürich, 2010. MR2603659

[16] Eckart Viehweg, *Quasi-projective moduli for polarized manifolds*, Ergebnisse der Mathematik und ihrer Grenzgebiete (3) [Results in Mathematics and Related Areas (3)], vol. 30, Springer-Verlag, Berlin, 1995. MR1368632

Department of Mathematics, UC Irvine, Irvine, California 92617
Email address: `zlu@uci.edu`

Department of Mathematics, Johns Hopkins University, Baltimore, Maryland 21218
Email address: `hxu@math.jhu.edu`

Contemporary Mathematics
Volume **735**, 2019
https://doi.org/10.1090/conm/735/14827

On orthogonal Ricci curvature

Lei Ni and Fangyang Zheng

ABSTRACT. In this paper we discuss some recent progresses in the study of compact Kähler manifolds with positive orthogonal Ricci curvature, a curvature condition defined as the difference between Ricci curvature and holomorphic sectional curvature. In the recent works by authors and the joint work of authors with Q. Wang the comparison theorems, vanishing theorems, and structural theorems for such manifolds have been proved. We also constructed examples of this type of manifolds, and give some classification results in low dimensions.

1. Orthogonal Ricci curvature

Let (M^n, g) be a Kähler manifold of complex dimension n. Its *orthogonal Ricci curvature* $Ric^{\perp}$ is defined by (cf. [**21**]):

$$Ric^{\perp}_{X\overline{X}} = Ric(X, \overline{X}) - R(X, \overline{X}, X, \overline{X})/|X|^2,$$

where X is a non-zero type $(1,0)$ tangent vector at a point $x \in M^n$. This curvature arises in the study of the comparison theorem for Kähler manifolds and the previous study of manifolds with so-called *nonnegative quadratic orthogonal bisectional curvature* (cf. [**4**], [**26**], [**16**], [**5**]). We refer the readers to [**21**] for a more detailed account on this topic. Clearly this curvature is closely related to Ricci curvature Ric and holomorphic sectional curvature H. It is natural to ask, what is the relationship between $Ric^{\perp}$ and Ric or H (other than the obvious one that $Ric^{\perp} + H = Ric$ for unit length tangent vectors), and what kind of compact complex manifolds M^n can admit Kähler metrics with $Ric^{\perp} > 0$ (or ≥ 0, or ≤ 0, or < 0, or $\equiv 0$) everywhere?

In this paper, we will focus on the curvature condition $Ric^{\perp}$ and pay particular attention to the class of compact Kähler manifolds with $Ric^{\perp} > 0$ everywhere, except in Section 2 where complete noncompact Kähler manifolds are also considered. Throughout this paper, we will assume that the complex dimension $n \geq 2$ unless stated otherwise, since $Ric^{\perp} \equiv 0$ when $n = 1$.

2010 *Mathematics Subject Classification.* 32L05, 32Q10, 32Q15, 53C55.

Key words and phrases. Compact complex manifolds, Kähler metrics, positive holomorphic sectional curvature, positive scalar curvature, projectivized vector bundles.

The research of LN is partially supported by NSF grant DMS-1401500 and the "Capacity Building for Sci-Tech Innovation-Fundamental Research Funds".

The research of FZ is partially supported by a Simons Collaboration Grant 355557.

©2019 American Mathematical Society

We start with the following observation. At a point $x \in M^n$, let us denote by $\mathbb{S}_x^{2n-1}$ the unit sphere of all type $(1,0)$ tangent vector at x of unit length. By a classic result of Berger, the average value of Ric or H over $\mathbb{S}_x^{2n-1}$ is $\frac{S(x)}{2n}$ or $\frac{S(x)}{n(n+1)}$, respectively, where $S(x)$ is the scalar curvature. Based on that, we get the following

LEMMA 1.1. *For a Kähler manifold (M^n, g) with $n \geq 2$, the average value of $Ric^\perp$ over the unit tangent sphere $\mathbb{S}_x^{2n-1}$ at any $x \in M^n$ is $\frac{(n-1)S(x)}{2n(n+1)}$. In particular, $Ric^\perp > 0$ (or ≥ 0, or < 0, or ≤ 0, or $\equiv 0$) implies $S(x) > 0$ (or ≥ 0, or < 0, or ≤ 0, or $\equiv 0$).*

So just like Ric or H, $Ric^\perp$ also dominates the scalar curvature S, in the sense that the sign of $Ric^\perp$ determines the sign of S. On the other hand, $Ric^\perp$ is clearly dominated by the bisectional curvature $B = R_{X\overline{X}Y\overline{Y}}$, where $|X| = |Y| = 1$, just like Ric or H. It is also dominated by the weaker curvature conditions *orthogonal bisectional curvature* $B^\perp$, which is defined by $R_{X\overline{X}Y\overline{Y}}$ for $|X| = |Y| = 1$ and $X \perp Y$, and the *quadratic orthogonal bisectional curvature* QB, which is defined in the following way:

The Kähler manifold (M^n, g) is said to have $QB > 0$ at $x \in M^n$, if for any unitary tangent frame $\{e_1, \ldots, e_n\}$ at x and any real numbers $a_1, \ldots, a_n$, not all equal to each other, it holds that $\sum_{i,j=1}^n R_{i\bar{i}j\bar{j}}(a_i - a_j)^2 > 0$.

This is a weaker curvature condition than $B^\perp > 0$, and yet by taking all but one of these a_i to be zero, we get

LEMMA 1.2. *A Kähler manifold (M^n, g) with $QB > 0$ (or ≥ 0, or < 0, or ≤ 0, or $\equiv 0$) will have $Ric^\perp > 0$ (or ≥ 0, or < 0, or ≤ 0, or $\equiv 0$).*

Some more elementary facts about orthogonal Ricci curvature. In complex dimension $n = 1$, one always have $Ric^\perp = 0$. For $n \geq 2$, the complex space forms $\mathbb{P}^n$, $\mathbb{C}^n$, and $\mathbb{H}^n$ respectively satisfies $Ric^\perp > 0, = 0$, or < 0. For product manifolds, we have the following:

LEMMA 1.3. *If both of the Kähler manifolds (M, g) and (N, h) satisfy $Ric^\perp > 0$ and $Ric \geq 0$, then the product manifold $(M \times N,\ g \times h)$ will have $Ric^\perp > 0$.*

This is because any tangent vector X of type $(1,0)$ on $M \times N$ can be uniquely written as $U + V$ where U is tangent to M and V is tangent to N, and

$$\begin{aligned} |X|^2 R_{X\overline{X}} - R_{X\overline{X}X\overline{X}} &= |X|^2(R_{U\overline{U}} + R_{V\overline{V}}) - (R_{U\overline{U}U\overline{U}} + R_{V\overline{V}V\overline{V}}) \\ &\geq \left(|U|^2 R_{U\overline{U}} - R_{U\overline{U}U\overline{U}}\right) + \left(|V|^2 R_{V\overline{V}} - R_{V\overline{V}V\overline{V}}\right). \end{aligned}$$

Here we used $R_{X\overline{Y}}$ to denote the Ricci tensor. In particular, $\mathbb{P}^{n_1} \times \cdots \times \mathbb{P}^{n_r}$ has $Ric^\perp > 0$ whenever all $n_1, \ldots, n_r \geq 2$.

There also exists an algebraic consideration viewing $Ric^\perp$ as the *holomorphic sectional curvature* of an algebraic curvature operator risen from the one acting on the two-forms via the Bochner formula. Recall the notations from the appendix of **[19]** and define an algebraic (Kähler) curvature operator

$$R_{Ric} = Ric \bar{\wedge}\, \mathrm{id},$$

where for any $A, B : T'_xM \to T'_xM$ Hermitian symmetric ($\bar{A}(X) = A(\bar{X}) = 0$)

$$\begin{aligned} \langle A \bar{\wedge} B(X \wedge \bar{Y}), \overline{Z \wedge \bar{W}} \rangle &\doteqdot \frac{1}{2}\Big(\langle \left(A \wedge \bar{B} + B \wedge \bar{A}\right)(X \wedge \bar{Y}), \overline{Z \wedge \bar{W}} \rangle \\ &\quad + \langle \left(A \wedge \bar{B} + B \wedge \bar{A}\right)(W \wedge \bar{Y}), \overline{Z \wedge \bar{X}} \rangle \Big). \end{aligned}$$

It is easy to check that $Ric^{\perp}(X,\overline{X}) = H_{R_{Ric}-R}(X)/|X|^2$. Here $H_{\widetilde{R}}(X)$ is the holomorphic sectional curvature of $\widetilde{R} = R_{Ric} - R$. From this it is easy to see that $Ric^{\perp} \equiv 0$ implies that $\widetilde{R} \equiv 0$. Hence $Ric^{\perp} \equiv 0$, via the decomposition of the curvature operators, induces that either $n=1$, or $n=2$ R is conformally flat, or $n \geq 3$ and R is flat. If $Ric^{\perp}(X,\overline{X}) = c|X|^2$ for a constant $c \neq 0$, similarly one can conclude that either $n=2$, R is conformally flat or R is a multiple if identity. Hence $Ric^{\perp}$-Einstein is a very special condition.

2. Comparison theorems

The Laplacian comparison theorem is a cornerstone in Riemannian geometry and global analysis. In the Kähler case, the Laplacian of the distance function decomposes as the sum of the so-called *holomorphic Hessian* and *orthogonal Laplacian* in a natural manner [**21**].

Let (M^n,g) be a Kähler manifold and let us fix a point $p \in M^n$. Denote by ρ the function on M which is the distance from p. Let $Z = \frac{1}{\sqrt{2}}(\nabla\rho - \sqrt{-1}J\nabla\rho)$ be the type $(1,0)$ unit tangent vector in the radial direction, then the orthogonal Laplacian is defined by

$$\Delta^{\perp}\rho = \Delta\rho - \nabla^2(Z,\overline{Z}),$$

and the second term on the right hand side is the holomorphic Hessian. As observed in [**21**], the comparison of orthogonal Ricci curvature will lead to comparison on orthogonal Laplacians:

THEOREM 2.1. *Let (M^n,g) be a complete Kähler manifold with $Ric^{\perp} \geq (n-1)\lambda$, where λ is a constant. Let $(\tilde{M},\tilde{g})$ be a complex space form of the same dimension with constant holomorphic sectional curvature 2λ. Fix $p \in M$ and $\tilde{p} \in \tilde{M}$, and denote by ρ, $\tilde{\rho}$ the distance function from p or $\tilde{p}$, respectively. Then for any $x \in M$ not in the cut locus of p, it holds*

$$\Delta^{\perp}\rho(x) \;\leq\; \Delta^{\perp}\tilde{\rho}|_{\tilde{\rho}=\rho(x)} \;=\; (n-1)\cot_{\frac{\lambda}{2}}(\rho).$$

Similarly, the comparison of holomorphic sectional curvature H leads to comparison on holomorphic Hessians. For distance function to points, this was proved by G. Liu in [**17**], using the argument of [**15**]. In [**21**], we generalized it to distance functions to complex submanifolds:

THEOREM 2.2. *Let (M^n,g) be a complete Kähler manifold with $H \geq 2\lambda$, and let $(\tilde{M},\tilde{g})$ be a complex space form of the same dimension with constant holomorphic sectional curvature 2λ. If $P \subset M$ and $\tilde{P} \subset \tilde{M}$ are complex submanifolds, and denote by ρ, $\tilde{\rho}$ the distance function from P or $\tilde{P}$, respectively. Then for any $x \in M$ not in the focal locus of P, it holds*

$$\nabla^2\rho(Z,\overline{Z})|_x \;\leq\; \nabla^2\tilde{\rho}(\tilde{Z},\overline{\tilde{Z}})|_{\tilde{\rho}=\rho(x)}.$$

In particular, when $\lambda = 0$ and $\tilde{P}$ is a point, it holds that

$$\nabla^2\rho(Z,\overline{Z})|_x \leq \frac{1}{2\rho(x)} \iff \nabla^2\log\rho(Z,\overline{Z}) \leq 0.$$

When the curvature assumptions in the above two theorems are both valid, then as in [**15**] one has the volume comparison theorem

Corollary 2.3. *Let (M^n, g) be a complete Kähler manifold with $Ric^\perp \geq (n-1)\lambda$ and $H \geq 2\lambda$, and $(\tilde{M}, \tilde{g})$ the complex space form of the same dimension with constant holomorphic sectional curvature 2λ. Then for any $x \in M$ and $\tilde{x} \in \tilde{M}$, it holds that $\Delta\rho(x) \leq \Delta\tilde{\rho}|_{\rho(x)}$, and for any $0 < r \leq R$, it holds*

$$\frac{Vol(B(x,R))}{Vol(B(x,r))} \leq \frac{Vol(\tilde{B}(\tilde{x},R))}{Vol(\tilde{B}(\tilde{x},r))},$$

where B and $\tilde{B}$ are geodesic balls in M and $\tilde{M}$, respectively. The equality holds if and only if $B(x,R)$ is holomorphically isometric to $\tilde{B}(\tilde{x},R)$.

Note that the lower bounds on $Ric^\perp$ and H gives the condition $Ric \geq (n+1)\lambda$, so there is volume comparison in the Riemannian setting. However, the above comparison in the Kähler setting is sharper.

Theorem 2.1 can be generalized to the case of complex hypersurfaces, which can be viewed as the Kähler version of Heintze-Karcher Theorem [**10**] with the assumption on Ricci curvature being replaced by $Ric^\perp$:

Theorem 2.4. *With M, $\tilde{M}$ in Theorem 2.1, if $P \subset M$ and $\tilde{P} \subset \tilde{M}$ are complex hypersurfaces and ρ, $\tilde{\rho}$ are distance functions to P, $\tilde{P}$, respectively, then for any x not in the focal locus of P,*

$$\Delta^\perp\rho(x) \leq \Delta^\perp\tilde{\rho}|_{\tilde{\rho}=\rho(x)} = (n-1)\tan_{\frac{\lambda}{2}}(\rho).$$

Similarly, one can generalize Theorem 2.2 to the orthogonal Hessian of the distance function. For any real value u on M, we will denote by $\nabla^{2\perp}u(X,\overline{X})$ the restriction of $\nabla^2 u(X,\overline{X})$ on the spaces of all type $(1,0)$ vectors X perpendicular to both ∇u and $J\nabla u$. We have the following:

Theorem 2.5. *Let (M^n, g) be a complete Kähler manifold with orthogonal bisectional curvature $B^\perp \geq \lambda$, and let $(\tilde{M}, \tilde{g})$ be a complex space form of the same dimension with constant holomorphic sectional curvature 2λ. Fix $p \in M$ and $\tilde{p} \in \tilde{M}$ and denote by ρ, $\tilde{\rho}$ the distance function from p or $\tilde{p}$, respectively. Then for any $x \in M$ not in the cut locus of p, it holds*

$$\nabla^{2\perp}\rho(x) \leq \nabla^{2\perp}\tilde{\rho}|_{\tilde{\rho}=\rho(x)}.$$

A similar argument as in the classical Bonnet-Myers Theorem case would imply that, for any complete Kähler manifold (M^n, g) with $Ric^\perp$ bounded from below by a positive constant, the diameter of M^n is bounded from above, hence M^n must be compact. As a consequence, we get the following:

Corollary 2.6. *Let (M^n, g) be a compact Kähler manifold with $Ric^\perp > 0$ everywhere. Then the fundamental group $\pi_1(M)$ is finite.*

It was conjectured [**21**] that such manifolds are all simply-connected, in fact, they all should be rationally-connected. But so far, we have only been successful in proving this for $n \leq 4$.

3. Vanishing theorems

In [**27**], it was shown that any compact Kähler manifold M^n with positive holomorphic sectional curvature must be projective, answering affirmatively a question raised in [**28**]. The proof was done by showing that the Hodge number $h^{2,0}$ vanishes, namely, there are no non-trivial global holomorphic 2-forms on M^n. It was actually proved that all the Hodge number $h^{p,0}$ vanishes for any $1 \leq p \leq n$, that is, any global holomorphic p-form on M^n must be identically zero.

The proof of this vanishing theorem used the form version of the Bochner identity

$$\partial\overline{\partial}|s|^2 = \langle \nabla s, \overline{\nabla s} \rangle - \tilde{R}(s, \overline{s}, \cdot, \cdot),$$

where s is any holomorphic p-form on M^n, i.e., any holomorphic section of $\bigwedge^p \Omega$, where Ω is the holomorphic cotangent bundle of M^n, and $\tilde{R}$ is the curvature of $\bigwedge^p \Omega$. Following the same approach, we were able to show that

THEOREM 3.1. *Let (M^n, g) be a compact Kähler manifold with $Ric^\perp > 0$ everywhere. Then $h^{p,0} = 0$ for $p = 1, 2, n-1, n$. In particular, M^n is always projective.*

Of course it is believed that such a manifold will have $h^{p,0} = 0$ for any $1 \leq p \leq n$, in fact, the manifold should be rationally connected. But so far we are not able to show that, as we don't know how to deal with the vanishing of holomorphic p-forms for $p \geq 3$. At present, we also don't know how to prove that any compact Kähler manifold with $Ric^\perp > 0$ must be simply-connected, except when the dimension is at most 4 which is a consequence of the above vanishing theorem.

Note that if we already know that $Ric^\perp > 0$ implies that $h^{p,0} = 0$ for all $1 \leq p \leq n$, then all such M^n must be simply-connected, by the following well-known argument. Note that the Euler characteristic of the structure sheaf $\mathcal{O}_M$ of M is given by

$$\chi(\mathcal{O}_M) = 1 - h^{1,0} + h^{2,0} - \cdots + (-1)^n h^{n,0},$$

so the vanishing of the Hodge numbers implies that $\chi(\mathcal{O}_M) = 1$. Let $\pi : \widetilde{M} \to M$ be the universal covering space. π is finite of degree d since $\pi_1(M)$ is finite. So we have $\chi(\mathcal{O}_{\widetilde{M}}) = d \cdot \chi(\mathcal{O}_M) = d$. On the other hand, since $\widetilde{M}$ with the pull back metric also has $Ric^\perp > 0$, thus $\chi(\mathcal{O}_{\widetilde{M}}) = 1$, so we must have $d = 1$, namely, M is simply-connected.

In a related recent work [**22**], we examined the vanishing theorems for a new set of curvature conditions, where we were able to achieve optimal results. Given a Kähler manifold (M^n, g), if Σ is a k-dimensional complex subspace of the tangent space $T_x^{1,0}M$ of M at $x \in M$, then we will denote by $S_k(x, \Sigma)$ the average value of the holomorphic sectional curvature function H, integrated over the unit sphere $\Sigma \cap \mathbb{S}_x^{2n-1}$.

S_k will be called the k-scalar curvature, which interpolates between $S_1 = H$ and $S_n = S$, the usual scalar curvature. We will say that M^n has positive k-scalar curvature, denoted as $S_k > 0$, if $S_k(x, \Sigma) > 0$ for any $x \in M$ and any k-dimensional subspace Σ at x. Clearly, $S_k > 0$ implies $S_l > 0$ for any $l > k$. So the strength of S_k deceases as k increases. In [**22**], it was proved that

THEOREM 3.2. *Let M^n be a compact Kähler manifold with $S_2 > 0$. Then $h^{p,0} = 0$ for any $2 \leq p \leq n$. In particular, M^n is always projective.*

Note that there are complex 2-tori that are not projective. By taking the product of such a torus with a complex projective space, we see that $S_3 > 0$ does not guarantee projectiveness. So the above result is sharp in some sense. For general k, we also have the following.

THEOREM 3.3. *Let M^n be a compact Kähler manifold with $S_k > 0$, where k is an integer between 2 and n. Then $h^{p,0} = 0$ for any $k \leq p \leq n$.*

Note that the statement is also true when $k = 1$, which is exactly the vanishing theorem of [**27**] for compact Kähler manifolds with positive holomorphic sectional curvature.

The curvature S_k is related to $Ric^{\perp}$ in the following way. For any k-subspace $\Sigma \subset T_x^{1,0}M$, we define

$$S_k^{\perp}(x,\Sigma) = k \cdot \fint_{Z\in\Sigma, |Z|=1} Ric^{\perp}(Z,\overline{Z})\, d\theta(Z)$$

where $\fint f(Z)\, d\theta(Z)$ denotes $\frac{1}{Vol(\mathbb{S}^{2k-1})}\int_{\mathbb{S}^{2k-1}} f(Z)\, d\theta(Z)$. We say $S_k^{\perp}(x) > 0$ if for any k-subspace $\Sigma \subset T_x^{1,0}M$, $S_k^{\perp}(x,\Sigma) > 0$. The following generalization of Theorem 3.1 can also be obtained.

THEOREM 3.4. *Let (M^n, g) be a compact Kähler manifolds such that $S_2^{\perp}(x) > 0$ for any $x \in M$. Then $h^{2,0} = 0$. In particular, M is projective.*

Observe that, if $\Sigma = \text{span}\{E_1, E_2, \cdots, E_k\}$, then we have

$$\begin{aligned}
&\frac{1}{k} S_k^{\perp}(x,\Sigma)\\
&= \fint_{Z\in\Sigma,\, |Z|=1} Ric^{\perp}(Z,\overline{Z})\, d\theta(Z) = \fint_{Z\in\Sigma,\, |Z|=1} \left(Ric(Z,\overline{Z}) - H(Z)\right) d\theta(Z)\\
&= \fint \frac{1}{Vol(\mathbb{S}^{2n-1})}\left(\int_{\mathbb{S}^{2n-1}} \left(nR(Z,\overline{Z},W,\overline{W}) - H(Z)\right) d\theta(W)\right) d\theta(Z)\\
&= \frac{1}{Vol(\mathbb{S}^{2n-1})}\int_{\mathbb{S}^{2n-1}} \left(\fint \left(nR(Z,\overline{Z},W,\overline{W}) - H(Z)\right) d\theta(Z)\right) d\theta(W)\\
&= \frac{1}{k}\left(Ric(E_1,\overline{E}_1) + Ric(E_2,\overline{E}_2) + \cdots + Ric(E_k,\overline{E}_k)\right) - \frac{2}{k(k+1)} S_k(x,\Sigma)
\end{aligned}$$

where $S_k(x,\Sigma)$ is the scalar curvature of R restricted to Σ defined in the above. The positivity of the partial sum $Ric(E_1,\overline{E}_1) + Ric(E_2,\overline{E}_2) + \cdots + Ric(E_k,\overline{E}_k)$ for any unitary frame is called the k-positivity of Ricci.

Given the results on the projectivity for compact Kähler manifolds with $Ric^{\perp} > 0$ and $S_2 > 0$, naturally questions could be asked for compact Kähler manifolds with $Ric^{\perp} < 0$, or $S_2 < 0$. For example, *are all compact Kähler manifolds with $Ric^{\perp} < 0$ projective? When (or if) the K_M of such a manifold is ample? Similarly one can ask when a compact Kähler manifold with $S_2 < 0$ is projective, and when K_M is ample.* Regarding the question for $S_2 < 0$ manifolds, there have been some recent progresses ([**25**], [**24**]) for $S_1 = H < 0$ case.

4. Examples: classical Kähler C-spaces with $b_2 = 1$

It is well known that any compact Hermitian symmetric space M^n will have bisectional curvature nonnegative everywhere, while its Ricci curvature and holomorphic sectional curvature are positive everywhere. It is verified in [**20**] that such M^n will have $Ric^{\perp} > 0$ if and only if M^n does not contain $\mathbb{P}^1$ as a factor:

THEOREM 4.1. *A compact Hermitian symmetric space M^n has $Ric^{\perp} > 0$ if and only if it does not have a $\mathbb{P}^1$ factor.*

More generally, the set of all compact Hermitian symmetric spaces is contained in the larger set of all Kähler C-spaces, which are the orbit spaces of the adjoint representations of compact simple Lie groups. Note that not all irreducible Kähler C-spaces of dimension at least 2 satisfy $Ric^{\perp} > 0$, for instance, we will see later that the flag threefold

$$M^3 = \{([z],[w]) \in \mathbb{P}^2 \times \mathbb{P}^2 \mid \sum_{i=0}^{2} z_i w_i = 0\}$$

cannot admit any Kähler metric with $Ric^{\perp} > 0$. However, this may be caused by the fact that its second betti number is bigger than one. We propose the following:

CONJECTURE 4.2. *Any Kähler C-spaces with $b_2 = 1$ and $n \geq 2$ will satisfy $Ric^{\perp} > 0$.*

Note that Kähler C-spaces with $b_2 = 1$ consist of the four classical sequences plus finitely many exceptional ones, and by using the computations by Itoh [**13**] and by Chau and Tam [**5**], we verified in [**20**] the following

THEOREM 4.3. *Any classical Kähler C-space with $b_2 = 1$ and $n \geq 2$ satisfies $Ric^{\perp} > 0$.*

To describe the story about the exceptional ones, let us recall that Kähler C-space with $b_2 = 1$ are characterized as $(\mathfrak{g}, \alpha_i)$ (see [**13**], [**5**], or [**16**] for more details). Here $\mathfrak{g}$ is a simple complex Lie algebra, and $\{\alpha_1, \ldots, \alpha_r\}$ is a fundamental root system with respect to the Cartan subalgebra $\mathfrak{h} \subset \mathfrak{g}$, with r the rank and $1 \leq i \leq r$. Simple complex Lie algebras are fully classified as the four classical sequences $A_r = \mathfrak{sl}_{r+1}$ $(r \geq 1)$, $B_r = \mathfrak{so}_{2r+1}$ $(r \geq 2)$, $C_r = \mathfrak{sp}_{2r}$ $(r \geq 3)$, $D_r = \mathfrak{so}_{2r}$ $((r \geq 4)$, plus the exceptional ones E_6, E_7, E_8, F_4 and G_2.

For the exceptional ones, $(E_6, \alpha_1) = (E_6, \alpha_6)$ is the compact Hermitian symmetric space of type V, which has dimension 16 and rank 2. (E_7, α_7) is the compact Hermitian symmetric space of type VI, which has dimension 27 and rank 3. Theses two form the only exceptional irreducible compact Hermitian symmetric spaces. Also, $(G_2, \alpha_1) = \mathbb{Q}^5$ is the quadratic hypersurface in $\mathbb{P}^6$ which is a type IV Hermitian symmetric space. In [**5**], it was proven that

THEOREM 4.4 (Chau-Tam [**5**]). *The following exceptional Kähler C-spaces with $b_2 = 1$ all have $QB > 0$, hence have $Ric^{\perp} > 0$:*

$$(G_2, \alpha_2),\ (F_4, \alpha_i)_{i=1,2,4},\ (E_6, \alpha_i)_{2,3,5},\ (E_7, \alpha_i)_{i=1,2,5},\ (E_8, \alpha_i)_{i=1,2,8}.$$

For the remaining exceptional Kähler C-spaces with $b_2 = 1$, namely, the following list

$$\mathcal{E}_0 = \{(F_4, \alpha_3),\ (E_6, \alpha_4),\ (E_7, \alpha_i)_{i=3,4,6},\ (E_8, \alpha_i)_{i=3,4,5,6,7}\},$$

Chau and Tam proved in [**5**] that each of them does not satisfy $QB \geq 0$. However, we do believe that each of them satisfy $Ric^{\perp} > 0$, which is a much weaker condition than $QB > 0$.

For Kähler C-spaces with $b_2 > 1$, it would be a very interesting question to determine the subset which satisfies $Ric^{\perp} > 0$.

5. Examples: projectivized vector bundles

We have seen that $\mathbb{P}^n \times \mathbb{P}^m$ has $Ric^{\perp} > 0$ for any $n, m \geq 2$. Now we would like to generalize that to projectivized vector bundles. Let (M^n, g) be a compact Kähler manifold and (E, h) be a holomorphic vector bundle of rank r over M, equipped with a Hermitian metric h. Let $\pi : P = \mathbb{P}(E^*) \to M$ be the projectivized bundle associated with E, that is, for any $x \in M$, the fiber $\pi^{-1}(x) = \mathbb{P}(E_x)$ is the projective space of all complex lines in E_x through the origin.

Denote by L be the holomorphic line bundle on P dual to the tautological subbundle. L is determined by the short exact sequence on P:

$$0 \to \mathcal{O}_P \to \pi^* E^* \otimes L \to T_{P|M} \to 0,$$

where $T_{P|M} = \ker(\, d\pi : T_P \to \pi^* T_M)$ is the relative tangent bundle. The metric h on E induces naturally a Hermitian metric $\hat{h}$ on L, whose curvature form is

$$C_1(L, \hat{h}) = \omega_{\mathrm{FS}} - \frac{\sqrt{-1}}{|v|^2} \Theta^h_{v\overline{v}}$$

at any point $(x, [v]) \in P$, where $x \in M$ and $0 \neq v \in E_x$. Here ω_{FS} is the Kähler form of the Fubini-Study metric on the fiber of π. For a positive constant λ, consider the closed $(1,1)$ form on P:

$$\omega_G = \lambda \pi^* \omega_g + C_1(L, \hat{h}).$$

Then for λ sufficiently large, G is a Kähler metric on P.

In [**11**], Hitchin showed that any Hirzebruch surface $\mathbb{F}_k$ admits Kähler metrics with positive holomorphic sectional curvature. Here $\mathbb{F}_k = \mathbb{P}(E^*)$ for $E = \mathcal{O} \oplus \mathcal{O}(k)$ over $\mathbb{P}^1$, where k is any nonnegative integer. In [**1**], this was generalized to any projectivized vector bundle over any compact Kähler manifold with positive holomorphic sectional curvature, namely, it was shown that when the base manifold (M^n, g) has positive holomorphic sectional curvature and when λ is sufficiently large, the above metric G always has positive holomorphic sectional curvature. Following this computation, in [**20**], we obtained the following:

THEOREM 5.1. *Let (M^n, g) be a compact Kähler manifold with $Ric^{\perp} > 0$, and (E, h) be a Hermitian vector bundle over M of rank $r \geq 3$ such that for any $x \in M$ and any $0 \neq v \in E_x$,*

$$Ric^{g\perp}_{X\overline{X}} + R(\det E)_{X\overline{X}} - \frac{r}{|v|^2} R^h_{v\overline{v}X\overline{X}} > 0 \tag{5.1}$$

for any tangent vector $0 \neq X \in T^{1,0}_x M$. Here $R(\det E)$ is the curvature of the determinant line bundle $\det E = \bigwedge^r E$ equipped with the metric induced by h. Then on the projectivized bundle $P = \mathbb{P}(E^*)$, the Kähler metric G with $\omega_G = \lambda \pi^* \omega_g + C_1(L, \hat{h})$ will have $Ric^{\perp} > 0$ everywhere when λ is sufficiently large.

We remark that the rank requirement $r \geq 3$ here is necessary, as we shall see later that any $\mathbb{P}^1$-bundle over any space can never admit a Kähler metric with $Ric^{\perp} > 0$. We also remark that the curvature condition (5.1) is independent of the scaling of metrics g or h, as well as tensoring of E by a line bundle.

When the dimension of the base manifold is 3 or higher, the above theorem gives non-trivial examples of manifolds with $Ric^{\perp} > 0$, including those which are not Kähler C-spaces. For instance, we have the following.

COROLLARY 5.2. *Consider $E = \mathcal{O}(a_1) \oplus \cdots \oplus \mathcal{O}(a_r)$ on $\mathbb{P}^n$ where $a_1 \geq a_2 \geq \cdots \geq a_r$ are integers. If $r \geq 3$, and $n-1 > (a_1 - a_2) + \cdots + (a_1 - a_r)$, then $P = \mathbb{P}(E^*)$ will admit a Kähler metric with $Ric^{\perp} > 0$.*

For instance, for $E = \mathcal{O}^{\oplus 2} \oplus \mathcal{O}(-1)$ over $\mathbb{P}^3$, the Fano fivefold $P^5 = \mathbb{P}(E^*)$ has $Ric^{\perp} > 0$. Note that it is not a Kähler C-space, as it contains a section with negative normal bundle.

Similarly, for any $n \geq 3$, one can check that the curvature condition (5.1) is satisfied for the holomorphic cotangent bundle $E = \Omega_{\mathbb{P}^n}$ over $\mathbb{P}^n$, so we have:

COROLLARY 5.3. *For any $n \geq 3$, the $(2n-1)$-dimensional manifold $\mathbb{P}(T_{\mathbb{P}^n})$ has $Ric^{\perp} > 0$.*

In contrast, when the base manifold is 2-dimensional, the theorem does not give much information. In fact, we have the following result which is in sharp contrast with the higher base dimensional cases:

THEOREM 5.4. *Let P be a holomorphic fiber bundle over a compact complex surface S with fiber $\mathbb{P}^m$, where $m \geq 2$. If P admits a Kähler metric with $Ric^{\perp} > 0$ everywhere, then S is biholomorphic to $\mathbb{P}^2$ and P is biholomorphic to $\mathbb{P}^2 \times \mathbb{P}^m$.*

So if we take any non-trivial $\mathbb{P}^m$-bundle over $\mathbb{P}^2$, for instance, $\mathbb{P}(T_{\mathbb{P}^3}|_{\mathbb{P}^2})$ or $\mathbb{P}(\mathcal{O}^{\oplus 2} \oplus \mathcal{O}(1))$ over $\mathbb{P}^2$, we know by the above theorem that the total spaces do not admit any Kähler metric with $Ric^{\perp} > 0$. So the total space of a holomorphic fiber bundle may not admit any Kähler metric with $Ric^{\perp} > 0$ even when both the fiber and the base do.

6. Structural results

In the two previous sections, we have seen some existence results. Now let us turn our attention to the obstruction or non-existence side, and use them to obtain some structural results. Our goal is to obtain differential and algebraic geometric consequences from the curvature condition $Ric^{\perp} > 0$.

In [**20**], a generalization of a theorem of Frankel [**7**] was obtained:

THEOREM 6.1. *Let M^n be a compact Kähler manifold with $Ric^{\perp} > 0$. If Y_1 and Y_2 are smooth compact complex hypersurfaces in M, then $Y_1 \cap Y_2 \neq \phi$.*

Note that when the hypersurfaces Y_1 and Y_2 in the above theorem are singular, the same conclusion holds. This can be proved by a slight modification of the proof given in [**20**].

As an immediate corollary, we know that manifolds with $Ric^{\perp} > 0$ cannot be the blowing up of a (smooth or singular) point, or a fiberation over a curve:

COROLLARY 6.2. *Let M^n be a compact Kähler manifold with $Ric^{\perp} > 0$. Then there exists no surjective holomorphic map from M^n onto a complex curve, and there exists no birational morphism $f : M \to Z$ onto a normal variety Z, where a smooth hypersurface in M is mapped to a (smooth or singular) point.*

A Lefschetz type theorem can also be proved for compact Kähler manifolds with $Ric^{\perp} > 0$, namely, for a pair of complex hypersurfaces (Y_1, Y_2) in M, or for a hypersurface Y in M. The key here is that for any pair of hypersurfaces Y_1, Y_2, one can consider the space Ω of all paths in M originating from Y_1 and ending in Y_2. The energy $\mathcal{E}(\gamma)$ of a path $\gamma \in \Omega$ is defined in the usual way, and it is well known that the critical points of the energy functional are normal geodesics, namely, geodesics which intersects Y_i orthogonally. The same argument as in the proof of the above theorem implies the following index estimate, which includes the theorem as a special case since the minimizers can be identified with $Y_1 \cap Y_2$ (cf. [**23**]).

COROLLARY 6.3. *Let γ be a nontrivial critical point (namely a nonconstant normal geodesic after [**21**]). Then the index $ind(\gamma) \geq 1$. In particular,*

$$\pi_0(\Omega, Y_1 \cap Y_2) = \{0\}, \quad \iota_* : \pi_1(Y_1, Y_1 \cap Y_2) \to \pi_1(M, Y_2) \text{ is surjective.} \tag{6.1}$$

When $Y_1 = Y_2 = Y$, this implies that $\pi_1(M, Y) = \{0\}$.

Note that in [**21**] it was conjectured that $\pi_1(M) = \{0\}$. The last statement of the corollary is clearly a consequence of an affirmative answer to the conjecture.

An important geometric property for manifolds with $Ric^{\perp} > 0$ is the following:

THEOREM 6.4. *Let (M^n, g) be a compact Kähler manifold with $Ric^{\perp} > 0$. Let C be an irreducible curve in M and $f : \tilde{C} \to C \subset M$ be its normalization. If we denote by K_M the canonical line bundle of M and let g be the genus of $\tilde{C}$, then we have*

$$K_M^{-1} C \geq 3 - 2g.$$

In particular, $K_M^{-1} C \geq 3$ for any rational curve C in M.

For a smooth rational curve $C \subset M$, we have the short exact sequence of vector bundles on C

$$0 \to T_C \to T_M|_C \to N_C \to 0$$

where N_C is the normal bundle of C in M. By taking their first Chern classes, we get

$$c_1(N_C) = c_1(T_M|_C) - c_1(T_C) = K_M^{-1} C - 2 > 0$$

by the above theorem. In other words, for any smooth rational curve in M^n with $Ric^{\perp} > 0$, the normal bundle must have positive first Chern class.

The above results already put severe restrictions to the class $\mathcal{M}_n^{\perp}$ of all compact complex manifolds of complex dimension n which admit Kähler metrics with $Ric^{\perp} > 0$ everywhere. For instance, $\mathcal{M}_2^{\perp}$ consists of $\mathbb{P}^2$ alone by result in [**9**] on manifold with positive orthogonal bisectional curvature $B^{\perp}$, as when $n = 2$, $Ric^{\perp}$ coincides with $B^{\perp}$.

By Mori's theory on extremal rays and the cone-contraction theorems, one can use the above numerical restriction for $Ric^{\perp} > 0$ manifolds to draw conclusions on low dimensional cases (see for instance [**18**], [**14**], [**6**]):

THEOREM 6.5. *Let (M^3, g) be a compact Kähler manifold of dimension 3 with $Ric^{\perp} > 0$. Then M^3 is isomorphic to either $\mathbb{P}^3$ or $\mathbb{Q}^3$.*

Here and below we will denote by $\mathbb{Q}^n$ the smooth quadric in $\mathbb{P}^{n+1}$. In dimension 4, one could use the results in Mori's program (see for instance [**3**], [**12**], [**2**]) to narrow things down to the following list:

THEOREM 6.6. *Let (M^4, g) be a compact Kähler manifold of dimension 4 with $Ric^{\perp} > 0$. Then M^4 is isomorphic to either $\mathbb{P}^2 \times \mathbb{P}^2$ or a Fano fourfold with $b_2 = 1$ and pseudo index $i(M) \geq 3$.*

Recall that the *pseudo index* of a Fano manifold M is defined to be the minimum of the intersection number $K_M^{-1}C$ with the anti-canonical line bundle, where C runs through all rational curves in M.

A key intermediate step in deriving the above theorem is to rule out the possibility of a class of fourfolds, including for instance a $\mathbb{P}^2$-bundles over a Barlow's surface. The result was stated as Theorem 5.4 in the previous section, and we need a slight modification of the vanishing theorem on holomorphic 2-forms to achieve that goal.

Recall that a *del Pezzo manifold* M^n is defined to be a Fano manifold with index $n-1$, where the index is the largest integer r such that $K_M^{-1} = rA$ for an ample divisor A. For $n \geq 3$, such manifolds were completely classified by Fujita in [**8**], arranged by their degree $d = A^n$:

- $d = 1$: $X_6^n \subset \mathbb{P}(1^{n-1}, 2, 3)$, a degree 6 hypersurface in the weighted projective space.
- $d = 2$: $X_4^n \subset \mathbb{P}(1^n, 2)$, a degree 4 hypersurface in the weighted projective space.
- $d = 3$: $X_3^n \subset \mathbb{P}^{n+1}$, a cubic hypersurface.
- $d = 4$: $X_{2,2}^n \subset \mathbb{P}^{n+2}$, a complete intersection of two quadrics.
- $d = 5$: Y^n, a linear section of $\mathbb{G}r(2,5) \subset \mathbb{P}^9$.
- $d = 6$: $\mathbb{P}^1 \times \mathbb{P}^1 \times \mathbb{P}^1$, or $\mathbb{P}^2 \times \mathbb{P}^2$, or the flag threefold $\mathbb{P}(T_{\mathbb{P}^2})$.
- $d = 7$: $\mathbb{P}^3 \# \overline{\mathbb{P}^3}$, the blowing up of $\mathbb{P}^3$ at a point.

We believe that by further application of the deep and rich results in algebraic geometry, one should be able narrow things down even more, and in particular in dimension 4, we would like to propose the following:

CONJECTURE 6.7. *A compact complex manifold M^4 of dimension 4 admits a Kähler metric with $Ric^{\perp} > 0$ if and only if M^4 is biholomorphic to $\mathbb{P}^4$, or $\mathbb{Q}^4$, or a del Pezzo fourfold: X_6^4, X_4^4, X_3^4, $X_{2,2}^4$, Y^4, or $\mathbb{P}^2 \times \mathbb{P}^2$.*

Based on the structural theorems we obtained so far, we see that in dimension $n \leq 4$, the $Ric^{\perp} > 0$ condition is quite restrictive, it means (assuming the above conjecture holds true) Fano manifolds with index 3 or higher. This of course is more restrictive than $Ric > 0$, which means Fano, or $H > 0$, which is known to be rationally connected but the exact subset is still quite unclear. In fact, even for $\mathbb{P}^2 \# 2\overline{\mathbb{P}^2}$, the blowing up of $\mathbb{P}^2$ at two points, it is still unknown whether it admits a Kähler metric with $H > 0$ or not.

For dimensions 5 or higher, however, there are more examples of $Ric^{\perp} > 0$ manifolds, for instance, there are examples of Fano manifold with index 1 that lies in the set $\mathcal{M}_n^{\perp}$ of compact Kähler n-manifolds with $Ric^{\perp} > 0$. It is unclear how the set $\mathcal{M}_n^{\perp}$ look like, or how is it related to the Fano or the $H > 0$ class. We don't know if all such manifolds are Fano, even though all the examples in $\mathcal{M}_n^{\perp}$ that we were able to construct so far are Fano, but we do believe that all manifolds in $\mathcal{M}_n^{\perp}$ are rationally connected. In any event, we think they should form an interesting class of projective manifolds, and perhaps worth some attention from algebraic geometers.

References

[1] Angelynn Alvarez, Gordon Heier, and Fangyang Zheng, *On projectivized vector bundles and positive holomorphic sectional curvature*, Proc. Amer. Math. Soc. **146** (2018), no. 7, 2877–2882, DOI 10.1090/proc/13868. MR3787350

[2] Marco Andreatta and Massimiliano Mella, *Morphisms of projective varieties from the viewpoint of minimal theory*, Dissertationes Math. (Rozprawy Mat.) **413** (2003), 72, DOI 10.4064/dm413-0-1. MR1997253

[3] Marco Andreatta and Gianluca Occhetta, *Special rays in the Mori cone of a projective variety*, Nagoya Math. J. **168** (2002), 127–137, DOI 10.1017/S0027763000008400. MR1942399

[4] R. L. Bishop and S. I. Goldberg, *On the second cohomology group of a Kaehler manifold of positive curvature*, Proc. Amer. Math. Soc. **16** (1965), 119–122, DOI 10.2307/2034011. MR0172221

[5] Albert Chau and Luen-Fai Tam, *Kähler C-spaces and quadratic bisectional curvature*, J. Differential Geom. **94** (2013), no. 3, 409–468. MR3080488

[6] Thomas Dedieu and Andreas Höring, *Numerical characterisation of quadrics*, Algebr. Geom. **4** (2017), no. 1, 120–135, DOI 10.14231/AG-2017-006. MR3592468

[7] Theodore Frankel, *Manifolds with positive curvature*, Pacific J. Math. **11** (1961), 165–174. MR0123272

[8] Takao Fujita, *Classification theories of polarized varieties*, London Mathematical Society Lecture Note Series, vol. 155, Cambridge University Press, Cambridge, 1990. MR1162108

[9] HuiLing Gu and ZhuHong Zhang, *An extension of Mok's theorem on the generalized Frankel conjecture*, Sci. China Math. **53** (2010), no. 5, 1253–1264, DOI 10.1007/s11425-010-0013-y. MR2653275

[10] Ernst Heintze and Hermann Karcher, *A general comparison theorem with applications to volume estimates for submanifolds*, Ann. Sci. École Norm. Sup. (4) **11** (1978), no. 4, 451–470. MR533065

[11] Nigel Hitchin, *On the curvature of rational surfaces*, Differential geometry (Proc. Sympos. Pure Math., Vol. XXVII, Part 2, Stanford Univ., Stanford, Calif., 1973), Amer. Math. Soc., Providence, R. I., 1975, pp. 65–80. MR0400127

[12] Andreas Höring and Carla Novelli, *Mori contractions of maximal length*, Publ. Res. Inst. Math. Sci. **49** (2013), no. 1, 215–228, DOI 10.4171/PRIMS/103. MR3030002

[13] Mitsuhiro Itoh, *On curvature properties of Kähler C-spaces*, J. Math. Soc. Japan **30** (1978), no. 1, 39–71, DOI 10.2969/jmsj/03010039. MR0470904

[14] János Kollár, *Extremal rays on smooth threefolds*, Ann. Sci. École Norm. Sup. (4) **24** (1991), no. 3, 339–361. MR1100994

[15] Peter Li and Jiaping Wang, *Comparison theorem for Kähler manifolds and positivity of spectrum*, J. Differential Geom. **69** (2005), no. 1, 43–74, DOI 10.4310/jdg/1121540339. MR2169582

[16] Qun Li, Damin Wu, and Fangyang Zheng, *An example of compact Kähler manifold with nonnegative quadratic bisectional curvature*, Proc. Amer. Math. Soc. **141** (2013), no. 6, 2117–2126, DOI 10.1090/S0002-9939-2013-11596-0. MR3034437

[17] Gang Liu, *Three-circle theorem and dimension estimate for holomorphic functions on Kähler manifolds*, Duke Math. J. **165** (2016), no. 15, 2899–2919, DOI 10.1215/00127094-3645009. MR3557275

[18] Shigefumi Mori, *Threefolds whose canonical bundles are not numerically effective*, Ann. of Math. (2) **116** (1982), no. 1, 133–176, DOI 10.2307/2007050. MR662120

[19] Lei Ni and Luen-Fai Tam, *Poincaré-Lelong equation via the Hodge-Laplace heat equation*, Compos. Math. **149** (2013), no. 11, 1856–1870, DOI 10.1112/S0010437X12000322. MR3133296

[20] L. Ni, Q. Wang, and F. Zheng, *Manifolds with positive orthogonal Ricci curvature*. arXiv:1806.10233

[21] Lei Ni and Fangyang Zheng, *Comparison and vanishing theorems for Kähler manifolds*, Calc. Var. Partial Differential Equations **57** (2018), no. 6, Art. 151, 31, DOI 10.1007/s00526-018-1431-x. MR3858834

[22] L. Ni and F. Zheng, *Positivity and Kodaira embedding theorem*. arXiv:1804.09696.

[23] Richard Schoen and Jon Wolfson, *Theorems of Barth-Lefschetz type and Morse theory on the space of paths*, Math. Z. **229** (1998), no. 1, 77–89, DOI 10.1007/PL00004651. MR1649314

[24] Valentino Tosatti and Xiaokui Yang, *An extension of a theorem of Wu-Yau*, J. Differential Geom. **107** (2017), no. 3, 573–579, DOI 10.4310/jdg/1508551226. MR3715350

[25] Damin Wu and Shing-Tung Yau, *Negative holomorphic curvature and positive canonical bundle*, Invent. Math. **204** (2016), no. 2, 595–604, DOI 10.1007/s00222-015-0621-9. MR3489705

[26] Damin Wu, Shing-Tung Yau, and Fangyang Zheng, *A degenerate Monge-Ampère equation and the boundary classes of Kähler cones*, Math. Res. Lett. **16** (2009), no. 2, 365–374, DOI 10.4310/MRL.2009.v16.n2.a12. MR2496750

[27] Xiaokui Yang, *RC-positivity, rational connectedness and Yau's conjecture*, Camb. J. Math. **6** (2018), no. 2, 183–212, DOI 10.4310/CJM.2018.v6.n2.a2. MR3811235

[28] Shing Tung Yau, *Problem section*, Seminar on Differential Geometry, Ann. of Math. Stud., vol. 102, Princeton Univ. Press, Princeton, N.J., 1982, pp. 669–706. MR645762

Lei Ni. Department of Mathematics, University of California, San Diego, La Jolla, California 92093

Email address: lni@math.ucsd.edu

Fangyang Zheng. Department of Mathematics, The Ohio State University, Columbus, Ohio 43210

Email address: zheng.31@osu.edu

Contemporary Mathematics
Volume **735**, 2019
https://doi.org/10.1090/conm/735/14828

The Anomaly flow on unimodular Lie groups

Duong H. Phong, Sebastien Picard, and Xiangwen Zhang

Abstract. The Hull-Strominger system for supersymmetric vacua of the heterotic string allows general unitary Hermitian connections with torsion and not just the Chern unitary connection. Solutions on unimodular Lie groups exploiting this flexibility were found by T. Fei and S.T. Yau. The Anomaly flow is a flow whose stationary points are precisely the solutions of the Hull-Strominger system. Here we examine its long-time behavior on unimodular Lie groups with general unitary Hermitian connections. We find a diverse and intricate behavior, which depends very much on the Lie group and the initial data.

1. Introduction

The Hull-Strominger system [**21**, **22**, **36**] is a system of equations for supersymmetric vacua of the heterotic string which generalizes the well-known Calabi-Yau compactifications found by Candelas, Horowitz, Strominger, and Witten [**7**]. It is also of considerable interest from the point of view of non-Kähler geometry and non-linear partial differential equations [**15**, **16**, **30**, **32**, **33**]. While more special solutions continue to be constructed (see e.g. [**2**, **9**–**17**, **26**, **29**] and references therein), the complete solution of the Hull-Strominger system still seems distant. In [**31**], a general strategy was proposed, namely to look for solutions as stationary points of a flow, called there the Anomaly flow. Despite its unusual original formulation as a flow of $(2,2)$-forms, the Anomaly flow turns out to have some remarkable properties, including some suggestive analogies with the Ricci flow [**34**], and the fact [**35**] that it can recover the famous solutions found in 2006 by Fu and Yau [**15**, **16**]. Even so, very little is known at this time about the Anomaly flow in general, and it is important to gain insight from more examples.

Part of the interest in the Hull-Strominger system lies in it allowing metrics on complex manifolds which have non-vanishing torsion. For such metrics, there is actually a whole line of natural unitary connections, namely the Yano-Gauduchon line [**19**] passing by the Chern unitary connection and the Bismut connection. The Hull-Strominger system can be formulated with any specific choice along this line [**22**], and not just the Chern unitary connection. This flexibility is particularly useful if we further impose the Hermitian-Yang-Mills equation on the connection,

2010 *Mathematics Subject Classification.* Primary 53C55, 53C44.

Work supported in part by the National Science Foundation under NSF Grant DMS-12-66033 and the Simons Collaboration Grant-523313.

©2019 American Mathematical Society

which is interesting from the point of view of physics as the solution then satisfies the heterotic equations of motion [**13, 24**].

Solutions to the Hull-Strominger system were found by Fei and Yau on unimodular complex Lie groups [**12**], using an ansatz originating from [**3**] and [**5**]. The advantage of unimodular Lie groups is that any left-invariant Hermitian metric is automatically balanced, so if we fix a holomorphic vector bundle with a Hermitian-Yang-Mills invariant metric, the Hull-Strominger will reduce to a single equation. A difficulty in using general Hermitian connections is that their Riemannian curvature tensor Rm will in general have components of all $(2,0)$, $(1,1)$, and $(0,2)$ types. But Fei and Yau discovered that, for unimodular Lie groups, the term $\mathrm{Tr}(Rm \wedge Rm)$ reduces to a $(2,2)$-form, and that solutions to the Hull-Strominger system can be found with any connection on the Yano-Gauduchon line, except for the Chern unitary connection and the Lichnerowicz connection.

The purpose of this paper is analyze the Anomaly flow on unimodular Lie groups. As had been stressed earlier, the study of the Anomaly flow has barely begun, and there is a strong need for good examples. Even in the case of toric fibrations analyzed in [**35**], the convergence of the flow was established only for a particular type of initial data, and a fuller understanding of the long-time behavior of the flow is still not available. Unlike the more familiar cases of the Kähler-Ricci and the Donaldson heat flows, which are known to converge whenever a stationary point exists, the Anomaly flow is expected to behave differently depending on the initial data. This can be traced back to the fact that the conformally balanced condition is weaker than the Kähler condition, and that the positivity and closedness of a $(2,2)$-form seem to carry less information than the positivity and closedness of a $(1,1)$-form. The case of Lie groups is particular appealing, as the flow reduces then to a system of ordinary differential equations, and its formalism can be readily extended to general connections on the Yano-Gauduchon line. Our main results are as follows.

Let X be a complex 3-dimensional Lie group. Fix a basis $\{e_a\}$ of left-invariant holomorphic vector fields on X, and let $\{e^a\}$ be the dual basis of holomorphic forms. Let c^d_{ab} denote the structure constants of X defined by this basis

$$[e_a, e_b] = \sum_d e_d\, c^d{}_{ab}. \tag{1.1}$$

The Lie group X is said to be unimodular if the structure constants satisfy the condition

$$\sum_a c^a{}_{ab} = 0 \tag{1.2}$$

for any b. This condition is invariant under a change of basis of holomorphic vector fields (see e.g. Section §3.1 for a proof). The basis $\{e_a\}$ also defines a holomorphic, nowhere vanishing $(3,0)$-form on X, given by

$$\Omega = e^1 \wedge e^2 \wedge e^3. \tag{1.3}$$

We define the norm of Ω with respect to any metric ω by

$$\|\Omega\|^2 = i\Omega \wedge \bar{\Omega} \left(\frac{\omega^3}{3!}\right)^{-1}. \tag{1.4}$$

THEOREM 1.1. *Let X be a unimodular complex 3-dimensional Lie group. Let $t \to \omega(t)$ be the Anomaly flow on X, as defined in (3.4) below. Set $\omega(t) = i g_{\bar{a}b}(t)\, e^b \wedge \bar{e}^a$. Then the Anomaly flow is given explicitly by*

$$(1.5) \qquad \partial_t g_{\bar{a}b} \;=\; \frac{1}{2\|\Omega\|_\omega} g^{d\bar{p}} \left(g_{\bar{i}s} \overline{c^i{}_{ap}} c^s{}_{bd} - \frac{\alpha'\tau}{4}\, g_{\bar{\ell}i} g^{n\bar{j}}\, \overline{c^m{}_{ap}}\, \overline{c^\ell{}_{mj}} c^i{}_{sn} c^s{}_{bd} \right).$$

Here α' is the string tension, and $\tau = 2\kappa^2(2\kappa - 1)$, where κ indicates the connection (2.13).

In three dimensions, as listed in Fei and Yau [**12**] and Knapp [**25**], the unimodular complex Lie groups consist precisely of those whose Lie algebra is isomorphic to that of either the Abelian Lie group $\mathbf{C}^3$, the nilpotent Heisenberg group, the solvable group of (complexfications of) rigid motions in $\mathbf{R}^2$, or the semi-simple group $SL(2, \mathbf{C})$. The Anomaly flow behaves differently in each case. Fixing the structure constants in each case as spelled out in §3, we have:

THEOREM 1.2. *Assume that $\alpha'\tau > 0$.*

(a) *When $X = \mathbf{C}^3$, any metric is a stationary point for the flow, and the flow is consequently stationary for any initial metric $\omega(0)$.*

(b) *When X is nilpotent, there is no stationary point. Consequently the flow cannot converge for any initial metric. If the initial metric is diagonal, then the metric remains diagonal along the flow, the lowest eigenvalue is constant, while the other two eigenvalues tend to $+\infty$ at a constant rate.*

(c) *When X is solvable, the stationary points of the flow are precisely the metrics with*

$$(1.6) \qquad g_{\bar{1}2} = \overline{g_{\bar{2}1}} = 0, \qquad \frac{\alpha'\tau}{4} g^{3\bar{3}} = 1.$$

The Anomaly flow is asymptotically instable near any stationary point. However, the condition $g_{\bar{1}2} = \overline{g_{\bar{1}2}} = 0$ is preserved along the flow, and for any initial metric satisfying this condition, the flow converges to a stationary point.

(d) *When $X = SL(2, \mathbf{C})$, there is a unique stationary point, given by the diagonal metric*

$$(1.7) \qquad g_{\bar{a}b} = \frac{\alpha'\tau}{2}\, \delta_{ab}.$$

The linearization of the flow at the fixed point admits both positive and negative eigenvalues. In particular, the flow is asymptotically instable.

Our main result shows that, as in the Ricci flow, the behavior of the Anomaly flow depends on the initial data, and singularities may occur. We hope that the singularities encode interesting geometric information, and that in general when solutions exist there is a geometric condition which gives a hypersuface of initial data where the flow is stable.

The paper is organized as follows. In §2, the terms in the main equation in the Hull-Strominger system, namely the anomaly cancellation condition, are worked out in detail for connections on the Yano-Gauduchon line for left-invariant metrics. While the arguments are directly inspired by those of Fei and Yau, we have adopted a more explicit component formalism that should make the proof as well as the resulting formulas more accessible and convenient for geometric flows. The Anomaly flow is analyzed in §3. Theorem 1.1 is proved in §3.2, and Theorem

1.2 is proved in §3.3, beginning in each case with the identification of the stationary points, and ending with the analysis of the Anomaly flow proper.

2. The curvature of unitary connections with torsion on Lie groups

In this section, we derive formulas for the curvature of an arbitrary connection on the Yano-Gauduchon line for Lie groups. They have been obtained before by Fei and Yau [**12**], but we need a formulation convenient for our subsequent study of the Anomaly flow.

2.1. Unitary connections with torsion. Let (X,ω) be a complex manifold of complex dimension n, equipped with a Hermitian metric ω. We shall denote by z^j, $1\leq j\leq n$, local holomorphic coordinates on X, and by ξ^α, $1\leq\alpha\leq 2n$, real smooth local coordinates on X. In particular $g_{\bar kj}$ denotes the metric in holomorphic coordinates, and $g_{\alpha\beta}$ the metric in real coordinates. We consider only unitary connections on X, that is connections ∇ on $T^{1,0}(X)$ which preserve the metric, $\nabla g_{\bar kj}=0$. Our notation for covariant derivatives is

$$\nabla_\alpha V^\gamma=\partial_\alpha V^\gamma+A^\gamma{}_{\alpha\beta}V^\beta. \tag{2.1}$$

and our conventions for the curvature $R_{\alpha\beta}{}^\gamma{}_\delta$ and the torsion tensors $T^\gamma{}_{\alpha\beta}$ are,

$$[\nabla_\beta,\nabla_\alpha]V^\gamma=R_{\alpha\beta}{}^\gamma{}_\delta V^\delta+T^\delta{}_{\alpha\beta}\nabla_\delta V^\gamma, \tag{2.2}$$

with similar formulas when using the complex coordinates $z^j,\bar z^j$, $1\leq j\leq n$. For example, in complex coordinates, the components of the torsion are

$$\begin{aligned} T^p{}_{\bar jm}&=A^p_{\bar jm}, & T^{\bar p}{}_{\bar jm}&=-A^{\bar p}_{m\bar j},\\ T^p{}_{jm}&=A^p_{jm}-A^p_{mj}, & T^{\bar p}{}_{\bar j\bar m}&=A^{\bar p}_{\bar j\bar m}-A^{\bar p}_{\bar m\bar j}. \end{aligned} \tag{2.3}$$

Among the unitary connections, of particular interest are the following three special connections:

(1) The Levi-Civita connection ∇^L on $T(X)$, characterized by the fact that its torsion is 0. In general, it does not preserve the complex structure, and in particular, it does not induce a connection on the complex tangent space $T^{1,0}(X)$. It preserves the complex structure J if and only if ω is Kähler.

(2) The Chern connection ∇^C, characterized by the fact that it preserves the complex structure, and that $\nabla^C_{\bar j}W^k=\partial_{\bar j}W^k$, i.e., $A^p_{\bar jm}=0$. By unitarity, it follows that $A^p_{jm}=g^{p\bar q}\partial_j g_{\bar qm}$. Thus the components of the torsion of the Chern connection in the first line of (2.3) vanish, and the torsion reduces to a section of the following bundle

$$\mathcal{T}=T+\bar T\in(\Lambda^{2,0}\otimes T^{1,0})\oplus(\Lambda^{0,2}\otimes T^{0,1}). \tag{2.4}$$

with

$$T=\frac{1}{2}T^\ell{}_{jk}\frac{\partial}{\partial z^\ell}\otimes(dz^k\wedge dz^j),\quad \bar T=\frac{1}{2}\bar T^{\bar\ell}{}_{\bar j\bar k}\frac{\partial}{\partial\bar z^\ell}\otimes(d\bar z^k\wedge d\bar z^j) \tag{2.5}$$

and

$$\bar T^{\bar\ell}{}_{\bar j\bar k}=\overline{T^\ell{}_{jk}}.$$

The torsion of the Chern connection is 0 exactly when ω is Kähler.

(3) The Bismut connection $\nabla^{\mathcal H}$. Connections obtained by modifying the Levi-Civita connection by a 3-form seem to have been written down first by Yano in

his book [**39**], which became the standard reference for it in the physics literature on supersymmetric sigma models (see e.g. [**18, 21, 23, 36**] and references therein). They were subsequently rediscovered by Bismut [**4**], who also identified among them the connection preserving the complex structure, which is now known as the Bismut connection. Thus the Bismut connection can be written in two different ways, depending on whether we use ∇^L or ∇^C as reference connection. To write it with ∇^L as reference connection, introduce the following real 3-form [1],

$$\mathcal{H} = \frac{1}{4}(H + \bar{H}) \in \Lambda^{2,1} \oplus \Lambda^{1,2} \subset \Lambda^3 \tag{2.6}$$

where

$$H = \frac{1}{2}\, g_{\bar{m}\ell}\, T^\ell{}_{jk}\, dz^k \wedge dz^j \wedge d\bar{z}^m, \qquad \bar{H} = \frac{1}{2}\, g_{\bar{\ell}m}\, \bar{T}^{\bar{\ell}}{}_{\bar{j}\bar{k}}\, d\bar{z}^k \wedge d\bar{z}^j \wedge dz^m,$$
$$H_{\bar{m}jk} = g_{\bar{m}\ell} T^\ell{}_{jk}, \qquad \bar{H}_{m\bar{j}\bar{k}} = g_{\bar{\ell}m} \bar{T}^{\ell}{}_{\bar{j}\bar{k}}. \tag{2.7}$$

Note that, in terms of the Hermitian form $\omega = i g_{\bar{k}j} dz^j \wedge d\bar{z}^k$, we have

$$H = i\partial\omega, \qquad \bar{H} = -i\bar{\partial}\omega. \tag{2.8}$$

Next, introduce the 1-form $S^{\mathcal{H}} = (S^{\mathcal{H}}_\alpha) d\xi^\alpha$ valued in the space of endomorphisms anti-symmetric with respect to the metric $g_{\alpha\beta}$ by setting [2]

$$\mathcal{H} \equiv \frac{1}{3!}\, \mathcal{H}_{\alpha\beta\gamma}\, d\xi^\gamma \wedge d\xi^\beta \wedge d\xi^\alpha,$$
$$(S^{\mathcal{H}}_\alpha)^\beta{}_\gamma = 2\mathcal{H}_{\alpha\rho\gamma}\, g^{\rho\beta}. \tag{2.9}$$

The Bismut connection is then defined by

$$\nabla^{\mathcal{H}} = \nabla^L + S^{\mathcal{H}}. \tag{2.10}$$

This formulation shows that it is unitary and that its torsion can be identified with a 3-form. If we use the Chern connection as the reference connection, we can also write

$$\nabla^{\mathcal{H}}_j W^p = \nabla^C_j W^p - T^p{}_{jk} W^k, \tag{2.11}$$

$$\nabla^{\mathcal{H}}_{\bar{j}} W^p = \nabla^C_{\bar{j}} W^p + g^{p\bar{q}} g_{\bar{m}r} \bar{T}^{\bar{m}}{}_{\bar{j}\bar{q}} W^r. \tag{2.12}$$

This formulation shows that the Bismut connection is unitary and that it manifestly preserves the complex structure. The equivalence of (2.10) and (2.11) can be found in [**4**].

The two connections ∇^C and $\nabla^{\mathcal{H}}$ determine a line of connections $\nabla^{(\kappa)}$ which are all unitary and preserve the complex structure. Indeed, in terms of an orthonormal frame, a unitary connection is characterized by a Hermitian matrix, and a real linear combination of Hermitian matrices is again Hermitian. Also a linear combination of connections on $T^{1,0}(X)$ is again a connection on $T^{1,0}(X)$, hence our assertion. We

---

[1] It is important to distinguish the form $H \in \Lambda^{2,1}$ from the torsion tensor $T \in \Lambda^{2,0} \otimes T^{1,0}$, although their coefficients are the same, and they can be recovered from one another. Perhaps a good analogue is the metric $g_{\bar{k}j}$ and the corresponding symplectic form $\omega = i g_{\bar{k}j}\, dz^j \wedge d\bar{z}^k$, which are distinct objects, although they can be recovered from one another, once the complex structure is fixed.

[2] An endomorphism $S^\beta{}_\gamma$ is antisymmetric with respect to a metric $g_{\rho\beta}$ if $\langle SW, U\rangle = -\langle W, SU\rangle$, which means that $g_{\beta\rho} S^\beta{}_\gamma = -S^\mu{}_\rho g_{\gamma\mu}$, which means that $S_{\rho\gamma}$ is anti-symmetric, if we raise and power indices using the metric $g_{\beta\rho}$.

refer to this line of connections as the Yano-Gauduchon line of canonical connections [**19**]. With ∇^C as reference connection, the connection $\nabla^{(\kappa)}$ can be expressed as

$$\nabla^{(\kappa)}_j W^p = \nabla^C_j W^p - \kappa\, T^p{}_{jk} W^k, \tag{2.13}$$

$$\nabla^{(\kappa)}_{\bar j} W^p = \nabla^C_{\bar j} W^p + \kappa\, g^{p\bar q} g_{\bar m r} \bar T^{\bar m}{}_{\bar j \bar q} W^r. \tag{2.14}$$

Clearly $\kappa = 0$ corresponds to the Chern connection and $\kappa = 1$ to the Bismut connection. The case $\kappa = 1/2$ is known as the Lichnerowicz connection.

2.2. Evaluation of curvature forms. Let (X, ω) be now a complex Lie group with left-invariant metric ω. Let $e_1, \ldots, e_n \in \mathbf{g}$ be an orthonormal frame of left-invariant holomorphic vector fields on X, and let $e^1, \ldots, e^n$ be the dual frame of holomorphic 1-forms, so that

$$\omega = i \sum_a e^a \wedge \bar e^a. \tag{2.15}$$

The structure constants of the Lie algebra $\mathbf{g}$ in this basis are defined in (1.1) and denoted by $c^d{}_{ab}$. They satisfy the Jacobi identity

$$c^q{}_{ir} c^r{}_{jk} + c^q{}_{kr} c^r{}_{ij} + c^q{}_{jr} c^r{}_{ki} = 0. \tag{2.16}$$

By Cartan's formula for the exterior derivative, we have

$$\partial e^a = \frac{1}{2} c^a{}_{bd}\, e^d \wedge e^b \tag{2.17}$$

which is the Maurer-Cartan equation.

Our goal is to determine the torsion and curvature of an arbitrary connection $\nabla^{(\kappa)}$ on the Yano-Gauduchon line in terms of the frame $e^1, \cdots, e^n$ and the structure constants $c^d{}_{ab}$. First, we express the connection forms for $\nabla^{(\kappa)}$ in terms of the structure constants of X:

LEMMA 2.1. *Let $\nabla^{(\kappa)}$ be a connection on the Yano-Gauduchon line for a left-invariant Hermitian metric ω on the Lie group X. Let $e^1, \cdots, e^n$ be a left invariant orthonormal basis for ω, and let $c^d{}_{ab}$ be the corresponding structure constants, as defined above. Then for any vector field $V = V^a e_a$,*

$$\nabla^{(\kappa)}_b V^a = \partial_{e_b} V^a + \kappa c^a{}_{bd} V^d, \quad \nabla^{(\kappa)}_{\bar b} V^a = \partial_{\bar e_b} V^a - \kappa \overline{c^d{}_{ba}} V^d. \tag{2.18}$$

Proof. We use the expression for $\nabla^{(\kappa)}$ with the Chern connection ∇^C as reference connection. For this, we need the torsion form H of the Chern connection. Taking the exterior derivative of ω and applying (2.17) gives

$$H = i\partial\omega = -\frac{1}{2} \sum c^d{}_{ab}\, e^b \wedge e^a \wedge \bar e^d. \tag{2.19}$$

Therefore

$$H_{\bar d ab} = (i\partial\omega)_{\bar d ab} = -c^d{}_{ab}. \tag{2.20}$$

Next, since the coefficients of the metric ω are constant in an orthonormal frame, the Chern connection reduces to the exterior derivative: $\nabla^C_j = \partial_j$ in this frame. The lemma follows now from the formula (2.13) giving $\nabla^{(\kappa)}$ in terms of ∇^C, H, and the parameter κ.

Henceforth, we shall denote $\nabla^{(\kappa)}$ by just ∇ for simplicity. We obtain

$$A^a{}_{bd} = \kappa c^a{}_{bd}, \quad A^a{}_{\bar b \bar d} = -\kappa \overline{c^d{}_{ba}}, \quad A^{\bar a}{}_{b\bar d} = -\kappa c^d{}_{ba}, \quad A^{\bar a}{}_{\bar b \bar d} = \kappa \overline{c^a{}_{bd}}. \tag{2.21}$$

LEMMA 2.2. *The components of the curvature of the connection* $\nabla^{(\kappa)}$ *in the orthonormal frame* $e^1, \cdots, e^n$ *are given by*

$$R_{kj}{}^p{}_q = (\kappa - \kappa^2) c^r{}_{kj} c^p{}_{rq}, \qquad R_{\bar k \bar j}{}^p{}_q = -(\kappa - \kappa^2)\overline{c^r{}_{kj} c^q{}_{rp}}$$
$$R_{\bar k j}{}^p{}_q = \kappa^2(-c^p{}_{jr}\overline{c^q{}_{kr}} + \overline{c^r{}_{kp}} c^r{}_{jq}). \tag{2.22}$$

Proof. The defining formula (2.2) for the curvature and the torsion $t^d{}_{ba}$ of the connection $\nabla^{(\kappa)}$ in coordinates becomes the following formula in the frame $\{e_a\}$,

$$[\nabla_a, \nabla_b] W^c = R_{ba}{}^c{}_d W^d + t^d{}_{ba} \nabla_d W^c + c^d{}_{ab} \nabla_d W^c \tag{2.23}$$

where the last term on the right hand side is the contribution of the commutator $[e_a, e_b]$. H We now compute the left hand side using the formulas of the previous lemma. We find

$$\begin{aligned} & R_{kj}{}^p{}_q W^q + t^q{}_{kj} \nabla_q W^p + c^q{}_{jk} \nabla_q W^p \\ = \ & [\nabla_j, \nabla_k] W^p \\ = \ & c^q{}_{jk} \partial_{e_q} W^p + \kappa^2(-c^p{}_{kr} c^r{}_{js} + c^p{}_{jr} c^r{}_{ks}) W^s + 2\kappa c^r{}_{kj} \nabla_r W^p. \end{aligned} \tag{2.24}$$

By (2.18) and the Jacobi identity (2.16), we see that the $(2,0)$ component is

$$\begin{aligned} R_{kj}{}^p{}_q &= -\kappa c^r{}_{jk} c^p{}_{rq} + \kappa^2(-c^p{}_{kr} c^r{}_{jq} + c^p{}_{jr} c^r{}_{kq}) \\ &= (\kappa - \kappa^2) c^r{}_{kj} c^p{}_{rq}. \end{aligned} \tag{2.25}$$

This proves the first formula in the lemma. Next, we have

$$\begin{aligned} & R_{\bar k \bar j}{}^p{}_q W^q + \bar t^{\bar q}{}_{\bar k \bar j} \nabla_{\bar q} W^p + \overline{c^q{}_{jk}} \nabla_{\bar q} W^p \\ = \ & [\nabla_{\bar j}, \nabla_{\bar k}] W^p \\ = \ & \overline{c^q{}_{jk}} \partial_{\bar e_q} W^p + \kappa^2(\overline{c^r{}_{jp} c^s{}_{kr} - c^r{}_{kp} c^s{}_{jr}}) W^s - 2\kappa \overline{c^r{}_{kj}} \nabla_{\bar r} W^p. \end{aligned} \tag{2.26}$$

We see that the $(0,2)$ component is

$$R_{\bar k \bar j}{}^p{}_q = \kappa \overline{c^r{}_{jk} c^q{}_{rp}} + \kappa^2(\overline{c^r{}_{jp} c^q{}_{kr} - c^r{}_{kp} c^q{}_{jr}}) = -(\kappa - \kappa^2)\overline{c^r{}_{kj} c^q{}_{rp}}. \tag{2.27}$$

which is the second formula in the lemma. Finally, we have

$$\begin{aligned} & R_{\bar k j}{}^p{}_q W^q - A^{\bar q}{}_{j \bar k} \nabla_{\bar q} W^p + A^q{}_{\bar k j} \nabla_q W^p \\ = \ & [\nabla_j, \nabla_{\bar k}] W^p \\ = \ & \kappa c^k{}_{jr} \nabla_{\bar r} W^p - \kappa \overline{c^j{}_{kr}} \nabla_r W^p + \kappa^2(-c^p{}_{jr} \overline{c^q{}_{kr}} + \overline{c^r{}_{kp}} c^r{}_{jq}) W^q. \end{aligned} \tag{2.28}$$

We see that the $(1,1)$ component is

$$R_{\bar k j}{}^p{}_q = \kappa^2(-c^p{}_{jr}\overline{c^q{}_{kr}} + \overline{c^r{}_{kp}} c^r{}_{jq}). \tag{2.29}$$

This completes the proof of the lemma.

We note our expressions satisfy the usual symmetries

$$R_{kj}{}^p{}_s = -R_{jk}{}^p{}_s, \quad R_{\bar k \bar j}{}^p{}_s = -R_{\bar j \bar k}{}^p{}_s \tag{2.30}$$

$$R_{\bar k \bar j}{}^p{}_q = -\overline{R_{kj}{}^q{}_p}, \quad R_{\bar k j}{}^p{}_q = \overline{R_{\bar j k}{}^q{}_p}. \tag{2.31}$$

Next, we compute $\mathrm{Tr}\,(Rm \wedge Rm)$ on the Lie group X. This computation was first done by Fei-Yau [**12**]. Their computation showed in particular that $\mathrm{Tr}\,(Rm \wedge$

Rm) is a $(2,2)$ form along the Yano-Gauduchon line. Here we recover their result using our formalism.

LEMMA 2.3. *Let $\nabla^{(\kappa)}$ be an arbitrary connection on the Yano-Gauduchon line, as before. Assume that* $\dim X = 3$. *Then* $\mathrm{Tr}\,(Rm \wedge Rm)$ *is a $(2,2)$-form, given explicitly by*

$$\mathrm{Tr}\,(Rm \wedge Rm)_{\bar k \bar\ell ij} \quad = \quad \tau\, \overline{c^r{}_{k\ell} c^s{}_{rp}} c^q{}_{ij} c^s{}_{qp}. \tag{2.32}$$

where we have set as before $\tau = 2\kappa^2(2\kappa - 1)$.

Proof. Because X is assumed to have dimension 3, the $(4,0)$ and $(0,4)$ components of $\mathrm{Tr}(Rm \wedge Rm)$ are automatically 0. Next, we show the vanishing of the $(3,1)$-component,

$$\mathrm{Tr}\,(Rm \wedge Rm)_{\bar\ell ijk} = 2(R_{ij}{}^p{}_s R_{\bar\ell k}{}^s{}_p + R_{ki}{}^p{}_s R_{\bar\ell j}{}^s{}_p + R_{jk}{}^p{}_s R_{\bar\ell i}{}^s{}_p). \tag{2.33}$$

Fixing an index (ijk,ℓ), we use the previously computed formulas for the curvature of a Lie group to obtain

$$R_{ij}{}^p{}_s R_{\bar\ell k}{}^s{}_p = (\kappa - \kappa^2)\kappa^2 c^r{}_{ij} c^p{}_{rs}(-c^s{}_{kq}\overline{c^p{}_{\ell q}} + \overline{c^q{}_{\ell s}} c^q{}_{kp}). \tag{2.34}$$

Applying the Jacobi identity (2.16),

$$\begin{aligned} & R_{ij}{}^p{}_s R_{\bar\ell k}{}^s{}_p \\ = \; & \kappa^3(1-\kappa)(-c^p{}_{ir} c^r{}_{js} c^s{}_{kq}\overline{c^p{}_{\ell q}} + c^p{}_{jr} c^r{}_{is} c^s{}_{kq}\overline{c^p{}_{\ell q}} \\ & - c^q{}_{kp} c^p{}_{jr} c^r{}_{is}\overline{c^q{}_{\ell s}} + c^q{}_{kp} c^p{}_{ir} c^r{}_{js}\overline{c^q{}_{\ell s}}) \\ = \; & \kappa^3(1-\kappa)(-c^p{}_{ir} c^r{}_{js} c^s{}_{kq} + c^p{}_{jr} c^r{}_{is} c^s{}_{kq} - c^p{}_{ks} c^s{}_{jr} c^r{}_{iq} + c^p{}_{ks} c^s{}_{ir} c^r{}_{jq})\overline{c^p{}_{\ell q}}. \end{aligned}$$

If we denote $F_{ijk}{}^p{}_q = c^p{}_{ir} c^r{}_{js} c^s{}_{kq}$, then

$$R_{ij}{}^p{}_s R_{\bar\ell k}{}^s{}_p = \kappa^3(1-\kappa)(-F_{ijk}{}^p{}_q + F_{jik}{}^p{}_q - F_{kji}{}^p{}_q + F_{kij}{}^p{}_q)\,\overline{c^p{}_{\ell q}}.$$

Upon cyclically permuting (ijk), we see that

$$R_{ij}{}^p{}_s R_{\bar\ell k}{}^s{}_p + R_{ki}{}^p{}_s R_{\bar\ell j}{}^s{}_p + R_{jk}{}^p{}_s R_{\bar\ell i}{}^s{}_p = 0. \tag{2.35}$$

Therefore

$$\mathrm{Tr}\,(Rm \wedge Rm)_{\bar\ell ijk} = 0 \tag{2.36}$$

as claimed. Next, we compute the $(1,3)$ component,

$$\mathrm{Tr}\,(Rm \wedge Rm)_{\bar i\bar j\bar k\ell} \quad = \quad 2(R_{\bar i\bar j}{}^p{}_s R_{\bar k\ell}{}^s{}_p + R_{\bar j\bar k}{}^p{}_s R_{\bar i\ell}{}^s{}_p + R_{\bar k\bar i}{}^p{}_s R_{\bar j\ell}{}^s{}_p). \tag{2.37}$$

Using the symmetric property (2.31) and the vanishing of the $(3,1)$ part, we see that the $(1,3)$ part also vanishes.

Finally, we evaluate the $(2,2)$-component,

$$\mathrm{Tr}\,(Rm \wedge Rm)_{\bar k\bar\ell ij} = 2R_{\bar k\bar\ell}{}^p{}_s R_{ij}{}^s{}_p + 2(R_{\bar kj}{}^p{}_s R_{\bar\ell i}{}^s{}_p - R_{\bar ki}{}^p{}_s R_{\bar\ell j}{}^s{}_p). \tag{2.38}$$

From the previously established formulas, we have

$$R_{\bar k\bar\ell}{}^p{}_s R_{ij}{}^s{}_p = -(\kappa - \kappa^2)^2\, \overline{c^r{}_{k\ell} c^s{}_{rp}} c^q{}_{ij} c^s{}_{qp}. \tag{2.39}$$

Next,

$$\begin{aligned} & R_{\bar kj}{}^p{}_s R_{\bar\ell i}{}^s{}_p \\ = \; & \kappa^4(-c^p{}_{jr}\overline{c^s{}_{kr}} + \overline{c^r{}_{kp}} c^r{}_{js})(-c^s{}_{iq}\overline{c^p{}_{\ell q}} + \overline{c^q{}_{\ell s}} c^q{}_{ip}) \\ = \; & \kappa^4(c^p{}_{jr}\overline{c^s{}_{kr}} c^s{}_{iq}\overline{c^p{}_{\ell q}} + c^q{}_{ip}\overline{c^r{}_{kp}} c^r{}_{js}\overline{c^q{}_{\ell s}} - c^p{}_{jr}\overline{c^s{}_{kr} c^q{}_{\ell s}} c^q{}_{ip} \\ & - c^s{}_{iq}\overline{c^p{}_{\ell q} c^r{}_{kp}} c^r{}_{js}). \end{aligned} \tag{2.40}$$

Some cancellation occurs and we are left with

$$\begin{aligned}
&(R_{\bar{k}j}{}^p{}_s R_{\bar{\ell}i}{}^s{}_p - R_{\bar{k}i}{}^p{}_s R_{\bar{\ell}j}{}^s{}_p) \\
= \;& \kappa^4(-\overline{c^s{}_{kr}c^q{}_{\ell s}}c^q{}_{ip}c^p{}_{jr} - \overline{c^p{}_{\ell q}c^r{}_{kp}}c^r{}_{js}c^s{}_{iq} + \overline{c^s{}_{kr}c^q{}_{\ell s}}c^q{}_{jp}c^p{}_{ir} \\
& + \overline{c^p{}_{\ell q}c^r{}_{kp}}c^r{}_{is}c^s{}_{jq}) \\
= \;& \kappa^4\left(\overline{c^s{}_{kr}c^q{}_{\ell s}}(-c^q{}_{ip}c^p{}_{jr} + c^q{}_{jp}c^p{}_{ir}) + \overline{c^p{}_{\ell q}c^r{}_{kp}}(-c^r{}_{js}c^s{}_{iq} + c^r{}_{is}c^s{}_{jq})\right) \\
= \;& \kappa^4(\overline{c^s{}_{kr}c^q{}_{\ell s}}c^q{}_{rp}c^p{}_{ij} - \overline{c^p{}_{\ell q}c^r{}_{kp}}c^r{}_{qs}c^s{}_{ij}) \\
= \;& \kappa^4(\overline{c^s{}_{kr}c^q{}_{\ell s}} - \overline{c^s{}_{\ell r}c^q{}_{ks}})c^q{}_{rp}c^p{}_{ij} \\
= \;& \kappa^4\overline{c^q{}_{rs}c^s{}_{k\ell}}c^q{}_{rp}c^p{}_{ij}.
\end{aligned}$$

Adding these two equations together, the terms of order κ^4 cancel and we obtain the desired formula.

Recall that we view the Lie algebra of X as generated by a given basis of left-invariant holomorphic vector fields e_a, with structure constants $c^d{}_{ab}$. So far we have considered only the metric $\omega = i\sum e^a \wedge \bar{e}^a$ defined by the condition that e_a be orthonormal. We consider now the general left-invariant metric given by

$$\omega = \sum g_{\bar{b}a}\, ie^a \wedge \bar{e}^b, \tag{2.41}$$

where $g_{\bar{b}a}$ is a positive-definite Hermitian matrix. We have

LEMMA 2.4. *Let X be a 3-dimensional complex Lie group, with a given basis of left-invariant holomorphic vector fields e_a with structure constants $c^d{}_{ab}$. Let ω be the metric given by (2.41), and let $\nabla^{(\kappa)}$ be the Hermitian connection on the Gauduchon line with parameter κ. Then*

$$\mathrm{Tr}\,(Rm \wedge Rm) = \frac{\tau}{4}\, g_{\bar{\ell}i}g^{n\bar{j}}\, \overline{c^m{}_{ab}}\,\overline{c^\ell{}_{mj}}c^i{}_{sn}c^s{}_{cd}\, e^d \wedge e^c \wedge \bar{e}^b \wedge \bar{e}^a. \tag{2.42}$$

and

$$\begin{aligned}
& i\partial\bar{\partial}\omega - \frac{\alpha'}{4}\mathrm{Tr}\,(Rm \wedge Rm) \\
= \;& \frac{1}{4}\left(g_{\bar{i}s}\overline{c^i{}_{ab}}c^s{}_{cd} - \frac{\alpha'\tau}{4}\, g_{\bar{\ell}i}g^{n\bar{j}}\, \overline{c^m{}_{ab}}\,\overline{c^\ell{}_{mj}}c^i{}_{sn}c^s{}_{cd}\right) e^d \wedge e^c \wedge \bar{e}^b \wedge \bar{e}^a.
\end{aligned} \tag{2.43}$$

Proof. These formulas for general metrics follow from the ones obtained earlier for the metric $i\sum_a e^a \wedge \bar{e}^a$ after performing a change of basis. More specifically, we let P be a matrix such that

$$\bar{P}^{\bar{a}}{}_{\bar{p}}g_{\bar{a}b}P^b{}_q = \delta_{\bar{p}q}. \tag{2.44}$$

Therefore, denoting $g^{a\bar{b}}$ to be the inverse of $g_{\bar{b}a}$, we have

$$g_{\bar{b}a} = (\bar{P}^{-1})^{\bar{r}}{}_{\bar{b}}(P^{-1})^r{}_a, \quad g^{a\bar{b}} = P^a{}_r\bar{P}^{\bar{b}}{}_{\bar{r}}. \tag{2.45}$$

We now perform a change of basis and define

$$f_i = e_r P^r{}_i, \quad [f_i, f_j] = f_r k^r{}_{ij}. \tag{2.46}$$

The induced transformation laws are

$$f_i = e_r P^r{}_i, \quad f^i = (P^{-1})^i{}_r e^r, \quad e_i = f_r(P^{-1})^r{}_i, \quad e^i = P^i{}_r f^r, \tag{2.47}$$

$$k^\ell{}_{ij} = (P^{-1})^\ell{}_s P^r{}_i P^q{}_j c^s{}_{rq}. \tag{2.48}$$

By construction, ω is diagonal in the basis $\{f_a\}$. From (2.32), we can compute the curvature in the basis $\{f^a\}$,

$$\mathrm{Tr}\,(Rm \wedge Rm) = \frac{\tau}{4}\,\overline{k^r{}_{ab}k^s{}_{rp}}k^q{}_{cd}k^s{}_{qp}\, f^d \wedge f^c \wedge \bar{f}^b \wedge \bar{f}^a. \tag{2.49}$$

Using the above transformation laws, $\mathrm{Tr}\,(Rm \wedge Rm)$ can be rewritten in terms of the $\{e^a\}$ basis. A straightforward computation gives the first formula in the lemma. To obtain the second formula, we begin by computing $i\partial\bar{\partial}\omega$ in the model case where the metric is $\sum_a ie^a \wedge \bar{e}^a$. We had already found $i\partial\omega$ in (2.19). Differentiating again gives

$$i\partial\bar{\partial}(i\sum_a e^a \wedge \bar{e}^a) = \frac{1}{4}\sum \overline{c^\ell{}_{ab}}\, c^\ell{}_{cd}\, e^d \wedge e^c \wedge \bar{e}^b \wedge \bar{e}^a. \tag{2.50}$$

Reverting to the general metric ω given by (2.41) and performing the same change of bases as before, we find

$$i\partial\bar{\partial}\omega = \frac{1}{4} g_{\bar{i}s}\overline{c^i{}_{ab}}c^s{}_{cd}\, e^d \wedge e^c \wedge \bar{e}^b \wedge \bar{e}^a. \tag{2.51}$$

Combining this formula with the one found previously for $\mathrm{Tr}(Rm \wedge Rm)$, we obtain the second formula stated in the lemma. Q.E.D.

3. Hull-Strominger systems and Anomaly flows

We come now to the study of the Anomaly flow on the complex Lie group X.

3.1. The Hull-Strominger system on unimodular Lie groups. First we recall the Hull-Strominger system [**21, 22, 36**]. Let X be a 3-dimensional complex manifold with a nowhere vanishing holomorphic $(3,0)$-form Ω. The Hull-Strominger system is a system of equations for a Hermitian metric ω on X and a holomorphic vector bundle $E \to X$ equipped with a Hermitian metric $H_{\bar{\alpha}\beta}$ satisfying

$$\begin{aligned} &F^{2,0} = F^{0,2} = 0, \qquad \omega^2 \wedge F^{1,1} = 0\\ &i\partial\bar{\partial}\omega - \frac{\alpha'}{4}\mathrm{Tr}(Rm \wedge Rm - F \wedge F) = 0\\ &d^\dagger\omega = i(\bar{\partial} - \partial)\log \|\Omega\|_\omega \end{aligned} \tag{3.1}$$

where $F^{p,q}$ are the components of the Chern curvature $F \in \Lambda^2 \otimes End(E)$ of the metric $H_{\bar{\alpha}\beta}$, and $\|\Omega\|_\omega$ denotes the norm of Ω with respect to the metric ω as defined in (1.4). The first equation in (3.1) is the familiar Hermitian-Yang-Mills equation, and its solution is well-known for given metric ω by the theorem of Donaldson-Uhlenbeck-Yau [**8, 37**], and its extension to the Hermitian setting [**6, 27**]. Thus the most novel aspects in the Hull-Strominger system resides in the other two equations. It has been pointed out by Li and Yau [**26**] that the last equation is equivalent to the following condition of "conformally balanced metric",

$$d(\|\Omega\|_\omega \omega^2) = 0 \tag{3.2}$$

which is a generalization of the balanced condition introduced in 1981 by Michelsohn [**28**].

We shall take the bundle $E \to X$ to be trivial, F to be 0, and restrict ourselves to left-invariant metrics. In this case, the first equation in (3.1) is trivially satisfied. Furthermore, the norm $\|\Omega\|_\omega$ is a constant function on X, and the conformally balanced condition reduces to exactly the balanced condition of Michelsohn [**28**], i.e., $d\omega^2 = 0$.

A key observation, also exploited earlier in the work of Fei and Yau [**12**], is that, on a unimodular complex Lie group, any left-invariant metric is balanced. This statement is well-known and to our knowledge first appeared in [**1**]. For the reader's convenience, we provide the brief argument. Recall that a complex Lie group is said to be unimodular if there exists a left-invariant basis of holomorphic vector fields with structure constants ${c^d}_{ab}$ satisfying the condition (1.2). In view of the transformation rule (2.48) for structure constants under a change of basis of left-invariant vector fields, this statement holds for all bases if and only if it holds for some basis. Let now ω be any invariant Hermitian metric on X, and express it in terms of any basis of holomorphic e^a forms as (2.41). A direct calculation using (2.17) gives

$$\partial\omega^2 = \sum \frac{1}{2}(g_{\bar p r}g_{\bar q c} - g_{\bar p c}g_{\bar q r}){c^r}_{ab}\, e^a \wedge e^b \wedge e^c \wedge \bar e^p \wedge \bar e^q. \tag{3.3}$$

Choosing e^a to be orthonormal with respect to ω, we can assume that $g_{\bar b a} = \delta_{ba}$, and we readily see that the condition $d\omega^2 = 0$ is equivalent to the unimodular condition (1.2).

Thus, on unimodular Lie groups, the Hull-Strominger system for a left-invariant metric reduces to the middle equation in (3.1).

3.2. Proof of Theorem 1.1. The Anomaly flow, introduced in [**31**], is a parabolic flow whose stationary points are solutions of the Hull-Strominger system. In the present setting, X is a 3-dimensional unimodular complex Lie group with a basis $\{e_a\}$ of left-invariant holomorphic vector fields, $\Omega = e^1 \wedge e^2 \wedge e^3$, and the bundle $E \to X$ is taken to be trivial. Then the Anomaly flow [**31**] is defined to be the following flow of $(2,2)$-forms,

$$\partial_t(\|\Omega\|_\omega \omega^2) = i\partial\bar\partial\omega - \frac{\alpha'}{4}\mathrm{Tr}\,(Rm^{(\kappa)} \wedge Rm^{(\kappa)}), \tag{3.4}$$

with any given initial data of the form $\|\Omega\|_{\omega_0}\omega_0^2$. Here $\kappa \in \mathbf{R}$ is fixed, and $Rm^{(\kappa)}$ is the curvature of the connection $\nabla^{(\kappa)}$ on the Yano-Gauduchon line for the evolving metric ω.

Theorem 1 of [**34**] shows how to rewrite this flow as a curvature flow for the Hermitian metric ω (see formula (2.46) in [**34**]). Setting $\Psi = i\partial\bar\partial\omega - \frac{\alpha'}{4}\mathrm{Tr}\,(Rm^{(\kappa)} \wedge Rm^{(\kappa)})$, we obtain

$$\partial_t g_{\bar a c} = -\frac{1}{2\|\Omega\|_\omega} g^{d\bar b}\Psi_{\bar a \bar b c d}. \tag{3.5}$$

Substituting in the formulas obtained in Lemma 2.4 for Ψ, we obtain Theorem 1.1.

3.3. Proof of Theorem 1.2. We discuss the unimodular Lie group case by case, as listed in Theorem 1.2.

3.3.1. *The Abelian Case.* In this case, we have for all a, b,

$$[e_a, e_b] = 0 \tag{3.6}$$

and all the structure constants ${c^d}_{ab}$ vanish. The flow (1.5) is static for all initial data, and part (a) of Theorem 1.2 is immediate.

3.3.2. *Nilpotent case.* By the classification theorem in [**12**], in this case we may assume the Lie algebra satisfies the commutation relations

$$[e_1, e_3] = 0, \ [e_2, e_3] = 0, \ [e_1, e_2] = e_3. \tag{3.7}$$

It follows that ${c^3}_{12} = 1$ and all other structure constants vanish. Substituting these structure constants into the flow (1.5) gives the following system

$$\partial_t g_{\bar 1 1} = \frac{g^{2\bar 2} g_{\bar 3 3}}{2\|\Omega\|}, \ \ \partial_t g_{\bar 1 2} = -\frac{g^{1\bar 2} g_{\bar 3 3}}{2\|\Omega\|}, \ \ \partial_t g_{\bar 2 2} = \frac{g^{1\bar 1} g_{\bar 3 3}}{2\|\Omega\|}, \ \ \partial_t g_{\bar p 3} = 0. \tag{3.8}$$

We see that there are no stationary points in this case. This proves part (b) of Theorem 1.2. In this case, we can also describe completely the flow for diagonal initial data. The diagonal property is preserved, and setting $g_{\bar b a}(t) = \lambda_a(t)\delta_{ab}$, we find that $\lambda_3(t) = \lambda_3(0)$ is constant in time, while $\lambda_1(t)$ and $\lambda_2(t)$ satisfy the ODE system,

$$\partial_t \lambda_1 = \frac{\sqrt{\lambda_3(0)}}{2}\sqrt{\frac{\lambda_1}{\lambda_2}}, \ \ \partial_t \lambda_2 = \frac{\sqrt{\lambda_3(0)}}{2}\sqrt{\frac{\lambda_2}{\lambda_1}}. \tag{3.9}$$

In particular,

$$\partial_t \frac{\lambda_1}{\lambda_2} = 0 \tag{3.10}$$

Thus the ratio $\lambda_1(t)/\lambda_2(t)$ is constant for all time. Substituting in the previous equation, we can solve explicitly for $\lambda_1(t)$ and $\lambda_2(t)$,

$$\lambda_1(t) = \lambda_1(0) + \frac{1}{2} t \sqrt{\frac{\lambda_1(0)\lambda_3(0)}{\lambda_2(0)}}, \qquad \lambda_2(t) = \lambda_2(0) + \frac{1}{2} t \sqrt{\frac{\lambda_2(0)\lambda_3(0)}{\lambda_1(0)}}$$

which tend both to ∞ as $t \to \infty$.

3.3.3. *Solvable case.* In this case, we may assume the Lie algebra satisfies the commutation relations

$$[e_3, e_1] = e_1, \ [e_3, e_2] = -e_2, \ [e_1, e_2] = 0 \tag{3.11}$$

that is, ${c^1}_{31} = 1$, ${c^2}_{32} = -1$, and all the other structure constants vanish. Substituting these structure constants into the flow (1.5) gives the following system

$$\partial_t g_{\bar 1 1} = \frac{1}{2\|\Omega\|}\Big(g^{3\bar 3} g_{\bar 1 1} - \beta g^{3\bar 3} g^{3\bar 3} g_{\bar 1 1} \Big), \tag{3.12}$$

$$\partial_t g_{\bar 1 2} = \frac{1}{2\|\Omega\|}\Big(g^{3\bar 3} g_{\bar 1 2} - \beta g^{3\bar 3} g^{3\bar 3} g_{\bar 1 2} \Big), \tag{3.13}$$

$$\partial_t g_{\bar 1 3} = \frac{1}{2\|\Omega\|}\Big(- g^{1\bar 3} g_{\bar 1 1} + g^{3\bar 2} g_{\bar 1 2} + \beta g^{1\bar 3} g^{3\bar 3} g_{\bar 1 1} + \beta g^{3\bar 2} g^{3\bar 3} g_{\bar 1 2} \Big), \tag{3.14}$$

$$\partial_t g_{\bar 2 2} = \frac{1}{2\|\Omega\|}\Big(g^{3\bar 3} g_{\bar 2 2} - \beta g^{3\bar 3} g^{3\bar 3} g_{\bar 2 2} \Big) \tag{3.15}$$

$$\partial_t g_{\bar 2 3} = \frac{1}{2\|\Omega\|}\Big(g^{1\bar 3} g_{\bar 2 1} - g^{2\bar 3} g_{\bar 2 2} + \beta g^{1\bar 3} g^{3\bar 3} g_{\bar 2 1} + \beta g^{2\bar 3} g^{3\bar 3} g_{\bar 2 2} \Big), \tag{3.16}$$

$$\begin{aligned} \partial_t g_{\bar 3 3} = \frac{1}{2\|\Omega\|}\Big(& g^{1\bar 1} g_{\bar 1 1} + g^{2\bar 2} g_{\bar 2 2} - g^{2\bar 1} g_{\bar 1 2} - g^{1\bar 2} g_{\bar 2 1} \\ & -\beta g^{1\bar 1} g^{3\bar 3} g_{\bar 1 1} - \beta g^{2\bar 2} g^{3\bar 3} g_{\bar 2 2} - \beta g^{2\bar 1} g^{3\bar 3} g_{\bar 1 2} - \beta g^{1\bar 2} g^{3\bar 3} g_{\bar 2 1} \Big). \end{aligned} \tag{3.17}$$

Here we set

$$\beta = \frac{\alpha'\tau}{4}. \tag{3.18}$$

We begin by identifying the stationary metrics. Let $g_{\bar{a}b}$ be a stationary metric. From (3.12) we see that $\beta > 0$ and $g^{3\bar{3}} = \beta^{-1}$. Substituting into (3.14), we obtain

$$2g^{3\bar{2}}g_{\bar{1}2} = 0. \tag{3.19}$$

If $g_{\bar{1}2} = 0$, the stationary point is of the desired form. Hence we must have $g^{3\bar{2}} = 0$. Similarly, substituting $g^{3\bar{3}} = \beta^{-1}$ into (3.16) leads to either $g_{\bar{2}1} = 0$ or $g^{1\bar{3}} = 0$. Hence we may assume that $g^{3\bar{2}} = g^{3\bar{1}} = 0$ which implies that $g_{\bar{3}1} = g_{\bar{3}2} = 0$ and $g_{\bar{3}3} = \beta$. Next, by (3.17), we conclude

$$0 = 2(g^{2\bar{1}}g_{\bar{1}2} + g^{1\bar{2}}g_{\bar{2}1}) = 4\,\mathrm{Re}\,\{g^{2\bar{1}}g_{\bar{1}2}\}. \tag{3.20}$$

By the formula for inverse matrices, we have

$$g^{2\bar{1}}g_{\bar{1}2} = -\frac{|g_{\bar{2}1}|^2 g_{\bar{3}3}}{\det g}. \tag{3.21}$$

Therefore $g_{\bar{1}2} = 0$ and the solution is of the desired form.

It follows that the stationary metrics are exactly the metrics which satisfy

$$g_{\bar{1}2} = 0, \quad g^{3\bar{3}} = \beta^{-1}. \tag{3.22}$$

Using $g^{3\bar{3}} = (\det g)^{-1}g_{\bar{1}1}g_{\bar{2}2}$, these equations can also be rewritten as

$$g_{\bar{1}2} = 0, \quad \frac{|g_{\bar{1}3}|^2}{g_{\bar{1}1}} + \frac{|g_{\bar{2}3}|^2}{g_{\bar{2}2}} = g_{\bar{3}3} - \beta. \tag{3.23}$$

In particular, there are stationary points $g_{\bar{b}a}$ which are not diagonal. For example, setting $g_{\bar{1}2} = g_{\bar{2}1} = 0$, $g_{\bar{3}3} = 2 + \beta$ and all other entries to 1 gives a stationary point. More generally, the moduli space of solutions requires locally two complex parameters $g_{\bar{1}3}$ and $g_{\bar{2}3}$, and two real parameters $g_{\bar{1}1}$, $g_{\bar{2}2}$.

Next, we examine the Anomaly flow. First, we consider the case of initial metrics with $g_{\bar{1}2}(0) = 0$. This condition is clearly preserved under the flow, and the flow for the other components of the metric becomes

$$\partial_t g_{\bar{1}1} = \frac{1}{2\|\Omega\|}g^{3\bar{3}}g_{\bar{1}1}(1 - \beta g^{3\bar{3}}), \tag{3.24}$$

$$\partial_t g_{\bar{2}2} = \frac{1}{2\|\Omega\|}g^{3\bar{3}}g_{\bar{2}2}(1 - \beta g^{3\bar{3}}), \tag{3.25}$$

$$\partial_t g_{\bar{3}3} = \frac{1}{2\|\Omega\|}(g^{1\bar{1}}g_{\bar{1}1}(1 - \beta g^{3\bar{3}}) + g^{2\bar{2}}g_{\bar{2}2}(1 - \beta g^{3\bar{3}})), \tag{3.26}$$

$$\partial_t g_{\bar{1}3} = \frac{1}{2\|\Omega\|}g^{1\bar{3}}g_{\bar{1}1}(-1 + \beta g^{3\bar{3}}), \tag{3.27}$$

$$\partial_t g_{\bar{2}3} = \frac{1}{2\|\Omega\|}g^{2\bar{3}}g_{\bar{2}2}(-1 + \beta g^{3\bar{3}}). \tag{3.28}$$

We shall use the following simple formulas for the entries of the inverse metric $g^{a\bar{b}}$,

$$g^{1\bar{1}} = \frac{g_{\bar{2}2}g_{\bar{3}3} - |g_{\bar{2}3}|^2}{\det g}, \quad g^{2\bar{2}} = \frac{g_{\bar{1}1}g_{\bar{3}3} - |g_{\bar{1}3}|^2}{\det g}, \quad g^{3\bar{3}} = \frac{g_{\bar{1}1}g_{\bar{2}2}}{\det g}$$

$$g^{1\bar{3}} = -\frac{g_{\bar{2}2}g_{\bar{1}3}}{\det g}, \quad g^{2\bar{3}} = -\frac{g_{\bar{1}1}g_{\bar{2}3}}{\det g}. \tag{3.29}$$

We note the following identities:

- $g_{\bar 2 2} = a g_{\bar 1 1}$ with $a = \frac{g_{\bar 2 2}}{g_{\bar 1 1}}(0)$. This follows directly from equation (3.24) and (3.25).
- $|g_{\bar 1 3}| = b g_{\bar 1 1}$ with $b = \frac{|g_{\bar 1 3}|}{g_{\bar 1 1}}(0)$. Indeed, putting $g^{1\bar 3}$ into equation (3.27) and $g^{3\bar 3}$ into (3.24), we obtain

$$\partial_t g_{\bar 1 3} = \frac{1}{2\|\Omega\|}\frac{g_{\bar 1 3} g_{\bar 2 2} g_{\bar 1 1}}{\det g}(1-\beta g^{3\bar 3}), \quad \partial_t g_{\bar 1 1} = \frac{1}{2\|\Omega\|}\frac{g_{\bar 1 1} g_{\bar 2 2} g_{\bar 1 1}}{\det g}(1-\beta g^{3\bar 3}).$$

Hence

$$\partial_t \ln |g_{\bar 1 3}| = \partial_t \ln g_{\bar 1 1} \tag{3.30}$$

and this implies the desired relation.

- $|g_{\bar 2 3}| = c g_{\bar 1 1}$ with $c = \frac{|g_{\bar 2 3}|}{g_{\bar 1 1}}(0)$. The proof is similar to the previous case.
- $g^{3\bar 3} = \frac{d}{g_{\bar 1 1}^2}$ with $d = g_{\bar 1 1}^2(0) g^{3\bar 3}(0)$. First, we compute

$$\begin{aligned}
\partial_t g^{3\bar 3} &= \frac{g_{\bar 2 2}\partial_t g_{\bar 1 1}}{\det g} + \frac{g_{\bar 1 1}\partial_t g_{\bar 2 2}}{\det g} - \frac{g_{\bar 1 1} g_{\bar 2 2}}{(\det g)^2}\partial_t \det g \\
&= \frac{1}{\|\Omega\|} g^{3\bar 3} g^{3\bar 3}(1-\beta g^{3\bar 3}) - \frac{g_{\bar 1 1} g_{\bar 2 2}}{(\det g)^2}\partial_t \det g. \qquad (3.31)
\end{aligned}$$

It follows from (3.27) and (3.28) that

$$\partial_t |g_{\bar 1 3}|^2 = \frac{1}{\|\Omega\|} g^{3\bar 3}(1-\beta g^{3\bar 3})|g_{\bar 1 3}|^2, \quad \partial_t |g_{\bar 2 3}|^2 = \frac{1}{\|\Omega\|} g^{3\bar 3}(1-\beta g^{3\bar 3})|g_{\bar 2 3}|^2. \tag{3.32}$$

Next,

$$\begin{aligned}
\partial_t \det g &= \partial_t (g_{\bar 1 1} g_{\bar 2 2} g_{\bar 3 3} - g_{\bar 1 1}|g_{\bar 2 3}|^2 - g_{\bar 2 2}|g_{\bar 1 3}|^2) \\
&= \partial_t g_{\bar 1 1} g_{\bar 2 2} g_{\bar 3 3} + g_{\bar 1 1}\partial_t g_{\bar 2 2} g_{\bar 3 3} + g_{\bar 1 1} g_{\bar 2 2}\partial_t g_{\bar 3 3} - \partial_t g_{\bar 1 1}|g_{\bar 2 3}|^2 \\
&\quad g_{\bar 1 1}\partial_t |g_{\bar 2 3}|^2 - \partial_t g_{\bar 2 2}|g_{\bar 1 3}|^2 - g_{\bar 2 2}\partial_t |g_{\bar 1 3}|^2 \\
&= \frac{1}{2\|\Omega\|}(1-\beta g^{3\bar 3})\Big(2 g^{3\bar 3} g_{\bar 1 1} g_{\bar 2 2} g_{\bar 3 3} + g_{\bar 1 1} g_{\bar 2 2}(g^{1\bar 1} g_{\bar 1 1} + g^{2\bar 2} g_{\bar 2 2})\Big) \\
&\quad - \frac{3}{2\|\Omega\|} g^{3\bar 3}(1-\beta g^{3\bar 3})(g_{\bar 1 1}|g_{\bar 2 3}|^2 + g_{\bar 2 2}|g_{\bar 1 3}|^2). \qquad (3.33)
\end{aligned}$$

Using (3.29) yields

$$\begin{aligned}
\partial_t \det g &= \frac{1}{2\|\Omega\|}(1-\beta g^{3\bar 3}) g^{3\bar 3}\Big(2 g_{\bar 1 1} g_{\bar 2 2} g_{\bar 3 3} + (\det g)(g^{1\bar 1} g_{\bar 1 1} + g^{2\bar 2} g_{\bar 2 2}) \\
&\quad -3 g_{\bar 1 1}|g_{\bar 2 3}|^2 - 3 g_{\bar 2 2}|g_{\bar 1 3}|^2\Big) \\
&= \frac{2}{\|\Omega\|}(1-\beta g^{3\bar 3}) g^{3\bar 3}\det g. \qquad (3.34)
\end{aligned}$$

Combining (3.31) and (3.34),

$$\partial_t g^{3\bar 3} = -\frac{1}{\|\Omega\|} g^{3\bar 3} g^{3\bar 3}(1-\beta g^{3\bar 3}). \tag{3.35}$$

Comparing this equation with (3.24), it follows that

$$\partial_t g^{3\bar{3}} = -2\frac{\partial_t g_{\bar{1}1}}{g_{\bar{1}1}} g^{3\bar{3}}, \quad \partial_t \ln g^{3\bar{3}} = 2\partial_t \ln \frac{1}{g_{\bar{1}1}}, \tag{3.36}$$

which implies $g^{3\bar{3}} = \frac{d}{g_{\bar{1}1}^2}$ with $d = g_{\bar{1}1}^2(0) g^{3\bar{3}}(0)$.

Denote $\lambda = g_{\bar{1}1}$. Using the previously derived formulas and solving for $g_{\bar{3}3}$ from the expression for $\det g$, it follows that

(3.37)

$$g_{\bar{1}1} = \lambda, \quad g_{\bar{2}2} = a\lambda, \quad |g_{\bar{1}3}| = b\lambda, \quad |g_{\bar{2}3}| = c\lambda, \quad \det g = \frac{a}{d}\lambda^4, \quad g_{\bar{3}3} = \frac{\lambda^2}{d} + \left(\frac{c^2}{a} + b\right)\lambda.$$

Putting the above relations into equation (3.24),

$$\partial_t \lambda = \frac{(\det g)^{1/2}}{2} g^{3\bar{3}} g_{\bar{1}1}(1 - \beta g^{3\bar{3}}) = \frac{1}{2}\sqrt{\frac{a}{d}}\lambda^2 \cdot \frac{d}{\lambda^2} \cdot \lambda(1 - \frac{\beta d}{\lambda^2}) \tag{3.38}$$

Then

$$\partial_t \lambda^2 = \sqrt{ad}\,(\lambda^2 - \beta d), \tag{3.39}$$

and hence

$$|\lambda^2(t) - \beta d| = |\lambda^2(0) - \beta d| \cdot e^{\sqrt{ad}\,t}. \tag{3.40}$$

Suppose $\beta > 0$. Then λ is uniformly bounded above and away from zero as $t \to -\infty$, hence from (3.37) we see that the metric $g_{\bar{b}a}$ is uniformly bounded and remains non-degenerate. Thus the flow exists for all time $t < 0$ and $g_{\bar{1}1}^2(t) \to \beta d$ as $t \to -\infty$ for any initial data. In particular, this implies that $g^{3\bar{3}} \to \beta^{-1}$. This completes the description of the flow for initial data satisfying the condition $g_{\bar{1}2}(0) = 0$.

We consider now the case of initial data with $g_{\bar{1}2} \neq 0$. In this case, we claim that the flow cannot converge to a non-degenerate metric. Indeed, if $\partial_t g_{\bar{b}a} \to 0$ then $g_{\bar{1}2} \to 0$. However, from (3.12) and (3.13), we deduce

$$\frac{\partial_t g_{\bar{1}1}}{g_{\bar{1}1}} = \frac{\partial_t |g_{\bar{1}2}|}{|g_{\bar{1}2}|}. \tag{3.41}$$

Therefore for all times,

$$|g_{\bar{1}2}| = C g_{\bar{1}1}, \quad C = \frac{|g_{\bar{1}2}(0)|}{g_{\bar{1}1}(0)}. \tag{3.42}$$

The convergence to 0 of $g_{\bar{1}2}$ implies then the convergence to 0 of $g_{\bar{1}1} \to 0$, which contradicts the requirement that the limit be a non-degenerate metric. In particular, given a stationary solution g_∞, if we perturb it in the $g_{\bar{1}2}$ direction and run the Anomaly flow, the flow will not take the metric back to g_∞. This completes the proof of part (c) of Theorem 1.2.

3.3.4. *The Semi-Simple Case.* This is the case of $X = SL(2, \mathbf{C})$, whose standard basis $\{e_a\}$ has structure constants $c^k{}_{ij} = \epsilon_{kij}$, the Levi-Civita symbol.

We begin by showing that the metrics $g_{\bar{a}b} = 2\beta\delta_{ab}$ are the only stationary points of the flow, where β is defined in (3.18). In particular, the existence of a stationary point requires $\beta > 0$.

Assume then that $g_{\bar{a}b}$ is a stationary point of the flow. Let P be a matrix such that $\bar{P}^{\bar{a}}{}_{\bar{p}} g_{\bar{a}b} P^b{}_q = \delta_{\bar{p}q}$, and perform a change of basis by defining $f_i = e_r P^r{}_i$,

$[f_a, f_b] = f_c k^c{}_{ab}$. As previously discussed (2.45), in this new frame $\{f_a\}$ we have $\omega = i f^a \wedge \bar{f}^a$. The fixed point equation for the flow (1.5) becomes

$$h_{\bar{a}b} = \beta\, \overline{k^m{}_{ap}} h_{\bar{m}s} k^s{}_{bp}, \tag{3.43}$$

where we defined

$$h_{\bar{a}b} = \overline{k^m{}_{ap}} k^m{}_{bp}. \tag{3.44}$$

If we let $\kappa_{\bar{a}b} = \overline{c^m{}_{ap}} c^m{}_{bp}$, by symmetry of the Levi-Civita symbol we obtain $\kappa_{\bar{a}b} = 2\delta_{ab}$. By using the transformation laws (2.48), we derive

$$k^\ell{}_{ij} = (P^{-1})^\ell{}_s P^r{}_i P^q{}_j \epsilon_{srq}, \quad h_{\bar{a}b} = \bar{P}^{\bar{q}}{}_{\bar{a}} P^p{}_b \kappa_{\bar{q}p} = 2\bar{P}^{\bar{r}}{}_{\bar{a}} P^r{}_b. \tag{3.45}$$

If we substitute (3.45) into (3.43) and use $g^{a\bar{b}} = P^a{}_r \bar{P}^{\bar{b}}{}_{\bar{r}}$, cancellation occurs and we are left with

$$h_{\bar{a}b} = 2\beta\, \bar{P}^{\bar{p}}{}_{\bar{a}} P^q{}_b g^{r\bar{s}} \epsilon_{\ell ps} \epsilon_{\ell qr}. \tag{3.46}$$

We note the formula

$$\sum_\ell \epsilon_{\ell ps} \epsilon_{\ell qr} = \delta_{pq}\delta_{sr} - \delta_{pr}\delta_{qs}. \tag{3.47}$$

Combining (3.46) and (3.47), we obtain

$$\begin{aligned} h_{\bar{a}b} &= 2\beta\, \bar{P}^{\bar{p}}{}_{\bar{a}} P^q{}_b g^{r\bar{s}} (\delta_{pq}\delta_{sr} - \delta_{pr}\delta_{qs}) \\ &= 2\beta\, (\bar{P}^{\bar{r}}{}_{\bar{a}} P^r{}_b) \mathrm{Tr}(g^{-1}) - 2\beta\, \bar{P}^{\bar{p}}{}_{\bar{a}} g^{p\bar{q}} P^q{}_b. \end{aligned} \tag{3.48}$$

Substituting the relations $h_{\bar{a}b} = 2\bar{P}^{\bar{r}}{}_{\bar{a}} P^r{}_b$ and $g^{a\bar{b}} = P^a{}_r \bar{P}^{\bar{b}}{}_{\bar{r}}$, it follows that

$$h_{\bar{a}b} = \beta\, h_{\bar{a}b} (P^m{}_r \bar{P}^{\bar{m}}{}_{\bar{r}}) - 2\beta\, \bar{P}^{\bar{p}}{}_{\bar{a}} P^p{}_m \bar{P}^{\bar{q}}{}_{\bar{m}} P^q{}_b = \frac{1}{2}\beta\, h_{\bar{a}b} \mathrm{Tr}\, h - \frac{1}{2}\beta\, h_{\bar{a}m} h_{\bar{m}b}.$$

Hence

$$h_{\bar{a}b}(\beta \mathrm{Tr} h - 2) = \beta h_{\bar{a}m} h_{\bar{m}b}. \tag{3.49}$$

By multiplying $h^{b\bar{a}}$ on both sides of the equation, we obtain $\beta\, h_{\bar{b}b} = \beta \mathrm{Tr} h - 2$. In particular, $\mathrm{Tr} h = 3\beta^{-1}$. Putting this back into equation (3.49),

$$h_{\bar{a}b} = \beta h_{\bar{a}m} h_{\bar{m}b}. \tag{3.50}$$

Thus, $\beta h_{\bar{a}b}$ is an invertible idempotent matrix. This implies that $\beta\, h = I$ and hence $\bar{P}^{\bar{r}}{}_{\bar{a}} P^r{}_b = \frac{1}{2} h_{\bar{a}b} = \frac{1}{2\beta}\delta_{ab}$. Therefore,

$$g_{\bar{a}b} = (\bar{P}^{-1})^{\bar{r}}{}_{\bar{a}} (P^{-1})^r{}_b = 2\beta\, \delta_{ab} \tag{3.51}$$

as was to be shown.

Next, we show that the stationary metric is asymptotically unstable [3]. For this, it suffices to show that the flow can be restricted to a submanifold of Hermitian metrics, and that restricted to this submanifold, the flow is asymptotically unstable. We choose this submanifold to be the submanifold of metrics $g_{\bar{a}b}$ which are diagonal with respect to the given basis of invariant holomorphic vector fields,

$$g_{\bar{a}b} = \lambda_b \delta_{ab}. \tag{3.52}$$

Using the explicit form (1.5) of the flow and the fact that the structure constants are given by ε_{abc}, it is easy to verify that the diagonal form of metrics is preserved

[3] Recall that a stationary point for a flow is said to be asymptotically stable if the flow will converge to the stationary point for any initial data in some neighborhood of the point. The flow is said to a asymptotically unstable if it is not asymptotically stable.

along the flow. The flow reduces to the following ODE system for the eigenvalues λ_a,

$$\begin{aligned}
\partial_t\lambda_1 &= \frac{(\lambda_1\lambda_2\lambda_3)^{1/2}}{2}\Big(\lambda_2^{-1}\lambda_3+\lambda_3^{-1}\lambda_2-2\beta\,\lambda_1^{-1}-\beta\,\lambda_1(\lambda_2)^{-2}-\beta\,\lambda_1(\lambda_3)^{-2}\Big),\\
\partial_t\lambda_2 &= \frac{(\lambda_1\lambda_2\lambda_3)^{1/2}}{2}\Big(\lambda_1^{-1}\lambda_3+\lambda_3^{-1}\lambda_1-2\beta\,\lambda_2^{-1}-\beta\,\lambda_2(\lambda_1)^{-2}-\beta\,\lambda_2(\lambda_3)^{-2}\Big),\\
\partial_t\lambda_3 &= \frac{(\lambda_1\lambda_2\lambda_3)^{1/2}}{2}\Big(\lambda_1^{-1}\lambda_2+\lambda_2^{-1}\lambda_1-2\beta\,\lambda_3^{-1}-\beta\,\lambda_3(\lambda_1)^{-2}-\beta\,\lambda_3(\lambda_2)^{-2}\Big).
\end{aligned}$$

The linearization of the flow at the stationary point $\lambda_1=\lambda_2=\lambda_3=2\beta$ is easily worked out,

$$\partial_t\lambda_a=\sum_b Q_{ab}(\delta\lambda_b) \tag{3.53}$$

where the matrix $Q=(Q_{ab})$ is given by

$$Q=\sqrt{\frac{\beta}{2}}\begin{bmatrix}0&1&1\\1&0&1\\1&1&0\end{bmatrix} \tag{3.54}$$

The matrix Q has eigenvalues $-\sqrt{\beta/2}$ and $2\sqrt{\beta/2}$ with multiplicities 2 and 1 respectively. By a classical theorem on ordinary differential equations (see e.g. Theorem 3.3 in [**38**]), the presence of an eigenvalue with strictly positive real part implies that the flow is asymptotically unstable. Part (d) of Theorem 1.2 has now been proved, completing the proof of the theorem.

3.3.5. *Remarks.* We conclude with several remarks.

• For simplicity, we have formulated Theorem 1.2 under the assumption that $\alpha'\tau>0$. The behavior of the Anomaly flow can be readily worked out as well by similar methods when $\alpha'\tau\le 0$. The arguments are in fact simpler in that case, because there is then no cancellation between the two terms on the right hand side (1.5) of the flow. We leave the details to the reader.

• In general, the sign of the right hand side in the Anomaly flow is dictated by the requirement that the flow be weakly parabolic. But in the case of Lie groups, the flow reduces to an ODE, and both signs are allowed. The opposite sign can be obtained from the sign we chose here simply by a time-reversal.

• The remaining remarks are about the semi-simple case $SL(2,\mathbf{C})$. The eigenvalues of the linerarized operator at the stationary point imply that there is a stable surface and an unstable curve near this point. The stable surface appears difficult to identify explicitly, but the unstable curve is easily found. It is given by the line of metrics proportional to the identity matrix, $g_{\bar a b}=\lambda\,\delta_{ab}$. This line is preserved under the flow, which reduces to

$$\partial_t\lambda=\lambda^{\frac{3}{2}}\left(1-\frac{2\beta}{\lambda}\right). \tag{3.55}$$

This equation can be solved explicitly by

$$\lambda(t)=2\beta\left(\frac{Ce^{\sqrt{2\beta}t}+1}{1-Ce^{\sqrt{2\beta}t}}\right)^2,\quad \text{if }\lambda(0)>2\beta, \tag{3.56}$$

$$\lambda(t) = 2\beta \left(\frac{1 - Ce^{\sqrt{2\beta}t}}{Ce^{\sqrt{2\beta}t} + 1} \right)^2, \quad \text{if } \lambda(0) < 2\beta, \tag{3.57}$$

where

$$C = \left| \frac{\sqrt{\lambda}(0) - \sqrt{2\beta}}{\sqrt{\lambda}(0) + \sqrt{2\beta}} \right| < 1. \tag{3.58}$$

This shows that the flow terminates in finite time at $T = \frac{1}{\sqrt{2\beta}} \log \frac{1}{C}$, with $\lambda(t) \to +\infty$ as $t \to T$ if $\lambda(0) > 2\beta$, and $\lambda(t) \to 0$ as $t \to T$ if $\lambda(0) < 2\beta$.

• More generally, if two eigenvalues are equal at some time, then they are equal for all time. This follows from rewriting the flow as

$$\partial_t \lambda_1 = -\frac{(\lambda_1\lambda_2\lambda_3)^{\frac{1}{2}}}{2} \left(\beta(\frac{2}{\lambda_1} + \frac{\lambda_1}{\lambda_2^2} + \frac{\lambda_1}{\lambda_3^2}) - \frac{\lambda_2^2 + \lambda_3^2}{\lambda_2\lambda_3} \right) \tag{3.59}$$

with similar formulas for $\partial_t \lambda_2$ and $\partial_t \lambda_3$. In particular, we have

$$\begin{aligned} \partial_t \log \frac{\lambda_1}{\lambda_2} &= -\frac{(\lambda_1\lambda_2\lambda_3)^{\frac{1}{2}}}{2} (\beta(\frac{1}{\lambda_1^2} - \frac{1}{\lambda_2^2}) - \frac{\lambda_2^2 - \lambda_1^2}{\lambda_1\lambda_2\lambda_3}) \\ &= -\frac{(\lambda_1\lambda_2\lambda_3)^{\frac{1}{2}}}{2} \frac{\lambda_2^2 - \lambda_1^2}{\lambda_1^2\lambda_2^2} (\beta - \frac{\lambda_1\lambda_2}{\lambda_3}). \end{aligned} \tag{3.60}$$

Let $[0, T)$ be the maximum time of existence of the flow. This equation implies that if any two eigenvalues are equal at some time t_0, then they are identically equal on the whole interval $[0, T)$. Indeed, by the Cauchy-Kowalevska theorem, the eigenvalues are analytic functions of t near any time where they are all strictly positive. The equation (3.60) implies that, if say λ_1 and λ_2 are equal at t_0, then all derivatives in time of λ_1 and λ_2 at t_0 are also equal, as we can see by differentiating the equation (3.60). By analyticity, they must be equal in a neighborhood of t_0. Thus the set where λ_1 and λ_2 coincide is both open and closed. This establishes our claim.

• It follows that if the eigenvalues at the initial time are ordered as

$$\lambda_1 \geq \lambda_2 \geq \lambda_3 \tag{3.61}$$

then this ordering is preserved by the flow. The configuration space can be divided into the invariant and mutually disjoint subsets $\{\lambda_1 > \lambda_2 > \lambda_3\}$, $\{\lambda_1 = \lambda_2 > \lambda_3\}$, $\{\lambda_1 > \lambda_2 = \lambda_3\}$, $\{\lambda_1 = \lambda_2 = \lambda_3\}$. We have already shown that the flow diverges to $+\infty$ on the last invariant subset. We shall next make a few remarks on the other sets.

• On each of the other invariant subsets, we have the following: (a) If $\lambda_1(0) < 2\beta$, then $\lambda_1(t)$ is monotone decreasing, and in particular less than $\lambda_1(0)$ for all time $t \in [0, T)$; (b) If $\lambda_3(0) > 2\beta$, then $\lambda_3(t)$ is monotone increasing, and in particular greater than $\lambda_3(0)$ for all time $t \in [0, T)$. To see this, we express the flow as

$$\partial_t \lambda_1 = \frac{(\lambda_1\lambda_2\lambda_3)^{\frac{1}{2}}}{2} \left(\frac{\lambda_2}{\lambda_3} + \frac{\lambda_3}{\lambda_2} - \beta(\frac{2}{\lambda_1} + \frac{\lambda_1}{\lambda_2^2} + \frac{\lambda_1}{\lambda_3^2}) \right). \tag{3.62}$$

We shall make use of the following two estimates

$$\begin{aligned}(3.63)\qquad & \frac{\lambda_1}{\lambda_2}+\frac{\lambda_2}{\lambda_1}-\beta(\frac{2}{\lambda_3}+\frac{\lambda_3}{\lambda_1^2}+\frac{\lambda_3}{\lambda_2^2}) \\ \geq\ & \frac{\lambda_1}{\lambda_2}+\frac{\lambda_2}{\lambda_1}-\beta(\frac{2}{\lambda_3}+\frac{\lambda_3}{\lambda_3^2}+\frac{\lambda_3}{\lambda_3^2})=\frac{\lambda_1}{\lambda_2}+\frac{\lambda_2}{\lambda_1}-4\beta\frac{1}{\lambda_3}\end{aligned}$$

and

$$\begin{aligned}(3.64)\qquad & \frac{\lambda_2}{\lambda_3}+\frac{\lambda_3}{\lambda_2}-\beta(\frac{2}{\lambda_1}+\frac{\lambda_1}{\lambda_2^2}+\frac{\lambda_1}{\lambda_3^2})\leq\frac{\lambda_2}{\lambda_3}+\frac{\lambda_3}{\lambda_2}-\beta\frac{2}{\lambda_2}-\beta\frac{2}{\lambda_3} \\ =\ & \frac{1}{\lambda_2}(\lambda_3-2\beta)+\frac{1}{\lambda_3}(\lambda_2-2\beta).\end{aligned}$$

We can now establish (a). First, we claim that $\lambda_1(t)<2\beta$ for any time $t\in[0,T)$. Otherwise, let t_0 be the first time when $\lambda_1(t_0)=2\beta$. Then $\lambda_1(t)<2\beta$ on the interval $[0,t_0)$. On the interval $[0,t_0]$, we have then $\lambda_3\leq\lambda_2\leq\lambda_1\leq 2\beta$, and the inequality (3.64) implies that $\partial_t\lambda_1<0$ on this interval. It follows that $\lambda_1(t_0)\leq\lambda_1(0)<2\beta$, which contradicts our assumption. But now that we know that $\lambda_1(t)<2\beta$ for all time t, the same inequality (3.64) shows that $\lambda_1(t)$ is a strictly monotone decreasing function of time.

Next, we establish (b). Again, let t_0 be the first time when $\lambda_3(t_0)=2\beta$. On the interval $[0,t_0)$, we can apply the inequality (3.63) and obtain

$$(3.65)\qquad \partial_t\lambda_3\geq\frac{(\lambda_1\lambda_2\lambda_3)^{\frac{1}{2}}}{2}\left(\frac{\lambda_1}{\lambda_2}+\frac{\lambda_2}{\lambda_1}-\beta\frac{4}{\lambda_3}\right)>0$$

where we used the inequality $\frac{\lambda_1}{\lambda_2}+\frac{\lambda_2}{\lambda_1}\geq 2$. It follows that $\lambda_3(t_0)>\lambda_3(0)>2\beta$, which is a contradiction. Thus $\lambda_3(t)$ is a strictly monotone increasing function of time.

Acknowledgments. The authors would especially like to thank Teng Fei for many stimulating discussions. We also thank the referee for many useful suggestions.

References

[1] E. Abbena and A. Grassi, *Hermitian left invariant metrics on complex Lie groups and cosymplectic Hermitian manifolds* (English, with Italian summary), Boll. Un. Mat. Ital. A (6) **5** (1986), no. 3, 371–379. MR866545

[2] Björn Andreas and Mario Garcia-Fernandez, *Solutions of the Strominger system via stable bundles on Calabi-Yau threefolds*, Comm. Math. Phys. **315** (2012), no. 1, 153–168, DOI 10.1007/s00220-012-1509-9. MR2966943

[3] Bjorn Andreas and Mario Garcia-Fernandez, *Note on solutions of the Strominger system from unitary representations of cocompact lattices of* $SL(2,\mathbb{C})$, Comm. Math. Phys. **332** (2014), no. 3, 1381–1383, DOI 10.1007/s00220-014-1920-5. MR3262629

[4] Jean-Michel Bismut, *A local index theorem for non-Kähler manifolds*, Math. Ann. **284** (1989), no. 4, 681–699, DOI 10.1007/BF01443359. MR1006380

[5] Indranil Biswas and Avijit Mukherjee, *Solutions of Strominger system from unitary representations of cocompact lattices of* $\mathrm{SL}(2,\mathbb{C})$, Comm. Math. Phys. **322** (2013), no. 2, 373–384, DOI 10.1007/s00220-013-1765-3. MR3077919

[6] N. P. Buchdahl, *Hermitian-Einstein connections and stable vector bundles over compact complex surfaces*, Math. Ann. **280** (1988), no. 4, 625–648, DOI 10.1007/BF01450081. MR939923

[7] P. Candelas, Gary T. Horowitz, Andrew Strominger, and Edward Witten, *Vacuum configurations for superstrings*, Nuclear Phys. B **258** (1985), no. 1, 46–74, DOI 10.1016/0550-3213(85)90602-9. MR800347

[8] S. K. Donaldson, *Anti self-dual Yang-Mills connections over complex algebraic surfaces and stable vector bundles*, Proc. London Math. Soc. (3) **50** (1985), no. 1, 1–26, DOI 10.1112/plms/s3-50.1.1. MR765366

[9] Teng Fei, *A construction of non-Kähler Calabi-Yau manifolds and new solutions to the Strominger system*, Adv. Math. **302** (2016), 529–550, DOI 10.1016/j.aim.2016.07.023. MR3545938

[10] Teng Fei, *Some torsional local models for heterotic strings*, Comm. Anal. Geom. **25** (2017), no. 5, 941–968, DOI 10.4310/CAG.2017.v25.n5.a3. MR3733796

[11] T. Fei, Z. Huang, and S. Picard, *A Construction of Infinitely Many Solutions to the Strominger System*, preprint, arXiv:1703.10067.

[12] Teng Fei and Shing-Tung Yau, *Invariant solutions to the Strominger system on complex Lie groups and their quotients*, Comm. Math. Phys. **338** (2015), no. 3, 1183–1195, DOI 10.1007/s00220-015-2374-0. MR3355812

[13] Marisa Fernández, Stefan Ivanov, Luis Ugarte, and Raquel Villacampa, *Non-Kaehler heterotic string compactifications with non-zero fluxes and constant dilaton*, Comm. Math. Phys. **288** (2009), no. 2, 677–697, DOI 10.1007/s00220-008-0714-z. MR2500995

[14] Ji-Xiang Fu, Li-Sheng Tseng, and Shing-Tung Yau, *Local heterotic torsional models*, Comm. Math. Phys. **289** (2009), no. 3, 1151–1169, DOI 10.1007/s00220-009-0789-1. MR2511664

[15] Ji-Xiang Fu and Shing-Tung Yau, *The theory of superstring with flux on non-Kähler manifolds and the complex Monge-Ampère equation*, J. Differential Geom. **78** (2008), no. 3, 369–428. MR2396248

[16] Ji-Xiang Fu and Shing-Tung Yau, *A Monge-Ampère-type equation motivated by string theory*, Comm. Anal. Geom. **15** (2007), no. 1, 29–75. MR2301248

[17] Mario Garcia-Fernandez, *Lectures on the Strominger system*, Travaux mathématiques. Vol. XXIV, Trav. Math., vol. 24, Fac. Sci. Technol. Commun. Univ. Luxemb., Luxembourg, 2016, pp. 7–61. MR3643933

[18] S. J. Gates Jr., C. M. Hull, and M. Roček, *Twisted multiplets and new supersymmetric nonlinear σ-models*, Nuclear Phys. B **248** (1984), no. 1, 157–186, DOI 10.1016/0550-3213(84)90592-3. MR776369

[19] Paul Gauduchon, *Hermitian connections and Dirac operators* (English, with Italian summary), Boll. Un. Mat. Ital. B (7) **11** (1997), no. 2, suppl., 257–288. MR1456265

[20] Gueo Grantcharov, *Geometry of compact complex homogeneous spaces with vanishing first Chern class*, Adv. Math. **226** (2011), no. 4, 3136–3159, DOI 10.1016/j.aim.2010.10.005. MR2764884

[21] C. M. Hull, *Superstring compactifications with torsion and spacetime supersymmetry*, Superunification and extra dimensions (Torino, 1985), World Sci. Publishing, Singapore, 1986, pp. 347–375. MR872720

[22] C. M. Hull, *Compactifications of the heterotic superstring*, Phys. Lett. B **178** (1986), no. 4, 357–364, DOI 10.1016/0370-2693(86)91393-6. MR862401

[23] C. M. Hull and E. Witten, *Supersymmetric sigma models and the heterotic string*, Phys. Lett. B **160** (1985), no. 6, 398–402, DOI 10.1016/0370-2693(85)90008-5. MR818681

[24] Stefan Ivanov, *Heterotic supersymmetry, anomaly cancellation and equations of motion*, Phys. Lett. B **685** (2010), no. 2-3, 190–196, DOI 10.1016/j.physletb.2010.01.050. MR2593876

[25] Anthony W. Knapp, *Lie groups beyond an introduction*, 2nd ed., Progress in Mathematics, vol. 140, Birkhäuser Boston, Inc., Boston, MA, 2002. MR1920389

[26] Jun Li and Shing-Tung Yau, *The existence of supersymmetric string theory with torsion*, J. Differential Geom. **70** (2005), no. 1, 143–181. MR2192064

[27] Jun Li and Shing-Tung Yau, *Hermitian-Yang-Mills connection on non-Kähler manifolds*, Mathematical aspects of string theory (San Diego, Calif., 1986), Adv. Ser. Math. Phys., vol. 1, World Sci. Publishing, Singapore, 1987, pp. 560–573. MR915839

[28] M. L. Michelsohn, *On the existence of special metrics in complex geometry*, Acta Math. **149** (1982), no. 3-4, 261–295, DOI 10.1007/BF02392356. MR688351

[29] Antonio Otal, Luis Ugarte, and Raquel Villacampa, *Invariant solutions to the Strominger system and the heterotic equations of motion*, Nuclear Phys. B **920** (2017), 442–474, DOI 10.1016/j.nuclphysb.2017.04.021. MR3656401

[30] D.H. Phong, S. Picard, and X.W. Zhang, *On estimates for the Fu-Yau generalization of a Strominger system*, arXiv:1507.08193, to appear in J. Reine Angew. Math.

[31] Duong H. Phong, Sebastien Picard, and Xiangwen Zhang, *Geometric flows and Strominger systems*, Math. Z. **288** (2018), no. 1-2, 101–113, DOI 10.1007/s00209-017-1879-y. MR3774405

[32] Duong Phong, Sebastien Picard, and Xiangwen Zhang, *A second order estimate for general complex Hessian equations*, Anal. PDE **9** (2016), no. 7, 1693–1709, DOI 10.2140/apde.2016.9.1693. MR3570235

[33] Duong H. Phong, Sebastien Picard, and Xiangwen Zhang, *The Fu-Yau equation with negative slope parameter*, Invent. Math. **209** (2017), no. 2, 541–576, DOI 10.1007/s00222-016-0715-z. MR3674222

[34] Duong H. Phong, Sebastien Picard, and Xiangwen Zhang, *Anomaly flows*, Comm. Anal. Geom. **26** (2018), no. 4, 955–1008, DOI 10.4310/CAG.2018.v26.n4.a9. MR3853932

[35] Duong H. Phong, Sebastien Picard, and Xiangwen Zhang, *The anomaly flow and the Fu-Yau equation*, Ann. PDE **4** (2018), no. 2, Art. 13, 60, DOI 10.1007/s40818-018-0049-9. MR3836294

[36] Andrew Strominger, *Superstrings with torsion*, Nuclear Phys. B **274** (1986), no. 2, 253–284, DOI 10.1016/0550-3213(86)90286-5. MR851702

[37] K. Uhlenbeck and S.-T. Yau, *On the existence of Hermitian-Yang-Mills connections in stable vector bundles*, Comm. Pure Appl. Math. **39** (1986), no. S, suppl., S257–S293, DOI 10.1002/cpa.3160390714. Frontiers of the mathematical sciences: 1985 (New York, 1985). MR861491

[38] Ferdinand Verhulst, *Nonlinear differential equations and dynamical systems*, Universitext, Springer-Verlag, Berlin, 1990. Translated from the Dutch. MR1036522

[39] Kentaro Yano, *Differential geometry on complex and almost complex spaces*, International Series of Monographs in Pure and Applied Mathematics, Vol. 49, A Pergamon Press Book. The Macmillan Co., New York, 1965. MR0187181

Department of Mathematics, Columbia University, 2990 Broadway, New York, New York 10027

Email address: `phong@math.columbia.edu`

Department of Mathematics, Columbia University, 2990 Broadway, New York, New York 10027

Current address: Department of Mathematics, Harvard University, One Oxford Street, Cambridge, Massachusetts 02138

Email address: `spicard@math.harvard.edu`

Department of Mathematics, University of California, 340 Rowland Hall, Bldg. 400, Irvine, California 92697

Email address: `xiangwen@math.uci.edu`

Contemporary Mathematics
Volume **735**, 2019
https://doi.org/10.1090/conm/735/14829

Pseudoconcave decompositions in complex manifolds

Zbigniew Slodkowski

ABSTRACT. We prove that the core of a complex manifold X is the union of pairwise disjoint pseudoconcave sets on which all uniformly bounded continuous plurisubharmonic functions on X are constant. Similarly, the minimal kernel of a weakly complete complex manifold decomposes into the union of compact pseudoconcave sets on which all continuous plurisubharmonic functions are constant. Versions of these results for standard smoothness classes are obtained. Analogous facts are discussed in the context of Richberg's regularization of continuous strongly plurisubharmonic functions.

0. INTRODUCTION

In this paper we discuss several situations in which pseudoconcave sets arise naturally as obstacles to constructions of plurisubharmonic functions (on complex manifolds) with certain properties and strictly (or strongly) plurisubharmonic on largest possible open set.

Thus in [**SlT04**], for a weakly complete manifolds (i.e. having a plurisubharmonic exhaustion function of some smoothness class) the *minimal kernel* was defined as the complement of the largest domain on which such exhaustion functions can be made strictly plurisubharmonic. Minimal kernels were shown to be pseudoconcave. Moreover, they can be sliced into compact pseudoconcave sets by level sets of exhaustion functions [**SlT04, MST15**]. This is an example of a pseudoconcave decomposition, which are a subject of this paper. One consequence is the following result (Theorem 2.1 below, an improved version of [**SlT04**, Corollary 3.7]:

A characterization of Stein manifolds. *If X is a manifold with a continuous plurisubharmonic exhaustion function and X does not contain any compact pseudoconcave set, then X is Stein.*

In this paper we obtain a complete decomposition of the minimal kernel into compact pseudoconcave *parts* such that all plurisubharmonic exhaustion functions of given smoothness are constant on each part (Theorem 5.2 below).

2010 *Mathematics Subject Classification.* Primary 32U05; Secondary 32F10, 32U35.
Key words and phrases. Plurisubharmonic functions; pseudoconcave sets.

©2019 American Mathematical Society

In similar spirit [**HST17a, HST17b, HST17c**] defined core of a complex manifold X as the complement of the largest domain on which bounded on X plurisubharmonic function (of a given smoothness class) can be strictly plurisubharmonic. The core was shown to be pseudoconcave, and the question was posed whether the core is the union of pairwise disjoint *irreducible components* on which every bounded, appropriately regular plurisubharmonic function is constant. In case $\dim_{\mathbb{C}} X = 2$ Harz et al [**HST17b**] have proved the the conjecture by using an earlier result of Shcherbina [**Shc93**]. We will present here the general proof of this conjecture, and of the corresponding (mentioned above) result for the minimal kernel, in all smoothness classes.

To avoid repetitions of similar arguments, and in view of possible further applications,we introduce in Section 4 the notion of an admissible class of plurisubharmonic functions. In this general context we define an analog of minimal kernel/core (Definition 4.2), prove a general pseudoconcave decomposition theorem for this set (Theorem 4.7), and then discuss its applications in Section 5. However, for the sake of clarity, we precede the general discussion with proofs two important special cases, Theorem 2.1 and 3.3, on which the more general proofs are patterned.

Finally, in Section 6 we introduce concept analogous to minimal kernel/core in the context of Richberg's uniform smoothing approximation of continuous plurisubharmonic functions, c.f. [**Ric68, Smi86**], and construct a similar pseudoconcave decomposition (Theorem 6.6). We hope that these concepts and results will be helpful in the study of uniform approximation of plurisubharmonic functions.

Section 1 presents necessary background information on pseudoconcave sets, and strongly plurisubharmonc functions.

The content of this paper is related to work of several mathematicians. We will make specific acknowledgments in 5.14, after presentation of the results.

Standing notations and conventions.A *manifold* will be always a complex manifold, countable at infinity, whether it is said explicitely or no.

A *ball*, always denoted B, will be a an open subset contained together with its closure in the coordinate patch of a complex coordinate system and biholomorphic to the standard ball in C^n.

An *exhaustion function* is always assumed plurisubharmonic.

1. Background on pseudoconcave sets

We will say that a set Z in a topological space X is *locally closed* if for every point z in Z there is a neighborhood V of z such that $\overline{V} \cap Z$ is closed in X. Clearly, if X is Hausdorff, Z is locally closed if and only if $\overline{Z} \setminus Z$ is closed in X.

1.1. Definition.[**Slo86**, Def 2.1] *Let Z be a locally closed subset of a complex manifold X. We say that Z has the local maximum property (is a local maximum set), if every point z_0 in Z has a relatively compact neighborhood V such that $\overline{V}$ is contained in a coordinate neighborhood of X, $\overline{V} \cap Z$ is closed, and for every plurisubharmonic function ϕ on a neighborhood of $\overline{V}$ it holds*

$$\max \phi | Z \cap \overline{V} \leq \max \phi | Z \cap bV$$

where $bV = \overline{V} \setminus V$.

1.2. Proposition.[**Slo86**, Prop. 2.3(iv)] *Let Z be a locally closed subset of a complex manifold X. Then Z is not a local maximum set if and only if there is a C^∞-smooth, strictly plurisubharmonic function ϕ on some open set V such $\phi|Z \cap V$ has strict maximum at some point of $Z \cap V$.*

1.3. Definition.[**Slo86**, Def. 4.1] *Let Z be a relatively closed subset of a domain W in an n-dimensional complex manifold X. We say that Z is 1-pseudoconcave, if Z can be covered by open subsets V_t of W, such that $W_t \setminus Z$ is (n-2)-pseudoconvex, in the sense of having an (n-2)-plurisubharmonic exhaustion function.*

The reader is referred to [**HST17a**] for a useful guide concerning terminological confusion regarding q-pseudoconcave and k-pseudoconvex sets.

1.4. Theorem.[**Slo86**, Th. 4.2] *A closed subset Z of a complex manifold X is 1-pseudoconcave, if and only if it is a local maximum set.*

[Below we use the terms *pseudoconcave* and *local maximum property* interchangeably, and always mean *1-pseudoconcave* by *pseudoconcave.*]

In this paper we will establish pseudoconcavity of various sets by using the concept of *maximal functions.* We collect here some basic facts. They follow from more general results in [**Bre59, Wal68**].

1.5. Definition.*Let $v : U \to \mathbb{R}$ be a continuous plurisubharmonic function on an open subset U of a complex manifold M. Then v is called a maximal function, if for every ball B in M such that $\overline{B} \subset U$, and for every uppersemicontinuous function $\phi : \overline{B} \to [-\infty, +\infty)$ which is plurisubharmonic on B and satisfies $\phi|bB \leqslant v|bB$, it holds $\phi|B \leqslant v|B$.*

1.6. Corollary.*If ϕ is a plurisubharmonic function in a domain U in a complex manifold X, and v is a maximal function in U, then for every ball B in U, such that $\overline{B} \subset U$,*

$$\max(\phi - v)|B \leqslant \max(\phi - v)|bB.$$

1.7. Theorem.*Let B be a ball in a complex manifold X, and $f : bB \to \mathbb{R}$ be a continuous function. Then there is a unique continuous function $u : \overline{B} \to \mathbb{R}$ such that $u|bB = f$ and $u|B$ is maximal. Furthermore, u can be constructed by the formula*

$$u(z) = \sup\{\phi(z) : \phi : \overline{B} \to [-\infty, +\infty) \text{ is usc}, \phi|B \text{ is psh }, \phi|bB \leq f\}.$$

The following two propositions are our basic tools for proving pseudoconcavity. They are both direct consequences of a more general [**Slo86**, Cor. 4.10]. Here we give a direct proof of the first, and a simple proof of Proposition 1.9 is in [**MST15**, Lem. 3.3].

1.8. Proposition.*Let v be a maximal function and ϕ be a plurisubharmonic function on a complex manifold U, such that $\phi(z_0) = v(z_0)$, and*

$$\phi(z) \leq v(z), \text{ for all } z \in U.$$

Then the set

$$Z := \{z \in U : \phi(z) = v(z)\}$$

has the local maximum property.

Proof. Suppose Z is not a local maximum set. By Proposition 1.2 there is a point p_0, a ball B containing p_0, with $\overline{B} \subset U$, and a smooth strictly plurisubharmonic function α, defined on a neighborhood of $\overline{B}$, such that $\alpha(p_0) = 0$ and

$$\alpha(p) < 0, \text{ for } p \in (\overline{B} \cap Z) \setminus \{p_0\}.$$

Denote $c := \max \alpha|(bB \cap Z$. Then $c < 0$. Consider now functions

$$\psi_n(z) = \alpha(z) + n\phi(z) - nv(z) \text{ , for } n = 1, 2, 3...$$

We claim that for some $n_0 > 0$,

$$\max \psi_{n_0}|bB < 0.$$

Let $W := \{z \in \text{domain}(\alpha) : \alpha(z) < 0\}$, and $K := bB \setminus W$. Then $K \subset \{\phi - v < 0\}$, and so if

$$\kappa := \max{(\phi - v)}|K,$$

then $\kappa < 0$, and

$$\max \psi_n|K \leq \max \alpha|K + n\kappa < 0, \text{ for } n \geq n_0.$$

Applying Corollary 1.6 to the plurisubharmonic function $\frac{1}{n}(\alpha + n\phi)$ and maximal function v, we obtain that $\psi_n = n(\frac{\alpha + n\phi}{n} - v)$ satisfies

$$0 = \psi_{n_0}(p_0) \leq \max \psi_{n_0}|bB < 0.$$

($bB \setminus K \subset W = \{\alpha < 0\}$, so $\psi_{n_0} < 0$ on $bB \setminus K$). This contradiction proves Prop. 1.8. □

1.9. Proposition. *Let Z be a local maximum set in a complex manifold X, and ϕ be a plurisubharmonic function defined in a neighborhood of Z. Suppose that $\phi|Z$ attains absolute maximum value at some point p_0. Then the set*

$$F := \{z \in Z : \phi(z) = \phi(z_0)\}$$

has the local maximum property.

1.10. Proposition. *Let Z_t, $t \in T$, be closed subsets of a complex manifold X. Assume that for every finite subset $\{t_1, t_2, ...t_n\} \subset T$, the intersection $Z_{t_1} \cap Z_{t_2} \cap ... \cap Z_{t_n}$ has the local maximum property. Then the set*

$$Z := \bigcap_{t \in T} Z_t$$

is a local maximum set, provided it is nonempty.

1.11. Corollary. *Let K be a compact local maximum set in a complex manifold X, and $\{\phi_t, t \in T\}$ a family of functions plurisubharmonic in some neighborhood of K. Then there is a compact pseudoconcave set F contained in K on which all functions of the family are constant.*

Proof. Assume that T is an interval of ordinal numbers. We will construct by transfinite induction compact pseudoconcave subset F_t of K such that ϕ_t is constant on F_t for every $t \in T$, and the family is monotonic decreasing. We can assume WLOG that ϕ_0 is a constant function 1, and let $F_0 = 1$. If $s = t + 1$, we let

$F_s := \{x \in K : \phi_{t+1}(x) = \max \phi_{t+1}|F_t\}$. This set, which is compact and nonempty, has local maximum property by Proposition 1.9. If s a limit ordinal, we let first $L := \bigcap_{t<s} F_t$. Since the family of sets F_t already constructed is monotonically decreasing and consists of compact pseudoconcave sets, L is a (nonempty) compact pseudoconcave set by Proposition 1.10. Now let $F_s := \{x \in L : \phi_s(x) = \max \phi_s|L\}$. By Proposition 1.9 F_s is a compact pseudoconcave set.

The inductive construction produces a monotonically decreasing family of compact local maximum sets. Let F be its nonempty compact intersection, which, by Proposition 1.10, has local maximum property. It is clear from the construction that all functions of the family are constant on F. □

1.12. Strongly plurisubharmonic functions. We follow the definition in [**Ric68**]. Let ϕ be a plurisubharmonic function on a complex manifold. We say it is *strongly* plurisubharmonic, if for every relatively compact neighborhood U contained with closure in some coordinate chart with the complex coordinates $z_1, ..., z_n$ there is a positive constant c_U, such that $dd^c\phi \geq c_U dd^c(|z_1|^2 + ... + |z_n|^2)$ (in the sense of positive currents). In this paper we will only use continuous strongly plurisubharmonic functions. We list next several of their properties that we will frequently use.

1.13. Properties of strongly plurisubharmonic functions. *Let X be a complex manifold. Then*

(i) *if ϕ is a strongly plurisubharmonic function on U, and ρ is a C^∞ smooth function with compact support in U, then there exist an $\epsilon > 0$, such that for every $t \in (-\epsilon, \epsilon)$, the function $\phi + t\rho$ is plurisubharmonic on U.*

(ii) *Let ϕ_n, $n = 1, 2, ...$, be continuous plurisubharmonic functions on X, with ϕ_n strongly plurisubharmonic on a domain U_n, $n = 1, 2, ...$ Assume that the series*

$$\sum_{n=1}^{+\infty} \phi_n$$

is uniformly convergent on compact sets in X. Then its sum is a continuous plurisubharmonic function on X and strongly plurisubharmonic on $\bigcup_n U_n$.

(iii) *Let $\phi_1, \phi_2, ...\phi_n$, be continuous plurisubharmonic functions on X, with ϕ_1 strongly plurisubharmonic on a domain $U \subset X$. Let $v : \mathbb{R}^n \to \mathbb{R}$ be smooth convex function, such that $\frac{\partial v}{\partial t_1} > 0$, and $\frac{\partial v}{\partial t_k} \geq 0$, for $k = 2, ..., n$, on the joint range of $(\phi_1, \phi_2, ...\phi_n)$. Then the function $\mu := v(\phi_1, \phi_2, ...\phi_n)$ is strongly plurisubharmonic on U (and continuous plurisubharmonic on X).*

1.14. Lemma. *For every $c \in \underline{R}^n$, $n = 1, 2, ...$, and for every $\delta > 0$, there is a convex function $v \in \mathbb{C}^\infty$ on $\mathbb{R}^n$ such that*

(i) *$v(t_1, t_2, ..., t_n) > t_1 + t_2 + ... + t_n$, for $t \in \mathbb{R}^n \setminus \{0\}$;*

(ii) *$v(c) = c_1 + c_2 + ... + c_n$;*

(iii) *$\frac{\partial v}{\partial t_i} > 0$ everywhere, for $i = 1, ..., n$;*

(iv) *$|v(t)| \leq const + (1 + \delta)|t|$, for $t \in \mathbb{R}^n$, i.e. v has linear growth.*

Proof. We first construct such function in dimension one, for $c = 0$. Let

$$g(t) = \int_0^t \arctan(s)ds, \text{ for } t \in \mathbb{R}.$$

We have, for $t \in \mathbb{R}$,

$$g'(t) = \arctan(t),$$
$$g''(t) = \frac{1}{t^2+1}.$$

Since $g''(t) > 0$ everywhere, v is convex on $\mathbb{R}$. Clearly $g(t) > 0$ on $(0, +\infty)$. As g is even ($g'(t)$ is odd), we have

$$g(t) > 0, \text{ for } t \neq 0, \text{ and } g(0) = 0.$$

Define now $v^0(t) := t + \epsilon g(t)$, for $t \in \mathbb{R}$, with $\epsilon \in (0, 2/\pi)$. Clearly, v^0 is convex and $\mathbb{C}^\infty$-smooth on $(\mathbb{R}^n)$, and

$$v^0(t) > t \text{ for } t \neq 0, \text{ and } v(0) = 0.$$

Also, $\frac{dv^0}{dt}(t) = 1 + \epsilon g'(t) = 1 + \epsilon \arctan(t) > 1 - (2/\pi)\epsilon > 0$. Furthermore, $g(t) \leq (\pi/2)\epsilon$, and so $v^0(t) \leq |t|(1 + \epsilon\pi/2)$. Thus v^0 satisfies all conditions (i)–(iv), when $c = 0$ and $n = 1$.

Now, for general n and arbitrary $c \in \mathbb{R}^n$, we let

$$v(t) := v^0(t_1 - c_1) + v^0(t_2 - c_2) + ... + v^0(t_n - c_n) + c_1 + c_2 + ... + c_n, \text{ for } t \in \mathbb{R}^n.$$

It is clear that v is a smooth convex function on $\mathbb{R}^n$ and satisfies all conditions (i)–(iv). □

2. A characterization of Stein manifolds

2.1. Theorem. *Let X be a complex manifold admitting a continuous plurisubharmonic exhaustion function. Then X is Stein if and only if it does not contain a compact pseudoconcave subset.*

Under the assumption that X has a C^k - smooth plurisubharmonic exhaustion function, $k \geq 2$, this fact was proven in [**SlT04**, Cor. 3.7]. Actually the proof in [**SlT04**] works also in the continuous category. Below we give a simplified proof of Theorem 2.1.

2.2. Definition.[**SlT04**, Sec 3.1] (i) *The minimal kernel, denoted Σ_X (which we will frequently shorten to Σ) is defined on a complex manifold with a contionuous plurisubharmonic exhaustion function as the complement of the union of all domains on which some such exhaustion function is strongly plurisubharmonic.*

(ii) *A minimal function for Σ_X is an exhaustion function (always plurisubharmonic on X) which is strongly plurisubharmonic on $X \setminus \Sigma_X$.*

2.3. Proposition.[**SlT04**, Lem. 3.1] *If X is complex manifold with a continuous exhaustion function, then it has a minimal function.*

The proof is the same as of Lemma 3.1 in [**SlT04**], but it uses Property 1.13(ii) above.

2.4. Lemma.(i) *Let $\phi : X \to \mathbb{R}$ be a continuous minimal exhaustion function, B an open ball intersecting Σ_X, and $u : \overline{B} \to \mathbb{R}$ a maximal function such that $u|bB = \phi|bB$. Then*

$$B \cap \Sigma_X = \{x \in B : \phi(x) = u(x)\}.$$

(ii) *Σ_X is a local maximum set.*

Proof. Denote

$$Z := \{x \in B : \phi(x) = u(x)\}.$$

Claim 1^0.

$$Z \cap (B \setminus \Sigma) = \emptyset.$$

Suppose there is a point $p \in Z \cap (B \setminus \Sigma)$. Choose a nonnegative, smooth function ρ on X with compact support such that $\overline{\text{supp}(\rho)} \subset B \setminus \Sigma$, and $\rho(p) = 1$. Since ϕ is minimal, it is strongly plurisubharmonic on $B \setminus \Sigma$, and so by Property 1.13(i) there is an $\epsilon > 0$ such that $\phi_\epsilon := \phi + \epsilon\rho$ is plurisubharmonic on $B \setminus \Sigma$. Since $\phi_\epsilon = \phi$ on $X \setminus \overline{\text{supp}(\rho)}$, ϕ_ϵ is plurisubharmonic on X, and $\phi_\epsilon|bB = \phi|bB = u|bB$. Hence $\phi_\epsilon(p) \leq u(p)$. However, as $p \in Z$, $u(p) = \phi(p) < \phi_\epsilon(p \leq u(p)$; contradiction.

Claim 2^0.

$$B \setminus Z \subset B \setminus \Sigma.$$

Consider any compact subset K of $B \setminus Z$. We will show that $K \cap \Sigma = \emptyset$. Choose $m > 0$ such that $(\phi + m)|K < u|K$. There is a smooth strictly plurisubharmonic function ψ on a neighborhood of $\overline{B}$ such that

$$u|\overline{B} < \psi|\overline{B} < (u + \frac{m}{2})|\overline{B}.$$

There is a neighborhood W of bB, such that ψ is defined on $W \cup \overline{B}$, and

$$\psi - \frac{m}{2} < \phi, \text{ on } W.$$

Define now a function $\tilde{\phi} : X \to \mathbb{R}$ by:

$$\tilde{\phi}(x) = \phi(x), \text{ for } x \in X \setminus \overline{B}$$

$$\tilde{\phi}(x) = \max\,(\psi(x) - \frac{m}{2}, \phi(x)), \text{ for } x \in \overline{B}.$$

Since for $x \in W$, $\psi(x) - \frac{m}{2} < \phi(x)$, it is clear that $\tilde{\phi}$ is a continuous plurisubharmonic exhaustion function on X. Now, if $x \in K$, then $\psi(x) - \frac{m}{2} > u(x) - \frac{m}{2} > \phi(x)$, and so $\tilde{\phi} = \psi - \frac{m}{2}$ in a neighborhood of K. Thus $\tilde{\phi}$ is strongly plurisubharmonic in a neighborood of K, hence $K \subset B \setminus \Sigma$. Since K was arbitrarilly chosen in $B \setminus Z$, Claim 2^0 follows.

Both claims mean $B \cap \Sigma = Z$ (statement(i)). By Proposition 1.8, the set $Z = \{u = \phi\}$ has the local maximum property, which means that Σ_X is pseudoconcave (statement(ii)). □

2.5. Lemma. *Let $\phi : X \to \mathbb{R}$ be a minimal continuous plurisubharmonic exhaustion function, and $c \in \phi(X)$. Then*

$$Z := \{x \in \Sigma : \phi(x) = c\}$$

is a compact set with local maximum property.

Proof. As Σ is closed and ϕ a continuous exhaustion function, Z is compact. Denote for $\epsilon > 0$,

$$v_\epsilon(t) := t + \epsilon(t - c)^2, t \in \mathbb{R},$$

$$\phi_\epsilon := v_\epsilon \circ \phi.$$

Observe that v_ϵ is strictly convex, and for small positive ϵ (i.e. such that $1 + 2\epsilon(\min\phi - c) > 0$), v_ϵ has positive derivative on the range of ϕ. By Property 1.13 (iii) ϕ_ϵ is strongly plurisubharmonic where ϕ was, hence a minimal function.

By definition of v_ϵ we have

$$\phi_\epsilon(x) > \phi(x), \text{ when } \phi(x) \neq c,$$

$$\phi_\epsilon(x) = \phi(x), \text{ when } \phi(x) = c.$$

Consider now any ball B intersecting Z; it is enough to show that $B \cap Z$ is a local maximum set. Let $u : \overline{B} \to \mathbb{R}$ be a continuous function which is maximal on B and $u|bB = \phi_\epsilon|bB$ (c.f.Theorem 1.7). By Lemma 2.4(i),

$$u(x) > \phi_\epsilon(x), \text{ if } x \in B \setminus \Sigma,$$

$$u(x) = \phi_\epsilon(x), \text{ if } x \in B \cap \Sigma,$$

$$u(x) = \phi_\epsilon(x) > \phi(x), \text{ if } x \in B \cap \Sigma \setminus \{\phi = c\},$$

$$u(x) = \phi_\epsilon(x) = \phi(x), \text{ if } x \in B \cap \Sigma \cap \{\phi = c\} = Z.$$

Combining these relations we obtain

$$u > \phi, \text{ on } B \setminus Z, , \text{ and } u = \phi, \text{ on } B \cap Z.$$

Thus, by Proposition 1.8, $B \cap Z$ has the local maximum property, and so Z does. □

2.6. Remark. *For a generalization of Lemma 2.5 for any family of continuous plurisubharmonic functions $\phi_t, t \in T$ see Theorem 5.2 below.*

Proof of Theorem 2.1. Assume X does not contain any compact pseudoconcave set. If the minimal kernel Σ_X is nonempty, then, by last lemma, for any minimal continuous exhaustion function ϕ, the sets $\Sigma_X \cap \{\phi(x) = c\}$, for any value c of ϕ, would be nonempty, compact, local maximum sets; contradiction. Thus the minimal kernel is empty, and so ϕ is a continuous strongly plurisubharmonic exhaustion function on X, which, by [**Ric68**], [**Smi86**] can be uniformly approximated on X by a smooth strictly plurisubharmonic function. This approximant has to be an exhaustion function, since ϕ is, and so X has to be Stein. The converse, that a Stein manifold cannot contain any (nonempty) *compact* set with local maximum property is well known. □

3. Decomposition of the core into pseudoconcave components

We recall first basic definitions and facts from [**HST17a**], see also [**PSh17**].

3.1. Definitions.(i) *The core $\mathbf{c}(X)$ of a complex manifold X is the complement of the union of all domains in X on which some bounded continuous plurisubharmonic function on X is strongly plurisubharmonic.*

(ii) *A minimal function for the core $\mathbf{c}(X)$ is a bounded continuous plurisubharmonic function on X which is strongly plurisubharmonic on $X \setminus \mathbf{c}(X)$.*

It was shown in [**HST17a**] that minimal functions exist (if $\mathbf{c}(X) \neq X$). See also Proposition 4.3 below.

3.2. Theorem.[**HST17a**] *In any complex manifold X the core $\mathbf{c}(X)$ is pseudoconcave.*

3.3. Theorem. *Let X be a complex manifold. Then its core $\mathbf{c}(X)$ can be decomposed into pseudoconcave subsets on which every bounded continuous plurisubharmonic function is constant.*

This theorem answers a cojecture posed in [**HST17b**]. A different proof was obtained independently by Poletsky and Shcherbina [**PSh17**]. See Subsection 5.14 for historical details.

In this section we consider core in the continuous category only. For other versions of the core see [**HST17a, HST17b**] and discussion in 5.3, 5.4 below.

The proof of Theorem 3.3 uses several lemmas. The first gives also a proof of Theorem 3.2, different from the one in [**HST17a**].

3.4. Lemma. *Let ϕ be a minimal function for $\mathbf{c}(X)$ (i.e. a bounded, continuous, plurisubharmonic function on X that is strongly plurisubharmonic on $X \setminus \mathbf{c}(X)$). Let B be a ball in X intersecting $\mathbf{c}(X)$, and let $u : \overline{B} \to \mathbb{R}$ be a continuous function which is maximal on B and $u|bB = \phi|bB$. Then*

$$B \cap \mathbf{c}(X) = \{x \in B : \phi(x) = u(x)\}.$$

Consequently $\mathbf{c}(X)$ is psudoconcave.

Proof. (*Sketch.*) The proof of analogous Lemma 2.4 carries over to the core situation, with no change except for replacing Σ_X by $\mathbf{c}(X)$. Specifically, if we let $Z := \{x \in B : \phi(x) = u(x)\}$, then we have to show that

$$Z \cap (B \setminus \mathbf{c}(X)) = \emptyset;$$
$$B \setminus Z \subset B \setminus \mathbf{c}(X).$$

corresponding respectively to claims 1^0 and 2^0 in the other proof. The reason it works here is that all changes done to function ϕ within the proof of Lemma 2.4 took place only in a relatively compact neighborhood of $\overline{B}$, so if ϕ is bounded on X, then the new functions ϕ_ϵ and $\tilde{\phi}$ are bounded on X (which is needed in the core situation). □

3.5. Lemma. *Let $\phi_0, \phi_1, ..., \phi_n$ be bounded, continuous, plurisubharmonic functions on X, and ϕ_0 be strongly plurisubharmonic on $X \setminus \mathbf{c}(X)$. Let $c_0, c_1, ..., c_n \in \mathbb{R}$. If the set*

$$Z := \mathbf{c}(X) \cap \bigcap_{j=0,1,...,n} \{x \in X : \phi_j(x) = c_j\}$$

is nonempty, then it has local maximum property.

Proof. By replacing ϕ_j by $\phi_j - c_j$ we can assume $c_j = 0$, i.e.

$$Z = \mathbf{c}(X) \cap \{\phi_0 = 0\} \cap \{\phi_1 = 0)\} ... \cap \{\phi_n = 0\}.$$

Let $\phi := \phi_0 + \phi_1 + ... + \phi_n$. Then ϕ is a minimal function for the core (i.e. is *bounded* on X and strongly plurisubharmonic on $X \setminus \mathbf{c}(X)$, by Property 1.13 (ii), since ϕ_0 is. Consider now function $v_\epsilon : \mathbb{R}^{n+1} \to \mathbb{R}$ defined by

$$v_\epsilon(t_0, t_1, ..., t_n) := t_0 + t_1 + ... + t_n + \epsilon(t_0^2 + t_1^2 + ... + t_n^2).$$

Then v_ϵ is convex on $\mathbb{R}^{n+1}$. Furthermore, if $m := \min\{\phi_j(x) : x \in X, j = 0, 1, ..., n\}$, then for small $\epsilon > 0$ such that $1 + 2\epsilon m > 0$ we have $\partial v_\epsilon / \partial t_j > 0$, for $j = 0, 1, ..., n$, on the joint range of $(\phi_0, \phi_1, ..., \phi_n)$. Then the function

$$\mu(x) := v_\epsilon(\phi_0(x), \phi_1(x), ..., \phi_n(x)), x \in X,$$

is a minimal function for the core (bounded on X, and strongly plurisubharmonic on $X \setminus \mathbf{c}(X)$, by Property 1.13(iii). By the above definitions,

$$\phi(x) \leq \mu(x), x \in X;$$

$$\{x \in X : \phi(x) = \mu(x)\} = \{x \in X : \phi_0(x) = \phi_1(x) = \ldots = \phi_n(x) = 0\}.$$

Hence

$$Z = \mathbf{c}(X) \cap \{\phi = \mu\}.$$

To show that Z is pseudoconcave, consider any ball intersecting Z, and let $u : \overline{B} \in \mathbb{R}$ be a continuous function, maximal on B, with $u|bB = \mu|bB$. Then, by Lemma 3.4 and the relation of ϕ and μ, we obtain

$$u(x) > \mu(x), \ , \text{for } x \in B \setminus \mathbf{c}(X),$$

$$u(x) = \mu(x), \ , \text{for } x \in B \cap \mathbf{c}(X),$$

$$u(x) = \mu(x) = \phi(x), \ , \text{for } x \in B \cap Z = B \cap \mathbf{c}(X) \cap Z,$$

$$u(x) = \mu(x) > \phi(x), \ , \text{for } x \in \mathbf{c}(X) \setminus Z$$

(because $Z \subset \mathbf{c}(X)$).

Combining these relations we obtain

$$\phi - u = 0 \ , \text{ in } Z \cap B,$$

$$\phi - u < 0 \ , \text{ in } B \setminus Z.$$

By Proposition 1.8 $Z \cap B$ is a local maximum set, and so Z is pseudoconcave. □

Proof of Theorem 3.3. Let $\{\phi_t : t \in T\}$ be the family of all bounded continuous plurisubharmonic functions on X. Consider $p \in \mathbf{c}(X)$. Let Z be the irreducible component of $\mathbf{c}(X)$ containing p. Then

$$Z := \mathbf{c}(X) \cap \bigcap_{t \in T} \{x \in X : \phi_t(x) = c_t\} \ , \text{ where } c_t := \phi_t(p).$$

Fix $t_0 \in T$ such that ϕ_{t_0} is a minimal function. Denote for any finite subset $\{t_1, t_2, ..., t_n\} \subset T$,

$$Z_{t_1, t_2, \ldots, t_n} := \mathbf{c}(X) \cap \bigcap_{j=0,1,\ldots,n} \{x \in X : \phi_{t_j}(x) = c_{t_j}\}$$

By Lemma 3.5 these sets have local maximum property. Since $Z = \bigcap_{t \in T} Z_t$ and $Z_{t_1, t_2, \ldots, t_n} = Z_{t_0} \cap Z_{t_1} \cap \ldots \cap Z_{t_n}$, we obtain by Proposition 1.10, that Z is pseudoconcave. □

Actually, in the formula for Z above, the core can be omitted, that is

$$Z = \bigcap_{t \in T} \{x \in X : \phi_j(x) = c_t\};$$

see 4.6 below, or discussion in [**HST17b**].

4. Admissible classes of plurisubharmonic functions

To extend the sample results given in Sections 2 and 3 to various regularity classes (for the minimal kernel and for the core) we introduce now a concept of *admissible* (certainly an overused word) class of plurisubharmonic functions and prove general theorems in this framework. Then the results for the minimal kernel and for the core will be immediately obtained as special cases (in Section 5). We found it natural do define the admissible classes using a variant of the concept of *presheaf* in which the classes of sections $\mathcal{F}$ are given only on open subsets U of the manifold X that are either relatively compact in X or cocompact. We will call such U *allowable.* The intuition behind it is that for relatively compact U the set of sections $\mathcal{F}(U)$ gives the regularity condition, while for cocompact U ("neighborhood of infinity") it captures the rate of growth of functions.

4.1. Definition. *Let be a complex manifold, and let for every allowable open $U \subset X$ a set $\mathcal{F}(U)$ of continuous plurisubharmonic functions on U be given. We call $\mathcal{F}$ an admissible class, if the following conditions are satisfied:*

(i) *for every sequence ϕ_n, in $\mathcal{F}(X)$, there is a sequence of positive numbers ϵ_n such that the series $\sum_{n=1}^{+\infty} \epsilon_n \phi_n$ converges uniformly on compact subsets of X to a function in $\mathcal{F}(X)$;*

(ii) *whenever $\phi \in \mathcal{F}(U)$ and $W \subset U$, then $\phi|W \in \mathcal{F}(W)$;*

(iii) *if $\phi : X \to \mathbb{R}$, $\{U_1, U_2, ..., U_n\}$ is a covering of X by allowable open sets, such that $\phi_i|U_i \in \mathcal{F}(U_i), i = 1, ..., n$, then $\phi \in \mathcal{F}(X)$;*

(iv) *for every allowable U in X, $\mathcal{F}(U)$ is a convex cone and contains all bounded C^∞-smooth plurisubharmonic functions on U;*

(v) *if $\phi_1, \phi_2, ..., \phi_n \in \mathcal{F}(U)$, (where U is allowable) and $v : \mathbb{R}^n \to \mathbb{R}$ is a C^∞-smooth convex function of at most linear growth, such that $\frac{\partial v}{\partial t_i} \geq 0$, $i = 1, 2, ..., n$, on the joint range of $(\phi_1, \phi_2, ..., \phi_n)$, then the function $\phi := v(\phi_1, \phi_2, ..., \phi_n)$ belongs to $\mathcal{F}(U)$;*

(vi) *if $\phi \in \mathcal{F}(U)$ and $\rho \in C^\infty(U)$, with $\overline{supp(\rho)} \subset U$, then there is $t > 0$, such that $\phi + t\rho \in \mathcal{F}(U)$.*

The notions of minimal kernel core, and minimal function extend naturally to the context of admissible classes of plurisubharmonic functions.

4.2. Definition. (i) *The singular locus of class $\mathcal{F}$, denoted $\Sigma^{\mathcal{F}}$, is the complement of the set of $x \in X$, such that some function $\phi \in \mathcal{F}(X)$ is strongly plurisubharmonic in a neighborhood of x.*

(ii) *$\phi \in \mathcal{F}(X)$ is called an $\mathcal{F}$-minimal function, if ϕ is strongly plurisubharmonic on $X \setminus \Sigma^{\mathcal{F}}$.*

Obviously, the singular locus $\Sigma^{\mathcal{F}}$ is a closed subset of X.

4.3. Proposition. *Under assumptions of Definition 4.1, $\mathcal{F}$-minimal functions exist.*

Proof. By our standing assumption that X is countable at ∞, there is a countable open covering U_n of the set $X \setminus \Sigma^{\mathcal{F}}$, and functions $\phi_n \in \mathcal{F}$, such that ϕ_n is strongly plurisubharmonic on U_n, for $n = 1, 2, ...$ By Definition 4.1 (ii) there is a sequence of positive numbers ϵ_n such that the function $\phi := \sum_{n=1}^{+\infty} \epsilon_n \phi_n$ is well defined and belongs to $\mathcal{F}(X)$. By Property 1.13(ii), ϕ is strongly plurisubharmonic on $\bigcup_{n=1}^{\infty} U_n$, and so is an $\mathcal{F}$-minimal function. □

4.4. Proposition. *For every admissible class $\mathcal{F}$, its singular locus $\Sigma^{\mathcal{F}}$ is pseudoconcave, provided it is nonempty.*

The proposition follows immediately from the next lemma.

4.5. Lemma. *Let $\mathcal{F}$ be an admissible class and $\phi : X \to \mathbb{R}$ an $\mathcal{F}$-minimal function. Let B be a ball in X intersecting $\Sigma^{\mathcal{F}}$, and let $u : \overline{B} \to \mathbb{R}$ be a continuous function which is maximal on B and $u|bB = \phi|bB$. Then*

$$B \cap \Sigma^{\mathcal{F}} = \{x \in B : \phi(x) = u(x)\}.$$

Consequently $B \cap \Sigma^{\mathcal{F}}$ is a local maximum set.

Proof. Denote $Z := \{x \in B : \phi(x) = u(x)\}$. The proof is modeled on that of Lemma 2.4. As there we have to show two claims:

Claim 1^0.

$$Z \cap (B \setminus \Sigma^{\mathcal{F}}) = \emptyset.$$

Claim 2^0.

$$B \setminus Z \subset B \setminus \Sigma^{\mathcal{F}}.$$

Proof of Claim 1^0 is essentially the same, but now we have to justify why $\phi_\epsilon \in \mathcal{F}(X)$, using "axioms" of admissible classes. If (in addition to other requirements in the proof of Lemma 2.4) U_1 is a neighborhood of p with closure in $\overline{B \setminus \Sigma^{\mathcal{F}}}$, and on which ϕ is strictly plurisubharmonic, and we assume that $\text{supp}(\rho) \subset U_1$, then for small enough $\epsilon > 0$ we have $\phi_\epsilon \in \mathcal{F}(U_1)$, by Definition 4.1 (vi), and using then (ii), (iii) we get $\phi_\epsilon \in \mathcal{F}(X)$.

Proof of Claim 2^0 is also similar to that in Lemma 2.4, but the gluing operation by which $\tilde{\phi}$ was defined there is more delicate now. As previously, take any compact set $K \subset B \setminus Z$. To show $K \cap \Sigma^{\mathcal{F}} = \emptyset$, choose $m > 0$ such that $\phi + m)|K < u|K$, and a C^∞-smooth *strictly* plurisubharmonic function ψ on a neighborhood of $\overline{B}$, such that

$$u|\overline{B} < \psi|\overline{B} < (u + \frac{m}{2})|\overline{B}.$$

There are now a $\delta > 0$ and a relatively compact neighborhood W of bB, such that

$$\psi(x) - \frac{m}{2} < \phi(x) - \delta \text{ , on } W,$$

$$\psi(x) - \frac{m}{2} - \delta > u(x - \frac{m}{2} > \phi(x) \text{ , for } x \text{ near } K.$$

We use now a well known *regularized maximum function.* That is, for every $\delta > 0$ there is a C^∞-smooth convex function $v_\delta = v : \mathbb{R}^2 \to \mathbb{R}$, with $\frac{\partial v}{\partial x}, \frac{\partial v}{\partial y} \geq 0$, and such that

$$v(x, y) = x, \text{ if } x - y > \delta,$$

$$v(x, y) = y, \text{ if } x - y < -\delta,$$

$$v(x, y) \geq \max(x, y),, \text{ for } (x, y) \in \mathbb{R}^2.$$

Define now $\tilde{\phi} : X \to \mathbb{R}$ as follows:

$$\tilde{\phi}(x) = \phi(x), \ , \text{ for } x \in X \setminus \overline{B},$$
$$\tilde{\phi}(x) = v(\psi(x) - \frac{m}{2}, \phi(x)), \text{ for } x \in \overline{B}.$$

Denote also

$$\alpha(x) = v(\psi(x) - \frac{m}{2}, \phi(x)), \text{ for } x \in W \cup \overline{B}.$$

By Def.4.1 (iv),(ii),(v), $\alpha \in \mathcal{F}(W \cup \overline{B})$ (note that v is of linear growth). By the choice of W, $\alpha|W = \phi|W$, and so $\tilde{\phi}|W = \phi|W$. Thus

$$\tilde{\phi}|(W \cup \overline{B}) = \alpha \in \mathcal{F}(W \cup \overline{B})$$
$$\tilde{\phi}|(X \setminus \overline{B}) = \phi|(X \setminus \overline{B}) \in \mathcal{F}(X \setminus \overline{B}).$$

By Def.4.1 (iii), $\tilde{\phi} \in \mathcal{F}(X)$. By the choice of m, δ, v we have, for x near K,

$$\tilde{\phi}(x) = v(\psi(x) - \frac{m}{2}, \phi(x)) = \psi(x) - \frac{m}{2}.$$

Hence $\tilde{\phi}$ is strongly plurisubharmonic on a neighborhood of K, i.e. $K \cap \Sigma^{\mathcal{F}} = \emptyset$. □

We extend now to the context of admissible classes the notions of irreducible components of the core. As in [**HST17b, PSh17**], we use an equivalence relation.

4.6. $\mathcal{F}$-components. *We say $x \sim y(\mathrm{rel}\mathcal{F})$, if $\phi(x) = \phi(y)$, for every $\phi \in \mathcal{F}(X)$. Like in* [**HST17b**] *the equivalence classes $[x]_{\mathcal{F}}$ are trivial when $x \notin \Sigma^{\mathcal{F}}$ (if $\phi(x) = \phi(y)$, and $x \neq y$, we can use Def.4.1 (vi) to change ϕ near x and not at y. So $\Sigma^{\mathcal{F}}$ is the union of equivalence classes of its elements. We call them $\mathcal{F}$-components.*

4.7. Theorem. *Let $\mathcal{F}$ be an admissible class on a complex manifold X. Then all $\mathcal{F}$-components of the singular locus $\Sigma^{\mathcal{F}}$ are pseudoconcave.*

The theorem is an easy consequence of the following finite intersection lemma.

4.8. Lemma. *Let $\phi_0, \phi_1, ..., \phi_n \in \mathcal{F}(X)$ and $c_0, c_1, ..., c_n \in \mathbb{R}$. Assume ϕ_0 is $\mathcal{F}$-minimal (i.e. strongly plurisubharmonic on $X \setminus \Sigma^{\mathcal{F}}$). Then the set*

$$Z := \Sigma^{\mathcal{F}} \cap \{x \in X : \phi_i(x) = c_i, i = 0, 1, ..., n\}$$

is pseudoconcave, provided it is nonempty.

Proof. Let $\phi := \phi_0 + \phi_1 + ... + \phi_n$. By Def. 4.1(iv), $\phi \in \mathcal{F}(X)$. Let $v : \mathbb{R}^{n+1} \to \mathbb{R}$ be a convex function corresponding to *point of contact* $c := (c_0, c_1, ..., c_n)$, given by Lemma 1.14. In particular

$v(t_0, t_1, ..., t_n) > t_0 + t_1 + ... + t_n$, for $t \neq c$;
$v(c) = c_0 + c_1 + ... + c_n$.

Since $\frac{\partial v}{\partial t_i} > 0$ everywhere, for $i = 1, ..., n$, and since v is of linear growth, the function

$$\mu := v(\phi_0, \phi_1, ...\phi_n)$$

belongs to $\mathcal{F}(X)$, by Def.4.1(v). By properties of v, we have

$$\mu(x) \geq \phi(x), \text{ for } x \in X;$$
$$\{x \in X : \mu(x) = \phi(x)\} = \{x \in X : \phi_i(x) = c_i, i = 0, 1, 2, ..., n\}.$$

Hence $Z = \Sigma^{\mathcal{F}} \cap \{\mu = \phi\}$.

From now on we can proceed like in the proof of Lemma 3.5, changing only $\mathbf{c}(X)$ there to $\Sigma^{\mathcal{F}}$ now, with functions ϕ and μ playing the same role. Like there, consider a maximal function u on B, with $u|bB = \mu|bB$ (where the ball B intersects $\Sigma^{\mathcal{F}}$ now). Since ϕ_0 is $\mathcal{F}$-minimal (strongly plurisubharmonic on $X \setminus \Sigma^{\mathcal{F}}$), and $\frac{\partial v}{\partial t_0} > 0$ everywhere, μ is strongly plurisubharmonic off $\Sigma^{\mathcal{F}}$ (by Property 1.13(iii)), i.e. it is $\mathcal{F}$-minimal, and so the relations

$$u(x) = \mu(x) \ , \text{ for } x \in B \cap \Sigma^{\mathcal{F}},$$

$$u(x) > \mu(x) \ , \text{ for } x \in B \setminus \Sigma^{\mathcal{F}}.$$

follow now from Lemma 4.5. With these adaptations, the ending procedure in the proof Lemma 3.5 gives the conclusion in our general situation. □

Proof of Theorem 4.7. The theorem is obtained from Lemma 4.8 in the same way, as Theorem 3.3 was from Lemma 3.5, with replacement of $\mathbf{c}(X)$ by $\Sigma^{\mathcal{F}}$, and with interpretation of $\{\phi_t : t \in T\}$ as $\mathcal{F}(X)$ now. □

5. Applications of admissible classes

5.1. Minimal kernels in weakly complete manifolds. Minimal kernels $\Sigma_k(X)$, $k = 0, 1, ... + \infty$ were defined in [**SlT04**] for manifolds X having a C^k-smooth (plurisubharmonic) exhaustion function. In terms of admissible classes, the definition in [**SlT04**] translates into $\Sigma_k(X) = \Sigma^{\mathcal{F}}$, where class $\mathcal{F}$ is the following. For a relatively compact open set U, $\mathcal{F}(U)$= the set of lower bounded, C^k-smooth,plurisubharmonic functions on U, and for cocompact U , we add the extra condition that for $\phi \in \mathcal{F}(U)$, the set $\{\phi \leq c\}$ is relatively compact in X for every $c \in \mathbb{R}$. Proposition 4.4 recovers (and extends to $k = 0$ and $k = 1$) Cor. 3.5 of [**SlT04**]. Theorem 4.7 implies

5.2. Theorem. *In a manifold X with a C^k-smooth plurisubharmonic exhaustion function (where $k = 0, 1, 2, ... + \infty$), the minimal kernel $\Sigma_k(X)$ is the union of pairwise disjoint compact pseudoconcave parts on which every C^k-smooth plurisubharmonic function is constant.*

We call *parts* $\mathcal{F}$-components in case of the minimal kernel. They are compact because exhauston functions are constant on them.

5.3. Cores $\mathbf{c}^{(k)}(X)$, $k = 0, 1, 2, ..., +\infty$. They were introduced and shown to be pseudoconcave in [**HST17a**],c.f. also [**PSh17**]. (Lemma 3.4 (and 4.5) above gives a different proof of their pseudoconcavity.) In terms of Section 4, $\mathbf{c}^{(k)}(X)$ is the singular locus of the admissible class $\mathcal{F}$ for which $\mathcal{F}(U)$ consists of all uniformly bounded, C^k-smooth plurisubharmonic function on U. In case of the core(s), the $\mathcal{F}$-components are called irreducible components in [**HST17b**]. The irreducible components are oftentimes noncompact, always in Stein manifods. Theorems 3.3, 4.7 above imply the folowing.

5.4. Theorem. *For every $k = 0, 1, 2, ..., +\infty$ the core $\mathbf{c}^{(k)}(X)$ is the disjoint union of closed pseudoconcave subsets on which every bounded plurisubharmonic C^k-smooth function on X is constant.*

5.5. Generalization of minimal kernels $\Sigma_k(X)$ to arbitrary complex manifolds. Consider any (countable at ∞) complex manifold X, and $k \in \{0, 1, 2, ..., +\infty\}$.

Consider an admissible class $\mathcal{F}$, where $\mathcal{F}(U)$ consists of all C^k-smooth plurisubharmonic functions on U. The singular locus $\Sigma^{\mathcal{F}}$ of this class is defined as the minimal kernel $\Sigma_k(X)$, in case of an arbitrary manifold.

5.6. Remark. *Assume that X has a C^k-smooth exhaustion function. Then the two definitions of $\Sigma_k(X)$ agree.*

Proof. We only need to verify that if p has a neighborhood U on which some plurisubharmonic function $\psi \in C^k(X)$ is strongly plurisuharmonic, then there is a C^k-smooth exhaustion funtion on X which is strongly plurisubharmonic on a neighborhood of p. Now if $d := \psi(p)$, take a smooth, lower-bounded convex fuction $v : \mathbb{R} \to \mathbb{R}$, such that $v(t) = t$, for $t \geq d-1$, and consider $v \circ \psi$. This function is C^k-smooth, plurisubharmonic, and lower bounded on X, and strongly plurisubharmonic near p. So if ϕ is any C^k-smooth exhaustion function on X, then $\phi + v \circ \psi$ is a C^k exhaustion function that is strongly plurisubharmonic near p, as needed. □

In the absence of an exhaustion function in $C^k(X)$ parts of the minimal kernel do not have to be compact. For example $X := \mathbb{P}\mathbb{C}^2 \setminus \{\text{a point}\}$ is a single part. The following weaker condition is obviously sufficient for compactness of $\mathcal{F}$-components.

5.7. Definition. *We say that functions in $\mathcal{F}$ separate points of X from infinity, if for every sequence of points (x_n), $n = 0, 1, ...,$ not relatively compact in X, there is a function ϕ in $\mathcal{F}$ such that $\phi(x_0) \neq \phi(x_n)$, for some $n > 0$.*

Applying again the results of Section 4 we obtain

5.8. Corollary. *For any complex manifold and $k = 0, 1, ..., +\infty$*

(i) *the minimal kernel $\Sigma_k(X)$ is pseudoconcave and it is the union of pairise disjoint pseudoconcave parts on which every $\mathbb{C}^k$ - smooth plurisubharmonic function is constant;*

(ii) *if, in addition, points of X are separated from infinity by $\mathbb{C}^k$-smooth plurisubharmonic functions then all parts are compact.*

We can formulate now a generalization of Theorem 2.1.

5.9. Corollary. *Let X be a complex manifold without compact pseudoconcave subsets. Assume that continuous plurisubharmonic functions separate points of X from infinity. Then there is a $\mathbb{C}^\infty$-smooth function strictly plurisubharmonic on X.*

Proof. In view of the assumptions Corollary 5.8 implies that $\Sigma(X)(= \Sigma^0(X)$) has to be empty. Thus any minimal function is strongly plurisubharmonic on X and by Richberg's theorem [**Ric68**] it can be uniformly approximated on X by smooth strictly plurisubharmonic function. □

(Note that X does not have to be Stein, e.g. $X = \mathbb{C}^2 \setminus \{0\}$ satisfies the assumptions of Corollary 5.9.)

5.10. Relations between cores and minimal kernels for different smoothness classes. Beyond the obvious inclusions

$$\mathbf{c}^{(0)}(X) \subset \mathbf{c}^{(1)}(X) \subset \ldots \subset \mathbf{c}^{(k)}(X) \subset ... \subset \mathbf{c}^{(\infty)}(X),$$

$$\Sigma_0(X) \subset \Sigma_1(X) \subset ... \subset \Sigma_k(X) \subset \ldots \subset \Sigma_\infty(X),$$

not much is known about the relations between these sets, in particular whether some or all of these inclusion might be proper.

5.11. Remark. *Let $k = 0, 1, ..., +\infty$.*

(i) *If one of the cores $\mathbf{c}^{(k)}(X)$, is nonempty, $k = 0, 1, ..., \infty$, then all are.*

(ii) *If one of the minimal kernels $\Sigma_k(X)$ is nonempty, $k = 0, 1, ..., +\infty$, then all are.*

Proof. (i) In view of the above inclusions, it is enough to observe that if $\mathbf{c}^{(0)}(X)$ is empty, then so is $\mathbf{c}^{(\infty)}(X)$. The former case means that the 0-minimal function is bounded and strongly plurisubharmonic on X. Using again Richberg approximation [**Ric68**], we obtain a *uniformly bounded* $\mathbb{C}^\infty$-smooth strictly plurisubharmonic function on X, i.e. $\mathbf{c}^{(\infty)}(X)$ is enpty, The proof of *(ii)* is analogous.

5.12. Inclusions for $\mathcal{F}$-components, for different classes $\mathcal{F}$. *Assume that $\mathcal{F} \supset \mathcal{F}^*$. Then*

(i) *every $\mathcal{F}$-component is contained in a unique $\mathcal{F}^*$-component;*

(ii) *whenever Z^* is a compact $\mathcal{F}^*$-component, then it contains some (compact) $\mathcal{F}$-component Z;*

(iii) *Every compact pseudoconcave sebset of X has a nonempty intersection with some $\mathcal{F}$-component Z.*

Proof. (i) is obvious. For (ii), apply Corollary 1.11 with $K := Z^*$ and $\{\phi_t : t \in T\} := \mathcal{F}(X)$. Then F constructed there is contained in a unique $\mathcal{F}$-component Z, which has to be contained in Z^*. (iii) also follows from Corollary 1.11 in the same way. □

It is not known, whether Z in (ii) is unique. Corresponding problem for (usually noncompact) core components is open.

For many interesting open problems concerning the cores we refer to [**HST17b**] and [**PSh17**].

5.13. Other admissible classes. There are many admissible classes defined in terms of (practically arbitrary) growth conditions and suitable regularity to which the results of Section 4 can applied.

5.14. Acknowledgements and historical comments. This paper centers around the conjecture, posed in [**HST17b**], that the irreducible components of the cores (defined in various smoothness classes) are pseudoconcave. In the case of two-dimensional complex manifolds Harz et al have proved the conjecture in [**HST17b**] by applying the theorem of Shcherbina about Levi flat graphs [**Shc93**]. The present author obtained a general proof both for cores $\mathbf{c}^{(k)}(X)$, and for minimal kernels, $\Sigma_k(X)$, for $k \geq 2$, (Theorems 5.4 and 5.2 above), by proving a general result for admissible classes of functions (Theorem 4.7 above), with assumption of C^k-smoothness of $k \geq 2$. These facts were presented in [**Slo17a**], a manuscript privately circulated in the Spring 2017. It could be mentioned that Lemma 4.8 above, which implies the above results, is an extension to the case of several functions of an earlier Lemma 3.4 from [**MST15**]. In the Summer 2017 the author noticed that the results of [**Slo17a**], as well as [**SlT04**, Cor. 3.7] extend with practically unchanged proofs to the continuous case ($\mathbf{c}^{(0)}(X)$ and $\Sigma_0(X)$. These improvements were presented (in particular Theorems 3.3 and 2.1 above), on seminars in Maryland and Bloomington [**Slo17b, Slo17c**]. Independently, Poletsky and Shcherbina [**PSh17**] obtained an entirely different proof of the continuous case of the conjecture (Theorem 3.3 above), together with a new characterization of the core $\mathbf{c}^{(0)}(X)$.

6. Pseudoconcave sets and Richberg's regularization

This section is inspired by Richberg's approximation/smoothing theorem [**Ric68**], but the focus is slightly shifted from the original. Given a continuous plurisubharmonic function $f : X \to \mathbb{R}$, X a complex manifold, we want to know what is the largest open subset U of X, such that there is a sequence of continuous plurisubharmonic functions $f_n : X \to \mathbb{R}$, uniformly converging to f on X, with $f_n|U \in C^\infty(U)$. But, due to Richberg's theorem, our task is reduced to finding f_n's such that $f_n|U$ is strongly plurisubharmonic, and, in fact, even one such function suffices, as shown by the next proposition.

6.1. Proposition. *Let f, ψ be continuous plurisubharmonic functions on X and $U \subset X$ open. Assume that $\psi|U$ is strongly plurisubharmonic and $\|f - \psi\|_\infty < +\infty$. Then there is a sequence of continuous plurisubharmonic functions $f_n : X \to \mathbb{R}$, such that $f_n|U$ are C^∞-smooth and stictly plurisubharmonic, and $\lim_{n\to\infty} \|f_n - f\|_\infty = 0$.*

Proof. Let $\psi_n := (1 - 1/n)f + (1/n)\psi$. Then $\|\psi_n - f\|_\infty = (1/n)\|\psi - f\|_\infty \to 0$. Since $(1 - 1/n)f$ is plurisubharmonic nd $(1/n)\psi|U$ is strongly plurisubharmonic, $\psi_n|U$ is strongly plurisubharmonic. By Richberg's theorem there are continuous plurisubharmonic functions f_n on X, C^∞-smooth on U, such that
$\|f_n - \psi_n\|_\infty < 1/n$. Thus $\|f_n - f\|_\infty \to 0$. $\square$

To determine properties of the largest such set U (hoping it exists) we define now, for a fixed f continuous and plurisubharmonic on X, and denote by $\mathfrak{A}(f)$, the class of all continuous plurisubharmonic functions $\psi : X \to \mathbb{R}$ such that $\|f - \psi\|_\infty < +\infty$. (The class $\mathfrak{A}(f)$ resembles in many aspects the admissible classes in Section 4, but it is not a cone.)

6.2. Proposition. *There is an open set $U^* \subset X$ and a function $\phi \in \mathfrak{A}(f)$, such that*
(i) *$\phi|U^*$ is strongly plurisubharmonic;*
(ii) *for every $\psi \in \mathfrak{A}(f)$, every neighborhood on which it is strongly plurisubharmonic is contained in U^*.*

Proof. For every $\psi \in \mathfrak{A}(f)$ denote by U_ψ the largest open set where ψ is strongly plurisubharmonic. Let $U^* := \bigcup\{U_\psi : \psi \in \mathfrak{A}(f)\}$. Selecting a countable subcovering of U^* (X countable at infinity) we obtain a sequence (ψ_n) in $\mathfrak{A}(f)$ such that $U^* := \bigcup U_{\psi_n}$. There is a sequence of positive ϵ_n such that $\sum_{n=1}^{+\infty} \epsilon_n = 1$, and

$$\sum_{n=1}^{+\infty} \epsilon_n \|\psi_n - f\|_\infty < +\infty.$$

Then the function $\phi := \sum_{n=1}^{+\infty} \epsilon_n \psi_n$ is well defined, $\phi \in \mathfrak{A}(f)$, and is strongly plurisubharmonic on U^*. (i.e. $\square$

6.3. Definition. *We denote $\mathfrak{R}(f) := X \setminus U^*$ and call it the rigid set of f. We will call any $\psi \in \mathfrak{A}(f)$ such that ψ is strongly plurisubharmonic on $X \setminus \mathfrak{R}(f)$ a good approximant (of f).*

6.4. Proposition. *Let f be a continuous plurisubharmonic function on a complex manifold X. Assume that $\mathfrak{R}(f)$ is nonempty. Then $\mathfrak{R}(f)$ is pseudoconcave. More precisely, if ϕ is a good approximant to f, B a ball intersecting $\mathfrak{R}(f)$, and $u : \overline{B} \to \mathbb{R}$ a continuous function with $u|bB = \phi|bB$ and maximal (plurisubharmonic) on B, then*

$$B \cap \mathfrak{R}(f) = \{x \in B : u(x) = \phi(x)\},$$

and so $B \cap \mathfrak{R}(f)$ has the local maximum property.

Proof. The proof is a small adaptation of that of Lemma 2.4. Denote, like there

$$Z := \{x \in B : u(x) = \phi(x)\}.$$

Since this set has the local maximum property (by Proposition 1.8), we just need to show that $B \cap \mathfrak{R}(f) = Z$. As in the proof of Lemma 2.4 we have to check two claims.

Claim 1^0

$$Z \cap (B \setminus \mathfrak{R}(f)) = \emptyset;$$

Claim 2^0

$$B \setminus Z \subset B \setminus \mathfrak{R}(f).$$

The proof of Claim 1^0 is simply the same as in Lemma 2.4. (of course with replacing there $\Sigma(X)$ by $\mathfrak{R}(f)$). As for Claim 2^0, the construction of functions ψ and $\tilde{\phi}$ carries over to our new setting. (Notice that good approximants play now the same role as minimal functions in Lemma 2.4.) We just have to check that $\|\tilde{\phi} - f\|_\infty < +\infty$, which is obvious since $\tilde{\phi} - \phi = 0$ on $X \setminus B$, thus $\tilde{\phi} \in \mathfrak{A}(f)$. □

6.5. Pseudoconcave decomposition of $\mathfrak{R}(f)$.

To decompose the rigid set $\mathfrak{R}(f)$ we consider again like in [**HST17b**] an equivalence relation on X but somewhat differently than in Subsection 4.6 above. We define $x \sim_f y$ if

$$\phi(x) - f(x) = \phi(y) - f(y), \text{ for every } \phi \in \mathfrak{A}(f).$$

6.6. Theorem. *Let f be a continuous plurisubharmonic function on a complex manifold X. Then for every $p \in \mathfrak{R}(f)$ the equivalence class $[p]_f$ is pseudoconcave.*

Like previously, the theorem will follow from the finite intersection lemma.

6.7. Lemma. *Let $\phi_1, \phi_2, ..., \phi_n \in \mathfrak{A}(f)$, and $c_1, c_2, ..., c_n \in \mathbb{R}$. Assume that $\phi_1|X \setminus \mathfrak{R}(f)$ is strongly plurisubharmonic. If the set*

$$Z := \{x \in \mathfrak{R}(f) : \phi_1(x) - c_1 = \phi_2(x) - c_2 = ... = \phi_n(x) - c_n\}$$

is nonempty, then it is pseudoconcave.

Proof. In general terms we will proceed similarly like in Lemma 3.5, but there are essential changes caused by the fact that $\mathfrak{A}(f)$ is not a cone, but just a convex set. Since $(\phi_i - c_i) \in \mathfrak{A}(f)$, we can assume that c_i's are zero, and so

$$Z = \mathfrak{R}(f) \cap \{\phi_1 = \phi_2 = ... = \phi_n\}.$$

Let $\phi := (1/n)(\phi_1 + \phi_2 + ... + \phi_n)$. Since $\mathfrak{A}(f)$ is convex, $\phi \in \mathfrak{A}(f)$, and is strongly plurisubharmonic on $X \setminus \mathfrak{R}(f)$ (by Property 1.13(iii). Fix $M \in \mathbb{R}$, $M > \max\{\|\phi_i - \phi_j\|_\infty : i, j = 1, ..., n\}$. Let

$$W := \{(t_1, ..., t_n) \in \mathbb{R}^n : |t_i - \frac{t_1 + ... + t_n}{n}| < M, i = 1, ..., n\}.$$

Then W is a convex domain that contains the joint range of $(\phi_1, \phi_2, ..., \phi_n)$. Define $v_\epsilon : W \to \mathbb{R}$ as

$$v_\epsilon(t_1, ..., t_n) := \frac{t_1 + ... + t_n}{n} + \epsilon \sum_{i=1}^{n} \left(t_i - \frac{t_1 + ... + t_n}{n}\right)^2.$$

Clearly v_ϵ is convex. One checks

$$\frac{\partial v_\epsilon}{\partial t_i} = \frac{1}{n} + 2\epsilon \left(t_i - \frac{t_1 + ... + t_n}{n}\right),$$

and so $\frac{\partial v_\epsilon}{\partial t_i} > 0$ on W, provided $0 < \epsilon < 1/(4nM)$. Fix such ϵ. Then

$$\mu(x) := v_\epsilon(\phi_1(x), \phi_2(x), ..., \phi_n(x)), x \in W,$$

defines a continuous plurisubharmonic function on X. By Property 1.13(iii) μ is strongly plurisubharmonic on $X \setminus \mathfrak{R}(f)$. To check that $\mu \in \mathfrak{A}(f)$, write $\mu = \phi + g$, where

$$g(x) := \epsilon \sum_{i=1}^{n} \left(\phi_i - \frac{\phi_1 + ... + \phi_n}{n}\right)^2.$$

Since

$$g(x) := \epsilon \sum_{i=1}^{n} \left(\sum_{i=j}^{n} \frac{\phi_i - \phi_j}{n}\right)^2,$$

and $\|\phi_i - \phi_j\|_\infty < M$, we get $\|g\|_\infty < \epsilon n M^2$. By triangle inequality $\|\mu - f\|_\infty < +\infty$, ie $\mu \in \mathfrak{A}(f)$. Thus μ is a good approximant to f, to which Proposition 6.4 can be applied.

To show that Z is pseudoconcave, we proceed now like in the proof of Lemma 3.5, with Z, ϕ, μ playing the same role as there, and $\mathfrak{R}(f)$ replacing now $\mathbf{c}(X)$. We take any ball B intersecting Z, and let $u : \overline{B} \to \mathbb{R}$ be a continuous function, equal μ on bB, and maximal on B. Since μ is strongly plurisubharmonic on $B \setminus \mathfrak{R}(f)$, Proposition 6.4 implies

$$B \cap \mathfrak{R}(f) = \{x \in B : u(x) = \mu(x)\}.$$

By construction of ϕ, v_ϵ, and μ,

$$\{x \in X : \phi(x) = \mu(x)\} = \{x \in X : \phi_1(x) = \phi_2(x) = ... = \phi_n(x)\},$$

and so

$$\{x \in \mathfrak{R}(f) \cap B : \phi(x) = \mu(x)\} = Z \cap B.$$

We also have inequalities

$$u(x) > \mu(x) \geq \phi(x), \text{ for } x \in B \setminus \mathfrak{R}(f),$$

$$u(x) \geq \mu(x) \ > \phi(x), \text{ for } x \in B \setminus \phi_1 = \phi_2 = \ldots = \phi_n.$$

Consequently

$$\phi - u = 0 \ , \text{ on } B \cap Z,$$
$$\phi - u < 0 \ , \text{ on } B \setminus Z,$$

and so, by Proposition 1.8, $B \cap Z$ has local maximum property. □

Proof of Theorem 6.6. Let $\{\phi_t : t \in T\} := \mathfrak{A}(f)$, and denote $c_t := \phi_t(p) - f(p)$. Then

$$[p]_f = \{x \in \mathfrak{R}(f) : \phi_t(x) - c_t = f(x)\} = \mathfrak{R}(f) \cap \bigcap_{t \in T} \{\phi_t - c_t = f\}.$$

Let $t_0 \in T$ be such that ϕ_{t_0} is strongly plurisubharmonic on $X \setminus \mathfrak{R}(f)$. For every finite $\{t_1, ..., t_n\} \subset T$ let

$$Z_{t_1,...,t_n} := \{x \in \mathfrak{R}(f) : \phi_{t_0}(x) - c_{t_0} = \phi_{t_1}(x) - c_{t_1} = ...\phi_{t_n}(x) - c_{t_n} = f(x)\}.$$

By Lemma 6.7 these sets have local maximum property, and since

$$[p]_f = \bigcap_{\{t_1,...,t_n\}} Z_{t_1,...,t_n},$$

Proposition 1.10 implies that $[p]_f$ is pseudoconcave. □

References

[Bre59] H. J. Bremermann, *On a generalized Dirichlet problem for plurisubharmonic functions and pseudo-convex domains. Characterization of Šilov boundaries*, Trans. Amer. Math. Soc. **91** (1959), 246–276, DOI 10.2307/1993121. MR0136766

[HST17a] Tobias Harz, Nikolay Shcherbina, and Giuseppe Tomassini, *On defining functions and cores for unbounded domains I*, Math. Z. **286** (2017), no. 3-4, 987–1002, DOI 10.1007/s00209-016-1792-9. MR3671568

[HST17b] Harz, T., Shcherbina, N., Tomassini, G.: *On defining functions and cores for unbounded domains II.* J. Geom. Anal. First Online: 21 April 2017.

[HST17c] Harz, T., Shcherbina, N., Tomassini, G.: *On defining functions and cores for unbounded domains III.* submitted for publication.

[MST15] Samuele Mongodi, Zbigniew Slodkowski, and Giuseppe Tomassini, *Weakly complete complex surfaces*, Indiana Univ. Math. J. **67** (2018), no. 2, 899–935, DOI 10.1512/iumj.2018.67.6306. MR3798861

[MST16] Samuele Mongodi, Zbigniew Slodkowski, and Giuseppe Tomassini, *Some properties of Grauert type surfaces*, Internat. J. Math. **28** (2017), no. 8, 1750063, 16, DOI 10.1142/S0129167X1750063X. MR3681124

[PSh17] Poletsky, E. A., Shcherbina, N.: *Plurisubharmonically separable complex manifolds.* arXiv:1712.02005 [math.CV].

[Ric68] Rolf Richberg, *Stetige streng pseudokonvexe Funktionen* (German), Math. Ann. **175** (1968), 257–286, DOI 10.1007/BF02063212. MR0222334

[Shc93] N. V. Shcherbina, *On the polynomial hull of a graph*, Indiana Univ. Math. J. **42** (1993), no. 2, 477–503, DOI 10.1512/iumj.1993.42.42022. MR1237056

[Slo86] Zbigniew Slodkowski, *Local maximum property and q-plurisubharmonic functions in uniform algebras*, J. Math. Anal. Appl. **115** (1986), no. 1, 105–130, DOI 10.1016/0022-247X(86)90027-2. MR835588
210 (2004), 125–147.

[SlT04] Zibgniew Slodkowski and Giuseppe Tomassini, *Minimal kernels of weakly complete spaces*, J. Funct. Anal. **210** (2004), no. 1, 125–147, DOI 10.1016/S0022-1236(03)00182-4. MR2052116

[Slo17a] Slodkowski, Z.: *Splitting minimal kernel/core.* Preliminary notes Spring 2017.

[Slo17b] Slodkowski, Z.: *Pseudoconcave decompositions in complex manifolds.* Talk at Hopkins-Maryland Complex Geometry Seminar. Oct.31.2017. Abstract on: www.math.umd/tdarvas/CGS.html.

[Slo17c] Slodkowski, Z.: *Pseudoconcave decompositions in complex manifolds.* Talk at Indiana University (Bloomington) Analysis Seminar. Nov.2017..
Ann. **273** (1986), 397–413.

[Smi86] Patrick A. N. Smith, *Smoothing plurisubharmonic functions on complex spaces*, Math. Ann. **273** (1986), no. 3, 397–413, DOI 10.1007/BF01450730. MR824430
Mech. **18** (1968) 143–148.

[Wal68] J. B. Walsh, *Continuity of envelopes of plurisubharmonic functions*, J. Math. Mech. **18** (1968/1969), 143–148. MR0227465

Department of Mathematics, University of Illinois at Chicago, 851 South Morgan Street, Chicago, Illinois 60607

Email address: zbigniew@uic.edu

SELECTED PUBLISHED TITLES IN THIS SERIES

735 **Yanir A. Rubinstein and Bernard Shiffman, Editors,** Advances in Complex Geometry, 2019

731 **Robert G. Niemeyer, Erin P. J. Pearse, John A. Rock, and Tony Samuel, Editors,** Horizons of Fractal Geometry and Complex Dimensions, 2019

730 **Alberto Facchini, Lorna Gregory, Sonia L'Innocente, and Marcus Tressl, Editors,** Model Theory of Modules, Algebras and Categories, 2019

729 **Daniel G. Davis, Hans-Werner Henn, J. F. Jardine, Mark W. Johnson, and Charles Rezk, Editors,** Homotopy Theory: Tools and Applications, 2019

728 **Nicolás Andruskiewitsch and Dmitri Nikshych, Editors,** Tensor Categories and Hopf Algebras, 2019

727 **André Leroy, Christian Lomp, Sergio López-Permouth, and Frédérique Oggier, Editors,** Rings, Modules and Codes, 2019

726 **Eugene Plotkin, Editor,** Groups, Algebras and Identities, 2019

725 **Shijun Zheng, Marius Beceanu, Jerry Bona, Geng Chen, Tuoc Van Phan, and Avy Soffer, Editors,** Nonlinear Dispersive Waves and Fluids, 2019

724 **Lubjana Beshaj and Tony Shaska, Editors,** Algebraic Curves and Their Applications, 2019

723 **Donatella Danielli, Arshak Petrosyan, and Camelia A. Pop, Editors,** New Developments in the Analysis of Nonlocal Operators, 2019

722 **Yves Aubry, Everett W. Howe, and Christophe Ritzenthaler, Editors,** Arithmetic Geometry: Computation and Applications, 2019

721 **Petr Vojtěchovský, Murray R. Bremner, J. Scott Carter, Anthony B. Evans, John Huerta, Michael K. Kinyon, G. Eric Moorhouse, and Jonathan D. H. Smith, Editors,** Nonassociative Mathematics and its Applications, 2019

720 **Alexandre Girouard, Editor,** Spectral Theory and Applications, 2018

719 **Florian Sobieczky, Editor,** Unimodularity in Randomly Generated Graphs, 2018

718 **David Ayala, Daniel S. Freed, and Ryan E. Grady, Editors,** Topology and Quantum Theory in Interaction, 2018

717 **Federico Bonetto, David Borthwick, Evans Harrell, and Michael Loss, Editors,** Mathematical Problems in Quantum Physics, 2018

716 **Alex Martsinkovsky, Kiyoshi Igusa, and Gordana Todorov, Editors,** Surveys in Representation Theory of Algebras, 2018

715 **Sergio R. López-Permouth, Jae Keol Park, S. Tariq Rizvi, and Cosmin S. Roman, Editors,** Advances in Rings and Modules, 2018

714 **Jens Gerlach Christensen, Susanna Dann, and Matthew Dawson, Editors,** Representation Theory and Harmonic Analysis on Symmetric Spaces, 2018

713 **Naihuan Jing and Kailash C. Misra, Editors,** Representations of Lie Algebras, Quantum Groups and Related Topics, 2018

712 **Nero Budur, Tommaso de Fernex, Roi Docampo, and Kevin Tucker, Editors,** Local and Global Methods in Algebraic Geometry, 2018

711 **Thomas Creutzig and Andrew R. Linshaw, Editors,** Vertex Algebras and Geometry, 2018

710 **Raphaël Danchin, Reinhard Farwig, Jiří Neustupa, and Patrick Penel, Editors,** Mathematical Analysis in Fluid Mechanics, 2018

709 **Fernando Galaz-García, Juan Carlos Pardo Millán, and Pedro Solórzano, Editors,** Contributions of Mexican Mathematicians Abroad in Pure and Applied Mathematics, 2018

For a complete list of titles in this series, visit the
AMS Bookstore at **www.ams.org/bookstore/conmseries/**.